# Physical and Chemical Properties of Aerosols

# Physical and Chemical Properties of Aerosols

Edited by

**I. COLBECK**
University of Essex
Colchester
UK

## BLACKIE ACADEMIC & PROFESSIONAL
An Imprint of Chapman & Hall
London · Weinheim · New York · Tokyo · Melbourne · Madras

**Published by**
**Blackie Academic & Professional, an imprint of Thomson Science,**
**2–6 Boundary Row, London SE1 8HN, UK**

Thomson Science, 2–6 Boundary Row, London SE1 8HN, UK

Thomson Science, 115 Fifth Avenue, New York, NY 10003, USA

Thomson Science, Suite 750, 400 Market Street, Philadelpha, PA 19106, USA

Thomson Science, Pappelallee 3, 69469 Weinheim, Germany

First edition 1998

© 1998   Thomson Science

Thomson Science is a division of International Thomson Publishing   I(T)P˙

Typeset in 10/12 Times by AFS Image Setters Ltd, Glasgow

Printed in Great Britain by St Edmundsbury Press, Bury St Edmunds, Suffolk

ISBN   0 7514 0402 0

A catalogue record for this book is available from the British Library

Library of Congress Catalog Card Number: 97-77011

∞ Printed on acid-free text paper, manufactured in accordance with ANSI/NISO Z39.48-1992 (Permanence of Paper).

# Contents

# 4   Diffusion and coagulation      153
## S.K. FRIEDLANDER

# 5   Electrical and thermodynamic properties      198
## C.F. CLEMENT

# Contributors

**U. Baltensperger**   Paul Scherrer Institute, CH-5232 Villingen PSI, Switzerland

**C.F. Clement**   QUANTISCI, 15 Witan Way, Wantage, Oxon, OX12 9EU

**I. Colbeck**   Department of Biological and Chemical Sciences, Central Campus, University of Essex, Colchester, Essex CO4 3SQ, UK

**S.K. Friedlander**   UCLA, Chemical Engineering Department, University of California, Los Angeles, CA 90095-1592, USA

**A.J. Hickey**   School of Pharmacy, University of North Carolina, Beard Hall, CB#7360, Chapel Hill, NC 27599-7360, USA

**D.B. Ingham**   Department of Applied Mathematical Studies, University of Leeds, Leeds LS2 9JT, UK

**S. Jain**   Center for Microengineered Materials, Department of Chemical Engineering, University of New Mexico, Albuquerque, NM 87131, USA

**T.T. Kodas**   Center for Microengineered Materials, Department of Chemical Engineering, University of New Mexico, Albuquerque, NM 87131, USA

**C.B. Lalor**   School of Pharmacy, University of North Carolina, Beard Hall, CB#7360, Chapel Hill, NC 27599-7360, USA

**D. Majumdar**   Center for Microengineered Materials, Department of Chemical Engineering, University of New Mexico, Albuquerque, NM 87131, USA

**D. Mark**        Institute of Occupational Health, University of
                  Birmingham, Edgbaston, Birmingham B15 2TT, UK

**J.P. Mitchell** Trudell Medical Group, London, Ontario N5Z 3M5,
                  Canada

**S. Nyeki**      Paul Scherrer Institute, CH-5232 Villingen PSI,
                  Switzerland

**J.I.T. Stenhouse** Department of Chemical Engineering, Loughborough
                  University, Loughborough, Leics. LE11 5TU, UK

# Preface

Aerosol science is gaining importance in a growing number of fields, from microelectronics to food-processing. Problems of interest fall into two categories: 'good' aerosols, useful for various products and processes; and 'bad' aerosols, those found in pollution. Whatever the usage, the fundamental rules governing the behaviour of aerosols remain the same. Hence aspects of physics, chemistry, material science, pharmacy and statistics are all touched upon in this book. The result, if we have done our job well enough, is more than the sum of the parts.

The following quote from an editorial in the *Aerosol Society Newsletter* (No. 28, August 1996) succinctly sums up the importance of aerosol science: 'Whether it's mechanics, electrostatics, optics or dental hygiene, the behaviour of aerosols can be used as a paradigm for many of the fundamental underlying processes. The person who seeks perfection in their understanding of aerosols can cope with anything the scientific world might throw at them.'

Recognizing the growing importance of aerosols in various disciplines, numerous national and international societies have been established. In Europe, for instance, the Gesellschaft für Aerosolforschung (GAeF) was founded in 1972, while in the UK the Aerosol Society was founded in 1986. These societies, together with those from Finland, France, the Netherlands and the Nordic countries, have recently established the European Aerosol Assembly, one of the aims of which is to encourage and facilitate the development of education in the field of aerosol science within the European Union. Elsewhere societies are flourishing in USA, Japan, India, China, Russia and Israel.

This book has its origins in a short course entitled 'Aerosol Properties and Instrumentation' which was held at the University of Essex in the late 1980s and early 1990s. The course was designed for scientists and engineers from industry and from other research and development institutions. The course, given by established experts in the field, covered the basic principles of aerosol science as well as introducing more advanced concepts.

The treatment should give the reader a basic idea of the scope of the subject and of its applications. A significant change in style will be noted when the more rigorous approach to diffusion, coagulation and computational fluid dynamics is presented. The use of certain mathematical techniques is, of course, essential for the treatment of aerosol science. The authors have strived to develop the mathematical treatment in a manner that is sufficiently rigorous, while trying to provide clear indications of the intermediate steps.

The book opens with a brief introduction to aerosol science. Chapter 2 describes generation and calibration methods, while sampling principles are discussed in Chapter 3. Chapters on diffusion and coagulation, electrical and thermodynamic properties, and filtration follow. Aerosols are ubiquitous in our environment, and Chapter 7 describes the various sources of atmospheric aerosols. The final three chapters concentrate on applications of aerosol science: the emerging importance of aerosol science in material synthesis is presented in Chapter 8; the significance of pharmaceutical aerosols is examined in Chapter 9; and computational fluid dynamics as a tool to enable an understanding of aerosol sampling is addressed in Chapter 10.

During the editing process I have tried to be reasonably consistent with nomenclature. Symbols are defined in the text, and chapters carry a nomenclature listing where necessary.

Many people have helped to prepare this volume, and heartfelt thanks are extended to them all. Bill Sturges, Dai Griffiths, Sarah Dunnett and Andrey Filippov read various parts of the manuscript in draft, each providing comments which have much improved on the initial efforts.

IC

# 1 Introduction to aerosol science

I. COLBECK

## 1.1 Introduction

Mention aerosols to most people and they instantly think of a spray can. Few know that the word **aerosol** has an alternative, more fundamental meaning. An aerosol is a dispersion of fine solid particles or liquid droplets in a gas. The term aerosol originates from military research during the First World War and is associated with Donnon (Whytlaw-Gray *et al.*, 1923), although first publication of the term was due to Schmauss (1920) who used it as an analogue to hydrosol, a stable liquid suspension of solid particles. The fundamental properties of aerosols have been studied for more than a hundred years. In the nineteenth century aerosol particles represented the smallest division of matter known. Many great scientists of the time, such as Faraday, Tyndall, Lister, Kelvin, Maxwell, Aitken and Einstein, contributed to our understanding of aerosols (Gentry, 1995; Gentry and Lin, 1996; Preining, 1996; Davis, 1997). The formulations of the equations of motion of fine particles in fluids by Stokes, the investigations into the scattering of light by Tyndall and Rayleigh, and the study of atmospheric nuclei by Aitken are examples of the work which provided the basic background to further developments. To a certain extent in the First World War, and to a very large extent in the Second, intensive research was undertaken on the physics and chemistry of clouds, especially smoke screens, with the result that rapid progress was made over a relatively short period. Spurny (1993) has classified the first half of the twentieth century as the classical period of aerosol science and the current period as the modern era. The latter has grown through the development of fast analytical techniques, advancements in electronics and the growth of computer power as well as the recognition of the important role of aerosols in the environment and as a route to producing high-technology materials such as superconductors and optical fibres. Although at first sight an esoteric topic, aerosol science is of practical importance in many industries, such as pharmaceuticals, nuclear power, petrochemicals and tobacco. Additionally, aerosol science has long been of importance in military defence, industrial hygiene, agriculture, meteorology and other fields.

Aerosols are at the core of environmental problems such as global warming (section 7.5), acid rain, photochemical smog, stratospheric ozone depletion (sections 5.6 and 7.3.2) and air quality. In fact, recognition of the effects of aerosols on climate can be traced back to 44 BC (Charlson, 1997) when an eruption from Mount Etna was linked to cool summers and poor harvests. People have been aware of the occupational health hazard of exposure to aerosols for many centuries (Jensen and O'Brien, 1993; Cantrell *et al.*, 1993). It is only relatively recently that there has been increased awareness of the possible health effects of vehicular pollution, and in particular submicrometre particles (Committee on the Medical Effects of Air Pollutants, 1995).

The particle size of interest in aerosol behaviour ranges from molecular clusters of 0.001 µm to fog droplets and dust particles as large as 100 µm – a variation of $10^5$ in size. The upper and lower limits are debatable. Particles much greater than 100 µm do not stay airborne long enough to be measured and observed as aerosols. The lower limit is controlled by the size of a cluster of half a dozen or so molecules: this is the smallest entity of the condensed phase that can exist.

There are various types of aerosol, which are classified according to physical form and method of generation. The commonly used definitions are:

- **dust** – a solid particle formed by mechanical disintegration of a parent material, such as crushing, grinding and blasting;
- **fume** – solids produced by physicochemical reactions such as combustion, sublimation or distillation;
- **smoke** – a visible aerosol resulting from incomplete combustion;
- **mist** and **fog** – liquid aerosol produced by the disintegration of liquid or the condensation of vapour;
- **bioaerosol** – a solid or liquid aerosol consisting of, or containing, biologically viable organisms.

Dusts range in size from the submicrometre to visible, fumes and smoke are typically below 1 µm, while mists and fogs range from submicrometre to approximately 20 µm, although these droplets may coalesce to form larger drops of about 100 µm. Figure 1.1 illustrates the typical size range of various aerosols.

## 1.2   Descriptions of particle size

Particle size is the most important parameter for characterizing the physical behaviour of aerosols. From the atmospheric perspective, those above 2 µm diameter are coarse while those below this size are fine. The fine particles are then subdivided into the **nucleation** ($d < 0.1$ µm) and **accumulation** ($0.1 < d < 2$ µm) modes. The term **ultrafine particles** is often used by some aerosol scientists. The 1979 Workshop on Ultrafine Aerosols

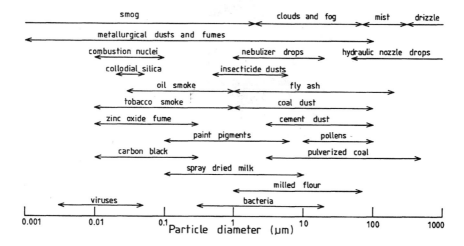

**Fig. 1.1** Particle size range for aerosols.

defined such particles as those with diameters below 100 nm (Liu *et al.*, 1982). The US Environmental Protection Agency (1996) used the term 'ultrafine particles in a biological context' to characterize particle size distributions with mass median diameter below approximately 100 nm. Nanostructured materials show improved or even new properties compared to those of micro- or macrostructured materials. The scientific and technological issue of nanostructured materials is on the verge of becoming a research and development field of high actuality. Far-reaching applications can be anticipated in life sciences, materials science, high-technology engineering and clean and durable industrial production processes (Chapter 8). The various disciplines (e.g. materials, molecular science, aerosol science) have used several definitions for nanometre particles. Pui and Chen (1997) have suggested that particles with diameters below 50 nm should be classified as nanoparticles.

Light microscopy was the first method used to determine the shape and size of particles. It permits direct measurement of particle size. The shape of a particle may provide information about its probable type and formation mechanism (McCrone and Delly, 1973). Particles may be resolved and observed around 0.25 µm, although size determination is most reliably achieved on particles larger than several micrometres in diameter. Figure 1.2 shows some of the earliest reported examinations, by microscope, of atmospheric particles (Sigerson, 1870a; 1870b). The figure shows particles collected in western Ireland (i.e. of maritime origin). Some of the particles were initially droplets, while others were collected dry. Sigerson also examined particles in 'country and urban air' and industrial

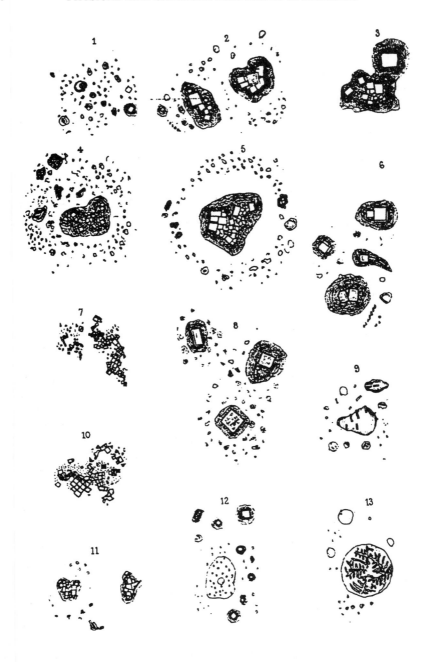

**Fig. 1.2** Drawings of atmospheric particles collected in western Ireland in 1870 (Sigerson, 1870a).

atmospheres. Interestingly, he speculates about the health hazards due to inhaling particles in urban air.

Particle size is based upon a two-dimensional projected image or silhouette. For spherical particles this is the diameter of the circular silhouette. However, airborne particles are rarely spherical and various diameters based on the geometry of the two-dimensional silhouette are used. If one is able to view the particles, it is possible to describe a diameter in the following terms:

- **Martin's diameter** – the length of the line that separates a particle into two equal portions. This diameter is often referred to as a 'statistical diameter' since the value depends upon the orientation of the particle, and only the mean value for all particle orientations is unique for a given particle. In practice, single measurements are made of each of numerous particles oriented randomly to a reference line.
- **Feret's diameter** – the maximum distance from edge to edge of a particle (all observations made in parallel planes).
- **Projected area diameter** – the diameter of a particle having the same projected area as the particle in question.

In general, Feret's diameter is the largest of these and Martin's diameter the smallest. Often it is not feasible to size large numbers of particles microscopically. In this case, other properties such as inertial or gravitational properties are used.

The above diameters are of little help when characterizing fibres. Assuming fibres of either cylindrical or prolate spheroidal shape, both length and diameter should be defined. However, real fibres typically do not satisfy this criterion. Many aerosols, especially those formed by combustion processes, exhibit complex structures, consisting of a large number of very much smaller primary particles. This property is described in section 1.7.

## 1.3  Sources

In order better to understand the properties of aerosols, it is useful to consider the characteristics of those from a number of different environments. Aerosols may be generated under both laboratory (Chapter 2) and natural conditions.

Microscopic particles are ubiquitous in our environment. They are produced by natural processes and by man, both intentionally and unintentionally. Wind-blown dust, smoke from forest fires and volcanic material are examples of naturally produced particles. Automobile exhaust and smoke from power generation are examples of anthropogenic particles.

Aerosol sources can be classified as primary or secondary. Primary sources are mainly of natural origin, whereas secondary sources (e.g. gas to particle conversion) are predominantly anthropogenic. The origins of aerosols are quite diverse and the sources vary both spatially and temporally (Colbeck 1995b; Pacyna, 1995). Emissions of aerosols from natural sources have altered in strength only slightly during the past centuries. However, anthropogenic emissions have grown dramatically, especially during the twentieth century. It is expected that the globally averaged aerosol concentration will stabilize due to emission controls, though their geographic distribution will change. Sources of atmospheric aerosol are discussed in detail in Chapter 7.

Once aerosols are in the atmosphere, their size, number and chemical compositions are changed by several mechanisms until they are ultimately removed by natural processes. Measurements of particle size distributions normally show up to three groups of particles (Figure 1.3):

- *Nucleation mode*. In this mode particles have recently been emitted from processes involving condensation of hot vapours, or freshly formed within the atmosphere by gas to particle conversion. Such particles

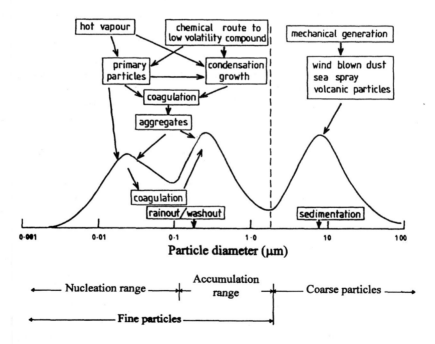

**Fig. 1.3** Schematic diagram of a typical size distribution and formation mechanisms for atmospheric particles.

account for the preponderance of particles by number, but because of their small size account for little of the total mass of airborne particles.

- *Accumulation mode.* These particles have grown from the nucleation mode by coagulation or condensation of vapours. They generally account for most of the aerosol surface area and a substantial part of the aerosol mass.
- *Coarse mode.* These particles are mainly formed by mechanical attrition processes, and hence soil dust, sea spray and many industrial dusts fall within this mode.

In many parts of the world anthropogenic sources of particulate matter dominate. Even with the introduction of emission control efforts, anthropogenic sources are still important. Any incombustible material present as inorganic impurities in fuel, which passes through the combustion process, will produce aerosol particles. Emissions from modern coal-fired power stations are an order of magnitude lower than those of a few decades ago as a result of more efficient particle removal systems. In many Third World countries the old technology is still in use and a significant percentage of the ash emitted there is from coal firing. Aerosols are also emitted in the workplace environment, with their characteristics being primarily determined by the nature of the industry and type of product.

Bioaerosols are particles of variable biological origin, among them pollen, fungal spores, bacterial cells, viruses, protozoa, excreta or fragments of insects, skin scales or hair of mammals, or other components, residues or products of organisms, such as bacterial saccharides. Bioaerosols are discussed in detail by Cox and Wathes (1996). Outdoors, airborne fungal spores originate from soil, vegetation and decaying plants and animals. Differences in quantity and quality of sources and meteorological conditions cause a strong temporal and spatial variation in airborne concentrations. For most genera, the highest numbers in the outdoor air are found during summer and autumn. During these seasons the outdoor air is the main source of fungi in the indoor air. Recent studies indicated that the outdoor air spores influence the presence of fungi in indoor environments, but indoor air spores are not a simple reflection of the presence of fungi in outdoor air.

In addition to the outdoor air, there are numerous sources of biological particulate matter such as foodstuffs, house plants and flowerpots, house dust, pets and their bedding material, textiles, carpets, wood material and furniture stuffing, from which spores of Alternaria, Aspergillus, Botrytis, Claclosporium, Penicillium, Scopulariopsis and yeast cells are occasionally released into the air. In non-industrial indoor environments, the most important source of airborne bacteria is the presence of humans. As a result, the concentration in air of normal human skin bacteria is often used as an indicator of indoor air quality. In industrial environments strong sources of bioaerosols may exist when organic material is handled, such as plants, hay, straw, wood chips, cereal grain and tobacco.

All natural and anthropogenic waters contain large numbers of microorganisms. Hence water or liquid droplets resulting from rain, splashes or bubbling processes may contain bioaerosols. These may remain airborne after the water evaporates. In non-industrial localities bioaerosol sources can develop due to microbial growth in a building's ventilation system or in the structure itself. The prerequisite for microbial growth is excessive and accessible moisture. Standing water is a good reservoir for microbial growth and a potential source of microbial aerosols when disturbed.

Indoor aerosols' concentrations are associated with both indoor and outdoor sources. The identification of sources and the assessment of their relative contribution can be a complicated process due to the presence of a number of indoor sources, which can vary from building to building. There are also uncertainties associated with estimating the impact of outdoor sources on the indoor environment. Many indoor activities generate pollutants. These include heating, cooking, cleaning, smoking, the use of a wide variety of consumer products, and the simple act of moving about and stirring up particles. Particles studied indoors have predominantly been those in the fine mode resulting from cigarette smoking or as emissions from combustion appliances. Coarse mode particles from re-entrainment of fibres, dust, animal and human dander and mould spores constitute the second most common form of indoor pollution.

## 1.4  Particle size distributions

Aerosols in which all particles are of the same size are termed **monodisperse**. These are, however, a very artificial and temporary phenomenon (Chapter 4), and aerosols typically contain a wide range of sizes and are termed **polydisperse**. Because of the presence of a range of sizes the problem arises of how to describe the size distribution.

In practice, the form of size distributions, although containing the same modes, can look very different according to whether the distribution is plotted as number distribution or mass distribution. The reason for this is that particles at the small end of the size distribution can be very abundant in number, but because mass depends upon the cube of diameter, such particles may contribute only a small amount of the total mass. Hence, a size distribution expressed in terms of the number of particles per size fraction will give far more emphasis to the smaller particles than a distribution expressed in terms of mass per size fraction.

The mode of presentation will probably depend upon the means of size measurement. If a size distribution is determined by a microscopic technique, the particles will have been quantified by number within given size ranges. If, however, a sample has been fractionated by impaction and each size fraction has been determined by weighing, a distribution by mass will have been generated.

Once an appropriate measure of particle size has been determined, information is still required about the size distribution. The number concentration or mass concentration distribution can be described in terms of a continuous function. The most popular size distribution of physical phenomena is the normal (Gaussian) distribution. This distribution can be described by the mean value ($d_{mean}$) and the standard deviation ($\sigma$):

$$d_{mean} = \frac{\sum n_i d_i}{\sum n_i},\tag{1.1}$$

$$\sigma = \left[\frac{\sum n_i (d_{mean} - d_i)^2}{(\sum n_i) - 1}\right]^{0.5}.\tag{1.2}$$

Such a distribution is rarely seen in aerosol science except in cases where the particles are virtually monodisperse, such as latex spheres used for calibration purposes (Chapter 2). However, most aerosols exhibit a skewed distribution with a long tail that extends out to relatively large particles. Such distributions are usually better described mathematically by the lognormal distribution. In this case the geometric mean diameter, $d_g$, and the geometric standard deviation, $\sigma_g$, are defined as follows:

$$\log d_g = \frac{\sum n_i \log d_i}{\sum n_i},\tag{1.3}$$

$$\log \sigma_g = \left[\frac{\sum n_i (\log d_g - \log d_i)^2}{(\sum n_i) - 1}\right]^{1/2},\tag{1.4}$$

in which $n_i$ is the number of particles having diameter $d_i$. For a perfectly monodisperse aerosol, $\sigma_g = 1$. The geometric mean and geometric standard deviation are usually obtained from count or mass distributions. The median diameter (which for a lognormal distribution equals the geometric mean) is termed the **count median diameter** (CMD) or **mass median diameter** (MMD).

It is frequently convenient to plot cumulative size data on lognormal probability paper (Figure 1.4). A linear plot indicates a lognormal distribution, the diameter corresponding to 50% frequency is the median or geometric mean diameter, and the geometric standard deviation is given by

$$\sigma_g = \frac{84.13\% \text{ diameter}}{50\% \text{ diameter}} = \frac{50\% \text{ diameter}}{15.87\% \text{ diameter}}.\tag{1.5}$$

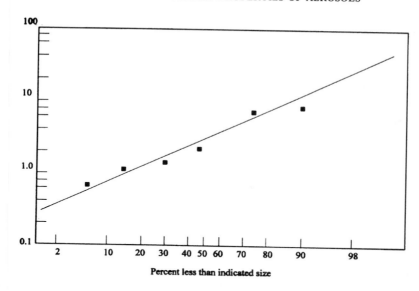

**Fig. 1.4** Log probability graph.

Provided a distribution is approximately lognormal, log probability graphs are the simplest method to determine the geometric mean and standard deviation. Details on the use of log probability plots are given in Hinds (1982). He recommends giving most weight to the central region (20–80%) and little weight to those points greater than 95% and less than 5%.

Perfect monodispersity is defined in a distribution with $\sigma_g = 1$, and so-called monodisperse distributions are classified as having $\sigma_g \leq 1.2$. One important property of the geometric standard deviation for an aerosol size distribution is that it is the same for a mass, surface area or number distribution of the same aerosol. This is useful because it is frequently necessary to measure one characteristic of the size distribution such as number distribution when what is really needed is some other characteristic such as the mass distribution. For lognormal distributions, the Hatch–Choate equations may be used to convert from number to mass or surface and vice versa (Hatch and Choate, 1929):

$$\text{MMD} = \text{CMD} \exp(3(\ln \sigma_g)^2), \tag{1.6}$$

$$\text{SMD} = \text{CMD} \exp(2(\ln \sigma_g)^2), \tag{1.7}$$

where SMD is the surface median diameter. In general,

$$d_p = \text{CMD} \exp(p(\ln \sigma_g)^2), \tag{1.8}$$

where $p$ is a parameter which serves to define the various possible diameters.

**Table 1.1** Conversion coefficients for equation (1.8) (after Reist, 1993)

| Required diameter | Value of $p$ |
| --- | --- |
| Mode | −1 |
| Count mean diameter | 0.5 |
| Diameter of average mass | 1.5 |
| Surface median diameter | 2 |
| Surface mean diameter | 2.5 |
| Mass median diameter | 3 |
| Mass mean diameter | 3.5 |

Values of $p$ and the associated diameter for a lognormal distribution are given in Table 1.1. Aerosols are rarely lognormal, so great care must be taken when calculating mass mean/median diameters based on count data, or vice versa. The exception to this is for cases of virtually monodisperse aerosols. Figure 1.5 shows the relative location of various diameters. The importance of large particles as far as mass distribution is concerned is clear from this figure: 50% of the aerosol mass is contained in particles of less than 30 μm diameter. On the other hand, in terms of numbers of particles, half the total number of particles are of less than 9 μm.

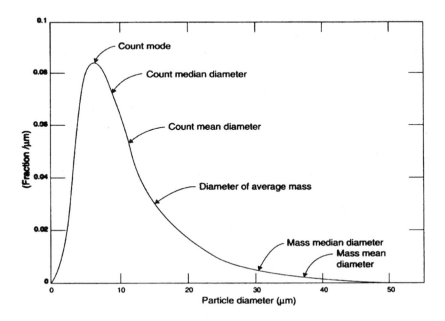

**Fig. 1.5** Arithmetic plot of size data for an aerosol of CMD = 9.0 μm and $\sigma_g = 1.89$ μm, showing various diameters.

## 1.5  Basic hydrodynamics

### 1.5.1  *Mean free path*

Before we can consider the motion of an aerosol, we have to consider the medium it is suspended in. The behaviour of the particles depends to a large extent on the properties and motion of the suspending gas. Only a brief outline will be given here. Full details may be found in Hinds (1982), Reist (1993) and Willeke and Baron (1993).

The ideal gas law relates pressure, $P$, volume, $v$, temperature, $T$, and number of moles, $n$, via the formula

$$Pv = nRT, \tag{1.9}$$

where $R$ is the molar gas constant. The gas density, $\rho_g$, is then

$$\rho_g = \frac{PM_{wt}}{RT}, \tag{1.10}$$

where $M_{wt}$ is the molecular weight. The effective molecular weight of air is $28.9\,\mathrm{g\,mol^{-1}}$. The **mean free path**, $\lambda$, is the mean distance that a gas molecule travels between collisions with other gas molecules:

$$\lambda = \frac{1}{\sqrt{2}n\pi d_m^2}, \tag{1.11}$$

where $d_m$ is the collision diameter of the molecule and $n$ the number of molecules per unit volume. Alternatively,

$$\lambda = \sqrt{\frac{\pi}{8}} \frac{\eta}{0.4987} \frac{1}{\sqrt{\rho_g P}}, \tag{1.12}$$

where $\eta$ is the viscosity of the gas. The mean free path in air as a function of temperature and relative humidity is given in Table 1.2. The amount of moisture present influences the gas density and hence is included in this table, although its effect is small. To maintain a gas of constant velocity, a constant force must be applied. Since this force results in a steady velocity, rather than acceleration, we can conclude from Newton's second

Table 1.2 Mean free path ($\times 10^{-8}$ m) of air as a function of relative humidity. Pressure $= 1.01325 \times 10^5$ Pa (after Jennings, 1988)

| $T$(K) | Relative humidity (%) | | |
|--------|------|------|------|
|        | 0    | 50   | 100  |
| 288.15 | 6.391 | 6.389 | 6.386 |
| 293.15 | 6.543 | 6.544 | 6.548 |
| 296.15 | 6.635 | 6.638 | 6.647 |
| 298.15 | 6.691 | 6.701 | 6.714 |

law of motion that there is an equal and opposing force in the system. Viscosity is a measure of this force. The temperature dependence of the gas viscosity is given by Sutherland's approximation

$$\eta = \eta_0 \sqrt{\frac{T}{T_0}} \left( \frac{1 + S/T_0}{1 + S/T} \right), \tag{1.13}$$

where $S$ is the Sutherland constant (110.4 K for air). Since $\eta$ is proportional to $T^{0.5}$, the mean free path is essentially inversely proportional to pressure and proportional to temperature:

$$\frac{\lambda}{\lambda_0} = \left( \frac{P_0}{P} \right) \left( \frac{T}{T_0} \right). \tag{1.14}$$

### 1.5.2  Reynolds number

If we consider the flow around geometrically similar shapes, it is possible to define the conditions when the flow will also be geometrically similar. Assuming the medium is incompressible and neglecting gravity, the main forces present are inertial and viscous forces. If the flow is similar, the components of the force of inertia must be proportional to the product $\rho V^2/l$ where $l$ is a characteristic length of the body and $V$ is the relative velocity between the fluid and the body. Frictional forces, on the other hand, are proportional to $\eta V/l^2$. For dynamic similarity $\rho V^2/l$ and $\eta V/l^2$ are in a fixed ratio. Hence $\rho Vl/\eta$ must be constant. This dimensionless number is called the **Reynolds number**. For spheres the characteristic length is the diameter and hence

$$Re = \frac{\rho d V}{\eta}. \tag{1.15}$$

At high $Re$ inertial forces dominate, while at low $Re$ viscous forces dominate. At low $Re$ the flow is laminar (often known as Stokes or viscous flow). It is not surprising that flows with high $Re$ and flows with low $Re$ have quite different characteristics.

### 1.5.3  Drag coefficient

The drag force, $F_D$, required to move a body at steady velocity through a viscous medium is a function of $l$, $V$ and the medium properties ($\rho$ and $\eta$). It is common practice to express the resistance exerted by a fluid in terms of a dimensionless coefficient obtained by dividing the resistance by $\frac{1}{2} \rho V^2$ and the area, $A$, of the body in question. Dimensional analysis leads to the following relationship for the **drag coefficient**, $C_D$:

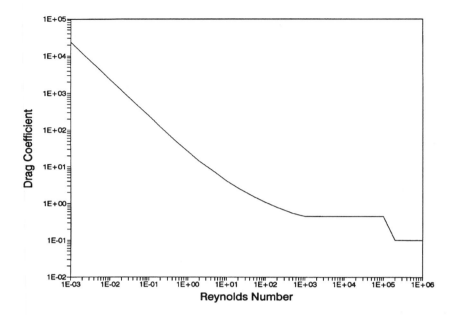

**Fig. 1.6** Drag coefficient versus particle Reynolds number for spherical particles.

$$C_D = \frac{F_D}{\frac{1}{2}\rho V^2 A}.\tag{1.16}$$

For spherical particles we have

$$F_D = C_D \rho V^2 d^2 \frac{\pi}{8}.\tag{1.17}$$

The drag coefficient relates the drag force to the velocity pressure. Its value is dependent upon the particle Reynolds number, as shown by the idealized curve in Figure 1.6. More details are given in Chapter 10. For high Reynolds number (1000 to 200 000), where inertial forces dominate, $C_D$ is virtually constant, with a value of approximately 0.44. Inserting this value into the above equation gives Newton's law.

## 1.6 Stokes' law

The natural laws of physics and chemistry which describe the behaviour of matter in the macro and molecular states also, of course, apply to aerosols. There are, however, some distinct differences between the basic properties of aerosols and those of liquid-based colloid systems which occur widely in industry. One major difference arises when considering the motion of a particle falling under the influence of gravity and the resistance offered by the medium to that motion. The general equation for the resistance force offered by the medium to the motion of a sphere was derived by Newton. Three kinds of resistance are associated with the motion, the predominant type depending upon the particle Reynolds number. For large $Re$, a given flow system may generally be treated as if it were inviscid. At large $Re$, frictional resistance and pressure drag predominate and the drag is primarily associated with the cross-sectional area normal to the flow. At small $Re$ ($Re < 1$), deformation drag is important and forces that act over the entire body surface must be taken into account. It is in the latter region that Stokes' law is applicable. There the resisting force on a moving particle is given by

$$F_D = 3\pi\eta Vd. \tag{1.18}$$

Hence

$$C_D = \frac{24}{Re}. \tag{1.19}$$

Various assumptions were made in the derivation of Stokes' law:

- incompressibility of medium;
- infinite extent of medium;
- medium is considered as continuous;
- rigid spherical particles;
- viscous medium.

Although air is compressible, compression is not important for aerosols in the Stokes region. The condition of infinite extent of the medium is never observed in practice, as there will always be some macroscopic body, for example the walls of a container, besides the particles. However, the extent of influence is so small it can be neglected without any significant error. The presence of other particles moving in close proximity will have the effect of reducing the resistance of the medium to the particle by setting the medium near the particle in motion. Hence, an ensemble of particles will settle faster than a single particle. Fuchs (1964) and Happel and Brenner (1965) have addressed more fully the implications of a non-infinite medium.

Liquid droplets may not meet the criteria of rigidity on two counts. First, large droplets may deform by the motion of the medium and will no longer be spherical. In fact they flatten out and offer greater resistance to falling than spherical particles. Fuchs (1964) states that, for water drops with diameters above 0.8 mm, appreciable deviation from the terminal velocity for equi-sized spheres occurs. Second, a circulation can develop within the moving droplet. This reduces friction at the surface and consequently the resistance offered by the medium. The resisting force then becomes

$$F_D = 3\pi\eta Vd\left(\frac{1 + (2\eta/3\eta_p)}{1 + (\eta/\eta_p)}\right) \tag{1.20}$$

where $\eta_p$ is the viscosity of the liquid of which the droplet is composed. Since the viscosity of gases is much less than the viscosity of liquids, this correction factor can be neglected.

In the intermediate region $(1 < Re < 1000)$ empirical relationships have been developed to extend Stokes' law (see Reist, 1993; Willeke and Baron, 1993; and references therein).

For aerosol particles falling freely in air, equation (1.18) is valid for $1 < d < 60\,\mu m$. Above $60\,\mu m$ diameter the drag is greater due to the motion induced in the air becoming sufficient to involve its inertia. Below $1\,\mu m$, air begins to act as a discontinuous medium, since the particle becomes comparable in size with the mean free path of the air molecules and the drag is less than that given by the equation above. When this condition occurs, the drag force is reduced by the so-called **Cunningham slip correction**, $C_c$, which depends on the mean free path of air. Hence:

$$F_D = \frac{3\pi\eta Vd}{C_c}, \tag{1.21}$$

where

$$C_c = 1 + Kn[\alpha + \beta \exp(-\gamma/Kn)]. \tag{1.22}$$

The **Knudsen number**, $Kn$, relates the gas molecular mean free path, $\lambda$, to the physical dimension of the particle $(Kn = 2\lambda/d)$. Various values for $\alpha$, $\beta$ and $\gamma$ have been reported (Table 1.3) although the actual differences in the computed values for $C_c$ are negligible. These parameters are generally

**Table 1.3** Values of the coefficients $\alpha, \beta$ and $\gamma$ required to calculate the Cunningham slip correction factor

| Author | Mean free path($\times 10^{-8}$ m) | $\alpha$ | $\beta$ | $\gamma$ |
|---|---|---|---|---|
| Hutchins *et al.* (1995)[a] | 6.73 | 1.2310 | 0.4695 | 1.1783 |
| Rader (1990) | 6.74 | 1.207 | 0.440 | 0.780 |
| Jennings (1988) | 6.635 | 1.252 | 0.399 | 1.100 |
| Allen and Raabe (1985)[a] | 6.73 | 1.142 | 0.558 | 0.999 |
| Allen and Raabe (1982) | 6.73 | 1.155 | 0.471 | 0.596 |
| Fuchs (1964) | 6.53 | 1.246 | 0.418 | 0.867 |
| Davies (1945) | 6.609 | 1.257 | 0.400 | 1.100 |
| Millikan (1923) | – | 1.209 | 0.406 | 0.893 |

[a] Data for solid particles, other data for oil droplets.

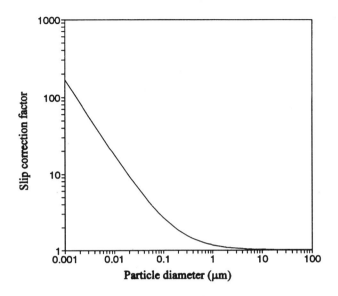

**Fig. 1.7** Slip correction factor as a function of particle size (20°C and standard atmospheric pressure).

based on aerosol droplets of oil. Because of differences in molecular interactions with the surfaces of solid particles and oil drops, different parameters should be appropriate for solid particles (Allen and Raabe, 1985; Hutchins *et al.*, 1995). It is important to use the mean free path with which these constants were determined. The mean free path is inversely proportional to pressure, and hence the value of the slip correction increases with altitude. The slip correction is plotted as a function of particle diameter in Figure 1.7. Note that for values of $d > 10\,\mu m$, $C_c = 1$ and that $C_c$ does not become large until $d < 1.0\,\mu m$. The slip correction represents the mechanism for transition from the continuum, $d > \lambda$, to the molecule region, $d < \lambda$. $Kn \ll 1$ indicates the **continuum regime** and $Kn \gg 1$ represents the **free molecular regime**. The intermediate range, approximately, between 0.1 and 10, is termed the **transition regime**.

In the previous discussion, particles were considered as spherical. While liquid droplets ($d < 1\,mm$) and some condensed vapours are spherical, many particles have irregular shapes, depending on how they were formed and the amount of agglomeration which has taken place. Another correction factor, this time the **dynamic shape factor**, must be applied to Stokes' law to account for the effect of shape on particle motion. Ignoring slip correction, we find

$$F = 3\pi\eta V d_v \chi, \tag{1.23}$$

**Table 1.4** Dynamic shape factors (after Davies, 1979; Wu and Colbeck, 1996)

| Shape | Dynamic shape factor |
|---|---|
| Sphere | 1.0 |
| Cube | 1.08 |
| Cluster of spheres | |
| 2-sphere chain | 1.12 |
| 3-sphere chain | 1.27 |
| 4-sphere chain | 1.32 |
| Dusts | |
| Bituminous coal | 1.05–1.11 |
| Quartz | 1.36–1.82 |
| Sand | 1.57 |
| $UO_2$ | 1.28 |
| Agglomerates | |
| Iron oxide | 13.2–17.7 |
| $(PuU)O_2$ | 1.96–2.85 |
| PtO | 1.1–3.6 |
| Carbonaceous smoke | 3.26–6.77 |
| MgO | 1.06–4.40 |
| Pb fume | 1.5–3.5 |

where $\chi$ is the dynamic shape factor and $d_v$ is the volume equivalent diameter (see later). The shape factor is always greater than or equal to one. This means that non-spherical particles settle more slowly than their equivalent volume spheres. The dynamic shape factor for an irregular particle will depend on the particle's orientation. Values for simple non-spherical (e.g. cube, cylinder) and more complex particles have been reported in the literature (Davies, 1979; Hinds, 1982; Reist, 1993). Table 1.4 shows some of these data. It is evident that compact spheres typically have values between one and two, while more extreme shapes may have larger values. Aggregates may have an effective density that is different from the bulk material density and hence a shape factor defined as a function of mass equivalent diameter may be more appropriate (Brockmann and Rader, 1990).

An important application of Stokes' law is the determination of the **gravitational settling velocity** of aerosol particles in still air. The gravitational force, $F_g$, is proportional to particle mass, $m$, and gravitational acceleration, $g$. As the particle begins to move, the surrounding gas exerts an equal and opposite drag force, which after a short period of acceleration equals the gravitational force and the particle reaches its terminal settling velocity, $V_{ts}$.

$$F_g = mg = \frac{(\rho_p - \rho_g)\pi d^3 g}{6} = \frac{3\pi\eta Vd}{C_c}, \qquad (1.24)$$

**Table 1.5** Particle parameters for unit-density particles under standard conditions

| Particle diameter (μm) | Slip correction factor | Settling velocity (cm s$^{-1}$) |
|---|---|---|
| 0.01 | 23.04 | $6.95 \times 10^{-6}$ |
| 0.1 | 2.866 | $8.65 \times 10^{-5}$ |
| 1 | 1.152 | $3.48 \times 10^{-3}$ |
| 10 | 1.015 | $3.06 \times 10^{-1}$ |
| 100 | 1.02 | $2.61 \times 10^{1}$ |

where $\rho_p$ and $\rho_g$ are the densities of the particle and gas respectively. The latter is included to account for the buoyancy effect, but this can usually be neglected because $\rho_p \gg \rho_g$. Hence

$$V_{ts} = \frac{\rho_p d^2 g C_c}{18\eta}. \tag{1.25}$$

For a particle Reynolds number $Re > 1.0$, inertial effects become significant in relation to viscous effects and an iterative or extended calculation procedure is necessary to calculate $V_{ts}$ (Hinds, 1982). This correction is not appreciable except for very large particles, and Stokes' law provides a reasonable estimate of settling velocity. Table 1.5 compares the slip correction factor and settling velocity for particles over a range of sizes.

The equivalent volume diameter, $d_v$, is the diameter of the sphere having the same volume as the irregular particle. In addition to this diameter, the Stokes diameter, $d_s$, and the aerodynamic diameter, $d_a$, find wide applications in aerosol technology. The Stokes diameter is the diameter of the sphere which has the same density and same falling velocity as the particle, while the aerodynamic diameter is the diameter of a sphere of unit density which has the same settling velocity as the particle. The aerodynamic diameter is particularly useful for characterizing filtration and respiratory deposition. Several instruments, such as impactors and elutriators, measure aerodynamic diameter. These diameters are related to each other via

$$V_{ts} = \frac{\rho_p d_v^2 C_c(d_v) g}{18\eta\chi} = \frac{\rho_b d_s^2 C_c(d_s) g}{18\eta} = \frac{\rho_0 d_a^2 C_c(d_a) g}{18\eta}, \tag{1.26}$$

where $\rho_0$ is unit density and $\rho_b$ is the bulk material of the particle. Ignoring slip correction, we obtain

$$d_a = d_v \left(\frac{\rho_p}{\rho_0\chi}\right)^{1/2} = d_s \left(\frac{\rho_b}{\rho_0}\right)^{1/2}. \tag{1.27}$$

The Stokes diameter is usually defined in terms of the density of the bulk material, thus removing the problem of defining the true density of the particle, which may be less than $\rho_b$ due to porosity or agglomerated structure. The equations are again only valid in the Stokes region and the slip correction must be applied to $d_v$, $d_a$ and $d_s$ for small particles (see equation (2.1)).

## 1.7  Fractal analysis

The advance of aerosol science has often necessitated the simplified treatment of aerosols as comprising spherical particles. Even rather simple non-spherical shapes such as ellipsoids, chains or fibres have not been easy to work with, either in theoretical treatments or in the laboratory. It has been shown in experiments (Forest and Witten, 1979) and simulations (Witten and Sander, 1981) that some aerosol particles, particularly those derived from combustion processes (i.e. grown by diffusion limited aggregation), are fractal-like (Figure 1.8). The most striking feature of a fractal is its scale invariance or self-similarity. If any part of a fractal is magnified, it appears similar to the object as a whole. The agglomerated

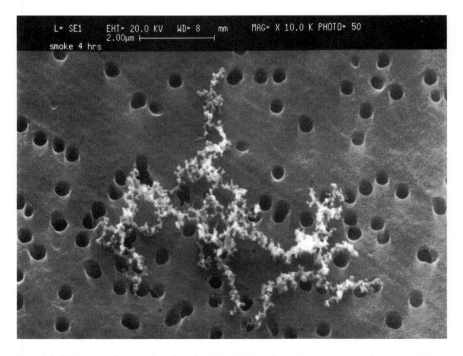

**Fig. 1.8** Electron micrograph of a fractal cluster of carbonaceous smoke on a 0.4 μm Nuclepore filter.

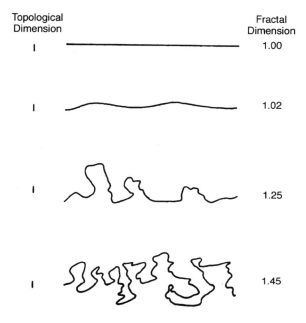

Topological
Dimension

Fractal
Dimension

| 1.00

| 1.02

| 1.25

| 1.45

**Fig. 1.9** The basic concept that the dimension of a physical quantity can be extended by adding fractional quantities related to the ruggedness of a system to the topological dimension. (Reproduced with permission from Kaye, 1989. Copyright 1989 VCH Publishers.)

particles mentioned above do not have strict geometrical similarity but do satisfy the criteria for fractals. An important characteristic of these structures is a power-law relationship between the mass, or number of primary particles, $N$, and the radius of gyration of the cluster, $R_g$, of the form

$$N \propto R_g^{D_f} \qquad (1.28)$$

where $D_f$ is the fractal dimension. The fractal dimension provides a quantitative measure of degree to which a structure fills the physical space beyond its topological dimension. Whereas a smooth line is confined to one dimension and a smooth surface to two, a fractal, with wrinkles on wrinkles, begins to infringe on other dimensions (Figure 1.9). For example, if the primary particles are aligned in a straight line then $D_f = 1$ and for particles on a regular two-dimensional array $D_f = 2$. For $D_f = 3$ a uniform spherical aggregate is implied. For a typical fractal cluster $D_f$ is lower than the space dimension and is usually non-integer, implying a structure intermediate between these idealized forms. The higher the fractal dimension, the more compact the structure described. Whereas all the

**Table 1.6** Fractal dimension for various aggregation models as a function of the spatial dimension (after Jullien, 1987)

|                            | 2-dimensional | 3-dimensional |
| -------------------------- | ------------- | ------------- |
| Particle–cluster           | 1.70          | 2.50          |
| Cluster–cluster            | 1.44          | 1.78          |
| Cluster–cluster (ballistic) | 1.51          | 1.91          |
| Cluster–cluster (chemical) | 1.55          | 2.04          |

various shape diameters vary as the particle size changes, the fractal dimension is constant over a range of particle sizes, provided the clusters form under similar conditions.

The formation process of agglomerates has been the subject of extensive simulation work (Smirnov, 1990), and many models have been postulated, among them the diffusion-limited particle–cluster, cluster–cluster, ballistic cluster–cluster and reaction-limited cluster–cluster aggregation models. The introduction of the cluster–cluster model allowed the simulation of realistic aggregate morphologies in which $D_f$ values were comparable to those obtained by experiment. In three-dimensional models, aggregate growth by diffusion-limited cluster–cluster aggregation results in a typical mean value $D_f \sim 1.80$, while growth by ballistic limited cluster–cluster aggregation gives $D_f \sim 1.95$. Table 1.6 summarizes the fractal dimension predicted by various aggregation models.

Many workers have found it convenient to work with what is defined as the **boundary fractal** of the agglomerate, obtained from the two-dimensional projection of the structure. Others have studied the internal structure of agglomerates and obtained the **mass** or **density fractal** dimension. With both the boundary and density fractal dimension some information is lost. However, each measurement method yields different details and there is diagnostic information on formation dynamics of the aerosol embedded in the fractal structure (Kaye, 1989).

First of all, if one looks at the ruggedness of the projected boundary of the agglomerate one can describe the structure of the boundary in terms of a boundary fractal dimension. This parameter is a measure of the ruggedness of the boundary. Such boundary fractal dimensions have been studied by various image analysis techniques, with the most familiar technique being described as the **structured walk** or **yardstick** method. The perimeter of the particle (Figure 1.10a) is bounded by straight incremental segments each of length $\Omega$ (Figure 1.10b). As the magnitude of the increment decreases, the perimeter increases as more fine structure is included. It is common practice to plot the logarithm of the perimeter as a function of the logarithm of the length scale. A graph of this kind is known for historic reasons as a **Richardson plot** (Figure 1.10c). It can be seen in this figure that there are two linear relationships and it can be shown that it is useful to describe the data line at coarse resolution as defining the structural boundary fractal dimension of the profile. The structural boundary fractal dimension is useful in describing the aerodynamic behaviour of an agglomerated aerosol and can also describe the way in which the profile physically interacts with its surroundings. The fractal dimension deduced from the data generated at high resolution is the textural fractal dimension.

However, for agglomerates formed from solid primary particles with a narrow size distribution, such as a typical combustion-generated aerosol, one is more concerned with the internal structure of the cluster. There are various techniques based on electron or optical microscopy, which use digitized images to determine this fractal dimension in two-dimensional space ($D_{2D}$) (Cleary *et al.*, 1990). Such methods are, however, only suitable for aggregates with $D_f < 2$. For those with $D_f > 2$ the projection onto a plane always results in $D_{2D} = 2$. The determination of the structure after deposition of the agglomerates on, for example, a filter can lead to erroneous results. By testing image analysis methods for model

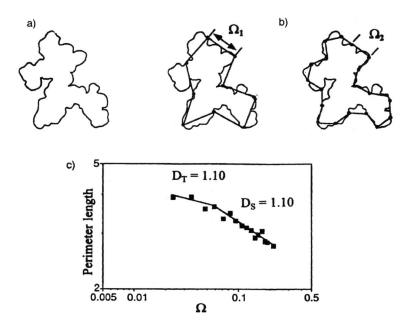

**Fig. 1.10** Concepts of fractal geometry: (a) profile of agglomerate; (b) structured walk explorations; (c) Richardson plot of a series of explorations of the agglomerate.

agglomerates it has been shown that the obtained value of the fractal dimension depends on the chosen evaluation method and that the true fractal dimension can be severely underestimated (Tence *et al.*, 1986; Rogak and Flagan, 1992). Hence *in-situ*, three-dimensional methods are generally favoured.

For example, one alternate technique for studying the morphology of aerosols is to probe the system using visible or ultraviolet radiation (Colbeck, 1995a; Colbeck and Nyeki, 1992; Bonczyk and Hall, 1992; Sorensen *et al.*, 1992). When one explores the structure of the agglomerates by such a technique one is not measuring the configuration of the profile in space, rather one is exploring the way that the subunits of the agglomerate are packed to occupy three-dimensional space. Thus fractal dimensions measured by light scattering studies should properly be called mass or density fractal dimensions. Various other *in-situ* methods have been reported, again measuring the density fractal. These methods include techniques based on the aerodynamic and mobility properties (Schmidt-Ott, 1988; Kütz and Schmidt-Ott, 1992), kinematic coagulation (Wu *et al.*, 1994), a modified Millikan cell (Nyeki and Colbeck, 1994; 1995) and a combination of differential mobility analysis with inductively coupled plasma optical emission spectrometry (Weber *et al.*, 1996).

Mathematical simulations may be used to generate models of agglomerates, and these may be compared with those that are actually generated in, for example, a flame so that one can establish formation dynamics from the comparison of the two structures. By varying the rules in which one simulates the growth of agglomerates, one can compare real agglomerates with simulated ones and gain information on the probable mechanisms of formation. Sutherland and Goodarz-Nia (1971) showed that the coagulation of particles that move on linear trajectories (i.e. ballistic monomer–cluster agglomeration) produces compact agglomerates. If the mean free path of the attaching monomers is comparable to or larger than the agglomerate size then one may assume linear trajectories. The fractal dimension for such agglomerates should be approximately 3. When $Kn < 1$ the diffusion-limited cluster–cluster aggregation model of Meakin and Skjeltorp (1993) is appropriate and the fractal dimension should be approximately 1.8. Therefore the Knudsen number should determine the upper limit for building close-packed agglomerates in the coagulation process. Tence *et al.* (1986) showed that different primary particle size distributions did not significantly influence the final fractal dimension, when agglomeration conditions remain unchanged. Therefore it must be concluded that the change of the mean free path of the carrier gas molecules is responsible for a change of the coagulation regime, and this has been confirmed by Weber (1992). Hence there is a value of agglomerate diameter where the fractal dimension of agglomerates changes from 3 to around 1.8 (free molecular to transition

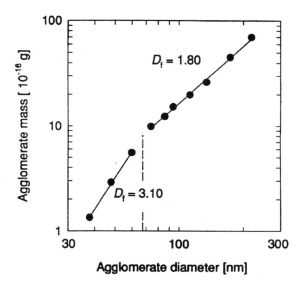

**Fig. 1.11** Double logarithmic plot of the mass per agglomerate and the mobility diameter. (Reprinted with permission from Weber *et al.* 1996. Copyright Elsevier Science Ltd.)

regime behaviour). Figure 1.11 shows agglomerate mass as a function of the mobility equivalent diameter. Two values of the fractal dimension are reported, with one relating to cluster–cluster agglomeration ($D_f \sim 1.8$) and the other to ballistic monomer–cluster agglomeration ($D_f \sim 3.0$). The change-over, in this case, is around 68 nm.

Fractal geometry is important in many areas of aerosol science. Kaye (1989) postulated that calculating the health hazard from the aerodynamic diameter can lead to serious underestimation of the actual hazard, and furthermore that the aerodynamic diameter can give a false impression of the physical magnitude of the aerosol when considering the design of respirators and filters. To illustrate these aspects he considered the sets of isoaerodynamic particles shown in Figure 1.12. Within each particle type some of the particles are going to be much more difficult to filter than one would anticipate from their aerodynamic diameter, whereas the more open agglomerated structures would be easier to filter. These latter particles have a surface area orders of magnitude bigger than those anticipated from their aerodynamic diameters and hence can carry large quantities of adsorbed chemicals into the lung. Since aerodynamic diameters underestimate the actual size of fractal clusters, it is often found that, for instance, diesel exhaust may be trapped by a relatively inefficient coarse filter compared to that predicted to be necessary based on aerodynamic size.

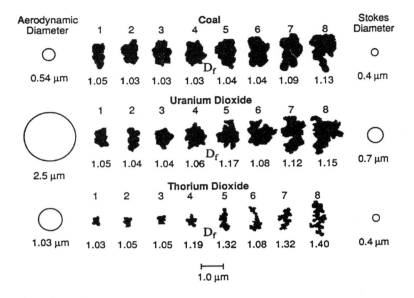

**Fig. 1.12** Variation in physical size and fractal dimension for coal dust, uranium dioxide aerosol and thorium dioxide aerosol. The structural fractal dimension is also reported. (Adapted from Kaye, 1989.)

The fractal characterization of aerosols has received increasing attention in recent years. Progress has been made in the study of such aspects as the hydrodynamic, diffusional, optical and elastic properties of fractal objects. However, little is known of the electrical properties of fractal aerosols. It is important from both a basic and a technological point of view to extend theories and experiments to fractals. The charging of fractal materials can give rise to interesting new phenomena which are not usually encountered in compact materials. The charging of low-density structures (e.g. smokes) could lead to their break-up; the possibility of restructuring them into more compact clusters could lead to a dramatic enhancement in their electrical mobilities. This result may have industrial applications in electrostatic precipitators for the removal of fractal clusters.

The mass fractal dimension of soot and other aerosols produced by combustion processes is becoming an important element in any models for use in predicting future climate changes. Climate modellers will need to know both the mass fractal dimension, to understand the scattering behaviour of the aerosols, and the boundary fractal dimensions, which governs the rate at which particles settle out of the atmosphere. Despite numerous algorithms and experimental techniques for determining the fractal dimension of aerosols, a quantitative description of atmospheric aerosols is rare. Those which have been reported have involved image analysis of ambient aerosols collected on a filter or transmission electron microscope grid (Katrinak et al., 1993; Xie et al., 1994; Kindratenko et al. 1994; Lee and Chou, 1994). Kindratenko et al. (1994) concluded that fractal analysis allows the unequivocal identification of particle source. Fractal analysis can also provide information on the formation processes of aerosols (Katrinak et al., 1993). A range of fractal dimensions were reported for carbonaceous aerosols in Phoenix, Arizona, indicating variations of diffusion-limited aggregation. Significantly, there was evidence that restructuring may occur (Katrinak et al., 1993). It appears likely that fractal analysis will continue to be developed and will provide an invaluable tool for the identification and apportionment of particles.

It is evident from Table 1.4 that the dynamic shape factors for fractal clusters are generally greater than those for simple non-spherical objects. This indicates that fractal clusters settle more slowly and hence have longer atmospheric lifetimes than simple non-spherical objects with the same mass. Theoretical studies by Berry (1989) have shown that clusters composed of 1000 individual spherules of 20 nm radius fall 10 times more slowly than a solid sphere of the same mass. Rogak et al. (1993) concluded that in the continuum regime the shape factor is related to the fractal dimension via

$$\chi = 0.9 \sqrt{\frac{D_f}{D_f + 2}} N^{1/D_f - 1/3} \tag{1.29}$$

For $D_f \leq 2$ the majority of the primary spherules in the free molecular regime are exposed to momentum exchange with the carrier gas. Hence the mobility diameter should be approximately equal to the area diameter and one would expect $\chi \propto N^{1/6}$, where $N$ is the number of primary particles. This agrees with equation (1.29) for $D_f = 2$ as well as with experimental results (Kops et al., 1975).

## References

Allen M.D. and Raabe O.G. (1982) Re-evaluation of Millikan's oil drop data for the motion of small particles in air. J. Aerosol Sci., 6, 537–547.

Allen M.D. and Raabe O.G. (1985) Slip correction measurements of spherical solid aerosol particles. Aerosol Sci. Technol., 4, 269–286.

Berry M.V. (1989) Falling fractal flakes. Physica D, 38, 29–31.

Bonczyk P.A. and Hall R.J. (1992) Measurement of the fractal dimension of soot using UV laser radiation. Langmuir, 8, 1666–1670.

Brockmann J.E. and Rader D.J. (1990) APS response to nonspherical particles and experimental determination of dynamic shape factor. Aerosol Sci. Technol., 13, 162–172.

Cantrell B.K., Williams K.L., Watts W.F. and Jankowski R.A. (1993) Mine aerosol management, in *Aerosol Measurement* (eds K. Willeke and P.A. Baron). Van Nostrand Reinhold, New York.

Charlson R.J. (1997) Direct climate forcing by anthropogenic sulfate aerosols: the Arrhenius paradigm a century later. *Ambio*, **26**, 25–31.

Cleary T.G., Samson R. and Gentry J.W. (1990) Methodology for fractal analysis of combustion aerosols and particle clusters. *Aerosol Sci. Technol.* **12**, 518–525.

Colbeck I. (1995a) Fractal analysis of aerosol particles. *Anal. Proc.*, **32**, 383–386.

Colbeck I. (1995b) Particle emission from outdoor and indoor sources, in *Airborne Particulate Matter* (eds T. Kouimtzis and C. Samara). Springer-Verlag, Berlin.

Colbeck I. and Nyeki S. (1992) Optical and dynamical investigations of fractal clusters. *Sci. Progress*, **76**, 149–166.

Committee on the Medical Effects of Air Pollutants (1995) *Non-biological Particles and Health, Department of Health*. HMSO, London.

Cox C.S. and Wathes C.M. (eds) (1996) *Bioaerosols Handbook*. Lewis Publishers, Boca Raton, FL.

Davies C.N. (1945) Definitive equations for the fluid resistance of spheres. *Proc. Phys. Soc.*, **57**, 259–270.

Davies C.N. (1979) Particle fluid interaction. *J. Aerosol Sci.*, **10**, 477–513.

Davis E.J. (1997) A history of single aerosol particle levitation. *Aerosol Sci. Technol.*, **26**, 212–254.

Forest S.R. and Witten T.A. (1979) Long range correlations in smoke particle aggregates. *J. Phys. A*, **12**, L109–L117.

Fuchs N.A. (1964) *The Mechanics of Aerosols*. Pergamon Press, Oxford.

Gentry J.W. (1995) The aerosol science contributions of Michael Faraday. *J. Aerosol Sci.*, **26**, 341–349.

Gentry J.W. and Lin J.-C. (1996) The legacy of John Tyndall in aerosol science. *J. Aerosol Sci.*, **27**, S503–S504.

Happel J. and Brenner H. (1965) *Low Reynolds Number Hydrodynamics*. Prentice Hall, Englewood Cliffs, NJ.

Hatch T. and Choate S.P. (1929) Statistical description of the size properties of non-uniform particulate substances. *J. Franklin Inst.*, **207**, 369–387.

Hinds W.C. (1982) *Aerosol Technology*. Wiley, New York.

Hutchins D.K., Harper M.H. and Felder R.L. (1995) Slip correction measurements for solid spherical particles by modulated dynamic light scattering. *Aerosol Sci. Technol.*, **22**, 202–218.

Jennings S.G. (1988) The mean free path in air. *J. Aerosol Sci.*, **19**, 159–166.

Jensen P.A. and O'Brien D. (1993) Industrial hygiene, in *Aerosol Measurement* (eds K. Willeke and P.A. Baron). Van Nostrand Reinhold, New York.

Jullien R. (1987) Fractal aggregates. *Comments Cond. Mat. Phys.*, **13**, 177–205.

Katrinak K.A., Rez P., Perkes P.R. and Buseck P.R. (1993) Fractal geometry of carbonaceous aggregates from an urban aerosol. *Environ. Sci. Technol.*, **27**, 539–547.

Kaye B.H. (1989) *A Random Walk through Fractal Dimensions*. VCH, Weinheim.

Kindratenko V.V., van Espem P.J.M, Treiger B.A. and van Grieken R.E. (1994) Fractal dimensional classification of aerosol particles by computer-controlled scanning electron microscopy. *Environ. Sci. Technol.*, **28**, 2197–2202.

Kops J., Dibbets G., Hermans L. and van de Vate J.F. (1975) The aerodynamic diameter of branched chain-like aggregates. *J. Aerosol Sci.*, **6**, 329–333.

Kütz S. and Schmidt-Ott A. (1992) Characterization of agglomerates by condensation-induced restructuring. *J. Aerosol Sci.* **23**, S357–S360.

Lee C. and Chou C.C.K. (1994) Application of fractal geometry in quantitative characterization of aerosol morphology. *Part. Part. Syst. Charact.*, **11**, 436–441.

Liu B.Y.H., Pui D.Y.H., McKenzie R.L., Agarwal J.K., Jaenicke R., Pohl F.G., Preining O., Reischl G., Szymanski W. and Wagner P.E. (1982) Intercomparison of different absolute instruments for measurement of aerosol number concentration. *J. Aerosol Sci.*, **13**, 429–450.

McCrone W.C. and Delly J.G. (1973) *The Particle Atlas*, 2nd edn, Ann Arbor Science, Ann Arbor, MI.

Meakin P. and Skjeltorp A.T. (1993) Application of experimental and numerical models to the physics of multiparticle systems. *Adv. Phys.*, **42**, 1–127.

Millikan R.A. (1923) Coefficients of slip in gases and the law of reflection of molecules from the surfaces of solids and liquids. *Phys. Rev.*, **21**, 217–238.

Nyeki S. and Colbeck I. (1994) The measurement of the fractal dimension of individual in-situ soot agglomerates using a Millikan cell technique. *J. Aerosol Sci.*, **25**, 75–90.

Nyeki S. and Colbeck I. (1995) Fractal dimension analysis of single, in-situ, restructured carbonaceous aggregates. *Aerosol Sci. Technol.*, **23**, 109–120.

Pacyna J.M. (1995) Sources, particle size distribution and transport of aerosols, in *Airborne Particulate Matter* (eds T. Kouimtzis and C. Samara). Springer-Verlag, Berlin.

Preining O. (1996) The many facets of aerosol science. *J. Aerosol Sci.*, **27**, S1–S6.

Pui D.Y.H. and Chen D.R. (1997) Nanometre particles: a new frontier for multidisciplinary research. *J. Aerosol Sci.*, **28**, 539–544.

Rader D.J. (1990) Momentum slip correction factor for small particles in nine common gases. *J. Aerosol Sci.*, **21**, 161–168.

Reist P.C. (1993) *Aerosol Science and Technology*, 2nd edn. McGraw-Hill, New York.

Rogak S.N. and Flagan R.C. (1992) Characterization of the structure of agglomerate particles. *Part. Part. Syst. Charact.*, **9**, 19–27.

Rogak S.N., Flagan R.C. and Nguyen H.V. (1993) The mobility and structure of agglomerates. *Aerosol Sci. Technol.*, **18**, 25–47.

Schmauss A. (1920) Die Chemie des Nebels der Wolken und des Regens. *Die Umschau*, **24**, 61–63.

Schmidt-Ott A. (1988) In-situ measurement of fractal dimensionality of ultra-fine particles. *Appl. Phys. Lett.*, **52**, 954–956.

Sigerson G. (1870a) Micro-atmospheric researches. *Proc. Roy. Irish Acad. Sci.*, **1**, 13–22.

Sigerson G. (1870b) Further researches on the atmosphere. *Proc. Roy. Irish Acad. Sci.*, **1**, 22–31.

Smirnov B.M. (1990) The properties of fractal clusters. *Phys. Reports*, **188**, 1–78.

Sorensen C.M., Cai J. and Lu N. (1992) Light scattering measurements of monomer size, monomers per aggregate and fractal dimension for soot aggregates in flames. *Appl. Optics*, **31**, 6547–6557.

Spurny K.R. (1993) Aerosol science of the early years. *J. Aerosol Sci.*, **24**, S1–S2.

Sutherland D.N. and Goodarz-Nia I. (1971) Floc simulation – effect of collision sequence. *Chem. Eng. Sci.*, **26**, 2071–2085.

Tence M., Chevalier J.P. and Jullien R. (1986) On the measurement of the fractal dimension of aggregated particles by electron microscopy: experimental method, corrections and comparison with numerical models. *J. Physique*, **47**, 1989–1998.

US Environmental Protection Agency (1996) *Air Quality Criteria for Particulate Matter*, EPA/600/P-95/001af. Office of Research and Development, Washington, DC.

Weber A.P. (1992) *Characterization of the Geometrical Properties of Agglomerated Aerosol Particles*, PSI Report 129. Paul Scherrer Institute, Villigen PSI, Switzerland.

Weber A.P., Baltensperger U., Gaggeler H.W. and Schmidt-Ott A. (1996) In situ characterization and structure modification of agglomerated aerosol particles. *J. Aerosol Sci.*, **27**, 915–929.

Whytlaw-Gray R., Speakman J.B. and Campbell J.H.P. (1923) Smokes part I – a study of their behaviour and a method of determining the number of particles they contain. *Proc. Roy. Soc. A*, **102**, 600–615.

Willeke K. and Baron P.A. (eds) (1993) *Aerosol Measurement*. Van Nostrand Reinhold, New York.

Witten R. and Sander L. (1981) Diffusion limited aggregation, a kinetic phenomenon. *Phys. Rev. Lett.*, **47**, 1400–1403.

Wu Z. and Colbeck I. (1996) Studies of the dynamic shape factor of aerosol agglomerates. *Europhys. Lett.*, **33**, 719–724.

Wu Z., Colbeck I. and Simons S. (1994) Determination of the fractal dimension of aerosols from kinetic coagulation. *J. Phys. D*, **27**, 2291–2296.

Xie Y., Hopke P.K., Casuccio G. and Henderson B. (1994) Use of multiple fractal dimensions to quantify airborne particle shape. *Aerosol Sci. Technol.*, **20**, 161–168.

## Nomenclature

| | |
|---|---|
| $A$ | area |
| $C_c$ | Cunningham slip correction factor |
| $C_D$ | drag coefficient |
| $d$ | particle diameter |
| $d_a$ | aerodynamic diameter |
| $d_g$ | geometric mean diameter |
| $d_m$ | collision diameter of molecule |
| $d_s$ | Stokes diameter |
| $d_v$ | volume equivalent diameter |
| $D_f$ | fractal dimension |
| $F_D$ | drag force |
| $g$ | gravitational acceleration |
| $Kn$ | Knudsen number |
| $l$ | characteristic length of a body |
| $M_{wt}$ | molecular weight |
| $N$ | number of particles |
| $P$ | pressure |
| $Re$ | Reynolds number |
| $R_g$ | radius of gyration |
| $S$ | Sutherland constant |
| $T$ | temperature |
| $v$ | volume |
| $V$ | velocity |
| $V_{ts}$ | terminal velocity |
| | |
| $\eta$ | viscosity |
| $\lambda$ | mean free path |
| $\rho$ | density |
| $\sigma$ | standard deviation |
| $\sigma_g$ | geometric standard deviation |
| $\chi$ | dynamic shape factor |

*Subscripts*

| | |
|---|---|
| g | gas |
| p | particle |
| b | bulk |

# 2 Aerosol generation and instrument calibration

J.P. MITCHELL

## 2.1 Scope

There is an increasing awareness of the need to verify that instruments measuring properties of aerosols, such as particle size and concentration, are operating correctly. Quality systems, such as ISO 9001 (ISO, 1994), emphasize the requirement that all instrumentation used for making measurements be calibrated on a regular basis as part of the method validation process. The term **calibration** in this context is taken to be the determination of bias associated with the measurement of an aerosol property under specified conditions in order to obtain meaningful results (BSI, 1993). This process ideally requires the setting up of a chain of traceability back to the 'absolute' internationally accepted standards of length and mass. At least four levels of measurement may be involved in this calibration chain:

1. measurement of unknown aerosol;
2. secondary calibration standards (in-house methods, test powders such as Arizona Road Dust);
3. primary calibration standards (particle-based certified reference materials (CRMs));
4. international standards of length and mass for particle size and concentration, respectively.

Calibrations are often performed at level 2 because of the limited availability of CRMs (especially in the submicrometre size range), their high cost if used regularly, and the difficulty in ensuring that they are dispersed adequately. This chapter is therefore concerned mainly with secondary methods; however, mention is also made of recent developments associated with primary standards, and suggestions are made as to how they may be used to generate test aerosols.

There are several excellent reviews on the topic of aerosol generation for calibration purposes, including those by Fuchs and Sutugin (1966); Liu (1974); Raabe (1976); and Willeke (1980). More recently, Chen (1993), Cheng and Chen (1995) and John (1993) have also examined the subject, adding information about appropriate calibration aerosol handling. Chen (1993) and Cheng and Chen (1995) have widened the topic to include the calibration of related equipment such as flow meters and pressure transducers as well as the use of chambers for sampling under calm air conditions and wind tunnels for sampling aerosols moving in a well-defined flow.

## 2.2 Basic concepts

Unlike many other analytes, an aerosol is not a single-phase system nor one component of a uniform mixture, but a two-phase dispersion of solid particles and/or liquid droplets in a gas (BSI, 1993). Aerosols are semi-stable, because once formed, processes such as gravitational sedimentation and phoretic forces are continually at work depleting the particle or droplet concentration. It follows that the process of calibrating an aerosol analyser must not only involve the means to generate the test aerosol, but also deliver the aerosol in a well-defined way to the point of measurement. Until recently, almost all information about aerosol generation for calibration purposes was based on the generation of particles of controlled size. It may be as important to control particle concentration if a meaningful calibration is to be achieved, especially when detecting biases, such as size-dependent inlet losses.

Calibration aerosols may be classified as either monodisperse or poly-disperse. The particle size distribution, if unimodal, can often be approximated by the lognormal function, where the degree of dispersity is given by the geometric standard deviation ($\sigma_g$). A perfectly monodisperse aerosol has $\sigma_g$ of unity. However, a practical definition of monodispersity where $\sigma_g$ is less than 1.2 is considered acceptable by many groups, and there are several types of aerosol generator that can meet this criterion, each operating within a well-defined size range (Figure 2.1). Mixtures comprising several sizes of monodisperse calibration particles can be used to span the operating range of an aerosol analyser. Such 'cocktails' are very useful at distinguishing bias from size-related inlet losses, as well as the magnitude of any spreading of size measurements by techniques that involve matrix inversion and/or correlation procedures to derive the size distribution from the raw data.

Polydisperse aerosols are effective where time is of the essence, as a single aerosol can be all that is required for a calibration. However, it is essential to know how well the aerosol is dispersed and sampled before assumptions are made about instrument performance. These requirements place a severe limit on their effective use, as it is necessary to measure the calibrant aerosol simultaneously by an independent (and preferably traceable) technique to verify its size distribution. Nevertheless, such aerosols are widely used to verify the performance of instruments such as aerosol mass monitors, where it is desirable to simulate 'real-life' aerosols,

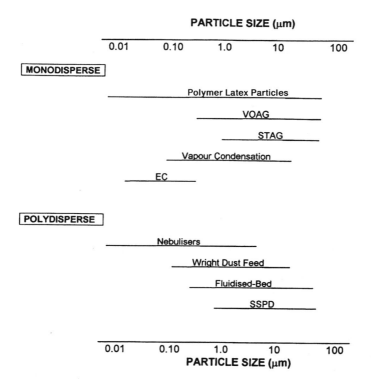

**Fig. 2.1** Calibration aerosol generators.

rather than work with purpose-made monodisperse particles, whose behaviour within the analyser may be different. As with monodisperse aerosols, the various methods for polydisperse aerosol generation operate within differing size ranges (Figure 2.1). It is important that conditions for either calm air or isokinetic sampling are achieved when using polydisperse aerosols, so that size-selective bias is not introduced between the aerosol source and the measurement zone of the instrument (Hinds, 1982).

Aerosol analysers are often calibrated in an enclosure (test chamber), which is connected to the aerosol generator in some way, and usually operated at room ambient pressure. Operation of the instrument within the test chamber may be preferred to minimize sampling losses, particularly when calibrating with polydisperse aerosols. However, even under these circumstances, consideration should be given to the possibility of biased sampling due to thermal convection currents near the inlet. Wind tunnels are often used for calibrations in which sampling efficiency of the analyser and inlet are being assessed together. It is normal to work with different sizes of monodisperse liquid droplets, or with solid particles if secondary effects, such as particle bounce and blow-off, are being investigated (John, 1993). The placement of the sampler in the working section of the wind tunnel is critical, and a second 'reference' sampler is usually located close to, but not obstructing the inlet to the analyser under calibration.

Chen (1993) categorized aerosol analysing instruments as follows: collection-and-analysis equipment (group A); and real-time, direct reading devices (group B). Each of these categories can be further subdivided into instruments that primarily determine number-weighted and those that measure mass- or volume-weighted size distributions (Table 2.1). Group A instruments fall into either classification. For instance, particles collecting on individual stages of a cascade impactor are weighed (assayed) to derive mass-based size distribution data. In contrast, number-weighted data are derived by filter collection microscopy. Group B instruments generally measure single particles on a one-by-one basis and therefore provide number-weighted data. However, laser diffractometers are an important exception, as they analyse the ensemble of particles simultaneously in the measurement zone to provide a volume-weighted size distribution. As a rule, monodisperse aerosols may be used to calibrate either number- or mass-based size analysers, as statistical errors introduced by transforming a number- to a mass-size distribution or vice versa are small. However, care is needed with polydisperse aerosols to establish traceability either by weighing (group A) or by counting microscopy (group B). It is better to avoid transforming size distributions from one weighting to another, as the few particles at one extreme of the distribution (e.g. large particles when transforming a number- to a mass-weighted size distribution) can exert a large influence on the final data.

It is often assumed that a calibration by spherical particles of uniform, known density provides information about analyser response in normal use with aerosols comprising non-spherical particles with unknown

**Table 2.1** Representative classes of aerosol analysers and the type of size distribution measured

| Instrument class | Group | Size distribution weighting |
|---|---|---|
| Impactors, impingers, cyclones | A | Mass |
| Inertial spectrometers | A | Mass |
| Real-time aerodynamic particle sizers | B | Number |
| Single particle optical counters | B | Number |
| Laser (phase) Doppler systems | B | Number |
| Laser diffractometers | B | Mass |
| Electrical mobility analysers | B | Number |
| Diffusion battery | A | Mass/number |
| Microscopy-image analysis | A | Number |

density. For aerodynamic particle size spectrometers, the aerodynamic size scale incorporates both particle density ($\rho_p$), unit (water droplet) density ($\rho_0$) and dynamic shape factor ($\chi$) through the relationship

$$d_a = d_v \left[\frac{\rho_p}{\chi\rho_0}\right]^{1/2} \left[\frac{C_c(d_v)}{C_c(d_a)}\right]^{1/2},  \qquad (2.1)$$

where $d_a$ and $d_v$ are aerodynamic and volume equivalent diameters respectively, and $C_c(d_a)$ and $C_c(d_v)$ are their Cunningham slip correction factors (Hinds, 1982). As long as the density and shape factor of the calibrant are known (the latter unity for spheres), particle size based on microscopy or another traceable method that provides volume equivalent diameter can be converted to aerodynamic diameter. Unfortunately this condition is not realized with calibrant particles that are porous or substantially non-spherical, because their preparation conditions are not well controlled. Assumptions then have to be made to link the two size scales, with a consequent loss of traceability. Even when the aerodynamic diameter is established, certain aerodynamic size analysers operate in the ultra-Stokesian regime where the simple relationship described by equation (2.1) no longer applies (section 2.5.4). Such instruments exhibit significant biases related to density (Wang and John, 1987) and shape (Marshall et al., 1991) in their size measurements.

The case of aerodynamic particle sizing instruments is illustrative of the need to be aware that secondary properties of calibrants may have a significant influence on the analyser response. A further example arises in the sizing of particles by methods that rely on scattered light intensity. Here, the refractive index of the calibrant may profoundly alter the relationship between size and light intensity associated with certain optical geometries, especially near-forward scattering. For this reason, it has been customary to calibrate such analysers with polymer latex microspheres, whose refractive index in air is well established (1.5905), and whose size, based on volume equivalent diameter, can be made traceable to the international length standard by microscopy.

The electrical charge associated with any test aerosol is a further issue that should be carefully considered. This is self-evident with the calibration of electrical mobility analysers, where charge modification to the aerosol is always undertaken prior to classification. However, a significant problem may arise with certain instruments in which adhesion between incoming particles and a collection surface is a prerequisite for accurate measurement. For instance, oscillating quartz crystal microbalances in which particles collect on an electrically insulating surface can be susceptible to charge build-up when sampling non-charge equilibrated aerosols, leading to loss of sensitivity as incoming particles are repelled by the accumulated electrostatic charge. It is therefore wise to ensure that calibrant aerosols are charge equilibrated (near zero net charge) by passing them through a bipolar ion generator such as a radioactive source (krypton-85 or polonium-210).

## 2.3 Calibration methods

### 2.3.1 *Monodisperse aerosols*

2.3.1.1 *Methods.* There are five commonly encountered sources of monodisperse aerosols: prefabricated particles, such as glass or polymer latex; controlled atomization; controlled vapour condensation; electrostatic size classification of a polydisperse aerosol; and naturally occurring uniform particles, such as pollen and fungal spores (not discussed further). A number of published standards contain detailed information about the first four of these, specifically German VDI standard 3491 Part 3 (VDI, 1980a) on polymer latex microspheres; Part 4 (VDI, 1980b) on the Sinclair–LaMer generator; Part 7 (VDI, 1987a) on the Rapaport–Weinstock generator; and Part 12 (VDI, 1987b) on centrifugal atomization. In addition, British Standard BS 3406 Part 7 (BSI, 1988) describes various methods of generating polymer latex aerosols for calibrating single particle optical counters.

2.3.1.2 *Prefabricated particles.* Several types of prefabricated particles are available for calibrating aerosol analysis equipment. Polymer (usually polystyrene) latex (PSL) microspheres are most frequently encountered, and are available as aqueous dispersions that require dilution before use in an atomizer- or nebulizer-based aerosol generator. Suppliers include Duke Scientific (Palo Alto, CA, USA), which provides a wide range of particles in the range from 0.01 to 30 μm volume equivalent diameter certified by methods traceable to the US National Institute of Standards and Technology (NIST); Dyno Industrier A/S, Lillestrøm, Norway, which makes highly uniform particles in the range from 2 to 20 μm volume equivalent diameter; Bangs Laboratories Inc., Carmel, IN, USA; Polysciences, Warrington, PA, USA; Seragen Diagnostics, Indianapolis, IN, USA; and Japan Synthetic Rubber Co. Ltd, Tokyo, Japan. A few of the larger sizes of PSL (over 5 μm volume equivalent diameter) are also obtainable as dry powders. Many sizes of PSL particles are available containing attached fluorescent dye which can be detected by fluorimetry, thereby providing a mass-based assay. However, care is needed to ensure that the dye does not leach from the particles into the suspending fluid before aerosol generation. Typically, PSL particles are highly monodisperse ($\sigma_g < 1.02$). However, particles larger than about 5 μm volume

**Table 2.2** Monodisperse PSL and polydisperse CRMs

(a) Monodisperse PSL CRMs

| Code | Source | Nominal mean particle diameter (μm) | Reference |
|---|---|---|---|
| SRM 1961 | NIST | 29.62 | Hartman *et al.* (1991) |
| SRM 1960 | NIST | 9.89 | Lettieri *et al.* (1991) |
| SRM 1962 | NIST | 2.977 | Hartman *et al.* (1992) |
| SRM 1690 | NIST | 0.895 | Mulholland *et al.* (1985) |
| SRM 1691 | NIST | 0.269 | Lettieri and Hembree (1989) |
| SRM 1693 | NIST | 0.107 | Kinney *et al.* (1991) |
| CRM 167 | BCR | 9.48 | Thom *et al.* (1985) |
| CRM 166 | BCR | 4.82 | Thom *et al.* (1985) |
| CRM 165 | BCR | 2.22 | Thom *et al.* (1985) |

(b) Polydisperse CRMs

| Code | Material | Source | Nominal particlesize[a] range (μm) | Reference |
|---|---|---|---|---|
| SRM 1003 | glass | NIST | 3–5 | |
| CRM 067 | quartz | BCR | 2.4–32 | Wilson *et al.* (1980) |
| CRM 070 | quartz | BCR | 1.2–20 | Wilson *et al.* (1980) |
| CRM 066 | quartz | BCR | 0.35–3.5 | Wilson *et al.* (1980) |

[a] Based on volume equivalent diameter.

equivalent diameter are often more polydisperse, the exception being those supplied by Dyno Industrier A/S, which are made by a polymer swelling process that preserves monodispersity.

Certain sizes of monodisperse PSL have been made available recently as CRMs by the European Bureau of Community Reference (BCR) and by NIST (Table 2.2a). These primary standards are expensive, and as a result their use is confined to the most critical applications.

In addition to PSL microspheres, uniform-sized glass beads of known density are available as calibrants from suppliers such as Duke Scientific, but the range of available sizes is limited, and does not extend to much finer than 0.15 μm volume equivalent diameter. Monodisperse, non-porous silica microspheres ($\rho_p = 1.96\,\mathrm{g\,cm}^{-3}$) are also available as an alternative to PSL in the range 0.15 to 5.0 μm volume equivalent diameter (Bangs Laboratories Inc.). Both glass and silica calibrants may be useful for calibrations at high temperatures, where PSL particles are unsuitable. However, particle counting is the main method of detection, as they are difficult to assay.

The most common method of generating PSL calibrants involves the use of a pneumatic medical nebulizer to form the aerosol by atomization of a dilute aqueous suspension containing the PSL particles

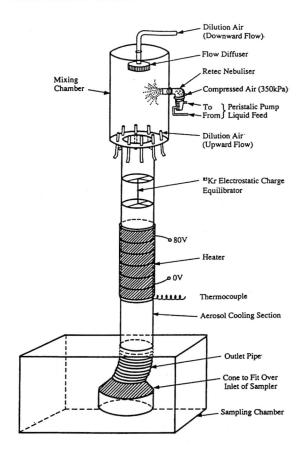

**Fig. 2.2** PSL calibration aerosol generator. Reprinted from Cox and Wathes (1995). Copyright Lewis Publishers, an imprint of CRC Press, Boca Raton, Florida, USA.

(section 2.3.2.2). The resulting droplets are subsequently dried and electrostatic charge equilibrated. A typical configuration of a PSL aerosol generator is shown in Figure 2.2, and other examples are given in BS 3406 Part 7 (BSI, 1988). This type of aerosol generator works well with PSL particles from about 0.1 to 5 µm volume equivalent diameter. Larger particles are inefficiently nebulized as their size exceeds that of most of the droplets formed by atomization.

There are two precautions that need to be considered when generating PSL aerosols by nebulization: multiplet formation; and the presence of spurious (residual) particles from contamination.

The formation of multiplet particles is brought about because the

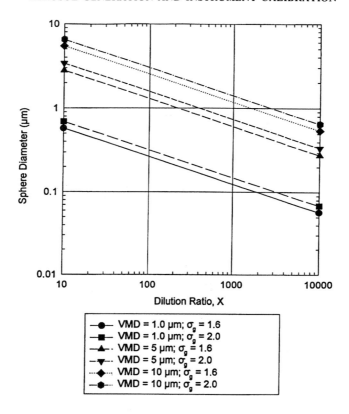

**Fig. 2.3** Dilution ratio required to generate a singlet ratio of 0.95 versus sphere diameter from stock of 10% w/v, for various droplet distributions formed by the nebulizer. (After Raabe, 1968.).

droplets initially produced by the nebulizer are mostly larger than the size of the PSL microspheres. Raabe (1968) described the relationship between the volume median diameter (VMD) and $\sigma_g$ of the water droplet aerosol produced by the nebulizer, to the dilution factor $(X)$ required to generate an aerosol comprising a known ratio of singlet particles $(R)$ from a stock suspension of known volumetric fraction of PSL $(F)$:

$$X = F(VMD)^3 \exp(4.5\ln^2 \sigma_g)\left[\frac{[1 - 0.5\exp(\ln \sigma_g)^2]}{(1 - R)d_v^3}\right], \qquad (2.2)$$

where $d_v$ is the volume equivalent diameter (mean size) of the PSL particles specified by the supplier. This relationship is valid for $R > 0.9$ and $\sigma_g < 2.1$, and is illustrated in Figure 2.3 for various water droplet size distributions, where $R$ is 0.95 (95%) and $F$ is 0.1 (10% w/v). Chen (1993)

**Table 2.3** Relative aerodynamic diameters of aggregates comprising up to eight singlet spherical particles ($\chi = 1.00$)

| Number of singlets | Configuration | Diameter |
|---|---|---|
| 1 | Singlet | 1.00 |
| 2 | Doublet | 1.19 |
| 3 | Chain triplet | 1.28 |
| 3 | Triangular triplet | 1.34 |
| 4 | Chain quadruplet | 1.38 |
| 4 | Triangular quadruplet | 1.42 |
| 5 | Chain quintuplet | 1.42 |
| 6 | Chain sextuplet | 1.45 |
| 4 | Tetrahedral quadruplet | 1.47 |
| 7 | Chain heptuplet | 1.48 |
| 5 | Triangular quintuplet (3 in line) | 1.50 |
| 6 | Triangular sextuplet (4 in line) | 1.52 |
| 8 | Chain octuplet | 1.52 |

tabulated values of VMD and $\sigma_g$ for well-known pneumatic nebulizers, and information for additional nebulizers can often be obtained from the manufacturer's promotional literature. A 10% w/v stock suspension containing 2 µm diameter PSL particles must be diluted 15 times to achieve a singlet ratio of 0.95, assuming that the VMD and $\sigma_g$ of the droplet distribution are 5 µm and 1.2, respectively. If $\sigma_g$ increases to 2.0, other parameters remaining constant, the dilution factor must increase to 80. In comparison, a stock suspension containing 0.2 µm diameter PSL particles must be diluted more than 15 000 times, assuming the same original water droplet distribution.

The presence of multiplets may be desirable for certain types of calibration, notably of aerosol spectrometers and real-time aerodynamic particle size analysers. This is because the ratios of aerodynamic sizes for the simpler aggregate shapes are well established in relation to the size of singlets (Stöber et al., 1969; Table 2.3). The probability of formation greatly decreases with increasing aggregate size, so it is likely that only aggregates comprising fewer than four singlets will be observed unless both the sensitivity and size resolution of the analyzer are high. These criteria are met with devices such as the spiral-duct aerosol centrifuge (SDC, section 2.5.3), and to illustrate an extreme case, Stöber and Flachsbart (1971) were able to resolve aggregates containing as many as twenty-three 1.8 µm diameter singlets under optimum conditions with their SDC.

The formation of submicrometre residual particles from empty droplets containing dissolved surfactant/stabilizer from the original PSL suspension is of major concern when attempting to produce aerosols comprising particles finer than 0.5 µm volume equivalent diameter. The use of surfactant-free PSL suspensions does not fully overcome the

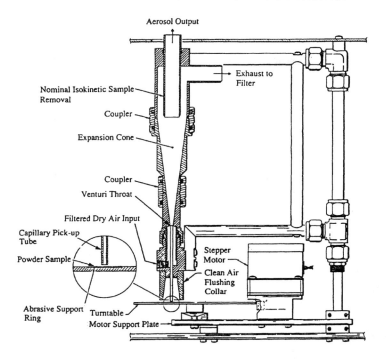

**Fig. 2.4** Small-scale powder dispenser (TSI Inc.) used to disperse larger PSL particles. Reprinted by permission of Particle Instruments Division, TSI Inc., St Paul, Minnesota, USA.

problem, as stabilizers are still present. An effective measure is to dilute the suspension supplied by the manufacturer with the purest water available (dissolved organic and inorganic species removed). The PSL particles can then be centrifuged and the supernatant liquid replaced with pure water. However, this process may need to be undertaken as much as 10–15 times to eliminate surfactant. Alternatively, as the contaminant particles are polydisperse, the electrical classifier method may be used to size-separate them from the uniform-sized PSL particles (section 2.3.1.5).

PSL particles larger than 5 μm volume equivalent diameter are better generated by dry dispersion, and the Small-Scale Powder Disperser (SSPD, TSI Inc.) has been developed with this application in mind (Figure 2.4). It can efficiently suspend particles as large as 30 μm in the upward direction (section 2.3.2.3). However, the resulting aerosol number concentration is much lower than that for nebulized micrometre-sized calibrants, and great care is needed to avoid severe losses due to impaction and gravitational deposition on surfaces between the SSPD outlet and the instrument being calibrated. Sharp bends should be avoided, with the

SSPD ideally located immediately beneath the analyser inlet; the analyser may need to be inverted to achieve this goal.

The use of 'cocktails' containing more than one size of monodisperse PSL particles is attractive, as they can provide calibration data at several sizes simultaneously. The process of producing such 'cocktails' is, however, arduous, as great care has to be taken to avoid significant coagulation in the blended aqueous suspension, and reference techniques to determine the relative number concentrations of each of the component sizes are in their infancy. At the present time, four different premixed 'cocktail' suspensions of PSL are about to become available in the UK as CRMs through the Office of Reference Materials of the Laboratory of the Government Chemist. These 'cocktails' were developed by AEA Technology plc in collaboration with Japan Synthetic Rubber Co. Ltd, Tokyo, Japan, as part of the recent UK Valid Analytical Measurement (VAM) initiative. The following nominal particle sizes (based on volume equivalent diameter) will be made available, with each component present at close to $10^8$ particles $cm^{-3}$:

- CRM AEA1004 – 0.1 μm, 0.2 μm, 0.5 μm;
- CRM AEA1005 – 0.2 μm, 0.5 μm, 1.0 μm;
- CRM AEA1006 – 0.5 μm, 1.0 μm, 2.0 μm;
- CRM AEA1007 – 1.0 μm, 2.0 μm, 5.0 μm.

These CRMs have been prepared from appropriate quantities of single-component PSL suspensions, paying attention to the differing dilution factor required for each component to produce approximately equal particle concentrations of each component. Particle concentration in each component before blending has been verified using a scanning electron microscope to count particles deposited from the liquid suspension onto a silicon wafer overlaid with a mask containing a small-diameter aperture to constrain the particles to a known region.

Single-component or 'cocktail' suspensions of PSL are potentially usable as aerosol concentration standards, as they can be nebulized from a well-defined stock suspension and delivered in a known volume of air (or other diluent gas). The Aeromaster-I® aerosol generator (Japan Synthetic Rubber Co. Ltd, Tokyo, Japan) is an example of a commercially available constant number concentration standard for use with PSL particles from about 0.1 to 1.0 μm volume equivalent diameter. The manufacturer has claimed that particle concentrations from $3.5 \times 10^{-1}$ to $3.5 \times 10^2 \, cm^{-3}$ are possible within a tolerance of ±10% during a period of 2 hours, using its proprietary brand of PSL microspheres (Airtex®). The particles are electrostatically charge equilibrated prior to delivery at the outlet, and a major application is the calibration of single particle counters for clean rooms. Table 2.4 is a guide to the number concentration of PSL particles ($N_{stock}, cm^{-3}$) of different sizes contained in undiluted 2.5% w/v and

**Table 2.4** Particle number concentration present in PSL suspensions supplied as 2.5% w/v and 10% w/v suspensions

| PSL diameter ($\mu$m) | 2.5% w/v suspension (cm$^{-3}$) | 10% w/v suspension (cm$^{-3}$) |
|---|---|---|
| 0.05 | $3.64 \times 10^{14}$ | $1.46 \times 10^{15}$ |
| 0.10 | $4.55 \times 10^{13}$ | $1.82 \times 10^{14}$ |
| 0.20 | $5.68 \times 10^{12}$ | $2.27 \times 10^{13}$ |
| 0.50 | $3.64 \times 10^{11}$ | $1.46 \times 10^{12}$ |
| 0.75 | $1.08 \times 10^{11}$ | $4.32 \times 10^{11}$ |
| 1.0 | $4.55 \times 10^{10}$ | $1.82 \times 10^{11}$ |
| 2.0 | $5.68 \times 10^{9}$ | $2.27 \times 10^{10}$ |
| 5.0 | $3.64 \times 10^{8}$ | $1.46 \times 10^{9}$ |
| 10.0 | $4.55 \times 10^{7}$ | $1.82 \times 10^{8}$ |

10% w/v suspensions. The particle number concentration at the outlet from a PSL aerosol generator ($N_{PSL}$, cm$^{-3}$), can be estimated from

$$N_{PSL} = \frac{N_{stock} R_a}{10^3 X Q_a} \eta(d_v), \qquad (2.3)$$

where $X$ is the dilution factor of the PSL suspension in the nebulizer compared with that in the stock suspension, $R_a$ is the rate of aerosolization of the liquid suspension in the nebulizer (cm$^3$ min$^{-1}$), $Q_a$ is the flow rate of air diluting the aerosol (1 min$^{-1}$) and $\eta(d_v)$ is the particle size-dependent fractional transport efficiency of the aerosol generator to the measurement point. The aerosol mass concentration can be derived from the mean particle size and density (1.05 g cm$^{-3}$ for PSL). It is therefore possible to achieve traceability for both number and mass concentration standard aerosols produced in this way through measurements of suspension feed rate, nebulizer concentration and particle size. However, internal losses after aerosol formation (affecting $\eta(d_v)$) cannot easily be quantified without undertaking a rigorous mass balance, severely limiting the validation of such aerosol sources, and explaining their lack of development.

### 2.3.1.3 *Atomization methods*

2.3.1.3.1 *Spinning top/disc.* The spinning top (May, 1949; Walton and Prewett, 1949) is an example of a centrifugal atomizer that is capable of producing monodisperse aerosols ($\sigma_g < 1.05$) of liquid droplets or solid particles in the size range from about 1 to 50 $\mu$m volume equivalent diameter. The commercially available version (STAG Mark-2, BIRAL, Portishead, UK; BGI Inc., Waltham, MA, USA) is based on refinements to the original design of May (1966). Solid particles are formed by delivering a solution or colloidal suspension to the centre of a flat-surfaced

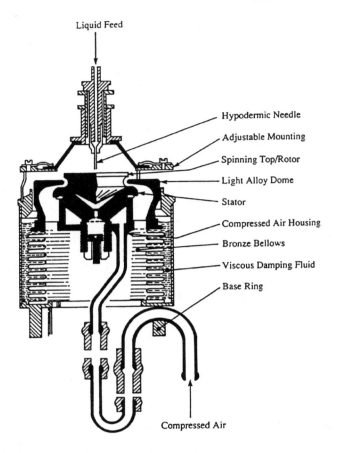

Liquid Feed

Hypodermic Needle

Adjustable Mounting

Spinning Top/Rotor

Light Alloy Dome

Stator

Compressed Air Housing

Bronze Bellows

Viscous Damping Fluid

Base Ring

Compressed Air

**Fig. 2.5** May spinning-top aerosol generator (STAG). Reprinted from May (1966), by permission of IOP Publishing Ltd, Bristol, UK.

spinning rotor through a hollow (hypodermic) needle (Figure 2.5). In the commercially available equipment, the rotor table is 2.54 cm in diameter and achieves rotation speeds in the range 700 to 1000 revolutions per second. A thin film of liquid spreads outwards from the centre of the rotor by centrifugal force. Film rupture occurs to form uniform-size primary droplets together with ligaments that eventually form smaller satellite droplets when centrifugal force overcomes surface tension. In the May apparatus, the satellites are separated from the primary droplets on the basis of their differing inertia by suction through a separate exhaust channel whose entrance forms part of the cover surrounding the rotor. The primary droplets travel beyond the influence of this suction and can be drawn off in either upwards or downwards direction for use.

The liquid droplet diameter ($d_1$, cm) can be predicted within a wide range of experimental conditions (Walton and Prewett, 1949):

$$d_1 = \frac{K}{\omega} \left[ \frac{T}{d_{\text{rot}} \rho_1} \right]^{1/2}, \qquad (2.4)$$

where $T$ is the liquid surface tension, $d_{\text{rot}}$ is the rotor table diameter, $\omega$ is the angular velocity of the rotor ($= 2\pi F_{\text{rot}}$ where $F_{\text{rot}}$ is rotation frequency), $\rho_1$ is the liquid density, and $K$ is a numerical constant reflecting several ill-determined parameters including the surface roughness of the rotor table and the degree of wetting by the liquid film. $K$ values can vary widely, and Mitchell (1984) reported a range from 2.3 to 9.4 for a range of aqueous liquids containing ferric oxide sol, methylene blue and sodium fluorescein. This limits the ability to predict droplet (particle) size from theoretical considerations (unlike the vibrating orifice aerosol generator (section 2.3.1.3.2)). Traceability must therefore be achieved using an independent technique, such as filter collection microscopy, to measure the sizes of several hundred droplets/particles.

The size of dried particles formed from the evaporation of colloidal suspensions or solutions is predictable assuming uniform surface mass transfer to the vapour phase from a spherical surface:

$$d_{\text{v}} = d_1 \left[ \frac{C}{\rho_{\text{s}}} \right]^{1/3}, \qquad (2.5)$$

where $C$ is the suspension (solution) mass concentration and $\rho_{\text{s}}$ is the solid particle density. Careful control of drying conditions is needed to produce smooth, spherical particles (Mitchell, 1984), essential if the aerodynamic diameter of the particles is being calculated from their volume equivalent diameter (equation (2.1)).

Several types of spinning top aerosol generator (STAG) have been developed for different applications. In one design, the primary droplets are drawn slowly upwards in a non-turbulent flow to permit time for partial evaporation to take place. This process may need to be completed by passing the particles through a trace-heated tube before charge equilibrating them prior to use (Mitchell, 1984). While highly spherical particles having excellent uniformity of size can be produced by this equipment, it suffers from the drawback that particles larger than $5\,\mu m$ aerodynamic diameter are difficult to transport efficiently. In an alternative design, the droplets were transported downwards from the point of generation to overcome this problem (Jenkins et al., 1985), and this equipment was able to produce particles as large as $15\,\mu m$ aerodynamic diameter efficiently.

Careful needle alignment, both centrally with respect to the rotor table and with the needle tip located at a fixed height above the rotor surface, is critical to obtain reliable results from a STAG. This is especially important when working with concentrated suspensions, where deposition of solid material on the rotor table causes it to settle, widening the gap to the needle tip. A mechanism was developed for use with the May spinning top equipment so that the needle tip could be raised or lowered from outside the enclosure housing the STAG (Mitchell and Stone, 1982). Smooth flow of liquid from needle to rotor table is diagnostic of good primary droplet production, whereas pulsations are associated with the formation of highly polydisperse droplets. Careful control of spent rotor drive air is also essential to avoid irregular operation of the rotor and more importantly to prevent satellite droplets contaminating the aerosol. It is important to ensure that a slight vacuum is present in the line conveying the spent rotor drive air out of the STAG enclosure. The purest reagents should always be used, and it is important to dry the droplets carefully to avoid the formation of non-spherical, porous particles whose aerodynamic size cannot be estimated readily. The optimum drying conditions must be determined for each STAG configuration and material being used, but as a general rule it has been found that the most common fault is the formation of misshapen doughnut-like particles when drying is too rapid. Charlesworth and Marshall (1960) and, more recently, Leong (1981) have established conditions for reliable formation of spherical particles from many solutions and suspensions.

2.3.1.3.2  *Vibrating orifice aerosol generator.*   If a thin stream of liquid is forced through a fine orifice under pressure, the flow will break up into discrete droplets by the interaction of external forces (gravity and surface tension). The formation of highly uniform droplets can be achieved by applying a periodic vibration to the orifice within narrow frequency ranges within which so-called **varicose** instability occurs. The most widely available aerosol generator based on this principle is the VOAG (Figure 2.6), developed by Berglund and Liu (1973) and available from TSI Inc. The heart of the system is the piezoceramic crystal surrounding the orifice disc, which may be mounted to generate droplets either upwards, during normal use (Vanderpool and Rubow, 1988) or downwards, for large particle production (Mitchell *et al.*, 1987). Like the spinning top generator, the VOAG can be used to form calibrant particles from either solutions or colloidal suspensions, and various orifice sizes are available that enable particles from about 0.5 to 50 μm volume equivalent diameter to be produced. The volume equivalent diameter of the dried particles can be predicted directly from the relationship

$$d_v = 10^4 \left[ \frac{Q_l C_f}{10\pi F_{\text{vib}}} \right]^{1/3}, \tag{2.6}$$

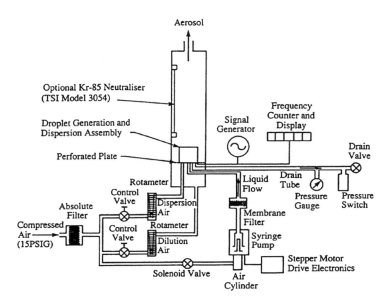

**Fig. 2.6** Vibrating orifice aerosol generator (VOAG). Reprinted by permission of Particle Instruments Division, TSI Inc., St Paul, Minnesota, USA.

where $Q_l$ is the liquid feed rate, $C_f$ is the fractional concentration of solute/suspension in the feed liquid and $F_{vib}$ is the orifice vibration frequency. The absolute nature of this relationship (with no empirical constant) makes the VOAG a primary calibration standard in terms of $d_v$, although it is recommended that an independent check be made of particle size and sphericity by a traceable method, such as microscopy. Aerosols produced by the VOAG can be highly monodisperse ($\sigma_g < 1.02$).

Significant care is required to operate the VOAG satisfactorily. In particular, the orifice is prone to plugging, and it is therefore recommended that the pressure in the liquid feed line be monitored. The VOAG should not produce satellite droplets if set up within the correct frequency range for the particular orifice and sufficient power (vibration amplitude) is available. However, slight variations outside the optimum operating range will result in the formation of poor-quality aerosols. For this reason, it is advised that the aerosol be monitored by an independently calibrated particle size analyser having adequate resolution, such as the Aerodynamic Particle Sizer® aerosol spectrometer (TSI Inc.). Careful control of drying conditions after droplet formation is also essential for the same reasons as described for the STAG. Particular conditions that produced highly spherical methylene blue microspheres with a VOAG have been described by Mitchell and Waters (1986).

The VOAG has the potential to be used as a particle concentration source from about 10 to 400 cm$^{-3}$, because of its extremely stable droplet production rate. However, care is needed to control and quantify not only the liquid and air feed rates, but also particle losses in pipework from the vibrating orifice to the point of measurement (Horton and Mitchell, 1992) and the spatial uniformity of particle concentration. The last requirement is a major limitation, as uniform mixing of particles larger than 5 μm aerodynamic diameter with carrier/dilution air is very difficult to achieve.

In principle, traceability can be achieved in terms of number concentration ($N$, particles cm$^{-3}$) through the relationship

$$N = \frac{0.06F}{Q_a}\eta(d_v),\tag{2.7}$$

where $Q_a$ is the flow rate of dilution gas (air) in l min$^{-1}$. In practice, there are always losses of particles after formation, and the particle size-dependent transport efficiency of the aerosol generator (expressed as a fraction) to the point of measurement ($\eta(d_v)$) has to be determined by an independent, non-invasive technique. In addition, a small percentage of doublet particles (typically less than 1%) are always formed by droplet collisions close to the dispersion orifice, reducing the particle number concentration.

The development and application of the VOAG as a particle concentration as well as a size standard necessitate precise control and regulation of liquid and air feed rates using devices such as an isocratic pump and gas flow controllers, respectively. It is a practical proposition only within a fairly narrow size range (0.5 to 5 μm aerodynamic diameter) where inertial impaction and gravitational deposition to internal surfaces are comparatively low.

2.3.1.4  *Heterogeneous vapour condensation.*  Sinclair and LaMer (1949) showed that the principle of heterogeneous vapour condensation could be applied to form monodisperse aerosols from about 0.1 to 10 μm volume equivalent diameter. In this process, a controlled concentration of polydisperse solid nuclei particles from 10 to 100 nm is exposed to saturated vapour of a suitable high-boiling-point liquid, such as di-2-ethylhexyl sebacate (DEHS) or dioctyl phthalate (DOP). Condensation of the vapour on the nuclei takes place as the vapour–nuclei mixture is cooled slowly during transportation to the sampling point. Under these conditions, condensation and growth are the same for each nucleus as it passes through the condensation region. Each droplet therefore grows to the same final size, which can be predicted (Hinds, 1982):

$$d_v = 10^4 \left[\frac{6C_{vap}}{\pi\rho_l N_{nucl}}\right]^{1/3}\tag{2.8}$$

assuming negligible condensation to the walls of the apparatus. $C_{vap}$ is the vapour mass concentration, $\rho_l$ is the liquid density and $N_{nucl}$ is the nuclei number concentration. Condensation aerosol generators typically produce aerosol particle concentrations in excess of $10^5\,cm^{-3}$, but some systems can attain $10^7\,cm^{-3}$, which is far higher than can be achieved with any of the other types of monodisperse aerosol generator.

The Sinclair–LaMer generator constructed by Muir (1965) is a more practical version of the original apparatus (Figure 2.7). A stream of oxygen-free nitrogen passes through a 2 litre flask containing the high-boiling-point liquid at a constant temperature maintained within ±0.5°C of a preset value (in the range 110 to 150°C for DEHS), and which is stirred continuously. The nuclei are generated from an electric arc across tungsten electrodes, but a nebulizer could also be used to form nuclei of materials, such as NaCl from very dilute solutions (section 2.3.2.2). In contrast with the original aerosol generator, the carrier gas is not bubbled through the liquid, and condensation to form the calibration particles takes place in a downward-facing chimney section.

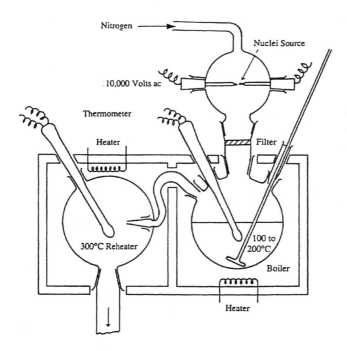

**Fig. 2.7** Sinclair–LaMer condensation aerosol generator. Reprinted from Muir (1965), with kind permission from Elsevier Science Ltd., The Boulevard, Langford Lane, Kidlington, Oxford OX5 1GB, UK.

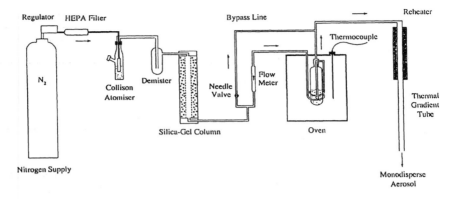

**Fig. 2.8** MAGE condensation aerosol generator. Reprinted from Horton *et al.* (1991), with kind permission from Elsevier Science Ltd., The Boulevard, Langford Lane, Kidlington, Oxford OX5 1GB, UK.

The Rapaport–Weinstock (RW) generator (Rapaport and Weinstock, 1955) was an attempt to attain more rapid changes to particle size than can be achieved by the Sinclair–LaMer generator. A polydisperse aerosol of the high-boiling-point liquid is generated by a pneumatic nebulizer located at the bottom of a vertical tube. These droplets are evaporated, leaving behind the nuclei that originate from impurities in the liquid. Condensation takes place near to the top of the tube as the vapour–nuclei mixture is cooled in a controlled thermal gradient. This aerosol generator takes only a few minutes to restabilize after changing operating conditions, and thermal decomposition of the bulk liquid (a problem with the Sinclair–LaMer apparatus) is avoided by continuously feeding fresh liquid to the heated section. Temperature control of the vaporizer section is less critical than required for the Sinclair–LaMer aerosol generator.

The monodisperse aerosol generator (MAGE) developed by Prodi (1972), commercially available from Physis S.R.L., Bologna, Italy, allows even more rapid changes to particle size by separating the nuclei into two streams, one of which bypasses the saturator vessel containing the high-boiling-point liquid (Figure 2.8). The streams are then recombined and passed through a reheater to ensure that premature condensation has not occurred before the vapour–nuclei mixture passes downwards through the cooling tube. Careful control of the ratio of bypass to main flows enables the droplet size to be adjusted and stabilized in seconds, because the thermal inertia of the system is so small. The MAGE uses a nebulizer to provide the condensation nuclei, which may be tagged with a fluorescent dye to provide droplets that can be conveniently assayed on a mass-weighted basis by fluorimetry (Horton *et al.*, 1992).

A MAGE was evaluated using DEHS as high-boiling-point liquid with NaCl nuclei from aqueous solution (Horton *et al.*, 1991), with the following observations:

- Incomplete drying of the nuclei does not influence monodispersity of the aerosol provided the reheater temperature is sufficient to ensure complete evaporation.
- Premature condensation between the saturator vessel and reheater must be avoided by trace-heating the linkage tube.
- The reheater temperature should be at least 50°C greater than that in the bubbler vessel.
- The level of liquid in the saturator vessel should be kept constant to avoid a small, but significant drift in particle size with time.

The condensation monodisperse aerosol generators (CMAGs) produced by TSI Inc. are similar in concept to the MAGE, but are based on the work of Altmann and Peters (1992) and Peters and Altmann (1993). The stream of nuclei can be partly diverted through a filter to reduce the concentration reaching the saturator vessel, thereby extending the upper size range of the aerosol generator. Further refinements involve improving monodispersity ($\sigma_g \simeq 1.1$) by selecting the central core of the flow emerging from the thermal gradient tube.

2.3.1.5 *Electrostatic classification.* Electrostatic size classification is a powerful technique for producing small quantities of calibrant particles from about 0.02 to 0.2 µm diameter from polydisperse source aerosols. For particles carrying one elementary unit of electric charge ($n = 1$), the electrical mobility ($Z_p$ (cm$^2$ statV$^{-1}$ s$^{-1}$)) is a unique function of the particle diameter ($d_v$):

$$Z_p = \left[\frac{neC_c}{3\pi\eta d_v}\right],\tag{2.9}$$

where $e$ is the elementary unit of charge ($4.8 \times 10^{-10}$ statC), $C_c$ is the Cunningham slip correction factor and $\eta$ is the gas viscosity. The aerosol produced is monodisperse, because almost all particles of this size range are singly charged during classification. The size distribution of the classified aerosol is triangular, with the count median diameter (CMD$_{em}$), based on electrical mobility diameter, being determined by the following parameters of the electrostatic classifier:

$$\text{CMD}_{em} = \frac{4 \times 10^7 ne\Lambda V C_c(d)}{3\eta(q_s - q_e)}\tag{2.10}$$

where $n = 1$ (singly charged particles), $V$ is the applied voltage, and $q_s$ and $q_e$ are the flow rates of sheath and excess air respectively (cm$^3$ s$^{-1}$) through the classifier. $\Lambda$ is the classifier dimension parameter given by

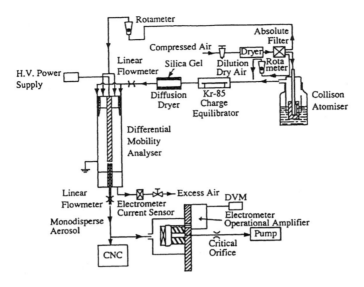

**Fig. 2.9** Electrostatic classifier as a source of monodisperse aerosols. Reprinted from Agarwal and Sem (1980), with kind permission from Elsevier Science Ltd., The Boulevard, Langford Lane, Kidlington, Oxford OX5 1GB, UK.

$L/\ln(r_2/r_1)$, where $L$ is the classifier length and $r_1$ and $r_2$ are the outer and inner radii of the classifier.

The electrostatic classifier (EC (TSI Inc.)) calibration method developed by Liu and Pui (1974), is a primary standard based on the above equation. Its operation is illustrative of the principles underlying the formation of these calibrant aerosols. A polydisperse aerosol is first formed by pneumatic nebulization of a dilute solution, such as sodium chloride (section 2.3.2.2). The dry submicrometre particles are then passed through a bipolar charge equilibrator before entering the EC (Figure 2.9). The incoming aerosol therefore comprises predominantly particles having zero electric charge, with progressively smaller numbers having ±1, ±2 charges, etc. The aerosol is introduced in laminar flow close to the walls of the cylindrical classifier, where the individual particles encounter a radial electric field maintained by the central cylindrical rod electrode that is at a high voltage with respect to the classifier wall. This rod is surrounded by a sheath of particle free air. If a negative potential is applied to the central electrode, particles that are positively charged will be drawn towards this electrode in well-defined trajectories that are a function of the internal geometry of the classifier, the flow rate, particle size and number of charges per particle (equation (2.10)). Other particles are either unaffected (no charge) or deflected towards the outer wall (negatively charged), and both groups are swept out of the EC to a filter. Only particles that have a narrow band of electrical mobility have trajectories that enable them to cross the sheath air flow, just missing the tip of the central electrode, passing through a slot below its base to emerge from the EC as 'monodisperse' aerosol. $\sigma_g$ values typically range from 1.04 to 1.10.

When generating particles larger than 0.04 μm diameter, there are always a small number of larger, multiply-charged particles that have the same electrical mobility. The proportion of such contaminant particles is minimized by ensuring that the $CMD_{em}$ of the original polydisperse aerosol is always smaller than the desired size of monodisperse aerosol, and by working in the size range below about 0.5 μm where the probability of multiply-charged particle formation is greatly reduced. It is this latter constraint as much as reduced electrical mobility with increasing particle size that defines the upper limit within which the EC method is effective. Some workers eliminate the multiply-charged particles by the use of a pre-impactor or by passing the near-monodisperse aerosol through a second EC.

The EC is usable in either under- or over-pressure modes, and is therefore highly effective at calibrating low-flow impactors by comparing inlet and outlet particle concentrations (section 2.5.2). In under-pressure mode, the aerosol is drawn off the 'monodisperse' port by the instrument being calibrated, and the particle concentration leaving the device may be monitored by a condensation nucleus counter (CNC). In the over-pressure mode, the flow at the 'monodisperse' port is generated within the EC and the instrument being calibrated is allowed to sample at atmospheric pressure at a vented coupling. In this configuration, the CNC measures the particle concentration at the inlet to the instrument being calibrated.

The EC can also be used as a number concentration standard for particles finer than 0.06 μm diameter, as almost all 'monodisperse' particles each carry unit positive charge that can be measured as an electric current ($I$) by a sensitive Faraday cup electrometer:

$$I = QeN, \tag{2.11}$$

where $N$ is the aerosol number concentration and $Q$ is the volumetric flow rate into the electrometer. This capability was used by Liu and Pui (1974) in their primary calibration of a condensation nucleus counter (section 2.5.8), and is still the method of choice for these devices.

### 2.3.2  *Polydisperse aerosols*

Polydisperse aerosols are less widely used as calibrants, mainly because it is more difficult to ensure that the aerosol has not been modified by size-selective processes. Despite this limitation, there are a few applications, such as the calibration of ambient dust monitors and laser diffractometers, where such aerosols are useful, provided that they are generated and transported to the analyser in a well-defined and reproducible manner.

2.3.2.1 *Prefabricated particles.* Like their monodisperse equivalents, polydisperse prefabricated particles require an aerosol generation system to disperse them adequately. This is usually a dry powder feed (section 2.3.2.3), especially when large quantities must be generated, but wet systems based on nebulizers (section 2.3.2.2) are also used. The aerosol generator may modify the original powder size distribution (especially with wet dispersion methods), and the calibrant should therefore always be independently size-analysed.

Polydisperse CRMs based on prefabricated particles are available from governmental certifying organizations (Table 2.2b), but few of these standards comprise particles finer than 10 μm volume equivalent diameter. The BCR CRMs have been certified both on a number-weighted (microscopy) and mass-weighted (gravitational sedimentation) basis (Wilson *et al.*, 1980). The latter provides particle size scaled in terms of Stokes diameter, which is related to aerodynamic diameter by the ratio of the bulk density of the particles to the density of water (Hinds, 1982), making these particles potentially useful for the calibration of size analysers based on inertial separation.

The main drawback to the use of the present range of CRMs is their high cost, combined with uncertainty concerning their secondary properties (particularly particle shape and refractive index in the case of the quartz particles from the BCR (Scarlett, 1985). For instance, these CRMs are of little use in the calibration of many optical aerosol analysers, because their irregular shape and variable refractive index result in an ill-defined relationship between the certified size distribution and the instrument response. The process could be simplified if new CRMs based on spherical particles of known density and refractive index are developed (Mitchell, 1992). Meanwhile, the use of well-characterized secondary standards, such as polydisperse glass microspheres (Whitehouse Scientific, Chester, UK or Gilson Co. Inc., Worthington, OH, USA (Table 2.5)) are acceptable. These particles have unimodal size distributions ranging from 0.1 to 1.0 μm, 0.5 to 3.0 μm, 2 to 10 μm and 5 to 30 μm volume equivalent diameter, with well-defined sphericity, density and refractive index.

Bulk powders, such as Arizona Road Dust and the (UK) Motor Industry Research Association (MIRA) fine powders (Table 2.5), may also be used as source materials for polydisperse calibrants. These powders are of particular use where a significant mass of aerosol is required, such as the calibration of environmental samplers using a wind tunnel (Chen, 1993).

2.3.2.2 *Nebulizers.* Pneumatic nebulizers are the simplest way to generate liquid droplet aerosols, typically using compressed air at a pressure from about 100 to 500 kPa to draw liquid from a reservoir to a

**Table 2.5** Polydisperse secondary standards suitable for use as calibration aerosols

| Material | Identification | Nominal size[a] range (μm) | Supplier |
|---|---|---|---|
| Soda glass | BS 170-172 | 0.1 to 1.0 | Whitehouse/Gilson[b] |
| Soda glass | BS 180-182 | 0.5 to 3.0 | Whitehouse/Gilson[b] |
| Soda glass | BS 190-193 | 2.0 to 10 | Whitehouse/Gilson[b] |
| Soda glass | BS 200-204 | 5 to 30 | Whitehouse/Gilson[b] |
| $SiO_2/Al_2O_3/Fe_2O_3$ | Arizona Road Dust | 73% mass < 20 μm 39% mass < 5 μm | AC Spark Plug Div.[c] |
| $Al_2O_3$ | MIRA 1 (BS 4552) | 2.5 to 9 | Powder Products[d] |
| $Al_2O_3$ | MIRA 2 (BS 4552) | 3 to 11 | Powder Products[d] |
| $Al_2O_3$ | MIRA 3 (BS 4552) | 6 to 21 | Powder Products[d] |
| $Al_2O_3$ | E8 (ISO 4020/1) | 3.5 to 10.5 | Powder Products[d] |
| $Al_2O_3$ | No. 2 (BS 2831) | 1 to 10 | Powder Products[d] |
| $Al_2O_3$ | No. 3 (BS 2831) | 8 to 32 | Powder Products[d] |

[a] Based on volume equivalent diameter.
[b] Other sizes available on request. Whitehouse Scientific Ltd, Waverton, Chester CH3 7PB, UK; Gilson Co. Inc., Worthington, OH 43085-0677, USA.
[c] AC Spark Plug Division, General Motors Corp., Flint, MI, USA.
[d] Powder Products Ltd, P.O. Box 5, Spondon, Derby, UK.

fine orifice as a result of the Bernouilli effect (Mercer *et al.*, 1968; Figure 2.10a). A wide range of these devices is available as they are extensively used in inhalation-based drug therapy, and Chen (1993) has listed the operating parameters of several nebulizers. Many of those intended for

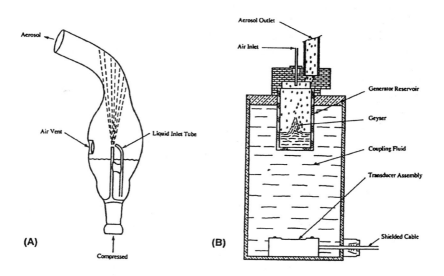

**Fig. 2.10** Pneumatic and ultrasonic nebulizer concepts. Reprinted from Cox and Wathes (1995). Copyright Lewis Publishers, an imprint of CRC Press, Boca Raton, Florida, USA.

medical use are optimized to operate at 50 pounds per square inch gauge (335 kPa), which is the normal air supply pressure in hospitals. Particles from pure non-volatile liquid (e.g. DEHS), dilute solutions of salts, or suspensions of insoluble particles may be generated by this method.

Ultrasonic nebulizers utilize the mechanical energy produced by a piezoelectric crystal vibrating in an applied a.c. electric field. The vibrations are transmitted through a coupling fluid to a vessel containing the liquid being nebulized (Figure 2.10b). Capillary waves formed on the surface break down rapidly into discrete droplets, whose sizes are similar to those produced by pneumatic nebulizers. Droplet concentrations are initially much higher than those produced by pneumatic nebulizers, as little or no air flow need be present above the liquid. Coagulation may therefore be an issue, and can be minimized by increasing the flow of carrier air.

When generating calibrants from suspensions, it is important to realize that the droplet size distribution produced by the nebulizer will influence the size distribution of the aerosol produced after evaporation of the suspending fluid (section 2.3.1.2). The size distribution of droplets formed by pneumatic nebulizers is typically close to lognormal, having a volume (mass) median diameter from 3 to 8 μm volume equivalent diameter with $\sigma_g$ values between 1.4 and 2.5. Limited control over droplet size can be exercised by varying the air pressure (finer droplets are formed at higher pressures). Nebulizers are unsuitable for suspending particles larger than about 5 μm aerodynamic diameter.

Nebulizers are frequently used to generate solid particles from dilute solutions. These particles may be of a fluorescent compound, making it possible to work with very small amounts of submicrometre calibrant particles having mass median diameters from about 0.05 to 0.2 μm volume equivalent diameter with $\sigma_g$ between 1.4 and 2.0, depending upon solute concentration in the nebulizer reservoir (Horton et al., 1989). They can also be used as nuclei in a heterogeneous condensation-type monodisperse aerosol generator (section 2.3.1.4). Alternatively, spherical particles containing entrapped radionuclides can be formed from suspensions by nebulization (Kanapilly et al., 1970), taking advantage of the very sensitive methods to detect emitted radiation to avoid the need for large quantities of calibrant.

Although nebulization is a simple process, precautions need to be observed, especially if the aerosol is being generated over a significant period of time. If the solvent evaporates rapidly after droplet formation, its continuous loss will cause a progressive increase in solute concentration with a resulting drift in size distribution of the dried particles to larger sizes. This problem can be overcome by circulating the solution from the reservoir of the nebulizer through a large auxiliary reservoir, cooling the nebulizer and/or prehumidifying the supply air. It may also be necessary to dehumidify the air containing the dry particles to preserve particle size if the substance being used to form the calibrant is hygroscopic (Horton *et al.*, 1989).

2.3.2.3  *Dry powder generators*.  There are several ways to disperse fine powders as aerosols (Hinds, 1980); commonly encountered methods are those based on direct dust-feed to a supply of air, fluidized-bed aerosol generators and the capillary-aspiration technique developed as the TSI Small-Scale Powder Disperser (SSPD (Figure 2.4)). Chen (1993) has summarized operating parameters for several widely used dry powder dispersers, including air and powder flow rates, aerosol mass concentration and optimum particle size range. The Wright (1950) dust feed (BGI Inc., Waltham, MA, USA) operates well with powders from 0.2 to 10 µm volume equivalent diameter, whereas the fluidized-bed system (FBAG (Marple *et al.*, 1978)) and SSPD (Blackford and Rubow, 1986) (both from TSI Inc.) are best used with powders 0.5 to 40 µm and 1 to 50 µm volume equivalent diameter, respectively. The choice of powder-aerosol generator depends mainly on the desired aerosol concentration. For instance, the SSPD produces very low-concentration aerosols ($300 \,\mu g \,m^{-3}$ to $40 \,mg \,m^{-3}$), whereas the FBAG is capable of providing aerosols of moderate mass concentration ($100 \,mg \,m^{-3}$ to $4 \,g \,m^{-3}$), and the Wright dust feed can generate aerosols having a mass concentration greater than $10 \,g \,m^{-3}$. Other dry powder systems are capable of producing higher mass concentrations, but their powder consumption rate precludes their use with CRMs or most standard test dusts.

The main advantages of dry systems are their ease of use and their ability in many cases to achieve near-perfect dispersion of the powder in aerosol form. However, they are all vulnerable to the influence of humidity on dispersion, especially with hydrophilic powders. In addition, particle shape can have a marked effect on performance; spherical particles are generally easier to disperse than plates and fibres. Correct dispersion involves matching powder properties with the correct air flow rate so that agglomerates are efficiently separated into single particles. The SSPD is particularly good at achieving well-dispersed aerosols, because there is a strong shear force encountered by the particles as they pass through a venturi. The FBAG is slower to respond to changes in operating conditions than other systems, but provides a very stable output once equilibration has been reached.

All dry powder systems are vulnerable to the build-up of electrostatic charge as a result of separating agglomerates and by triboelectrification from particle–surface interactions. These processes may result in enhanced internal deposition, resulting in a decrease in output. Aerosols should therefore be charge equilibrated as soon as possible after formation.

2.3.2.4 *Other methods.* Multimodal or wide size range polydisperse aerosols may be preclassified in order to select a single mode in the size distribution or to restrict the size distribution within a narrower range than that provided by the original generation method. An important application is the removal of multiplet particles formed when generating PSL particles, already described in relation to the use of the electrostatic classifier (section 2.3.1.5).

Polydisperse particles from about 1 to 10 µm aerodynamic diameter can be size-classified to form a narrower size distribution by inertial separation, usually by means of two virtual impactors in series (Pilacinski *et al.*, 1990). Flow rates through the system are adjusted so that the cut-point of the second virtual impactor is at a smaller size than that of the first device. The fine fraction passing the first impactor is therefore split into middle and fine subfractions by the second impactor, and the middle fraction is used as the calibrant. Romay-Novas and Pui (1988) have produced narrow distribution calibrants from 0.1 to 1.0 µm aerodynamic diameter using a single-stage microorifice impactor to remove the coarse fraction, and passing the fine fraction through an electrostatic classifier to extract a narrow, well-defined subfraction.

Finally, a polydisperse aerosol may be collected in an aerosol spectrometer where the size-segregated particles are deposited at different distances on a collection substrate depending on their aerodynamic size (Timbrell, 1972). Particles of the desired narrower size range are removed from the substrate and resuspended to produce either monodisperse or narrow-range polydisperse aerosols (Kotrappa and Moss, 1971; Marshall *et al.*, 1991; Booker and Mitchell, 1991). The technique is limited by the small amount of material that can be generated as usable aerosol, but it is valuable when the source material is expensive or hard to manufacture.

## 2.4  Particle shape standards

The response of many aerosol analysers is dependent not only on particle size but also on shape. Until recently shape-related effects were largely ignored, because of the difficulty of defining three-dimensional shape as a single size-related parameter, and the non-availability of reference particles having well-defined shapes. At present, the most useful application of shape standards is in the calibration of inertial aerosol analysers, where the relationship between aerodynamic diameter and volume equivalent diameter incorporates the dynamic shape factor ($\chi$) through equation (2.1). The dynamic shape factor also expresses the influence of particle shape on various mobility-based sizes, including the electrical mobility diameter. Particle shape influences the three-dimensional scattered light intensity pattern created when particles interact with light in optical aerosol analysers. Many systems assume the particles to be spherical, as solutions to the Lorenz–Mie equations that describe particle–light interaction are at present only available for spheres (and cylinders) in uniform illumination (van de Hulst, 1981).

The development of techniques for controlled single crystal growth by forced hydrolysis techniques (Matijevic, 1985) has enabled a variety of uniform size particles to be formed that have ellipsoidal, cubic, spindle and rod shapes (Gowland and Wilshire, 1992). These particles are available within narrow size limits between 1 and 20 μm volume equivalent diameter, and the precise shape and size range depend on the chemistry of the particle formation process.

Marshall and Mitchell (1992) showed how to relate volume equivalent diameter to aerodynamic diameter with monodisperse truncated cube-shaped particles from 2 to 20 μm volume equivalent diameter formed from natrojarosite ($NaFe_3(SO_4)_2(OH)_6$) by controlled crystal growth. The aerodynamic size of each batch was determined under Stokesian conditions in a Timbrell spectrometer (Timbrell, 1972). Electron microscopy was used to determine the count mean size, based on maximum particle length, from which the mean volume equivalent diameter could be determined with a knowledge of particle geometry. The dynamic shape factor was obtained by rearranging equation (2.1), neglecting the Cunningham slip correction factors, as the particles were larger than 2 μm volume equivalent diameter:

$$\chi = \frac{d_v^2 \rho_p}{d_a^2 \rho_0} \qquad (2.12)$$

$\chi$ was found to be remarkably uniform at $1.29 \pm 0.16$ within most of the size range, so that these particles could be used to assess shape-related effects in aerodynamic particle size analysers independently of particle size (Marshall et al., 1991; Cheng et al., 1993a).

Single particle shape standards can also be made by a process similar to that used to manufacture integrated circuit microchips (silicon micro-machining). The chosen two-dimensional particle profile (fibre, disc, etc.) is replicated millions of times by means of a mask that is used in the fabrication of silicon dioxide particles by photolithographic etching (Kaye et al., 1991). These particles are highly monodisperse and can be made in most regular shapes larger than 1 μm; however, their thickness is limited by the etching process to about 1 μm. Three particle shape CRMs prepared using this technology are about to become available in the UK as CRMs through the Office of Reference Materials of the Laboratory of the Government Chemist. These particle shape standards were developed by AEA Technology plc in collaboration with the University of Hertford-shire, UK, as part of the recent UK Valid Analytical Measurement

(VAM) initiative. Their specifications (mean $\pm 1$ standard deviation) are as follows:

(a) CRM AEA1001: $3.09 \pm 0.10\,\mu m$ long $\times 1.67 \pm 0.08\,\mu m$ wide $\times$
    $0.96 \pm 0.09\,\mu m$ deep ($1.00 \times 10^7$ particles per vial),
(b) CRM AEA1002: $7.51 \pm 0.22\,\mu m$ long $\times 1.72 \pm 0.11\,\mu m$ wide $\times$
    $1.02 \pm 0.06\,\mu m$ deep ($1.44 \times 10^7$ particles per vial),
(c) CRM AEA1003: $12.13 \pm 0.22\,\mu m$ long $\times 1.70 \pm 0.04\,\mu m$ wide $\times$
    $1.00 \pm 0.07\,\mu m$ deep ($2.74 \times 10^7$ particles per vial).

Condensation aerosol generators (section 2.3.1.4) can be used with substances that sublimate at high temperature to prepare comparatively large quantities of monodisperse particles having a uniform shape. Vaughan (1990) described the preparation of highly uniform 1 to $3\,\mu m$ long elongated cuboid 'fibre-like' particles from caffeine using a MAGE. Longer particles were produced when the caffeine was maintained at a higher temperature in the saturator vessel.

Finally, Cheng *et al.* (1993b) intentionally used concentrated PSL suspensions to form micrometre-sized multiplet particles, as a means of working with particles having well-defined dynamic shape factors (Table 2.3). The aggregates were collected using a point-to-plane electrostatic precipitator for sizing by electron microscopy in terms of volume equivalent diameter. The aerosol was also sampled by an Aerodynamic Particle Sizer® aerosol spectrometer (TSI Inc.) to determine shape-related bias in this instrument. This technique is very useful, as it provides size-related information for several values of $\chi$ in a single measurement, and it can also be extended to the submicrometre size range, which is not easily attainable by any of the other techniques. PSL aggregates are also useful in the investigation of shape-related bias with optical particle spectrometers; however, it should be noted that bias with such instruments is highly shape-specific.

## 2.5   Calibration of aerosol analysers

### 2.5.1   *Good calibration practice*

The following seven principles underlie a good calibration. Those that are related to the instrument influence its performance and should therefore be followed in normal use, as well as during calibration.

- Aim for traceability with choice of calibrant, and choose the properties of the particles (e.g. size, density, refractive index, shape) to match those of the aerosol(s) being sampled.
- Opt for the simplest generation technique producing sufficient calibrant.

- Minimize electrostatic charge associated with both calibrant and the equipment being tested.
- Avoid long sample lines and minimize bends and constrictions.
- Beware of secondary effects, such as particle bounce in impactors or shape effects with optical particle analysers.
- Calibrate auxiliary devices that affect the performance of instruments (e.g. flow meters used with impactors).
- Ensure that the conditions under which the calibration is taking place (temperature, pressure etc.) reflect the conditions of use.

### 2.5.2 Size-fractionating analysers

This large group of aerosol analysers operates by classifying the original aerosol into discrete fractions that are physically separated after analysis. Impactors, liquid impingers and gas cyclones are the most commonly encountered size fractionators. Calibration establishes the characteristic curve (Figure 2.11) of collection efficiency (grade efficiency) as a function of particle size for each stage (in a multi-stage device) or for the stage in question (in a single-stage system). The variation in collection efficiency is a smooth transition from 0 to 100% within a limited size range, which is in general narrower for impactors than with either impingers or gas cyclones.

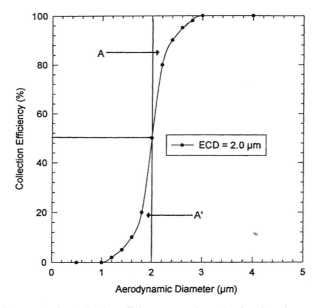

**Fig. 2.11** Representative collection efficiency curve for a size-fractionating aerosol analyser.

- The size corresponding to 50% collection efficiency is termed the effective cut-off diameter (ECD), the calibration constant for the separation stage.
- It is assumed that the inefficient collection of particles larger than the ECD is exactly balanced by the over-efficient collection of finer particles (the two areas $A$ and $A'$ in Figure 2.11 are equal).
- In multi-stage devices, it is assumed that stages operate independently of each other. In well-designed impactors, this assumption is probably valid. However, independent operation of stages is not always the case, especially with cyclones, but corrections for stage interactions are seldom undertaken.

Size fractionating devices are most frequently calibrated with monodisperse particles, despite the amount of work involved. For instance, the calibration of the eight-stage Andersen Mk-II cascade impactor (Graseby Andersen, Smyrna, GA, USA) can require more than 24 sizes of particles from 0.2 to 12 μm aerodynamic diameter, as at least three data points should be obtained per stage in the range between 20% and 80% collection efficiency, where the curve is steepest.

The choice of calibrant and aerosol generator depends on the particle sizes required (Figure 2.1). The following precautions should always be taken, whichever system is used.

- Ensure that the device being calibrated is leak-tight and operating at the desired flow rate.
- Avoid particle bounce and blow-off in impactors by greasing the collection surfaces appropriately (Marple and Liu, 1975; Rao and Whitby, 1978b). A thin coating of high-vacuum silicone grease is effective for most applications. It can be made up in $n$-hexane and applied as a 1% w/v solution to avoid an excessive application. Particle bounce is evident when the collection efficiency curve reaches a plateau at a value less than 100% before decreasing as particle size increases. Bounce is not a problem with virtual impactors, as the particles in both coarse and fine fractions are removed from the zone of separation whilst still gas-borne.
- Avoid overloading the collection surfaces, especially with multi-stage impactors. This difficulty is readily avoided by means of calibrant particles that are dye-based or contain a fluorescent marker, as accurate analytical assays can be undertaken with a few micrograms of material. Overloading causes changes to the stage collection efficiency curve similar to those brought about by particle bounce.
- Work with electrostatically charge equilibrated aerosols. Impactors that detect the mass of deposited particles by changes in frequency of an oscillating quartz crystal collection surface are particularly prone to large errors in both calibration and use, unless precautions are taken to minimize electrostatic charge (Horton et al., 1992).

The collection efficiency curve can most easily be determined on a mass-weighted basis, with the advantage that traceability is possible through the calibration of the analytical equipment used to measure the mass of tracer compound in the calibrant. Microscopy can be used to provide traceability in terms of particle size. The collection efficiency ($E_i$ (%)) for each stage ($i$) of an $n$-stage size fractionating analyser is given by

$$E_i(d_a) = 100 \frac{[M_i(d_a)]}{\left[ \sum_{k=i}^{n} M_k(d_a) \right] + [M_f(d_a)]}, \qquad (2.13)$$

where $M_i$, $M_k$ and $M_f$ are the masses of material collecting on stage $i$, from stage $i$ to the bottom of the device, and on any back-up filter, respectively. The determination of the ECD for a particular stage is simplified by approximating the S-shaped collection efficiency–size curve in the region close to 50% efficiency as a straight line (logarithmic size scaling) between two sizes, $d_1$ and $d_2$, where:

$$\text{ECD} = \exp\left[ \left( \frac{50 - E_1}{E_1 - E_2} \right) \ln\left( \frac{d_1}{d_2} \right) \right] + \ln(d_1) \qquad (2.14)$$

in which $E_1$ and $E_2$ are the collection efficiency values (%) corresponding to the chosen sizes. The VOAG (Figure 2.6) and the MAGE (Figure 2.8) are ideally suited to this application in the size ranges from about 1 to 15 μm and 0.1 to 5 μm aerodynamic diameter, respectively. Alternatively, monodisperse fluorescent-tagged PSL particles can be used as calibrants in the submicrometre size range, dissolving the polystyrene latex in 50% v/v $N$-methyl-2-pyrollidone/methanol to release the fluorescent dye for assay.

It is important to realize that the method described above yields no information about internal wall losses within devices such as cascade impactors, in which the collection surfaces are separate from other internal surfaces in contact with the aerosol during size separation. If the interior surfaces associated with each collection stage are washed and assayed, a modified collection efficiency ($E^*(d_a)$ (%)) is obtained that includes wall losses:

$$E_i^*(d_a) = 100 \frac{[M_i(d_a)]}{\left[ \sum_{k=i}^{n} M_k(d_a) \right] + [M_f(d_a)] + \left[ \sum_{k=i+1}^{n} L_k(d_a) \right]}, \qquad (2.15)$$

where the wall loss term is the sum of the wall losses from the stage below the one in question to the bottom of the device (Franzen and Fissan, 1979; Mitchell *et al.*, 1988). Wall losses as a result of parasitic inertial deposition to the walls can become significant at sizes in excess of about 5 μm aerodynamic diameter, depending on the design of the fractionator. Similar losses caused by a combination of gravitational sedimentation and diffusional deposition may also be a problem at smaller particle sizes with certain low-flow impactors, especially where there is significant dead volume between stages (Horton *et al.*, 1992).

Although less widely practised today, size-fractionating analysers may also be calibrated on a number-weighted basis, using PSL particles and measuring the particle concentration upstream $(N_u(d_a))$ and downstream $(N_d(d_a))$ of the size fractionating stage with a single particle counter. This technique is particularly appropriate with single-stage devices (Rao and Whitby, 1978a). However, it can also be undertaken with multi-stage analysers, provided the stages below the one of interest are removed to permit the particle counter to obtain a representative downstream sample. $E_i(d_a)(\%)$ is determined from

$$E_i(d_a) = 100\left[1 - \frac{N_d(d_a)}{N_u(d_a)}\right]. \tag{2.16}$$

The particle concentration measurement is comparative, but not strictly traceable unless the particle monitor has been calibrated using an aerosol number concentration standard. Traceability to particle size is by microscopy.

Virtual impactors require a slightly different approach to their calibration, and Chen *et al.* (1985a) described an approach that involves counting monodisperse particles of selected sizes in both minor and major flows from this type of separator (Figure 2.12). At high flow rates, $E_i(d_a)(\%)$ is given by:

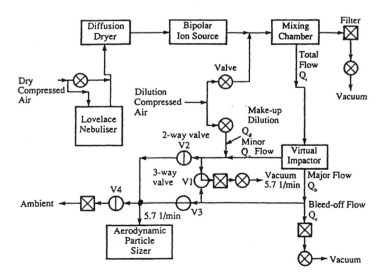

**Fig. 2.12** Calibration method with PSL particles for a virtual impactor. Reprinted from Chen *et al.* (1985a), with kind permission from Elsevier Science Ltd., The Boulevard, Langford Lane, Kidlington, Oxford OX5 1GB, UK.

$$E_i(d_a) = 100 \frac{N_a(d_a)}{N_a(d_a) + \left[\dfrac{N_b(d_a)Q_b}{Q_b - Q_c}\right]}, \tag{2.17}$$

where $N_a(d_a)$ and $N_b(d_a)$ are the number concentrations measured in the minor and major outlet flows respectively, and $Q_b$ and $Q_c$ are the volumetric flow rates in the major flow and particle counter bleed-off flow respectively. At low flow rates, $Q_c$ is replaced by a flow of dilution air to satisfy the need of the particle counter and expression (2.16) becomes:

$$E_i(d_a) = 100 \frac{N_a(d_a)}{N_a(d_a) + N_b(d_a)}. \tag{2.18}$$

### 2.5.3   Inertial/sedimentation spectrometers

This group of aerosol analysers preserves the particles or droplets after size separation on a collection substrate, and includes devices such as spiral-duct centrifuges (SDCs), various inertial spectrometers and sedimentation spectrometers. Calibration establishes the relationship between particle location in the spectrum of deposited material as a function of size, usually scaled as aerodynamic diameter, as this class of analysers relies on differing particle inertia or sedimentation under Stokesian conditions to achieve particle size separation. The most commonly encountered procedure involves the use of monodisperse PSL particles, especially those produced from concentrated aqueous suspensions (section 2.3.1.2) which contain several sizes of multiplets, whose relationship in terms of aerodynamic particle size to the singlet particles is well established (Table 2.3). Stöber and Flachsbart (1971) described the calibration of an SDC of their design, and Mitchell and Nichols (1988) described the rigorous calibration of an Inspec inertial spectrometer developed by Prodi et al. (1979). Marshall et al. (1990) undertook a similar calibration of a sedimentation-based Timbrell aerosol spectrometer (Timbrell, 1972). These studies demonstrated that aerosol spectrometers offer the highest size resolution when the ratio of aerosol to winnowing gas flow rates is kept as small as possible. Traceability is achieved through a knowledge of the mean PSL particle size (based on volume equivalent diameter) by microscopy, and the calibration of the travelling microscope or size-measuring device used to determine the precise distance along the surface where the particles collected.

A more rapid method of calibration involves sampling a stable polydisperse aerosol, and measuring the size of particles at predetermined locations by microscopy. For instance, in the calibration of a Timbrell spectrometer (Marshall et al., 1990), a polydisperse spray of polyvinyl acetate (PVA) particles was created by nebulization of a dilute solution (2.7% w/v) in ethanol, and the dried particles sized at different distances

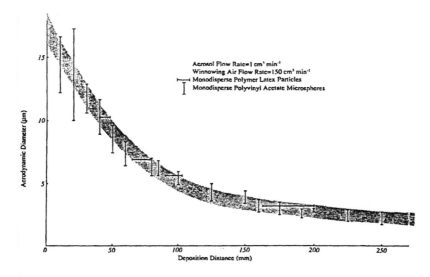

**Fig. 2.13** Representative calibration curve for an aerosol spectrometer. Reprinted from Marshall *et al.* (1990), with kind permission from Elsevier Science Ltd., The Boulevard, Langford Lane, Kidlington, Oxford OX5 1GB, UK.

along the collection surface (Figure 2. 13). Note that both monodisperse PSL and polydisperse PVA particles were required to achieve the required traceability in terms of particle size. A knowledge of both particle shape (spheres) and density was also required to scale the particle size axis in terms of aerodynamic diameter.

### 2.5.4 *Real-time aerodynamic particle size analysers*

The Aerodynamic Particle Sizer® spectrometer (APS, TSI Inc.) and the Aerosizer® spectrometer (API Inc., Hadley, MA) are the most widely used instruments in this class of aerosol analysers. They both operate as single particle detectors, constructing the size distribution from time-of-flight (TOF) measurements of many particles as they pass across two laser beams, following acceleration in a well-defined flow field (Remiarz *et al.*, 1983; Dahneke, 1973). In the case of the APS, TOF (transit time between laser beams) is obtained as a function of aerodynamic diameter (Figure 2.14) from the accumulator bins representing TOF intervals, and a similar situation exists with the Aerosizer®. Software in both instruments converts the number-TOF data to particle size distributions.

Calibration of these instruments is best undertaken with monodisperse, spherical particles of known density, and 8 to 10 different sizes of PSL

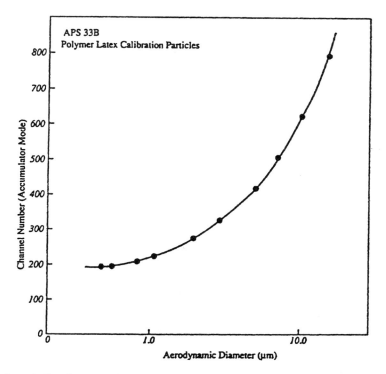

**Fig. 2.14** Calibration curve for a real-time aerodynamic particle size analyser. Reprinted by permission of Particle Instruments Division, TSI Inc., St Paul, Minnesota, USA.

microspheres covering the range of interest are most frequently used for this purpose. Both nebulization and dry dispersion techniques are required to cover the entire range of the APS (0.5 to 30 μm aerodynamic diameter) or the same size range with the Aerosizer®.

Careful control of instrument operating conditions is essential for the most precise work with all real-time aerosol spectrometers. For example, the APS is sensitive to variations in ambient pressure (Chen *et al.*, 1985b), and the aerosol flow rate to the Aerosizer® can vary significantly during instrument warm-up. Calibrations with gases other than air and at other than sea-level pressure have been undertaken with the APS, using a universal calibration curve based on the ratio of gas to particle velocity as a function of Stokes number (Rader *et al.*, 1990). In normal use, both aerosol and sheath air flow rates should be checked using a calibrated mass flowmeter.

The use of monodisperse calibration particles generated by controlled atomization (section 2.3.1.3), although frequently undertaken as an alternative to calibration with PSL particles, is not the preferred technique, as apart from the increased difficulty in generating the test aerosols in the first place, the sphericity and density of the particles cannot be readily verified. VOAG-produced solid particles can, however, be useful for calibrations where sizes in addition to those available with PSL microspheres are needed. It is inadvisable to generate liquid droplet calibrants for this purpose, as distortion occurs in the measurement zone, due to the acceleration experienced in flow field immediately prior to measurement (Griffiths *et al.*, 1986).

Real-time aerodynamic size analysers can also be calibrated with polydisperse, spherical particles of known density. This technique is especially useful to determine if the sampling conditions are optimized for the instrument, as the inlets of both the APS and Aerosizer® are not ideally suited to sampling moving air streams. Attachments, such as the AeroBreather® for the Aerosizer®, may also impose their own size-related sampling bias. Until recently, the main difficulty has been the location of suitably well-characterized calibration particles. However, soda-glass microspheres (section 2.3.2.1) are now available in the range from 0.5 to 20 µm with a known density of 2.4 g cm$^{-3}$. Their number size distribution is scaled in terms of volume equivalent diameter independently measured fully dispersed in liquid suspension by means of the electrical sensing zone (Coulter) method. The main issue is their effective dispersion in aerosol form, and this process may be conveniently accomplished by means of the SSPD (section 2.3.2.3). An Aerosizer® was recently calibrated effectively with these particles in the range 2 to 10 µm aerodynamic diameter. In this case, the microspheres were generated using a modified dry powder medical inhaler, as these devices are optimized to deliver fully dispersed powder particles in this size range (Mitchell and Nagel, 1996).

The performance of both the Aerodynamic Particle Sizer® and Aerosizer® is significantly influenced by particle density and shape because they size-separate particles under ultra-Stokesian conditions (Wang and John, 1987; Cheng *et al.*, 1993a; 1993b). It therefore follows that the density of calibration particles should be well defined and preferably independent of particle size. It is normal to ensure the calibrant particles are spherical, although the use of particle shape standards (section 2.4) is an important way in which to quantify shape-related bias with this class of instruments.

### 2.5.5   *Optical particle counters (OPCs)*

Multi-channel OPCs size-analyse individual particles on the basis of the relative intensity of light scattered to a detector as they pass through an illuminated measurement zone. In its simplest form, the scattered light collected at the detector is converted to a voltage, and this signal is then assigned to one of several channels based on its magnitude. Like real-time aerodynamic particle size analysers (section 2.5.4), multi-channel OPCs construct a complete size distribution from the total individual particle measurements after signal processing and data reduction to a manageable number of size classes. However, the particle size that is measured is not scaled to volume equivalent or aerodynamic diameter by virtue of the measurement technique, but depends on the optical properties (refracting and absorbing components of refractive index) to a greater or lesser extent, depending on particle size range and optical geometry (Cooke and Kerker, 1975). It follows that OPCs should be calibrated with spherical particles having a known chemical composition, size and refractive index if a meaningful size scale is to be obtained. Monodisperse PSL microspheres have therefore become the calibrant of choice for OPCs, since these properties are exceptionally well defined (section 2.3.1.2). Typically, the calibration curve for the OPC relates a property based on scattered light intensity at the detector (pulse height (voltage)/size channel number) to volume equivalent diameter (Szymanski and Liu, 1986; Figure 2.15).

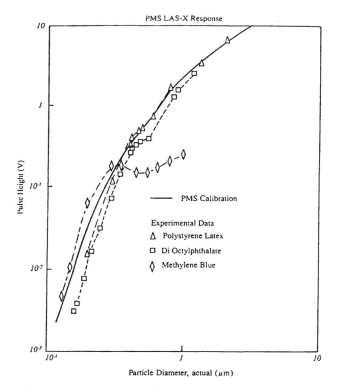

**Fig. 2.15** Calibration curve for a multi-channel optical particle counter. Reprinted from Symanski and Liu (1986), with kind permission from VCH Verlagsgesellschaft GmbH, Weinheim, Germany.

Calibration with monodisperse non-volatile oil (DEHS) droplets produced by the VOAG technique (section 2.3.1.3) is useful, particularly where it is necessary to generate test particles larger than 5 μm in diameter to establish if the instrument has any sampling- or measurement-related biases. Such a study was undertaken to quantify the size-related magnitude of the so-called 'border-zone error' in the size range from about 6 to 36 μm volume equivalent diameter with a range of OPCs whose measurement volumes are optically defined (Mitchell *et al.*, 1990). Border-zone error results in a systematic under-sizing of the small proportion of particles that graze the measurement volume of such instruments.

A PSL-based calibration curve is of limited value if OPCs are being used to detect aerosols containing particles or droplets which have substantially different optical properties (e.g. compare the curve for methylene blue with that for PSL in Figure 2.15). Alternative strategies are therefore required to obtain meaningful results. One of these is to utilize the size fractionating characteristics of a single-stage impactor to preseparate an aerosol of the material being worked with at a known ECD (section 2.5.2). The shapes of size distributions measured by the multi-channel OPC with and without the preseparator are compared on a channel-by-channel basis to arrive at the size channel at which 50% of the aerosol is removed by the impactor. This channel is then directly related to a fixed aerodynamic size (Marple and Rubow, 1976). When applied with several single-stage impactors in turn, each preseparator having a different ECD, a calibration curve of OPC size channel versus aerodynamic diameter can be constructed within a specified size range. Such a strategy made it possible to calibrate a Polytec HC-15 multi-channel OPC (Polytec GmbH, Waldbronn, Germany) with polydisperse water droplets from 2 to 20 μm aerodynamic diameter (Marshall *et al.*, 1988; Mitchell *et al.*, 1989).

OPCs can also be used to measure total particle concentration without size discrimination. Such devices are widely employed to monitor particles larger than about 0.1 μm diameter in clean rooms. The calibration of such devices is commonly undertaken by comparing the measured particle concentration against the number concentration calculated from counting the number of particles of a test aerosol (usually monodisperse PSL microspheres) within the specified size range, collected on a known surface area of a membrane filter located near to the OPC. Ideally this filter should sample the same aerosol as the OPC. In practice, it is located adjacent to the OPC, and samples the aerosol with the same entry nozzle configuration operating at the same flow conditions. Fissan *et al.* (1993) have recommended that OPCs used in clean rooms be calibrated in a regime that includes checks on flow rate, size resolution, counting efficiency (especially near the lower size limit of detection) and false count rate.

### 2.5.6 *Laser diffractometers*

Laser diffractometers are becoming widely used to measure aerosols, especially those produced by drug delivery devices, because of the simplicity and non-invasive nature of the measurement procedure. Diffractometers determine the ensemble light scattering from all particles in the measurement zone at a fixed time, to derive volume-, rather than number-weighted size distributions as the primary measurement. At present, performance verification is by means of reference reticles in which two-dimensional arrays of particle outlines are deposited in a chrome-on-glass matrix (Hirleman *et al.*, 1984). The calibration reticle supplied for the MasterSizer® series of diffractometers (Malvern Instruments, Malvern, UK) is a unimodal, near-lognormal distribution in the range from 10 to 100 μm volume equivalent diameter. Unfortunately, reticles cannot as yet be manufactured to test the performance below this range, which is where they are most needed by those measuring aerosols. Calibration by means of monodisperse PSL microspheres is currently the only practical solution. However, the nature of the conversion of the scattered light intensity data to the particle size distribution inevitably introduces some spread to the data, making it only meaningful to compare the peak of the measured PSL size distribution with the stated particle size. The arrival of good-quality polydisperse standards having well-defined optical properties (section 2.3.2.1) offers a better alternative procedure, provided that they can be dispersed in sufficient concentration to be detectable.

### 2.5.7 *Electrical mobility analysers and diffusion batteries*

Electrical mobility analysers and diffusion batteries size aerosol particles finer than approximately 0.5 μm volume equivalent diameter on the basis of their electrical charge and the influence of Brownian motion of surrounding gas molecules on the particles, respectively. Both types of analyser are frequently calibrated with monodisperse particles, either directly with nebulized PSL particles (section 2.3.1.2) or by the electrostatic classifier method (section 2.3.1.5). If PSL microspheres are generated by nebulization of dilute suspensions, contaminant particles formed from 'empty' solution droplets are a major limiting factor. However, such aerosols may be 'cleaned' by passing through an electrostatic classifier to eliminate particles outside the narrow range of sizes corresponding with the count mean diameter of singlet PSL particles. Calibration with polydisperse aerosols is useful, particularly for the electrical aerosol analyser (EAA), which tends to spread data from monodisperse particle measurements due to the presence of a charge distribution within each electrical mobility (particle size) range (Pui and Liu, 1979). Ultrafine particles from about 0.02 to 0.5 μm diameter can be formed by nebulization of dilute solutions (section 2.3.2.2), and this approach was used in a study comparing an EAA with a differential mobility particle sizer (DMPS (both TSI Inc.)), using sodium chloride particles (Horton *et al.*, 1989).

Bias in the measurements made by electrical mobility analysers is often caused by inaccurate setting and calibration of auxiliary components, such as flowmeters. Attention to the calibration of flowmeters is especially important if these devices are to be used where the ambient pressure is different from that at sea level (Yeh and Cheng, 1982).

Multi-stage diffusion batteries are also frequently calibrated with monodisperse particles, and considerations similar to those for electrical mobility analysers apply, except that the problem is to compare the penetration data for each size of calibrant as a function of stage number. Again, it is important to clean up any contaminant particles from 'empties', if PSL microspheres are being generated by nebulization. The flow rate, ambient temperature, pressure and mechanical properties of the diffusion battery stages (screen wire diameter or tube length) all influence the calibration (Cheng, 1989; Chen *et al.*, 1991).

### 2.5.8 *Condensation nuclei counters*

CNCs are widely used for the detection of the number concentration of submicrometre particles, often augmenting OPCs which are used to monitor larger particles. Their calibration relies on the production of nuclei of known number concentration, and this assessment can be made by collecting the particles on a filter for a given time at a known flow rate and counting particles deposited on known areas of the filter by electron microscopy. However, the process is time-consuming and unreliable with particles finer than 5 nm. The Nolan–Pollak manually-operated CNC (Pollak and Metnieks, 1959) has been used to calibrate other CNCs, as the decrease in light transmitted across the expansion chamber is directly related to the number of particles present. However, it is highly user-sensitive, making it an imperfect reference instrument. In the original calibration by Pollak and Metnieks (1959), the light transmission signal was calibrated against the aerosol concentration measured simultaneously by means of a photographic-based CNC. Their tables relating light transmission to particle concentration were validated using monodisperse particles generated by the more consistent electrostatic classifier method (Liu *et al.*, 1975).

Continuous-flow CNCs of the type developed by Agarwal and Sem (1980) have a very wide dynamic range, from $< 10^{-5}$ to $10^7 \, cm^{-3}$, and can operate in two modes. In the model 3020, 3022 and 3025 CNCs from TSI Inc., individual particles are counted at concentrations less than $10^3$, $10^4$ and $10^5 \, cm^{-3}$, respectively. At higher concentrations, the total light scattered in the view volume is used to determine particle concentration. These instruments are most conveniently calibrated with monodisperse aerosols produced by the electrostatic classifier method, diluting the 'monodisperse' aerosol from the electrostatic classifier with excess particle-free air to reduce the particle concentration to the required range.

### 2.5.9 *Mass concentration monitors*

Total- and respirable-mass concentration monitors based on the piezo-electric balance or beta-attenuation principle are most frequently calibrated using polydisperse secondary standards, such as Arizona Road Dust (Table 2.5), using a dry powder disperser (section 2.3.2.3). Considerable care has to be taken to ensure that the calibration aerosol is uniformly dispersed throughout the region in which the mass concentration monitoring equipment is being calibrated, especially if more than one unit is being simultaneously tested. A test chamber developed by Sem *et al.* (1977) for calibrating several piezoelectric microbalance mass monitors at a time was used with various polydisperse aerosols, including oil mist, welding fume and tobacco smoke. Each test aerosol was fed in turn to the chamber in which an equal number of filter samplers were located between each Piezobalance mass monitor (TSI Inc.).

Photometers are also used to measure aerosol mass concentration by light extinction. These instruments are sensitive to particle shape and refractive index, as well as size. The choice of calibration aerosol should therefore reflect the eventual use of the mass monitor (e.g. calibration with carbon black aerosol may be appropriate if the mass monitor is to be used for the measurement of aerosols from combustion processes). Similar considerations also apply for aerosol mass monitors that operate by measuring the total light scattered by the ensemble of particles in the measurement zone.

The most convenient, traceable way to calibrate aerosol mass concentration monitors is by means of filter collection of a known volume (flow rate × sample time) of the aerosol followed by gravimetric/chemical assay. Care is required in the choice of filter substrate, as absorption of ambient moisture will lead to erroneous results. PTFE or polycarbonate membrane filters are very useful as long as the aerosol mass is low to moderate ($< 10\,\mathrm{mg\,m^{-3}}$). For the most accurate work, it is essential to avoid electrical charging of the filter when weighing. Several hours of continuous sampling may be required to obtain sufficient mass on the filter(s) to weigh or assay accurately, if the aerosol mass concentration is less than $0.1\,\mathrm{mg\,m^{-3}}$.

In the case of mass monitors that accumulate particles on the collection surface, such as the Piezobalance or TEOM® system (Rupprecht and

Patashnick Co. Inc., Albany, NY, USA), the uptake or loss of ambient moisture by material collected since the last cleaning may cause erroneous results, and careful control of ambient relative humidity during calibration and use is therefore important.

Aerosol mass monitors can also be calibrated with monodisperse aerosols between 0.1 and 10 μm diameter, specifically in order to establish size-dependent changes in performance. PSL microspheres are very suitable for this purpose; however, the maximum mass concentration that can be generated in a practical way with the nebulizer method (section 2.3.1.2) is about 10 mg m$^{-3}$. Again, care is required to achieve uniform spatial dispersion of these calibrant particles, and filter collection followed by gravimetric assay is essential to verify the mass concentration that is present.

## Acknowledgement

The author would like to acknowledge the helpful suggestions and improvements provided by Gilmore J. Sem (Particle Instruments Division, TSI Inc.) during the preparation of this chapter.

## References

Agarwal J.K. and Sem G.J. (1980) Continuous-flow, single-particle-counting condensation nucleus counter. *J. Aerosol Sci.*, **11**, 343–357.
Altmann J. and Peters C. (1992) The adjustment of the particle size of a Sinclair–LaMer-type aerosol generator. *J. Aerosol Sci.*, **23**, S277–S280.
Berglund R.N. and Liu B.Y.H. (1973) Generation of monodisperse aerosol standards. *Environ. Sci. Technol.*, **7**, 147–153.
Blackford D.B. and Rubow K.C. (1986) A small-scale powder disperser. *Proc. NOSA Symposium*, Solna, Sweden, pp. 13–17. Available from TSI Inc., St Paul, MN, USA.
Booker D.R. and Mitchell J.P. (1991) Production of ceramic reference particles by inertial classification, in *Proc. 5th Ann. Conf. Aerosol Society*, Aerosol Society, Bristol, UK, pp. 77–82.
BSI (1993) *BS 2955: Glossary of Terms Relating to Particle Technology*. British Standards Institution, London.
BSI (1988) *Determination of Particle Size Distribution: Recommendations for Single Particle Light Interaction Methods*, BS 3406 Part 7: British Standards Institution, London.
Charlesworth D.H. and Marshall J.R. (1960) Evaporation from drops containing dissolved solids. *Am. Inst. Chem. Eng. J.*, **6**, 9–23.
Chen B.T. (1993) Instrument calibration, in *Aerosol Measurement: Principles, Techniques and Applications* (eds K. Willeke and P.A. Baron). Van Nostrand Reinhold, New York, pp. 493–520.
Chen B.T., Yeh H.C. and Cheng Y.S. (1985a) A novel virtual impactor – calibration and use. *J. Aerosol Sci.*, **16**, 343–354.
Chen B.T., Cheng Y.S. and Yeh H.C. (1985b) Performance of a TSI Aerodynamic Particle Sizer. *Aerosol Sci. Technol.*, **4**, 89–97.
Chen B.T., Cheng Y.S., Yeh H.C., Bechtold W.E. and Finch G.L. (1991) Test of the size resolution and sizing accuracy of the Lovelace parallel-flow diffusion battery. *Am. Ind. Hyg. Assoc. J.*, **52**, 75–80.

Cheng Y.-S. (1989) Diffusion batteries and denuders, in *Air Sampling Instruments* (ed. S.V. Hering). American Conference of Governmental Industrial Hygienists, Cincinnati, OH, pp. 405–419.

Cheng Y.-S., Barr E.B., Marshall I.A. and Mitchell J.P. (1993a) Calibration and performance of an API aerosizer. *J. Aerosol Sci.*, 24, 501–514.

Cheng Y.-S., Chen B.T., Yeh H.C., Marshall I.A., Mitchell J.P. and Griffiths W.D. (1993b) Behavior of compact non-spherical particles in the TSI aerodynamic particle sizer model APS33B: Ultra-Stokesian drag forces. *Aerosol Sci. Technol.*, 19, 255–267.

Cheng Y.-S. and Chen B.T. (1995) Aerosol sampler calibration, in *Air Sampling Instruments* (eds B.S. Cohen and S.V. Hering). American Conference of Governmental Industrial Hygienists, Cincinnati, OH, pp. 165–186.

Cooke D.D. and Kerker M. (1975) Response calculations for light scattering aerosol particle counters. *Appl. Optics*, 14, 734–739.

Cox C.S. and Wathes C.M. (1995) *Bioaerosols Handbook*. Lewis Publishers, Boca Raton, FL.

Dahneke B. (1973) Aerosol beam spectrometry. *Nature Phys. Sci.*, 244, 54–55.

Fissan H.J., Schmitz W. and Trampe A. (1993) Clean-room measurements, in *Aerosol Measurement* (eds K. Willeke and P.A. Baron). Van Nostrand Reinhold, New York, pp. 747–767.

Franzen H. and Fissan H. (1979) The separation behaviour of the Andersen 'non-viable' and 'stack' samplers used with glass fibre collection plates. *Staub Reinhalt. der Luft*, 39, 50.

Fuchs N.A. and Sutugin A.G. (1966) Generation and use of monodisperse aerosols, in *Aerosol Science* (ed. C.N. Davies). Academic Press, New York, pp. 1–30.

Gowland R.J. and Wilshire B. (1992) Standard powders for particle size analysis, in *Particle Size Analysis* (eds N.G. Stanley-Wood and R.W. Lines). Royal Society of Chemistry, Cambridge, pp. 99–107.

Griffiths W.D., Iles P.J. and Vaughan N.P. (1986) The behaviour of liquid droplet aerosols in an APS 3300. *J. Aerosol Sci.*, 17, 921–930.

Hartman, A.W., Doiron T.D. and Hembree G.G. (1991) Certification of NIST SRM 1961: 30 μm diameter polystyrene spheres. *J. Res. Natl. Inst. Stand. Technol.*, 96, 551–563.

Hartman A.W., Doiron T.D. and Fu J. (1992) Certification of NIST SRM 1962: 3 μm diameter polystyrene spheres. *J. Res. Natl. Inst. Stand. Technol.*, 97, 253–265.

Hinds W.C. (1980) Dry dispersion aerosol generator, in *Generation of Aerosols and Facilities for Exposure Experiments* (ed. K. Willeke). Ann Arbor Science, Ann Arbor, MI, pp. 171–188.

Hinds W.C. (1982) *Aerosol Technology*. Wiley, New York.

Hirleman E.D., Oechsle V. and Chigier N.A. (1984) Response characteristics of laser diffraction particle size analysers. *Opt. Eng.*, 23, 610–419.

Horton, K.D. and Mitchell J.P. (1992) Development of a number concentration standard for micron-sized particles. *J. Aerosol Sci.*, 23, S341–S344.

Horton K.D., Mitchell J.P. and Nichols A.L. (1989) Experimental comparison of electrical mobility aerosol analysers. *TSI J. Particle Instrum.*, 4, 3–19.

Horton K D., Miller R.D. and Mitchell J.P. (1991) Characterisation of a condensation-type monodisperse aerosol generator (MAGE). *J. Aerosol Sci.*, 22, 347–363.

Horton K.D., Ball M.H.E. and Mitchell J.P. (1992) The calibration of a California Measurements PC-2 quartz crystal cascade impactor (QCM). *J. Aerosol Sci.*, 23, 505–524.

ISO (1994) *Quality Systems – Model for Quality Assurance in Design, Production, Installation and Servicing*, ISO 9001. International Standards Organization, Geneva.

Jenkins R.A., Mitchell J.P. and Nichols A.L. (1985) Monodisperse microspheres for the calibration of aerosol analysers that operate at high temperatures, in *Particle Size Analysis 1985* (ed. P.J. Lloyd). Wiley, Chichester, pp. 197–209.

John W. (1993) The characteristics of environmental and laboratory-generated aerosols, in *Aerosol Measurement: Principles, Techniques and Applications* (eds K. Willeke and P.A. Baron). Van Nostrand Reinhold, New York, pp. 54–76.

Kanapilly G.M., Raabe O.G. and Newton G.J. (1970) A new method for the generation of aerosols of insoluble particles. *J. Aerosol Sci.*, 1, 313–323.

Kaye P.H., Hirst E. and Clark J.M. (1991 ) The manufacture of standard, non-spherical particles by silicon micromachining, in *Proc. 5th Ann. Conf. Aerosol Society*, Aerosol Society, Bristol, UK, pp. 223–228.

Kinney P.D., Pui D.Y.H., Mulholland G.W. and Bryner N.P. (1991) Use of the electrostatic classifier method to size 0.1 µm SRM particles – a feasibility study. *J. Res. Natl. Inst. Stand. Technol.*, **96**, 147–176.

Kotrappa P. and Moss O.R. (1971) Production of relatively monodisperse aerosols for inhalation experiments by aerosol centrifugation. *Health Phys.*, **21**, 531–535.

Leong K.H. (1981) Morphology of aerosol particles generated from the evaporation of solution droplets. *J. Aerosol Sci.*, **12**, 417–435.

Lettieri T.R. and Hembree G.G. (1989) Dimensional calibration of the NBS 0.3 µm diameter particle sizing standard. *J. Colloid Interface Sci.*, **127**, 566–572.

Lettieri T.R., Hartman A.W., Hembree G.G. and Marx E. (1991) Certification of SRM 1960: nominal 10 µm diameter polystyrene spheres ('space beads'). *J. Res. Natl. Inst. Stand. Technol.*, **96**, 669–691.

Liu B.Y.H. (1974) Laboratory generation of particulates with emphasis on submicron aerosols. *J. Air Pollut. Control Assoc.*, **24**, 1170–1172.

Liu B.Y.H. and Pui D.Y.H. (1974) A sub-micron standard and the primary, absolute calibration of the condensation nucleus counter, *J. Colloid Interface Sci.*, **47**, 155–171.

Liu B.Y.H., Pui D.Y.H., Hogan A.W. and Rich T.A. (1975) Calibration of the Pollak counter with monodisperse aerosols. *J. Appl. Meteorol.*, **14**, 46–51.

Marple V.A. and Liu B.Y.H. (1975) Inertial impactors; theory, design and use, in *Fine Particles* (ed. B.Y.H. Liu). Academic Press, New York, pp. 411–446.

Marple V.A. and Rubow K.L. (1976) Aerodynamic particle size calibration of optical particle counters. *J. Aerosol Sci.*, **7**, 425–433.

Marple V.A., Liu B.Y.H. and Rubow K.L. (1978) A dust generator for laboratory use. *Am. Ind. Hyg. Assoc. J.*, **39**, 26–32.

Marshall I.A. and Mitchell J.P. (1992) The preparation and characterisation of particle shape standards, in *Particle Size Analysis* (eds N.G. Stanley-Wood and R.W Lines). Royal Society of Chemistry, Cambridge, pp. 81–90.

Marshall I.A., Mitchell J.P., Nichols A.L. and Van Santen A. (1988) Calibration studies of a Polytec HC-15 optical aerosol analyser with water droplets, in *Particle Size Analysis* (ed. P.J. Lloyd). Wiley, Chichester, pp. 289–301.

Marshall I.A., Mitchell J.P. and Griffiths W.D. (1990) The calibration of a Timbrell aerosol spectrometer. *J. Aerosol Sci.*, **12**, 969–975.

Marshall I.A., Mitchell J.P. and Griffiths W.D. (1991) The behaviour of non-spherical particles in a TSI aerodynamic particle sizer. *J. Aerosol Sci.*, **22**, 73–90.

Matijevic E. (1985) Production of monodisperse colloidal particles. *Ann. Rev. Mater. Sci.*, **15**, 483–516.

May K.R. (1949) An improved spinning top homogeneous spray apparatus. *J. Appl. Phys.*, **20**, 932–938.

May K.R. (1966) Spinning-top homogeneous spray aerosol generator with shockproof mounting. *J. Sci. Instrum.*, **43**, 841–842.

Mercer T.T., Tillery M.I. and Chow H.Y. (1968) Operating characteristics of some compressed-air nebulisers. *Am. Ind. Hyg. Assoc. J.*, **29**, 66–78.

Mitchell J.P. (1984) The production of aerosols from aqueous solutions using the spinning top generator. *J. Aerosol Sci.*, **15**, 35–45.

Mitchell J.P. (1992) Certification and characterisation of a new range of Community Bureau of Reference polydisperse, spherical certified reference materials. *Anal. Proc. Roy. Soc. Chem.*, **29**, 508–509.

Mitchell J.P. and Nagel M.W. (1996) An assessment of the API Aerosizer for the real time measurement of medical aerosols from pressurized metered-dose inhalers. *Aerosol Sci. Technol.*, **25**, 411–424.

Mitchell J.P. and Nichols A.L. (1988) Experimental assessment and calibration of an inertial spectrometer. *Aerosol Sci. Technol.*, **9**, 15–28.

Mitchell J.P. and Stone R.L. (1982) Improvements to the May spinning-top aerosol generator. *J. Phys. E*, **15**, 565–567.

Mitchell J.P. and Waters S. (1986) Improvements to the vibrating orifice aerosol generator for the production of methylene blue particles. *J. Aerosol Sci.*, **17**, 556–560.

Mitchell J.P., Snelling K.W. and Stone R.L. (1987) Improvements to the vibrating orifice aerosol generator. *J. Aerosol Sci.*, **18**, 231–243.

Mitchell J.P., Costa P.A. and Waters S. (1988) An assessment of an Andersen Mark-II cascade impactor. *J. Aerosol Sci.*, **19**, 213–221.

Mitchell J.P., Nichols A.L. and Van Santen A. (1989) The characterisation of water droplet aerosols by Polytec optical aerosol analysers. *Part. Charact. J.*, **6**, 119–123.

Mitchell J.P., Ashcroft J., Fromentin A., Holmes R., Marsault P., McAughey J.J., Patel A. and Phillips H. (1990) Laboratory intercomparison of Polytec optical aerosol analysers, in *Aerosols: Science, Industry, Health and Environment* (eds S. Masuda and K. Takahashi). Pergamon Press, Oxford, pp. 643–646.

Muir D.C.F. (1965) The production of monodisperse aerosols by a LaMer–Sinclair generator. *Ann. Occup. Hyg.*, **8**, 233–240.

Mulholland G.W., Hartman A.W., Hembree G.G., Marx E. and Lettieri T.R. (1985) Development of a one-micrometer diameter particle size standard reference material. *J. Res. Natl. Bur. Stds.*, **90**, 2–25.

Peters C. and Altmann J. (1993) Monodisperse aerosol generator with rapid adjustable particle size for inhalation studies *J. Aerosol Med.*, **6**, 307–315.

Pilacinski W., Ruuskanen J., Chen C.C., Pan M.J. and Willeke K. (1990) Size-fractionating aerosol generator. *Aerosol Sci. Technol.*, 450–458.

Pollak L.W. and Metnieks A.L. (1959) New calibration of photo-electric nucleus counters. *Geofis. Pura Applicata*, **43**, 285–301.

Prodi V. (1972) A condensation aerosol generator for solid, monodisperse particles, in *Assessment of Airborne Particles* (eds T.T. Mercer, P.E. Morrow and W. Stöber). C.C. Thomas, Springfield, IL, pp. 169–181.

Prodi V., DeZaiacomo T., Melandri C. and Formigniani M. (1979) An inertial spectrometer for aerosol particles. *J. Aerosol Sci.*, **10**, 411–419.

Pui D.Y.H. and Liu B.Y.H. (1979) Electrical aerosol analyser: calibration and performance, in *Aerosol Measurement* (eds D.A. Lundgren, F.S. Harris, W.H. Marlow, M. Lippman, W.E. Clark and M.D. Durham. University Presses of Florida, Gainesville, pp. 384–399.

Raabe O.G. (1968) The dilution of monodisperse suspensions for aerosolisation. *Am. Ind. Hyg. Assoc. J.*, **29**, 439–443.

Raabe O.G. (1976) The generation of fine particles, in *Fine Particles* (ed. B.Y.H. Liu). Academic Press, New York, pp. 57–110.

Rader D.J., Brockmann J.E., Ceman D.L. and Lucero D.A. (1990) A method to employ the APS factory calibration under different operating conditions. *Aerosol Sci. Technol.*, **13**, 514–521.

Rao A.K. and Whitby K.T. (1978a) Non-ideal collection characteristics of inertial impactors – single-stage impactors and solid particles. *J. Aerosol Sci.*, **9**, 77–86.

Rao A.K. and Whitby K.T. (1978b) Non-ideal collection characteristics of inertial impactors – cascade impactors. *J. Aerosol Sci.*, **9**, 87–100.

Rapaport E. and Weinstock S.E. (1955) A generator for homogeneous aerosols. *Experimentia*, **11**, 363–364.

Remiarz R.J., Agarwal J.K., Quant F.R. and Sem G.J. (1983) Real-time aerodynamic particle size analyzer, in *Aerosols in the Mining and Industrial Work Environment* (eds V.A. Marple and B.Y.H. Liu). Ann Arbor Science, Ann Arbor, MI, pp. 879–895.

Romay-Novas F.J. and Pui D.Y.H. (1988) Generation of monodisperse aerosols in the 0.1 to 1.0 μm diameter range using a mobility classification–inertial impaction technique. *Aerosol Sci. Technol.*, **9**, 123–131.

Scarlett B. (1985) Measurement of particle size and shape: some reflections on the BCR reference material programme. *Part. Charact. J.*, **2**, 1–6.

Sem G.J., Tsurubayashi K. and Homma K. (1977) Performance of the piezoelectric microbalance respirable aerosol sensor. *Am. Ind. Hyg. Assoc. J.*, **38**, 580–588.

Sinclair D. and LaMer V.K. (1949) Light scattering as a measure of particle size in aerosols. *Chem. Rev.*, **44**, 245–267.

Stöber W. and Flachsbart H. (1971) High resolution aerodynamic size spectrometry of quasi-monodisperse latex spheres with a spiral centrifuge. *J. Aerosol Sci.*, **2**, 103–116.

Stöber W., Berner A. and Blaschke R. (1969) The aerodynamic diameter of aggregates of uniform spheres. *J. Colloid Interface Sci.*, **29**, 710.

Szymanksi W.W. and Liu B.Y.H. (1986) On the sizing accuracy of laser optical particle counters. *Part. Charact. J.*, **3**, 1–7.

Timbrell V. (1972) An aerosol spectrometer and its applications, in *Assessment of Airborne Particles* (eds T.T. Mercer, P.E. Morrow and W. Stöber). C.C. Thomas, Springfield, IL.

Thom R., Marchandise H. and Colinet E. (1985) The Certification of Monodisperse Latex Spheres in Aqueous Suspensions with Nominal Diameter 2.0 µm, 4.8 µm and 9.6 µm. Commission of the European Communities Report EUR 9662-EN.

Van de Hulst H.C. (1981) *Light Scattering by Small Particles*. Dover Press, New York.

Vanderpool R.W. and Rubow K.L. (1988) Generation of large, solid, monodisperse calibration aerosols. *Aerosol Sci. Technol.*, **9**, 65–69.

Vaughan N.P. (1990) The generation of monodisperse fibres of caffeine. *J. Aerosol Sci.*, **21**, 453–462.

VDI (1980a) *Generation of Latex Aerosols Using Nozzle Atomisers*, VDI 3491 Part 3. Verein Deutscher Ingenieure, Düsseldorf.

VDI (1980b) *Generation of Test Aerosols: Sinclair–LaMer Generator*, VDI 3491 Part 4. Verein Deutscher Ingenieure, Düsseldorf.

VDI (1987a) *Generation of Test Aerosols: Rapaport–Weinstock Generator*, VDI 3491 Part 7. Verein Deutscher Ingenieure, Düsseldorf.

VDI (1987b) *Generation of Test Aerosols Using Centrifugal Atomisers*, VDI 3491 Part 12. Verein Deutscher Ingenieure, Düsseldorf.

Walton W.H. and Prewett W.C. (1949) The production of sprays and mists of uniform drop size by means of spinning disc type sprayers. *Proc. Phys. Soc.*, **62**, 341–350.

Wang H.-C. and John W. (1987) Particle density correction for the Aerodynamic Particle Sizer. *Aerosol Sci. Technol.*, **6**, 191–198.

Willeke K. (1980) *Generation of Aerosols*. Ann Arbor Science, Ann Arbor, MI.

Wilson R., Leschonski K., Alex W., Allen T., Koglin B. and Scarlett B. (1980) *Certification Report on Reference Materials of Defined Particle Size*, Commission of the European Communities Report EUR 6825-EN.

Wright B.M. (1950) A new dust feed mechanism. *J. Sci. Instrum.*, **27**, 12–14.

Yeh H.C. and Cheng Y.S. (1982) Electrical aerosol analyser: an alternative method for use at high altitude or reduced pressure. *Atmos. Environ.*, **16**, 1269–1270.

## Nomenclature

| | |
|---|---|
| $C$ | suspension mass concentration |
| $C_c$ | Cunningham slip correction factor |
| $C_f$ | fractional mass concentration |
| $C_{vap}$ | vapour mass concentration |
| $d_a$ | aerodynamic diameter |
| $d_l$ | liquid droplet diameter |
| $d_v$ | volume equivalent diameter |
| $d_{rot}$ | rotor table diameter |
| $e$ | elementary unit of charge |
| $E$ | collection efficiency |
| $F$ | volumetric fraction |
| $F_{rot}$ | rotation frequency |
| $F_{vib}$ | orifice vibration frequency |
| $I$ | current |
| $L$ | classifier length |

| $M_i$ | mass of material on stage $i$ |
|---|---|
| $N$ | particle number concentration |
| $N_{nucl}$ | nuclei number concentration |
| $Q$ | flow rate |
| $Q_a$ | flow rate (air dilution) |
| $q_e$ | flow rate of excess air |
| $q_s$ | flow rate of sheath air |
| $R$ | ratio of singlet/total particles |
| $r_1$ | outer radius of classifier |
| $r_2$ | inner radius of classifier |
| $R_a$ | rate of aerosolization |
| $T$ | surface tension |
| $V$ | voltage |
| $X$ | dilution factor |
| $Z_p$ | electrical mobility |

| $\eta$ | viscosity |
|---|---|
| $\eta(d_v)$ | transport efficiency |
| $\Lambda$ | classifier dimension |
| $\rho$ | density |
| $\sigma_g$ | geometric standard deviation |
| $\chi$ | dynamic shape factor |

# 3 Transport and sampling of aerosols

D. MARK

## 3.1 Introduction

In a book describing the physical and chemical properties of aerosols it is very often assumed that the sample of particles collected for analysis is representative of those airborne particles from which the sample was taken. For gases, this assumption is reasonable provided buoyancy (and density) considerations are taken into account during the sampling, and the collected samples are kept under appropriate temperature conditions. However, particles do not behave in the same way as gas molecules when dispersed in air – they deposit under gravity, impact on bends due to particle inertia, are deposited on internal surfaces by molecular and turbulent diffusion, and are affected by thermal, electrostatic and acoustic forces. In addition, the efficiency with which particles are sampled is governed mainly by inertial and sedimentation forces, with other forces playing a role dependent upon the nature and size of the particles and the sampler, and upon the environmental conditions during which the samples were taken.

In this chapter, a brief insight into the transport properties of aerosols, which govern the influence that they have on human health and the environment, is given. This is followed by a description of the methodology employed to sample aerosols in workplaces, chimney stacks and the ambient environment. Within the limitations of a chapter it has not been possible to include all aspects of aerosol sampling, and subjects such as the measurement of the particle size distributions of aerosols, the measurement of bioaerosols in all environments and the measurement of deposited dust in the ambient atmosphere are not covered. For these topics, and others that are not covered, the reader is referred to other texts given under the 'Further Reading' heading at the end of the chapter.

## 3.2 Basic transport properties of aerosols

The transport of aerosols in the workplace and the ambient atmosphere, and their effects on human health and the environment, are strongly governed by a series of simple physical processes and particle properties.

These differ for each aerosol particle according to its size, shape and chemical composition, with particle size being the most important (and dealt with comprehensively elsewhere in the book). The main properties of aerosol particles in this context are the aerodynamic properties (involving the motion of particles in air) which determine whether, and for how long, particles remain airborne. This affects the atmospheric concentration and the distance travelled before deposition onto surfaces or to ground. These same properties determine whether particles enter the mouth and/or nose during breathing, and how far they penetrate into the respiratory system (see below). A brief outline of these physical properties is given here. A fuller explanation can be found in a number of well-known texts (e.g. Fuchs, 1964; Hinds, 1982; Vincent, 1989).

### 3.2.1 *The motion of airborne particles*

3.2.1.1 *Drag force on a particle.* When a particle moves relative to the air, it experiences forces associated with the resistance by the air to its relative motion. For very slow, creeping air flow over the particle, the drag force $F_D$ is given by the well-known Stokes' law

$$F_D = -3\pi d\eta v, \qquad (3.1)$$

where $d$ is the geometric diameter of the particle, $\eta$ is the air viscosity, and $v$ is the mean air velocity; the minus sign indicates that the drag force is acting so as to oppose the motion of the particle. This law is upheld only under certain limited conditions of particle size and aerodynamic conditions where the Reynolds number for the particle, defined as:

$$Re_p = \frac{dv\rho_a}{\eta}, \qquad (3.2)$$

is very small ($Re < 1$), where $\rho_a$ is the density of the air. Strictly, equation (3.1) should be modified by three factors. The first of these is the Cunningham slip correction factor ($C_c$), which takes account of the fact that the air surrounding the particle is not continuous but is made up of individual molecules. Very small particles (smaller than the mean free path between the gas molecules, $\sim 0.06\,\mu m$ for air at standard temperature and pressure) may slip between successive collisions with the molecules. Secondly, for large values of $Re_p$ the drag coefficient tends to become constant, a situation known as the **non-Stokesian** regime. Finally, particles are generally not spherical, and particle motion may be affected by their orientation.

Nevertheless, Stokes' law may – to a first approximation – be considered as a reasonable working assumption for understanding the motion of airborne particles.

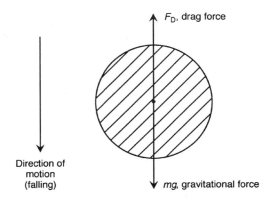

**Fig. 3.1** Forces acting on a particle of mass $m$ falling under gravity.

The starting point for all considerations of particle transport is the again well-known Newton's second law (mass × acceleration = net force acting). For this situation the net forces comprise the drag force (described above) and external forces (gravity, electrical, or some combination of forces). The effect of these forces is to generate and sustain particle motion, and provided that the particle is in motion relative to the air, the drag force will remain finite. For aerosol sampling, however, the nature of the drag force is the predominating influence since the external forces are usually of secondary importance. When the motion of the air, the particle and the forces acting are considered in three dimensions, the equations required to predict particle motion can become quite complicated.

3.2.1.2  *Motion under the influence of gravity.* A simple but very important example of particle motion under the influence of external forces can be found in the motion of particles under the influence of gravity. The forces acting on the particle are shown in Figure 3.1. The equation of motion for a spherical particle moving in the vertical ($y$) direction is given by:

$$m\frac{dv_y}{dt} = 3\pi\eta dv_y - mg \qquad (3.3)$$

where $v_y$ is the particle velocity in the $y$-direction, $m$ is the particle mass and $g$ is the acceleration due to gravity. This can be reorganized to give

$$\frac{dv_y}{dt} + \frac{v_y}{\tau} - g = 0, \qquad (3.4)$$

where $\tau$ is known as the particle relaxation time, given by

$$\tau = \frac{d^2 \rho_p}{18\eta} \qquad (3.5)$$

and $\rho_p$ is the particle density. Equation (3.4) is a simple first-order linear differential equation from which it can be shown that particle velocity under the influence of gravity tends exponentially towards a terminal value, known as the **sedimentation** or **falling speed**, given by:

$$v_{ts} = g\tau. \qquad (3.6)$$

For example, the falling speeds of spherical particles with the same density as water ($10^3 \, \text{kg m}^{-3}$) can be obtained as follows: for $d = 1 \, \mu\text{m}$, $v_{ts} = 0.003 \, \text{mm s}^{-1}$; $d = 5 \, \mu\text{m}$, $v_{ts} = 0.8 \, \text{mm s}^{-1}$; $d = 10 \, \mu\text{m}$, $v_{ts} = 3 \, \text{mm s}^{-1}$; $d = 100 \, \mu\text{m}$, $v_{ts} = 240 \, \text{mm s}^{-1}$, etc.

This is the simple case where slip, non-Stokesian conditions and particle non-sphericity are neglected. More generally:

$$v_{ts} = \left(\frac{C_c}{\chi}\right)\left(\frac{24}{C_D Re_p}\right) g\tau. \qquad (3.7)$$

Sedimentation is the main mechanism for the deposition of large particles onto horizontal surfaces such as cars, soil and vegetation. The main complicating factor is the turbulent fluctuations in wind speed near to the ground which can include upward motions of velocity as high as $500 \, \text{mm s}^{-1}$. However, these are rare occurrences and the particles will soon deposit again some time later.

### 3.2.1.3 *Motion without external forces.*

The concept of particle motion without the application of an external force is also helpful in understanding the behaviour of airborne particles. For example, consider the simplest case where the air is stationary and a spherical particle is projected into it with finite initial velocity in the $x$-direction. Motion is described by the equation

$$m\frac{\mathrm{d}v_x}{\mathrm{d}t} = -3\pi\eta d v_x. \qquad (3.8)$$

This has the simple solution

$$v_x = v_{x0} \exp\left(-\frac{t}{\tau}\right), \qquad (3.9)$$

where $v_{x0}$ is the initial particle velocity relative to the fluid at time $t = 0$. This expression has particular relevance to moving air, since it describes how a particle, initially injected into an airflow with zero velocity, is progressively pulled along by the drag force exerted by the fluid until it eventually catches up with it. The particle is then moving at the same velocity as the air and may be considered to be airborne.

Further integration gives the distance travelled by the particle relative to the air before it catches up with the airflow, or before it comes to rest if the air is stationary. This is known as the **particle stop distance** ($s$), and is given by

$$s = v_{x0}\tau. \tag{3.10}$$

This concept is particularly important when considering how particles behave within moving air that is changing direction – for instance, in the distorted flows entering the human respiratory system and passing through the lung airways or entering aerosol samplers.

For particles moving in these distorted flows, their behaviour may be scaled to give the dimensionless quantity known as the **Stokes number** ($St$),

$$St = \frac{d^2 \rho_p U}{18\eta D}, \tag{3.11}$$

where $D$ and $U$ are characteristic dimensional and velocity scales, respectively. For particles in the vicinity of aerosol samplers, $U$ is normally the external wind speed and $D$ the dimension of the orifice, the flow into which is responsible for the distortion. Combining (3.5) with (3.11) gives

$$St = \frac{\tau}{D/U} = \frac{\tau}{\tau_d} \tag{3.12}$$

where $\tau_d$ is equivalent to a time-scale which is characteristic of the flow distortion. For a very small particle with small particle relaxation time ($\tau$), $St$ will be small, indicating that the particle will tend to respond quickly to changes in the flow and tend to follow the airflow closely. A large particle, with large $\tau$, and correspondingly larger $St$ will tend to respond less effectively to the changing direction. A very large particle will therefore not follow the changes in flow direction and velocity, but will tend to follow the direction of its original motion.

This behaviour can also be followed by combining (3.10) and (3.12):

$$St = \frac{s}{D} \tag{3.13}$$

where $St$ is now expressed as the ratio of particle stop distance ($s$) to the characteristic dimension of the flow distortion ($D$). Here the particle will tend to follow the airflow when $s$ is small compared to $D$, and will not follow the airflow when $s$ is large compared to $D$.

This discussion has led us to understand that $St$ is an important measure of the ability of an airborne particle to respond to the movement of the air around it, and that particle trajectory patterns may differ to an extent dictated largely by the magnitude of $St$. We therefore have the concept of particle **inertia**, which is a function of both the particle and the airflow in

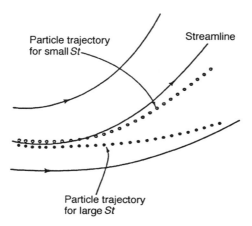

**Fig. 3.2** The concept of particle inertia.

which it is moving. Figure 3.2 illustrates this concept, which is of great importance in understanding the behaviour of aerosol samplers.

3.2.1.4 *Particle aerodynamic diameter.* One of the most useful concepts in particle transport is that of the **aerodynamic diameter** ($d_a$) of the particle. This is defined as the diameter of an equivalent spherical particle of density $10^3\,\mathrm{kg\,m^{-3}}$ with the same falling speed as the particle in question. It is related to the particle geometrical diameter $d_v$ with a knowledge of the particle density and shape from the following relation:

$$d_a = d_v \left\{ \frac{\rho C_c Re_p^*}{\rho^* C_c^* Re_p \chi} \right\} \qquad (3.14)$$

where $\rho_p$ is the particle density, $C_c$ is the Cunningham slip correction factor for the particle, $Re_p$ is the particle Reynolds number, and $\chi$ is the dynamic shape factor. The starred terms refer to the spherical water droplet.

### 3.2.2 *Impaction and interception*

Consider what happens when particles are transported in a distorted airflow such as around a bend or about a bluff obstacle (Figure 3.3). The air itself diverges to pass around the outside of the obstacle. Very small (inertialess) particles would do the same, but larger particles, because of the inertial behaviour described above, would tend to leave the flow streamlines and continue to travel in the direction of their initial motion. This tendency is greater the more massive the particle, the greater its

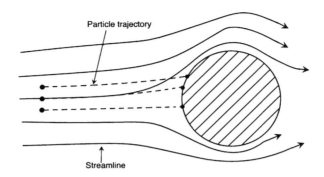

**Fig. 3.3** The impaction of particles.

approach velocity and the more sharply the flow diverges. This behaviour is consistent with a dependence on the Stokes number (3.11), where $D$ is the obstacle dimension and $U$ is the approach velocity. It can be seen in Figure 3.3 that some particles impact onto the surface of the obstacle, and assuming that they all stick to the surface, the **impaction efficiency** ($E$) is defined as:

$$E = \frac{\text{Number of particles arriving by impaction}}{\text{Number of particles geometrically incident on the body}}. \quad (3.15)$$

In addition, the particle trajectories will be dependent upon the Reynolds numbers of both the particle $Re_p$ and the flow around the obstacle ($Re$), such that

$$E = f(St, Re_p, Re). \quad (3.16)$$

Impaction is particularly important as a removal mechanism for particles close to the ground as they are carried by the wind around leaves, trees, grass, irregularities in the soil surface, as well as buildings and other human-made obstacles. It is also the main mechanism in the sampling of airborne particles whether by aspiration or by the use of impaction surfaces. In addition, it is used as a means of separating particles according to their aerodynamic diameters in cascade impactors.

Particles which are of the same order of size as the obstacle itself may be collected by **interception** if the particle trajectory passes close enough to the surface of the obstacle so that they touch, as shown in Figure 3.4. This is not an efficient collection process but is enhanced in practice by the presence of fine surface structure (leaf hairs, surface roughness of soil and building materials, etc.). For particle sampling, however, the phenomenon of interception may be largely disregarded.

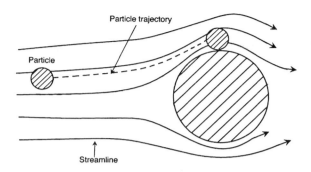

**Fig. 3.4** The collection of particles by interception.

### 3.2.3  *Diffusion*

3.2.3.1  *Molecular diffusion.*  So far we have discussed the transport of airborne particles by processes that are assumed to be well ordered and, in theory, deterministic. In practice, however, small particles in still or smooth airflows are seen to move randomly and erratically. This motion results from collisions with the surface of the particle of gas molecules, which are in thermal motion as described by the classical kinetic theory of gases. This movement is independent of any convection associated with the air itself and is known as **molecular** or **Brownian diffusion**. The result of this phenomenon is a net migration of particles from regions of high concentration to regions of low concentration. The resultant local net flux of particles by this process is described by Fick's law of classical diffusion, which in the simplest, one-dimensional case is given by:

$$-D_B \frac{dc}{dx}, \tag{3.17}$$

where $c$ is the local particle concentration, and $D_B$ is the coefficient of Brownian diffusion. For small particles in the Stokes regime, the latter is given by

$$D_B = \frac{k_B T}{3\pi \eta d_v}, \tag{3.18}$$

where $T$ is the air temperature (in kelvin) and $k_B$ is the Boltzmann constant $(1.38 \times 10^{-23} \, \text{J K}^{-1})$. $D_B$ therefore embodies the continual interchange of thermal energy between the gas molecules and the particles and vice versa. The effect of diffusion on particle transport becomes more significant for smaller particles, although even for a particle of diameter 1 μm in air, $D_B$ is only of the order of $10^{-11} \, \text{m}^2 \, \text{s}^{-1}$. From (3.16), the resultant rate of change of local particle concentration is therefore

$$\frac{dc}{dt} = D_B \frac{d^2 c}{dx^2} \tag{3.19}$$

$$c(x, t) = \frac{N_0}{(2\pi D_B t)^{0.5}} \exp\left(\frac{-x^2}{4 D_B t}\right) \tag{3.20}$$

for the concentration distribution along the $x$-direction at time $t$. The root mean square displacement of particles from their origin at time $t$ is

$$x' = (2 D_B t)^{0.5}. \tag{3.21}$$

The phenomenon of diffusion is important not only in how particles move from one point to another, but also in how they move in relation to one another. It is responsible for collisions between particles which result in the coagulation of small particles to form larger ones.

3.2.3.2  *Turbulent diffusion.*  The mixing of particles in a turbulent airflow may be thought of as a form of diffusion over and above the molecular variety described above. In most cases, the flux associated with turbulent diffusion may be described in terms of an expression which is directly analogous to Fick's law as given in (3.17). The only difference is that the turbulent diffusivity of the particles, $D_{pt}$, replaces $D_B$. The particle's ability to respond to the eddying, distorted turbulent motions of the surrounding air is dependent upon inertial considerations similar to those discussed earlier. We may describe an inertial parameter ($K_{pt}$) similar to the Stokes number already defined:

$$K_{pt} = \frac{\tau u'}{L}, \tag{3.22}$$

where $\tau$ is the particle relaxation time, and $u'$ and $L$ are the characteristic turbulence parameters. This leads to:

$$\frac{D_{pt}}{D_{ft}} = f(K_{pt}), \tag{3.23}$$

where $D_{ft}$ is the turbulent diffusivity for the fluid. As $K_{pt}$ increases (i.e. larger particles, greater turbulence intensity, smaller turbulence length scale), the particle responds less well to the fluctuations and so its diffusivity falls. For very large particles or very small length scales the particle does not see the turbulent motions as it travels with the mean flow.

### 3.2.4  *Aspiration*

**Aspiration** concerns the process by which particles are withdrawn from ambient air through an opening in an otherwise enclosed body. It is therefore relevant to aerosol sampling systems, and to the inhalation of

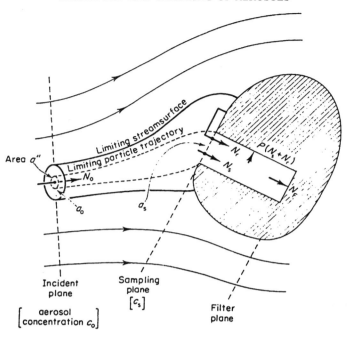

**Fig. 3.5** The aspiration of particles into an entry and the different sampler performance parameters.

aerosols by humans through the nose and/or mouth during breathing. Figure 3.5 shows a body of arbitrary shape placed in a moving airstream. It has a single orifice positioned at arbitrary orientation to the wind through which air is drawn at a fixed volumetric flow rate by a sampling pump. Particles brought to the vicinity of the sampler by the wind experience two competing flow regimes, and their behaviour in these regimes is dependent upon the inertial forces as the flow changes direction. Firstly, the external part diverges to pass around the outside of the body, and the particles undergo impaction on to the body, as described earlier. Secondly, the particles impact onto the plane of the orifice as they experience the convergent flow into the orifice. For more details, see Vincent (1989).

**Aspiration efficiency** ($A$) may be defined for a given particle aerodynamic ($d_a$) as

$$A = \frac{\text{Concentration of particles in the air actually entering the orifice}}{\text{Concentration of particles in the undisturbed upstream air}},$$

$$(3.24)$$

provided that the airflow and aerosol upstream of the sampler are uniformly distributed in space. From considerations of particle impaction from one region of the flow to another, a system of equations may be developed which can, in principle, provide estimates for $A$. Generally

$$A = f(St, U/U_s, \delta/D, \theta, B), \tag{3.25}$$

where $St\,(= d_a^2 \tau^* U/18\eta D)$ is a characteristic Stokes number for the aspiration system, $U$ is the external windspeed, $U_s$ is the mean sampling velocity, $\delta$ is the orifice dimension, $D$ is the body dimension, $\theta$ is the orientation with respect to the wind, and $B$ is an aerodynamic bluffness factor.

In practice, however, sampler performance needs to be described by more than one efficiency parameter. The aspiration efficiency, $A$, describes the purely aerodynamic part of sampler performance and is given by

$$A = \frac{N_s}{N_0} \tag{3.26}$$

where (referring again to Figure 3.5) $N_0$ is the number of particles passing through the limiting stream surface sufficiently far upstream to be undisturbed by the presence or action of the sampler, and $N_s$ is the number of particles passing directly through the sampling plane. The **entry** or **inlet efficiency**, $A_I$, describes the efficiency with which particles enter the sampler plane and includes those particles $N_r$ that have passed through the sampling plane after rebounding off the external walls of the sampler

$$A_I = \frac{N_s + N_r}{N_0}. \tag{3.27}$$

Finally, **the overall sampling** efficiency, $A_s$, takes account of those particles lost to the external walls during transport from entry plane to filter:

$$A_s = \frac{N_f}{N_0} = \frac{P(N_s + N_r)}{N_0} \tag{3.28}$$

where $N_f$ is the number of particles deposited on the filter (or collection surface) and $P$ is the fractional penetration of particles through the transmission section.

### 3.2.5  *Practical examples of the transport of aerosols*

3.2.5.1  *The lifetime of aerosols in the ambient atmosphere.*  In the ambient atmosphere aerosol particles arise from a whole range of sources: wind-raised dust from spoil heaps, construction sites and open fields, sea spray, industrial activity, emissions from animals and plants (microbial and fungal spores), traffic, volcanoes, forest fires and combustion

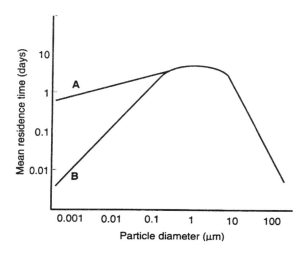

**Fig. 3.6** The residence time of aerosols in the atmospheric boundary layer. (Adapted from Jaenicke, 1993).

processes, residues from the evaporation of sprays, mists and fogs, and photochemical conversion of gas to particles (mainly sulphates and nitrates). Details of the particle size distributions of these aerosols have been given in Chapters 1 and 7.

The length of time that the particles are likely to remain airborne in the atmosphere is dependent mainly upon the size of the particle and the size-dependent removal mechanisms (see Figure 3.6). The smallest particles (1 nm diameter) last for only about 10 minutes due to agglomeration with other particles to form larger particles. These particles, from 100 nm to about 2 μm, may then be removed from the atmosphere by the scavenging action of rain droplets as they fall to ground. Main capture mechanisms by the rain droplets are diffusion, interception and impaction. Rain removes particles from the lower atmosphere in about 10 days, by which time they may have travelled several thousand kilometres in the lower troposphere, where the mean windspeed is about 7 m s$^{-1}$.

Larger particles (over 2 μm in size) are likely to remain airborne for just 10 to 20 hours and to travel distances of 20 to 30 km, before removal by dry deposition. The combination of diffusion, interception, impaction and sedimentation in removing the particles can be clearly seen in Figure 3.7. Near the surface of the earth the air is mixed rapidly by turbulent motions, generated by the friction of the wind at the surface. This turbulent mixing distributes particles uniformly throughout the 1 km high boundary layer. When the air is close to the Earth it flows parallel to the surface, and

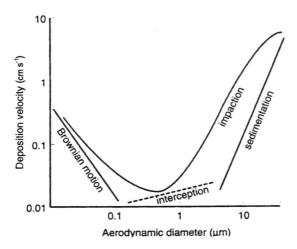

**Fig. 3.7** The mechanisms for the dry deposition of aerosols.

particles close to the surface are removed mechanisms associated with their aerodynamic diameters.

### 3.2.5.2    *The transport of particles in the human respiratory system.*    The inhalation of particles into the mouth and nose and their subsequent penetration into the various regions of the respiratory system is controlled by the transport mechanisms described in section 3.1 above. To aid the discussion, consider the simple representation of the human respiratory system given in Figure 3.8. Particles enter the mouth and/or nose when air is breathed in. The efficiency of entry (**sampling efficiency**, otherwise known as **inhalability**) is described in (3.25) and (3.27). This is governed by inertial impaction as described by the Stokes number, the velocity ratio, the bluffness of the body and the orientation of the mouth to the wind. Once entered, the largest particles may be too large to penetrate the passages between nasal hairs or the nasal turbinates and will be collected by interception. The air velocities in this nasopharyngeal region are high and so particles are deposited mainly by impaction as the air is constrained to change direction rapidly. Penetration of large particles through the larynx and into the tracheobronchial tree is dependent upon the inhalation flow rate and whether inhalation occurs through the nose or mouth.

Within the tracheobronchial region the fall in air velocity, and the transition from turbulent to laminar flow, reduces the importance of impaction as the principal mode of deposition, except close to the bifurcations where the airways divide into two and particles deposit on the intersections between the two pipes (known as carina). As we go deeper

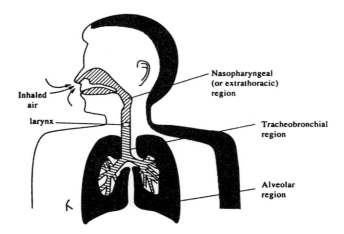

**Fig. 3.8** Simple representation of the human respiratory system.

into the lung the conducting airways branch successively to form conducting and terminal bronchioles, and finally to the respiratory bronchioles, the alveolar ducts and sacs where the gas exchange takes place. Table 3.1 gives details of the structure of the human lung, including the degree of branching, the dimensions of successive branches, the air velocities and the associated residence times. As airway size decreases, the distances which particles must travel to reach the walls are shorter and the mean air velocities are lower. The probability of particle deposition by

**Table 3.1** Summary of data on the structure of the human lung, including the degree of branching, the dimensions of successive branches, the air velocities and the associated residence times (from Vincent, 1995)

| Region | Airway | Generation | Number per generation | Area of duct $(cm^2)$ | Air velocity $(m\,s^{-1})$ | Residence time (ms) |
|---|---|---|---|---|---|---|
| Tracheo- | Trachea | 0 | 1 | 2.5 | 3.9 | 30 |
| bronchial | Bronchi | 1 | 2 | 2.3 | 4.3 | 11 |
| region | | 11 | $2 \times 10^3$ | 20 | 0.52 | 7.4 |
| | | 13 | | | | |
| | Bronchioles | 14 | $16 \times 10^3$ | 69 | 0.14 | 16 |
| | | 16 | $66 \times 10^3$ | 180 | 0.05 | 31 |
| Alveolar | Respiratory | 17 | | | | |
| region | Bronchioles | 18 | $26 \times 10^4$ | 530 | 0.03 | 60 |
| | Alveolar ducts | 21 | $2 \times 10^6$ | $3 \times 10^3$ | 0.003 | 210 |
| | Alveolar sacs | 23 | $8 \times 10^6$ | $1 \times 10^4$ | 0.001 | 550 |

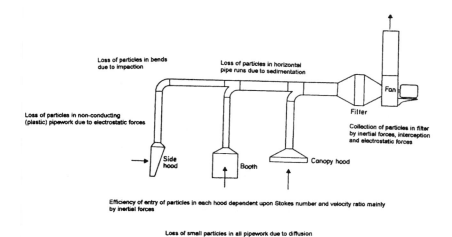

**Fig. 3.9** Schematic of a simple ventilation system showing where particles deposit.

*sedimentation* and *diffusion* increases. In the alveolar region, air velocities are very low and residence times high, and so the probability of deposition of particles which sediment and/or diffuse rapidly is very high.

3.2.5.3 *The transport of particles in ventilation ducts.* Once an aerosol particle has been captured by the inlet of a ventilation system, the aim is to transport the particles with maximum effectiveness to a part of the system where they can be either separated from the flow or discharged safely. During transport through the ventilation ducts, particles may be deposited by a combination of *sedimentation, inertial impaction and turbulent diffusion* (with Brownian diffusion expected to be negligible in the relatively fast-moving air that usually prevails). Such deposition can cause build-up of particles in certain parts of the ductwork, particularly in horizontal sections, and near bends, contractions, expansions, baffles and fittings (as shown in Figure 3.9). As a general rule of thumb, to minimize particle deposition in ventilation ducts, the layout should be as simple and undistorted as possible, with mean air velocity in the duct maintained well above about $15\,\mathrm{m\,s^{-1}}$.

A similar effect occurs with pipework connecting particle entries to particle monitors. Very often the pipework is long, has a number of sharp bends and is made from non-conducting plastic material. Considerable particle losses may be experienced in this pipework, resulting in a distorted and unrepresentative aerosol sample being presented to the detection system.

## 3.3 The sampling of aerosols

### 3.3.1 *Criteria for practical aerosol sampling*

There are three main sets of factors justifying interest in the sampling of aerosols: technological, ecological and health; see Figure 3.10 (Vincent, 1989).

***Technological*** justifications refer to the investigation of aerosol properties directly relevant to industrial processes where neither ecological, environmental nor health considerations are the primary focus of attention. This may include aerosol sampling for monitoring the emissions of particles from industrial processes and monitoring the effectiveness of subsequently installed aerosol control systems. It may overlap with the areas of ecology and/or health if the aerosol to be sampled is emitted to the outdoor atmosphere via exhaust ducts or fugitive leaks, or to indoor environments by lack of suitable control systems.

***Ecological*** justifications relate to the sampling of aerosols that are present in the outdoor ambient atmospheres for which human health is not the major concern. This includes: the sampling of aerosols for meteorological purposes associated with the formation of clouds and the effect of aerosols on radiative forcing (as a negative contribution to global warming) and atmospheric visibility; and the sampling of deposited particles for their effects on plants and on the soiling of buildings. The latter area may overlap with health in the case of particles deposited on plants that are subsequently eaten by humans or by animals that humans then eat.

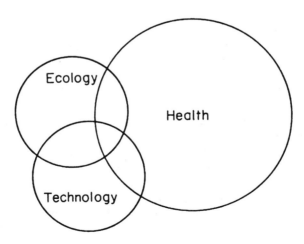

**Fig. 3.10** Reasons for sampling aerosols (Vincent, 1989).

*Health* justifications include all aspects of aerosol sampling where human health is directly the reason for the sampling. This area is the main reason for aerosol sampling and has received the most scientific development work in the second half of the twentieth century.

Before undertaking a sampling programme, it is essential that a number of simple questions are answered. *Why* do you need to sample? *How* will the results be used? The answers to these questions should provide a sampling rationale relevant to the practical situation concerned, and the most appropriate sampling equipment and strategy can then be chosen.

3.3.1.1 *Criteria for technological and ecological sampling of aerosols.* In the technological area, aerosol measurement is generally based on the criterion that all airborne particles are of interest and that a representative sample of total aerosol concentration is obtained. Ideally, this means that samplers must have a sampling efficiency of unity for all ranges of particle sizes present. This is feasible when sampling aerosols contained within chimney stacks and exhaust ducts by using thin-walled sampling tubes and isokinetic sampling (see below), but is almost impossible once the aerosol has been released into the atmosphere, especially for the larger particles. It is important, therefore, when interpreting sampling data describing the emission of airborne particles from a process to assess the likely efficiency and validity of the sampling equipment and strategy employed.

In the ecological area, we are concerned with aerosols in the ambient atmosphere, which are very variable in both size distribution and chemical composition. Ambient aerosols in urban environments are generally considered to exhibit a trimodal size distribution, with the three modes varying significantly in origin, chemical composition and residence time in the atmosphere (see section 3.2.5.1). Briefly, they comprise: the *nucleation* mode, consisting of particles with diameters from 0.05 to 0.1 μm, formed mostly by gas-to-particle conversions and combustion processes; the *accumulation* mode, consisting of particles with diameters from about 0.1 to 2 μm, formed as a result of the coalescence of particles in the nucleation mode; and the *coarse* mode, containing particles from about 2 to 100 μm, generated by a wide range of naturally occurring and human-made processes.

This wide diversity of particles and their differing effects on the environment means that there is not one sampling criterion that can be relevant to all particles. For atmospheric visibility studies, it is the number concentration of the finer particles that is the relevant criterion, since these particles are more numerous and have a higher coefficient of light scattering than the coarser ones. Such fine particles are also important in relation to aerosol effects on climate such as radiative forcing. Aerosol sampling relevant to these effects should therefore use instrumentation that operates according to this criterion. The other main ecological effect of airborne particles is associated with the deposition of coarse particles onto surfaces such as plants, buildings, cars, washing, etc. Generally, these effects are aesthetic in nature, but are nevertheless a major source of complaint from members of the general public. Health effects may be relevant due to the deposition of toxic particles (such as radioactive materials) onto plants, which are either directly eaten or are eaten by animals which humans then eat. For these nuisance dusts and those associated with health effects via secondary pathways, it is usually deposition to the ground or the flux of particles past a point that is of interest. Instruments used for this purpose are generally passive collection devices, where the mass of material deposited in a given area (or passing through a given area) per unit time is determined. The instruments used for this will be described briefly later in this chapter.

3.3.1.2 *Health-related aerosol sampling criteria.* For health effects that are suspected to have arisen from particles entering the body through the nose and mouth during breathing, one must use a sampler whose performance mimics the efficiency with which particles enter the nose and mouth and penetrate to the region in the body where the harmful effect occurs. Workers in the occupational hygiene field have realized this for some years and defined the respirable fraction for those particles that penetrate to the alveolar region of the lung and cause diseases such as pneumoconiosis, silicosis and asbestosis.

Since the early 1980s an *ad hoc* working group of the International Standards Organization (ISO/TC146/SC2 and SC3) and a working group of the European Committee for Standardization (CEN/TC137/WG3) have been formulating health-related sampling conventions for airborne dusts both in the ambient atmosphere and in the workplace. The final agreed conventions are now available as EN 481 (CEN, 1993) and ISO 7708 (ISO, 1995). They are defined in Figure 3.11 and comprise four main fractions.

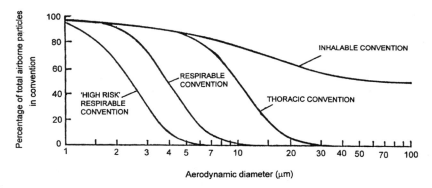

**Fig. 3.11** Health-related aerosol sampling conventions (ISO/CEN/ACGIH).

- The *inhalable fraction* $(E_I)$ is defined as the mass fraction of total airborne particles which is inhaled through the nose and/or mouth. It was derived from wind-tunnel measurements of the sampling efficiency of full-size tailor's mannequins and replaces the very loosely defined 'total' aerosol fraction used previously. It is given by

$$E_I = 0.5(1 + \exp[-0.06d_a]) + 10^{-5}U^{2.75}\exp(0.05d_a), \qquad (3.29)$$

where $d_a$ is the aerodynamic diameter of the particle (see (3.14)), and $U$ is the wind speed (up to $10\,\mathrm{m\,s^{-1}}$).
- The *thoracic fraction* $(E_T)$ is defined as the mass fraction of inhaled particles penetrating the respiratory system beyond the larynx. As a function of total airborne particles, it is given by a cumulative lognormal curve, with a median aerodynamic diameter of $10\,\mu m$ and geometric standard deviation of 1.5.
- The *respirable fraction* $(E_R)$ is defined as the mass fraction of inhaled particles which penetrates to the unciliated airways of the lung (alveolar region). As a function of total airborne particles, it is given by a cumulative lognormal curve with a median aerodynamic diameter of $4\,\mu m$ and a geometric standard deviation of 1.5.
- The *'high-risk' respirable fraction* is a definition of the respirable fraction for the sick and infirm, or for children, and is intended for use in the non-occupational environment. As a function of total airborne particles, it is given by a cumulative lognormal curve with a median aerodynamic diameter of $2.5\,\mu m$ and a geometric standard deviation of 1.5.

These conventions provide target specifications for the design of health-related sampling instruments, and give a scientific framework for the measurement of airborne dust for correlation with health effects. For example, the inhalable fraction applies to all particles that can enter the body, and is specifically of relevance to those coarser toxic particles that deposit and dissolve in the mouth and nose. The respirable fraction, on the other hand, relates to those diseases of the deep lung, such as the pneumoconioses, while the thoracic fraction may be relevant to incidences of bronchitis, asthma and upper airways diseases.

In the occupational field, these conventions have been adopted world-wide, and occupational exposure standards are being revised to specify aerosol sampling to be carried out with instruments designed to meet their requirements. For example, the 1997 revision of the UK MDHS 14 (HSE, 1997) refers to the new conventions. Until recently however, this philosophy has not been taken on board by the environmental community for health-related sampling. In the USA, measurement of an aerosol fraction in the ambient atmosphere specifically related to health effects has been carried out since the late 1980s. The US Environmental Protection Agency produced an earlier definition of the thoracic aerosol fraction

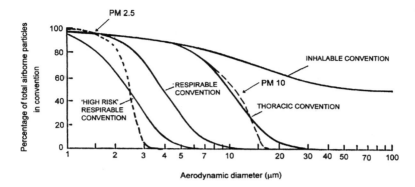

**Fig. 3.12** Definitions of PM10 and PM2.5 in relation to the ISO health-related sampling conventions.

(EPA, 1987), known as PM10, which differs from the ISO definition in that it has a 50% penetration at aerodynamic diameter 10.6 µm and zero penetration at 16 µm, as shown in Figure 3.12. It has been systematically measured with validated instruments in the USA since the late 1980s, and has been widely adopted in other countries during the 1990s, as the main criterion for ambient aerosol sampling for health effects. In practice, however, the difference between the ISO and PM10 definitions of the thoracic fraction is not significant because most instruments validated for sampling the PM10 fraction will also reliably sample the ISO thoracic fraction. At the time of writing (summer 1997), a further health-related aerosol convention is being discussed for future implementation in the USA for better correlation with some health effects. This is a finer fraction with a median aerodynamic diameter of about 2.5 µm, and separates particles in the accumulation mode from those human-made and wind-blown particles in the coarse mode. The shape of the proposed curve is somewhat sharper than that of the 'High risk' respirable fraction of ISO 7708, although they both seek to protect those whose health is most at risk from inhalation of airborne particles – children and those with respiratory or cardiovascular disease. Regulatory bodies in Britain are keeping a close eye on these developments with a view to introducing them in the future, especially as a possibly more reliable indicator of levels of vehicle-derived particles which have been shown to be mainly composed of submicrometre particles.

Fibrous aerosol particles – those with long aspect ratio, such as asbestos and human-made mineral fibres – have, historically, been considered separately by the scientific community. The definition of what is *respirable* for such particles is based not only on the aerodynamic factors that govern the deposition of fibres in the lung after inhalation, but also on their known dimension-associated health risk. For example, in the case of asbestos, long, thin fibres are thought to be more hazardous to health than short, fat ones. This is because they are capable of penetrating deep into the alveolar region of the lung, but the normal lung defence mechanisms are less able to eliminate long particles than isometric ones of similar aerodynamic diameter. Selection of the respirable fraction of the airborne fibres is therefore carried out, after they have been collected on filters, by sizing and counting under the microscope. Unlike for isometric particles, it is the number not the mass of particles that is measured.

The internationally agreed criterion for respirable asbestos fibres is that they should have an aspect ratio greater than 3:1, length greater than 5 μm and diameter less than 3 μm. This was agreed in 1979 by the Asbestos International Association (AIA, 1979) and is still widely used today. Both sampling and microscopy procedures are prescribed in this document, many versions of which have been proposed in different countries. While this prescriptive approach should lead to consistency in fibre measurements, it is not consistent with the more general conventions described above which specify instrument performance and not particular instrument designs. This method is currently under review, and research work is in progress to produce a thoracic aerosol inlet suitable for fibrous aerosols.

There are currently no internationally agreed definitions for the sampling of biological aerosols (bioaerosols), and work is in progress in a number of countries to resolve the situation. However, although not explicitly stated in the document EN 481, bioaerosols that are harmful to the various regions of the body should be sampled using instruments that select particles according to the health-related sampling conventions described in EN 481. The additional problems posed by bioaerosols are that some of the particles only cause problems to the human body when alive. These particles (bacteria, viruses, moulds, etc.) are detected by culturing them on media such as agar and so they must be kept alive and unharmed during the sampling process. For these particles it is the number rather than the mass concentration that is determined. This area is changing rapidly and it is hoped that agreed conventions and methodology will be available soon. These methods will not be described here, and the reader is referred to a number of documents such as the ACGIH Handbook on Air Sampling Instruments (ACGIH, 1994) and Griffiths and DeCosemo (1994) for details.

### 3.3.2    *The basics of an aerosol sampling system*

An aerosol sampling instrument (or sampler) always comprises a number of components which contribute to the overall accuracy with which a sample is taken. These components are the sampling inlet, the trans-

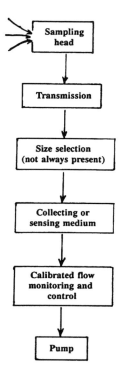

**Fig. 3.13** The basics of an aerosol sampling system.

mission section, the particle size selector (not always present), the collecting or sensing region, calibrated flow monitoring and control, and the pump. A simple schematic diagram of these essential components is given in Figure 3.13.

3.3.2.1 *The sampling inlet.* Many people involved in aerosol sampling have, in the past, overlooked the choice of sampling inlet. Instead they have concentrated on the protection of the filter from the rain and snow (for ambient aerosols), the performance of the pump, the choice of filter and the analytical technique to be employed. For the workplace, they have used any filter holder of suitable size, without knowing its particle sampling efficiency, which can have widely varying values dependent upon its design, and the particle size and wind speed in the workplace. This would lead to wide differences in the mass concentrations reported. In the ambient environment, the use of any cover which in effect defines the sampling inlet (as in the well-known high-volume sampler) may lead to large errors in the mass concentration measured, due to variable and inefficient particle sampling. This is especially neglected in the use of cascade impactors to obtain particle size distribution information about the airborne particles. Very often large particles are significantly under-represented in the particle size distributions reported. An inlet should be chosen that meets the requirements of the sampling strategy chosen. For both environments, the situation is improving because with acceptance of the new health-related sampling conventions described above, instrument manufacturers must, and users must, demonstrate that their equipment meets the relevant sampling convention if the result is to be used for regulatory purposes.

3.3.2.2 *Transmission section.* The transmission of sampled particles from the inlet to the particle size selector (if present) and on to the collecting or sensing region is an important part of the sampling system that is again often overlooked. The main problem is to avoid particle losses to the internal walls of the transmission pipe or channel. In samplers where the sampling inlet, size selector and collecting filter are built into a sampling head, care has normally been taken with the design of the sampler to minimize this problem. Losses of up to 100% of the sampled particles have been reported for some non-conducting personal sampling heads. However, for instruments where the inlet is remote from the sensing region (continuous particle counters and particle size analysers) it is essential to minimize the length of the connecting pipework and to keep sharp bends and horizontal surfaces to a minimum. Wall losses can occur over the whole size spectrum of particles, ranging from inertial and sedimentation losses for the large particles (aerodynamic diameters greater than 5 μm) to diffusional losses for particles of diameters less than 100 nm. The use of non-conducting plastic piping is not recommended due to possible enhanced deposition from electrostatic forces.

3.3.2.3 *Size selection.* For health-related sampling we wish to select aerosol fractions that relate to particle deposition in various regions of the respiratory system. For this, some form of particle size selector is used to select the relevant portion of the sampled aerosol. Particles are generally selected by aerodynamic means using physical processes similar to those involved in the deposition of particles in the respiratory system. Gravitational sedimentation processes are used to select particles in horizontal and vertical elutriators, centrifugal sedimentation is used in cyclones, inertial forces are used in impactors, while porous foams employ a combination of both sedimentation and inertia forces.

Owing to their size, and the requirement to be accurately horizontal or vertical for correct operation, elutriators are only used in some workplace static samplers. Personal samplers for the thoracic and respirable fractions are available using cyclones, impactors and foams for size selection.

Generally, in samplers for environmental aerosols, impactors or cyclones are used for particle size selection, although porous foams have been used in some prototype samplers.

3.3.2.4 *Filters*. A filter is the most common means of collecting the aerosol sample in a form suitable for assessment. That assessment might include gravimetric weighing on an analytical balance before and after sampling to obtain the sampled mass. It might also include visual assessment using an optical or electron microscope and/or a whole range of analytical and chemical techniques. The choice of filter type for a given application depends greatly on how it is proposed to analyse the collected sample. Many different filter materials, with markedly different physical and chemical properties, are now available. These include fibrous (e.g. glass, quartz), membrane (e.g. cellulose nitrate, polycarbonate, Teflon) and sintered (e.g. silver) filters. Membrane filters have the advantage that they can retain particles effectively on their surface (good for microscopy), whereas fibrous filters have the advantage of providing in-depth particle collection and hence a high load-carrying capacity (good for gravimetric assessment).

Such filters are available in a range of dimensions (e.g. from 25 to 140 mm diameter) and pore sizes (e.g. from 0.1 to 10 μm). Collection efficiency is usually close to 100% for particles in most size ranges of interest, although sometimes some reduction in efficiency might be traded against the lower pressure drop requirements of a filter with greater pore size. For some types of filter, electrostatic charge can present aerosol collection and handling problems – in which case, the use of a static eliminator may (but not always) provide a solution. For other types, weight variations due to moisture absorption can cause difficulty, especially when being used for the gravimetric assessment of low masses. It is therefore recommended that the stabilization of filters overnight in the laboratory should be carried out before each weighing, together with the use of blank 'control' filters to establish the level of variability. It is preferable that temperature and humidity control in the balance room be provided, especially when collected particle weights are low.

The chemical requirements of filters depend on the nature of the analysis which is proposed. As already mentioned, weight stability is important for gravimetric assessment. If particle counting by optical microscopy is required, then the filters used must be capable of being rendered transparent (i.e. cleared). Direct on-filter determination of elemental composition (e.g. by scanning electron microscope and energy-dispersive X-ray analyses, X-ray fluorescence, neutron activation analysis and particle induced X-ray emission) is often required. For this, filters must allow good transmission of the radiation in question, with low background scatter. Collected samples may also be extracted from the filter prior to analysis, using a range of wet chemical methods, ultrasonication, ashing, etc., each of which imposes a range of specific filter requirements.

3.3.2.5  *Pumps.*   Most samplers require a source of air movement so that particulate-laden air can be aspirated into the instrument. For personal samplers, the flow rates are low (usually from 1 to $4\,l\,min^{-1}$), the main limiting factor being the weight, which must be low enough for the sampler to be worn on the body without inconvenience to the wearer. Generally, because aerosol concentrations in the ambient environment are low, flow rates must be high to collect sufficient particles to weigh or to analyse chemically. Flow rates range from 16.7 to $1200\,l\,min^{-1}$, with one sampler (the WRAC) much higher. The actual volumetric flow rate will depend first on sampling considerations (e.g. entry conditions to provide the desired performance), and then on the amount of material to be collected for accurate assessment, analytical requirements, etc. Internal flowmeters, usually of the rotameter type or digital counters, are incorporated into some sampling pumps, but these must always be calibrated against a calibrated flow rate standard (wet gas meter, orifice plate, etc.) placed over the sampler inlet. It should also be noted that the flow rate may vary with the resistance imposed by the filter and its collected aerosol mass. For this reason, flow rates should be checked periodically during sampling and adjusted if necessary. Nowadays, however, most pumps employ some form of flow control where a sensor (e.g. pressure or velocity) is built into a feedback loop to eliminate the need for such regular attention during sampling.

### 3.3.3  *Sampling in stacks and ducts*

3.3.3.1  *Basic considerations.*   Sampling in chimney stacks and exhaust ducts is usually carried out for the purpose of assessing the emission of aerosols from industrial processes to the ambient atmosphere. This can be for determining compliance with air pollution emission standards, evaluating the performance of air pollution control equipment, establishing process material balances, or determining the 'source term' for predicting ground-level aerosol concentrations using atmospheric plume dispersion models. Because flows of air (more generally gas) in stacks and ducts are constrained by the duct walls, they are usually well defined in terms of direction and velocity. This enables isokinetic sampling using thin-walled tubes to provide the main basis for the sampling methodology. It is the only area of aerosol sampling where there is sufficient theoretical understanding of the sampling process to enable *true total aerosol* to be specified as the sampling criterion and the method of taking the practical measurements to be proposed.

Aerosol sampling is isokinetic when the inlet to the sampler, which should be a thin-walled tube, is aligned parallel to the air (or gas) streamlines and the air velocity ($U_s$) entering the sampler is identical to the free-stream velocity ($U$) in the duct approaching the sampler inlet. In this way there is no distortion of the streamlines in the vicinity of the inlet, and all particle sizes are sampled with an efficiency of unity. However, if $U_s$ does not equal $U$, or the sample is not aligned to flow, then the flow streamlines become distorted and curved, large particles leave the flow because of their inertia, and the resultant sample differs both in particle size distribution and concentration from that of the aerosol in the duct. Figure 3.14 shows isokinetic sampling and three conditions of anisokinetic sampling. In Figure 3.14b the sampler is not aligned to the airflow and large particles with high inertia will not be able to turn quickly enough into the inlet and will miss it, leading to under-sampling. In Figure 3.14c, where $U_s$ is greater than $U$, under-sampling also takes place because the air streamlines must converge into the sampler inlet and high inertia particles miss it. Finally, in Figure 3.14d, where $U_s$ is less than $U$, the streamlines diverge around the sampler and excess high-inertia particles enter the inlet. The basic limitations of isokinetic sampling have been studied in detail by a number of researchers, and are reviewed by Vincent (1989). From these considerations and papers such as Vincent et al. (1985), the extent of the errors in sampling due to anisokineticity can be calculated from the following equation:

$$A = 1 + \left[ 1 + \frac{1}{1 + G(\theta)St(\cos\theta + 4R^{1/2}\sin^{1/2}\theta)} \right](R\cos\theta - 1), \qquad (3.30)$$

where $R$ is the velocity ratio $U/U_s$, $G$ is an empirical constant, and $\theta$ is the orientation of the sampler entry to the free-stream wind.

Most of the gasflows in stacks and ducts are highly turbulent, so it is fortunate that the fluctuating flow appears to have very little effect on the sampling efficiency of thin-walled tubes facing the wind (Vincent et al. 1985). However, studies by Wiener et al. (1988) show that particle losses inside the entry probe are affected by the level of free-stream turbulence. Internal wall losses are potentially a major source of error in stack sampling and need to be carefully considered. We will deal with these later.

In practice, however, there are a number of problems in obtaining reliable aerosol samples from stacks and ducts, and the methods developed over 25 years ago are still undergoing improvement and standardization. The latest methodology is given in the ISO standard that has recently been approved, ISO 9096 (ISO, 1992). This is an update of the various national standards that have been widely used for some considerable time (e.g. US EPA, 1971 (revised 1987), and BS 3405, 1971 (revised 1983)). Similar CEN standards are being produced and will be published soon. In addition, a

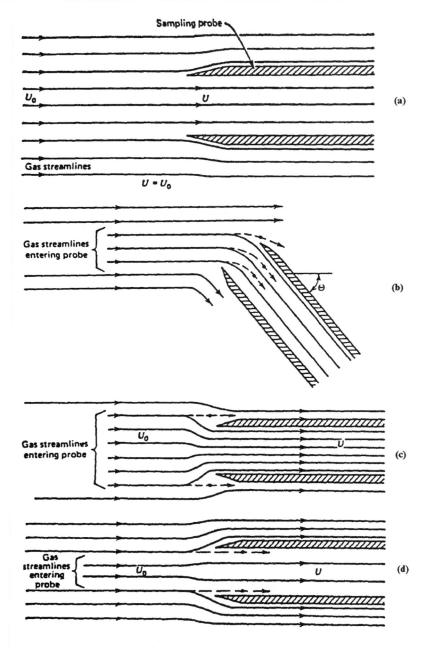

**Fig. 3.14** An explanation of the conditions for isokinetic sampling of aerosols: (a) true isokinetic conditions, $U_s = U, \theta = 0$; (b) non-isokinetic, $U_s = U$, but $\theta \neq 0$; (c) super-isokinetic, $U_s > U$; (d) sub-isokinetic, $U_s < U$.

very readable summary of the principles and methods used in the USA is given in ACGIH (1994).

The main difficulty involves the selection of the sampling station in the stack or duct in order to minimize possible interfering factors. In particular, close to bends, sudden changes in cross-section or some other form of flow disturbance, the nature of the gasflow – both upstream and downstream – may become non-uniform, unstable, highly turbulent, and subject to flow separations with regions of reverse flow. The concentration of aerosol particles in these flows will therefore be very unpredictable both spatially and temporally and difficult to measure. To overcome this problem, a sampling strategy has been developed that involves multi-point sampling across the duct at a location down the duct that is sufficiently distant from the flow disturbances mentioned above, such that their effects on the measured concentration are minimized. The sampling points are chosen such that the exhaust duct cross-section is divided into equal areas and the samples are taken at the centroids of each area. This is illustrated in Figure 3.15 for rectangular and circular ducts. Sampling is carried out

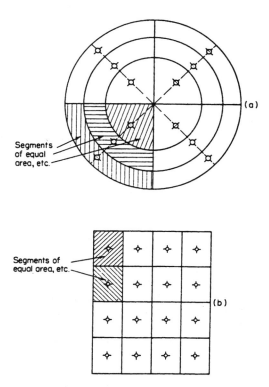

**Fig. 3.15** Location of sampling points over the cross-sections of ducts, (a) circular, (b) rectangular.

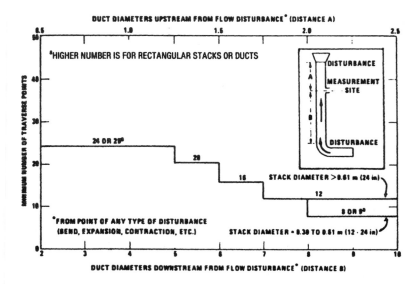

**Fig. 3.16** Minimum number of traverse points as a function of distance downwind from disturbance.

by the successive deployment of the same probe in traverses across the duct, with identical times being spent at each sampling point. The number of sampling points required depends upon the distance between the sampling station and the up- or downstream disturbance, with the number increasing as the distance decreases. This is illustrated in Figure 3.16, which is a diagram taken from the US EPA Method (1987) for both rectangular and circular ducts. It can be seen that smaller numbers of sampling points are required for small duct diameters (less than 610 mm).

3.3.3.2 *Practical methods for stack sampling.* The simplest method of sampling aerosols in ducts includes a first step which is to measure the distribution of gas velocity over the cross-section of the duct at each sampling point of the chosen sampling station. This is usually achieved using a pitot-static tube, which measures the velocity pressure at the sampling point from which the gas velocity can be calculated by application of Bernoulli's equation. The velocity at each sampling point is thus defined so that the aspiration flow rate required to achieve isokinetic conditions at each point can be determined.

The design of the sampling probe itself is governed by the considerations discussed in section 3.3.3.1 to ensure that 'true total aerosol' is measured. Most sampling probes in their simplest forms are thin-walled tubes with their leading edges chamfered at an angle of less than 30° to produce sharp

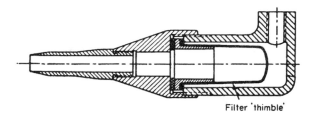

Filter 'thimble'

**Fig. 3.17** Example of simple standard sampling probes for aerosols.

edges with minimal aerodynamic blockage. An example of a standard probe is given in Figure 3.17. It has thin walls which expand slowly externally to connect to a filter holder housing a thimble-type filter. The internal diameter of the probe has two small expansions to the filter. The standard probe has the advantage of simplicity, but suffers from the fact that before sampling can begin, the velocity traverse must be carried out and the flow conditions must be assumed to remain constant during the sampling period.

This drawback is overcome in some probes by the incorporation of a gas velocity sensor into the thin-walled sampling tube. This enables simultaneous measurement to be made of the local free-stream velocity during the sampling run, thus eliminating the need for a separate velocity traverse. Sampling flow rate can then be adjusted during the run to accommodate changes in gas velocity. Systems have been developed using a number of different gas velocity sensors including, for example, pitot-static tubes and hot wire anemometers, and in modern systems adjustment of sampling flow rate is automatic using feedback loop circuits. A typical velocity sensing probe is shown in Figure 3.18.

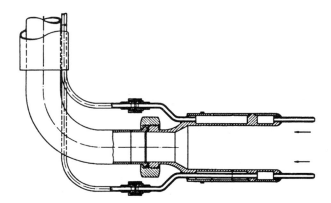

**Fig. 3.18** Example of typical velocity-sensing isokinetic sampling probe of the type described by Bohnet (1978).

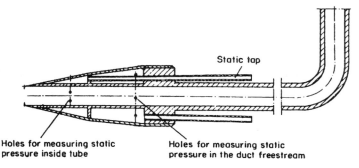

**Fig. 3.19** Example of typical null-type sampling probe of the type described by Dennis *et al.* (1957).

An extension to the combined probe approach, with the aim of eliminating the separate velocity traversing stage is seen in the 'null-type' sampling probe. It is based on the principle that, for an idealized thin-walled probe, if the gas velocity is the same inside the probe as outside it (necessary for isokinetic sampling), then the static pressure at the inside and outside walls should be the same. The null-static pressure balance is achieved when zero static pressure differential is developed between the inside and outside static pressure taps. While a number of these probes have been made, Dennis *et al.* (1957) found that they require calibration under the conditions of use to produce accurate results. A typical null-type probe is shown in Figure 3.19.

There are two basic configurations for these probes: in-stack and out-of-stack. In the in-stack configuration the collection filter is normally of the thimble type held in a filter holder attached directly to the probe entry, as shown in Figure 3.17. This has the advantage that the whole unit is located inside the stack or duct and particle losses inside the walls of the sampling tube are thereby minimized, and all of the sampled aerosol is available for assessment and analysis. However, particles collected on the filter are subjected to the temperature within the duct and for high temperatures (incinerators, etc.) volatilization and oxidation may take place, resulting in a loss of particulate matter. Examples of in-stack methods are US EPA Reference Method 17 (EPA, 1987) and British Standard Method BS803 (1983), shown in Figure 3.20.

Out-of-stack methods employ the filter or collection medium outside the stack. This has the advantage of allowing the collected sample to be maintained at a lower temperature, thereby minimizing the loss of collected material by volatilization and allowing a long-term sample to be collected. It also allows easier filter removal and the use of higher flow rates. However, it suffers from the disadvantage of considerable particle losses on the internal walls between probe entry and collector. This loss will be variable and dependent upon particle size, upon the length,

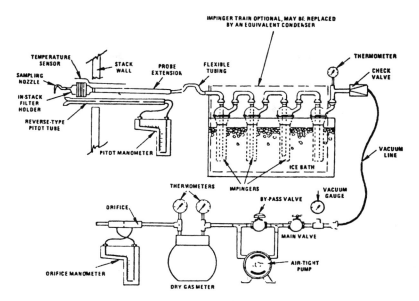

**Fig. 3.20** Schematic of in-stack sampling method (US EPA Reference Method 17).

diameter and configuration of conducting pipe and upon the temperature of that pipe. It is essential, therefore, that the internal walls of the conducting pipe are washed and the deposited particles added to those on the filter. Examples of out-of-stack methods include US EPA Method 5 (US EPA, 1987), shown in Figure 3.21.

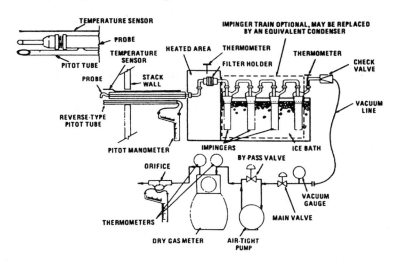

**Fig. 3.21** Schematic of out-of-stack sampling method (US EPA Reference Method 5).

3.3.3.3 *Direct-reading stack monitoring instruments.* All of the methods so far described have involved the extraction of a sample of gas from the stack and the separation of the particles onto some form of collection medium (filter). This process generally involves a day of sampling, and the result is not available until the next day at the earliest. If it is required continuously to monitor particle levels in stacks to ensure that emissions are kept to within legislative limits, or as a means of keeping process under control, then a real-time monitoring instrument is required. A number of such devices have been developed, the majority operating on the principle of the attenuation of a light beam as it traverses the particle-laden gas stream in the stack or duct. This measurement of opacity is dependent upon the size distribution, shape and refractive index of the particles and requires calibration *in situ* if conversion to mass concentration is required. Instruments have also been developed based upon other physical properties, including optical scintillation, triboelectric charging, etc., and a development of the vibrating element mass monitor for use in stacks and ducts is expected soon.

### 3.3.4 *Samplers for aerosols in the workplace*

3.3.4.1 *Sampling strategy.* The most important question to ask before setting out to develop a sampling strategy is whether it is necessary to sample. In the UK, the Control of Substances Hazardous to Health (COSHH) Regulations state clearly that an assessment of the likely risk to health at the workplace should be carried out first. Only if the estimated risk may be significant is it recommended that a sampling programme be instigated. There are then many questions to address before a reliable sampling strategy can be achieved: *what to measure; how to sample; whose exposure to measure; where to position the sampler; how long to sample; how many measurements to take; and how often to sample.*

 If sampling is to be carried out to assess the true exposures of individual workers (or of groups of workers), one of the most important questions to answer is whether a personal or static sampling strategy is to be used (or a combination of both). In static (or area) measurements, the chosen instrument is located in the workplace atmosphere and provides a measurement of aerosol concentration which is (hopefully) relevant to the workforce as a whole. For the case of personal measurements, the chosen instrument is mounted on the body of the exposed subject and moves around with that person at all times.

 When choosing one or other of these, some important considerations need to be taken into account. For a few workplaces (e.g. some working groups in long-wall mining), it has been shown that reasonably good comparison may be obtained between suitably placed static instruments and personal samplers. More generally, however, static samplers have been found to perform less well, tending to give aerosol concentrations which are consistently low compared to those obtained using personal samplers. One advantage with static samplers is that a relatively small number of instruments may be used to survey a whole workforce. If this can be shown to provide valid and representative results, it is a simple and cost-effective exercise. Furthermore, the high flow rates that are acceptable for static samplers mean that, even at very low aerosol concentrations, a relatively large sample mass can be collected in a short sampling period. The use of personal samplers, however, is more labour-intensive. More instruments are deployed and this leads to greater effort in setting them up and in recovering and analysing the samples afterwards. By definition, personal sampling involves the direct cooperation of the workers themselves. Also, for such samplers, it is inevitable that the capacities of the pumps used will be limited by their portability. So flow rates will usually be low (rarely greater than $4 \, l \, min^{-1}$). However, personal aerosol sampling is the only reliable means of assessing the true aerosol exposures of individual workers, so it is by far the most common mode of aerosol measurement in workplaces.

A combination of both static and personal measurements should provide the most cost-effective and comprehensive sampling strategy. Personal samplers can be used to provide the detailed individual exposure information for regulatory purposes on, say, one shift every month, while coverage of the other shifts may be achieved by using a strategically placed static monitor providing continuous assessment. Provided that the work process is relatively stable, an alarm monitor may be employed, set to trigger when the level is reached at which personal exposure is expected to exceed the occupational exposure limit. This system would require calibration of the sort described above.

### 3.3.4.2 *Practical samplers for inhalable aerosols*

3.3.4.2.1 *Static (or area) samplers.* Over many years, static samplers have been developed for the sampling of coarse aerosol in workplace atmospheres. The simplest are open-filter arrangements mounted on the box which contains the pump or systems in which the same open-filter holder is mounted independently. These samplers are widely used in a number of industries in Britain (e.g. nuclear and cotton). Similar devices have been used elsewhere, both in workplace and in ambient air sampling. However, their sampling performances have very rarely been tested, and the few experiments that have been carried out reveal that none of the samplers comes close to matching the inhalability criterion.

The only static aerosol sampler for workplaces designed from the outset to match the inhalability criterion is the $3 \, l \, min^{-1}$ Institute of Occupational Medicine (IOM) static inhalable aerosol sampler (Mark *et al.*, 1985), shown

**Fig. 3.22** IOM static inhalable aerosol sampler: (a) complete; (b) measured sampling efficiency.

in Figure 3.22. It incorporates a number of novel features. The sampler contains a single sampling orifice located in a head which, mounted on top of the housing containing the pump, drive and battery pack, rotates slowly about a vertical axis. The entry orifice forms an integral part of an aerosol collecting capsule which is located mainly inside the head. This capsule also houses the filter, and the whole capsule assembly (tare weight of the order of a few grams) is weighed before and after sampling to provide the full mass of aspirated aerosol. This system eliminates the possibility of errors associated with internal wall losses. When the capsule is mounted in the sampling head, the entry itself projects about 2 mm out from the surface of the head, creating a 'lip' around the orifice itself. This has the effect of preventing the secondary aspiration of any aerosol particles which strike the outside surface of the head and fail to be retained. The performance of this sampler, also shown in Figure 3.22, is in good agreement with the inhalability curve for particles with aerodynamic diameter up to about 100 μm. At present this is the only static sampler designed specifically for the inhalable fraction which is commercially available.

3.3.4.2.2 *Personal samplers.* For reasons outlined above, personal sampling is generally the preferred approach for workplace aerosols. Here, for coarse aerosol, a large number of different devices have been used world-wide, again originating historically for the purpose of sampling for 'total' aerosol. They include simple 25 and 37 mm filter holders in both open-face and single-orifice closed-face arrangements, single-orifice conical samplers and multi-orifice (seven-hole) samplers. A selection of these samplers is shown in Figure 3.23. Most of the samplers shown are intended for use at the sampling flow rate of $2 \, l \, min^{-1}$.

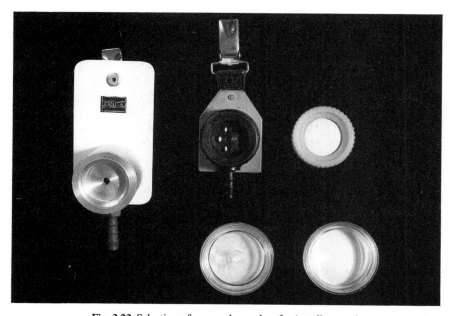

**Fig. 3.23** Selection of personal samplers for 'total' aerosol.

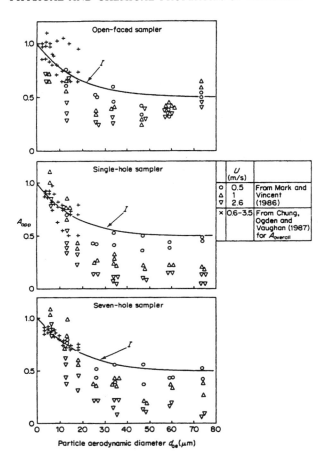

**Fig. 3.24** Early results for the sampling efficiency of three personal samplers for 'total' aerosol.

Over the past 10 years a number of experiments have been conducted to compare their performances with the inhalability curve (Mark and Vincent, 1986) culminating in a collaborative European study (Kenny *et al.,* 1997; Mark *et al*; 1997). All these tests were carried out with each sampler mounted on a life-size torso (e.g. of a mannequin), because it cannot be assumed that, if such samplers were to be tested independently, they would necessarily provide the same results. This is because the aerodynamic conditions governing the airflow around the sampler would be quite different. It follows, therefore, that devices designed as personal samplers should not be used in the static mode. Unfortunately, this common-sense guideline is widely ignored.

The earlier tests shown in Figure 3.24 led to the development of the only personal sampler specifically designed to meet the inhalable criterion –

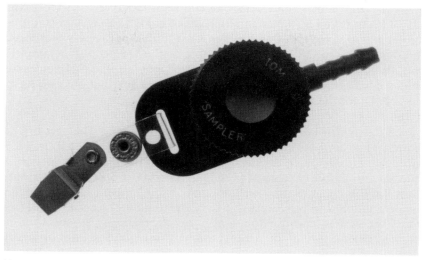

(a)

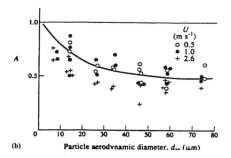

(b)

**Fig. 3.25** IOM personal inhalable aerosol sampler: (a) complete; (b) measured sampling efficiency.

the $2 \, \mathrm{l \, min^{-1}}$ IOM personal inhalable aerosol sampler (Mark and Vincent, 1986) shown in Figure 3.25. It features a 15 mm diameter circular entry which faces directly outwards when the sampler is worn on the torso. Like the IOM static inhalable aerosol sampler in Figure 3.22, the entry is incorporated into an aerosol collecting capsule which, during sampling, is located behind the face-plate. Use of this capsule ensures that the overall aspirated aerosol is always assessed. In addition, as for the static sampler, the lips of the entry protrude outwards slightly from the face-plate in order to prevent oversampling associated with particle blow-off from the external sampler surfaces. Experimental data obtained during its development, given in Figure 3.25, show a good match with the inhalability curve.

All the above-mentioned personal samplers, together with the French CIP10, were tested in the major EU-funded study. The results confirmed the validity of using the IOM sampler for inhalable particles in workplaces and demonstrated that one other sampler – the German GSP sampler – may also be used. The other samplers tested may also be used, but under limited conditions of wind speed and particle size distributions (see Kenny *et al.*, 1997, for the outcome of the study).

3.3.4.3 *Practical samplers for respirable aerosol.* The history of sampling fine aerosols in workplaces began with the respirable fraction, in particular with the emergence in the 1950s of the BMRC respirable aerosol criterion. A number of types of sampling device have since been developed. Most have in common the fact that they first aspirate a particle fraction which is assumed to be representative of the total workplace aerosol, from which the desired fine fraction is then aerodynamically separated inside the instrument, using an arrangement whose particle size-dependent penetration characteristics match the desired criterion. It is the fraction which remains uncollected inside the selector and passes through to collect onto a filter (or some other collecting medium) which is the fine fraction of interest.

3.3.4.3.1 *Static samplers.* A variety of static samplers for respirable aerosol have been built and successfully used in practical occupational hygiene. Some achieve particle size selection by the mechanism of horizontal gravitational elutriation. One example is the British 2.5 l min$^{-1}$ MRE Type 113A sampler (Dunmore *et al.*, 1964) shown in Figure 3.26; another is the similar 100 l min$^{-1}$ Hexhlet (Wright, 1954). For such devices, their penetration characteristics can be easily tailored using the elutriator theory developed by Walton (1954) to match closely the BMRC respirable aerosol curve (itself originally derived from elutriator theory). However, recent wind-tunnel studies (Mark *et al.*, 1993) have shown that the overall sampling efficiency (inlet and size selection) is affected by varying wind speed, as shown in Figure 3.26.

Other static respirable aerosol samplers have been designed to operate on the principles of cyclone selection. One example is the German 50 l min$^{-1}$ TBF50 sampler (Stuke and Emmerichs, 1973); another the French 50 l min$^{-1}$ CPM3 (Fabries and Wrobel, 1987). Although such cyclones can be designed having well-defined penetration characteristics, prediction of performance from theory is more complicated than for horizontal elutriators.

3.3.4.3.2 *Personal samplers.* Horizontal gravitational elutriators are very satisfactory for static respirable aerosol sampling, but are inevitably rather bulky and not conducive to miniaturization. Therefore horizontal

(a)

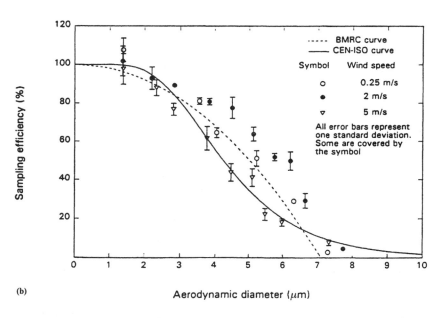

(b)

**Fig. 3.26** MRE Type 113A static respirable aerosol sampler: (a) complete; (b) measured sampling efficiency.

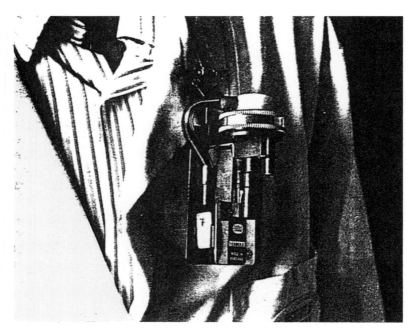

**Fig. 3.27** SIMPEDS personal respirable aerosol sampler.

gravitational elutriation is not an option for personal respirable aerosol samplers. On the other hand, cyclones are ideally suited for such purposes, and have found wide application. Well-known examples are the British $1.9 \, l \, min^{-1}$ cyclone derived from the SIMPEDS (Harris and Maguire, 1968) and shown in Figure 3.27; and the American 1.7 to $2.1 \, l \, min^{-1}$ 10 mm cyclone (Lippmann and Harris, 1962), whose selection characteristics have been shown to be in good agreement with the original BMRC and ACGIH respirable curves, respectively. To meet the new ISO/CEN respirable aerosol convention the SIMPEDS cyclone should be operated at a flow rate of $2.2 \, l \, min^{-1}$.

One further device has some interesting and unusual features and so deserves special mention. This is the French CIP10 (Courbon *et al.*, 1988) shown in Figure 3.28. Although this instrument is aimed primarily at collecting a finer (respirable) aerosol fraction, it is capable of also providing the concentration of the inhalable and thoracic fractions by the inclusion of special inserts in place of the size-selecting foam. It is particularly interesting because the instrument incorporates its own built-in pumping unit, consisting of a battery-driven, rapidly rotating polyester foam plug. The aerosol is aspirated through a downwards-facing annular entry and is collected efficiently by filtration (by a combination of mainly gravitational and inertial forces) in two static, coarse-grade foam plugs

**Fig. 3.28** CIP10 personal respirable aerosol sampler.

located inside the entry as well as on the finer-grade rotating one. As a result of the low pressure drop characteristics of such foam filtration media, a very high flow rate – by personal sampler standards – can be achieved, up to $10 \, l \, min^{-1}$.

### 3.3.4.4 *Practical samplers for thoracic aerosols*

3.3.4.4.1 *Static samplers.* Methodology for the sampling of thoracic aerosol in the occupational context was not widely considered prior to the establishment of the ISO and ACGIH criteria. For workplaces, the only aerosol standard that approaches the thoracic fraction is in the US cotton industry where a criterion was established in 1975 by the US National Institute of Occupational Safety and Health (NIOSH, 1975) based on a selection curve which falls to 50% at $15 \, \mu m$ (compared with $11.64 \, \mu m$ in the new CEN/ISO/ACGIH thoracic fraction). This implies recognition of the role of particle deposition in the large airways of the upper respiratory

**Fig. 3.29** MSP personal environmental monitor for thoracic aerosols.

tract in cotton workers' byssinosis. The recommended static sampling method employs the concept of vertical elutriation.

3.3.4.4.2 *Personal samplers.* The MSP Personal Environmental Monitor (Buckley *et al.*, 1991), shown in Figure 3.29, uses a single-stage impactor designed to select the thoracic fraction according to the earlier American PM10 definition. With a flow rate of $4 \, \mathrm{lmin}^{-1}$, it has a very sharp sampling curve which excludes some of the large particles allowed in the ISO/CEN/ACGIH thoracic convention. This sampler has been widely used in the USA for monitoring personal exposures in non-occupational situations such as homes and public places and has formed part of a major study to investigate aerosol exposures both indoors and outdoors.

There are a number of other samplers currently under development for the thoracic fraction. One is a modification of the CIP10 respirable aerosol sampler (described above) in which the foam size selector has been replaced by an inertial particle selection device. Another is a prototype sampler which uses a porous foam size selector behind the IOM inhalable aerosol entry (Mark *et al.*, 1988). Finally, new cyclones are being developed for sampling the thoracic fraction. These samplers should soon be commercially available.

3.3.4.5 *Direct-reading instruments.* In all of the instruments described above, the sampled aerosol is collected on a filter or some other substrate which may be assessed separately after sampling has been completed. Such instrumentation is suitable when time-averaged measurement can be justified. However, there are occasions where short-term (or even real-time) measurement is required – for example, when investigating the major sources of aerosol emission from an industrial process and the subsequent efficiency of control procedures introduced to minimize that emission. They may also form the basis of an alarm monitoring system of the type described above. For this purpose direct-reading instruments are used, which are based on three main principles: optical techniques, beta particle attenuation and oscillating element microbalances.

3.3.4.5.1 *Optical devices*.   Optical techniques, based on the principles of light extinction and scattering, provide an effective means by which aerosol can be assessed in real time. They have the great advantage that measurement can be made without disturbing the aerosol – provided of course that the particles can be introduced into the sensing zone of the instrument without loss or change. Their disadvantage is that interactions between light and airborne particles are strongly dependent on particle size and type, and so results are frequently difficult to interpret.

In the workplace, optical instruments operating on the basis of the detection of the scattered light are widely used. The most successful have been those designed for the monitoring of aerosol fractions within specific particle size ranges – in particular, the respirable fraction. They mainly involve devices which detect light (laser or infrared) scattered in the near-forward direction, using either horizontal elutriators or cyclones to select the respirable particles from aspirated samples. Examples include the RAM monitors, the latest of which is the DataRAM, and the SKC Hazdust. There is also a family of light-scattering instruments where the aerosol enters the sensing region by direct convection without the aid of a pump. These 'passive' devices rely on the optical response curve (which is more sensitive for the finer particles) to simulate the respirable size selection. The German TM-Digital is the main static instrument of this type commercially available. The lack of a pump means that the weight of the instruments can be minimized and there are designs such as the MINIRAM which purport to be personal. Similar principles are used in the hand-held wand-type instruments such as the AMS950 produced by Casella (London) Ltd (Figure 3.30), and the Hand-Held Aerosol Monitor produced by ppm Inc. These are widely used in walk-through surveys of workplaces to determine which processes are the major dust sources, and for this purpose the variation in response of the instrument with particle size and refractive index must not be forgotten.

The second type of instrument is based on the interaction between a focused light beam and each individual single particle. Such instruments are referred to as optical particle counters. From light-scattering principles, if an individual particle can be detected and registered electronically, it can be not only counted but also sized (i.e. placed into a given size band or 'channel' based on the magnitude of signal arising from

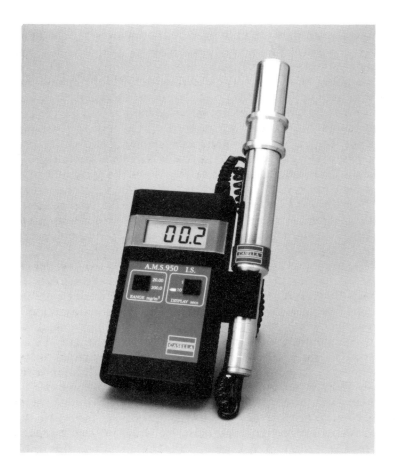

**Fig. 3.30** Example of portable light scattering direct-reading instrument – Casella AMS950.

the scattered light). By such means, instruments can be designed capable either of counting particles within specified size ranges or of providing an overall particle size distribution. As with aerosol photometers, many practical instruments have evolved within this category and have been widely used in research both in laboratories and in workplaces. The Grimm Model 1.105 dust monitor is basically an optical particle counter, which gives number concentrations in a number of particle size bands from 0.3 to 20 μm. The Fibrous Aerosol Monitor (FAM) is a version that sets out to provide counts of fibrous particles conforming to the 'respirable' fibre definition (as discussed earlier), even in the presence of non-fibrous particles.

3.3.4.5.2  *Beta attenuation devices.*  The beta attenuation concept is where the mass of particulate material deposited on a filter or some other surface is determined from the reduction in intensity of beta particles passing through the accumulated layer. In such instruments, the change in attenuation reflects the rate at which particles are collecting on the filter and hence on the concentration of the sampled aerosol. One advantage of this approach over optical instruments is that the attenuation of beta particles is directly dependent on particulate mass, and is almost independent of aerosol type or particle size distribution. Examples include the filter tape arrangements in the BAM, etc., and the small portable instruments produced by the GCA Inst. Inc.

3.3.4.5.3  *Vibrational mass balances.*  Another class of devices is what might be referred to as 'vibrational mass balances', the most common of which is the piezobalance. It is well-known that the frequency of mechanical oscillation of a piezoelectric crystal (e.g. quartz) is directly proportional to the mass of the crystal. Change in effective mass of the crystal, such as that due to the deposition of particles on its surface, is reflected in the change in its mechanical resonant frequency. The main instrument used in workplaces is the piezobalance, manufactured by TSI. While this device is highly sensitive, it suffers from particle overload at the collection site on the crystal, leading to particle deposition away from the central area of the crystal, and from poor coupling between the particles and the crystal. A recent development of this principle (more commonly used for environmental monitoring), the Tapered Element Oscillator Microbalance (TEOM), involves the use of a tapered glass tube which is fixed at the large end and supports a filter at the narrow end. The tube and filter are oscillated and again the deposition of particles on the filter causes a change in the resonant frequency of the tube.

An interesting and potentially very useful development has been the combination of direct-reading instruments and video recording of the operation. In this procedure the operator wears a direct-reading personal monitor and the contaminant levels are superimposed on the video film of his operational process. Although mostly used for the sampling of gases and vapours, some success has been obtained using the MINIRAM light-scattering device as the monitor. It has proved to be a very useful tool in demonstrating to the operator ways in which he can reduce his personal exposure.

A selection of the most common devices is given in Table 3.2.

### 3.3.5  *Sampling in the ambient environment*

3.3.5.1  *Samplers for 'total' aerosol.*  If information is required about the characteristics of all the particles that comprise the environmental

**Table 3.2** Examples of direct-reading instruments for workplace aerosols

| Name | Measurement technique | Flow rate ($l\,min^{-1}$) | Particle fraction | Concentration range[a] ($\mu g\,m^{-3}$) | Comments |
|---|---|---|---|---|---|
| TEOM Series 1100 Particle Mass Monitor | Tapered element oscillating microbalance | 3 | Total mass | $6–1.5 \times 10^3$ | Particles collected on filter – output directly related to mass. |
| TSI Portable Piezoelectric Respirable Aerosol Mass Monitor | Quartz crystal oscillating microbalance | ~2 | Respirable with 10 mm nylon cyclone | | Particles collected on crystal surface – output directly related to mass but problems with overloading limits application. |
| Casella AMS950 hand-held dust monitor | Light-scattering photometer | Passive | Nominally fine | $0.1–10^3$ indicated | Hand-held passive device suitable for walk-through surveys. Response dependent upon refractive index and size of particles. |
| SKC Hazdust | Light-scattering photometer | ~2 variable | Inhalable, thoracic or respirable from different inlets | $10–2 \times 10^5$ indicated | Calibrated with Arizona Road Dust. Response dependent upon refractive index and size of particles. Will not detect larger particles within inhalable convention. Has detachable sensor to give personal sampling. |
| DataRAM Portable Real-Time Aerosol Monitor | Light-scattering photometer | 2 | 'Total' and respirable from different inlets | $0.1–10^3$ indicated | Optical device calibrated with AC fine test dust. May need on-site calibration to give reliable mass measurements as response is dependent upon refractive index and size of particles. |
| TSI Model 8520 DUSTTRAK Aerosol Monitor | Light-scattering photometer | 1.4–2.4 | Particle size range 0.1 to 10 µm | $1–10^5$ indicated | Calibrated with Arizona Road Dust. Response is dependent upon refractive index and size of particles. |
| GRIMM Model 1.105 dust monitor | Optical particle counter | 1.2 | 'Total' and respirable from different inlets and size distribution in 8 channels | $1–10^5$ indicated | Optical particle counter with in-built filter for on-site calibration, as mass response may be dependent upon refractive index and size of particles. Gives number concentrations also. |

[a] Manufacturers' figures.

aerosol, then a sample must be collected that represents all particle sizes and types in the same relative proportions that they are present in the atmosphere. A sampler achieving this goal would have a sampling efficiency of unity, for all particles provide a true sample of 'total aerosol'. Only then can chemical and physical analysis of the sampled aerosol particles be considered as an unbiased characterization of atmospheric aerosols. In reality, however, there are no samplers commercially available that are capable of meeting this requirement for all particle sizes and environmental conditions found in the ambient atmosphere.

There are many technical problems involved in designing samplers for 'total' aerosol in the ambient atmosphere. They include: the low particle levels found, requiring very high sampling flow rates; the wide range of wind speeds over which they must operate; the requirement to have performance independent of direction to the wind (i.e. omnidirectional entry); adverse weather conditions (rain, snow, mist) in which they must operate; and the potentially wide range of particle sizes to be sampled.

European instrument designers in the early 1980s concentrated their efforts on protecting their samplers from the ingress of rain, snow, etc., and they produced similar samplers with flow rates from 4.5 to 45 $1 \text{min}^{-1}$, featuring downwards-facing omnidirectional entries, protected by cowls and rain shields. A selection of these samplers is shown in Figure 3.31. Their sampling efficiencies are given in Figure 3.32, and it can be seen that performance varies strongly with both particle aerodynamic diameter and wind speed in a manner that is broadly consistent with theory based on vertical elutriation modified to include the effects of cross-winds. It is to be expected that any other sampler with similar features would have similar sampling efficiency. Evidently none is satisfactory for sampling true total aerosol.

At the same time in the USA, the well-known Hi-vol total suspended particulate sampler was produced. Sampling at flow rates of 1200 $1 \text{min}^{-1}$, it features aerosol collection onto a large upwards-facing rectangular filter located inside a large, weatherproof housing (see Figure 3.33). Its performance was broadly similar to those for the samplers featured in Figure 3.31. In addition, due to the rectangular cross-section of the sampler, its performance is also dependent upon orientation to the wind. However, in a recent study of the performance of the Hi-vol PM10 sampling head (see below) it was found that if the internal impactor size selector is removed, the sampling efficiency is close to that of the Hi-vol TSP sampler, but without the orientation dependence. At this stage, it is important to mention the performance of the entry of the widely used Andersen cascade impactor (Andersen, 1988). Both versions of the device (viable and general particles) have a single upwards-facing circular sampling orifice as shown in Figure 3.34. The performance (also shown in Figure 3.34) is very similar to those for other samplers mentioned, and is

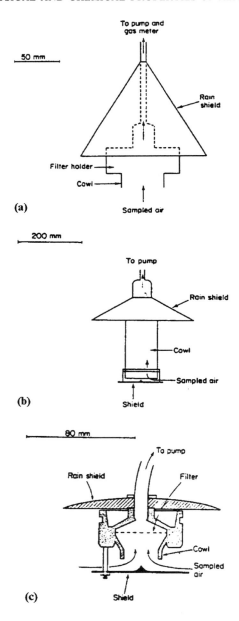

**Fig. 3.31** Early samplers for ambient aerosol concentrations: (a) schematic of the head of the Italian $20 \, \mathrm{l \, min^{-1}}$ ISTISAN sampler; (b) schematic of the head of German $242 \, \mathrm{l \, min^{-1}}$ (minimum) LISP; (c) schematic of the head of German $45 \, \mathrm{l \, min^{-1}}$ Kleinfiltergerät GS050/3 sampler for 'total' SPM (VDI, 1981b); (d) schematic of the French $25 \, \mathrm{l \, min^{-1}}$ Type PPA 60 sampler for 'total' SPM; (e) schematic of the British $4.4 \, \mathrm{l \, min^{-1}}$ M-type sampler for 'total' SPM.

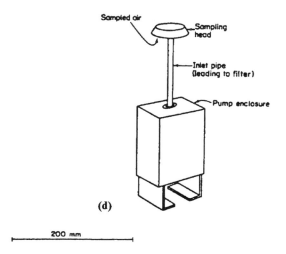

**(d)**

200 mm

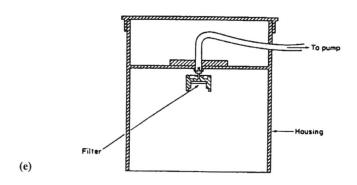

**(e)**

very dependent upon wind speed and particle size. This means, therefore, that the accuracy of the particle size distribution measurements obtained with this instrument is dependent upon the environmental conditions.

However, there are a small number of instruments that have been designed with the specific aim of sampling true total aerosol in the ambient atmosphere. The Wide Ranging Aerosol Classifier (WRAC) (Burton and Lundgren, 1987) makes use of large physical dimensions relative to the stop distance of ambient particles to ensure that particles up to at least 60 μm are sampled with high efficiency independently of wind speed. It has a central inlet of diameter 0.6 m, which is surrounded by a 1.6 m cylindrical shroud designed to act as a wind shield and produce calm-air conditions for the inlet. It is protected with a 1.6 m diameter rain cap, as shown in Figure 3.35. The sampling flow rate is 41 667 l min$^{-1}$, and there are four single-stage impactors with cut-points at 10, 20, 40 and 60 μm and

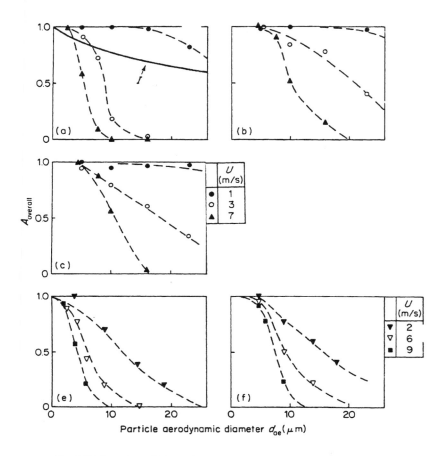

**Fig. 3.32** Sampling efficiency for each of the samplers shown in Fig. 3.31.

total particles filter sampling the air isokinetically inside the central inlet duct. Limited field testing by Hollander (1991) has shown that it may sample large particles with high efficiency, but full characterization in the controlled conditions in wind tunnels is required before definitive statements on its performance can be made. It is currently recommended as the reference instrument in a field equivalence testing protocol proposed by the European Union, but because of its large size and power requirements it is not suitable for general use.

A much smaller and simpler device was first introduced by May *et al.* (1976). Subsequently referred to as the 'aerosol tunnel sampler', the sampler has been developed further by Hofschreuder and Vrins (1986). Shown in Figure 3.36, it comprises a 150 mm diameter tube through which

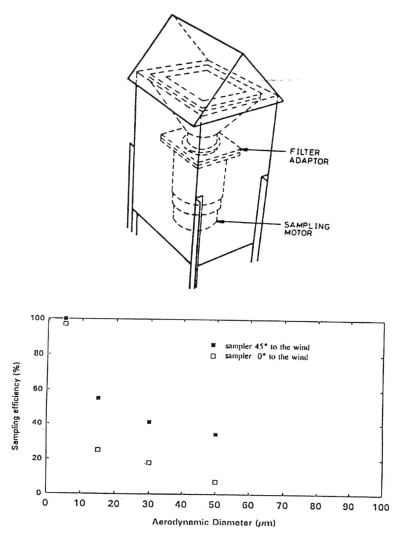

**Fig. 3.33** 'Hi-vol' sampler for total suspended particulate matter.

air is drawn by means of an axial fan to produce a mean air velocity of
$9\,\mathrm{m\,s^{-1}}$. Just in front of a flow-straightening honeycomb is the 10 mm
diameter thin-walled isokinetic sampler, located axially. This takes a
representative sample of the aerosol concentration inside the main tube,
which is designed to have sampling efficiency close to unity. The net effect
is that the aerosol collected in the isokinetic probe should be representative
of that in the ambient air outside. Results published in Vrins (1991) show

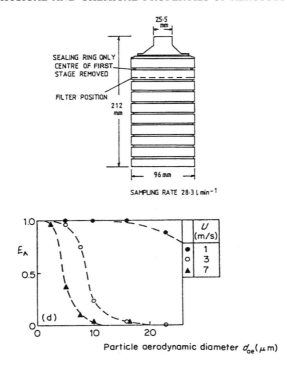

**Fig. 3.34** Andersen cascade impactor widely used for particle size distribution measurements of ambient aerosols.

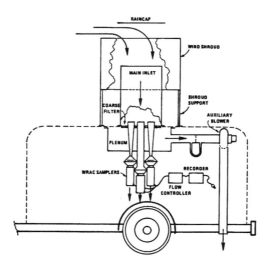

**Fig. 3.35** Wide Ranging Aerosol Classifier for ambient aerosols.

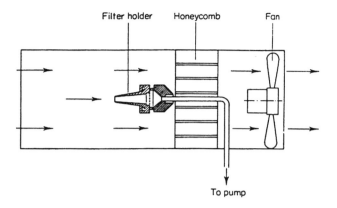

**Fig. 3.36** Schematic of the 'aerosol tunnel sampler' for the measurement of true total aerosols in the ambient atmosphere.

this to be the case. The whole arrangement is mounted on a pivot with a wind vane so that, under actual sampling conditions, it is always oriented with the tube mouth facing the wind. It differs from the other sampler mentioned above therefore in that it is not omnidirectional.

### 3.3.5.2 Samplers for human health effects

3.3.5.2.1 *Sampling strategy.* When formulating a strategy for the health-related sampling of aerosols in the ambient atmosphere it is necessary, first of all, to decide what the information is to be used for. There are two main types of programme:

- Long-term studies at a number of fixed sites nation-wide to monitor pollution trends. At these sites a whole range of pollutants (including aerosols) are measured by continuous monitors with the results being sent by telemetry to a central data acquisition centre. Besides providing information to meet guide and limit values in government and European directives, these programmes also provide invaluable data for public information and epidemiological studies.
- Shorter-term studies designed to address a specific pollution problem. In these studies a large number of samplers are concentrated in a small area around a particular aerosol source, and measurements are normally taken before and after a specific operation has taken place. For instance, the construction of a new road or a major industrial plant could result in significant dust exposures to the general public during construction but not before and after.

Recommended criteria for the siting of monitoring stations have been developed to cover all situations where human exposure to aerosols could arise and a useful, brief guide to the types of location to be included is given by Ott (1977). A number of considerations need to be addressed which will affect the aerosol measurement when selecting the location of outdoor monitoring sites. These include: the proximity of point sources, which could result in highly variable concentration gradients; obstructions or changes to airflow caused by tall buildings, trees, etc., and abrupt changes in terrain, which could introduce localized separations and swirls into the airflow, again causing highly variable concentration gradients; and the height of the sampler entry above the ground, which is a compromise between siting the entry at head height and ensuring that it does not get damaged or receive any extraneous, non-sampled material. For some samplers the sampling head must be aligned vertically.

Additionally, for research studies where attempts are being made to correlate the symptoms of disease with exposure to aerosols, it may not be sufficient just to monitor outdoor concentrations. In a recent study carried out in the USA (Ferris and Spengler, 1985) respirable aerosol levels indoors were found to be nearly twice as high as those measured outdoors.

Ideally, to ensure accurate estimates of the likely exposures of humans to aerosols in the ambient atmosphere, measurements should be taken continuously for the whole period of the study. However, this would be prohibitively expensive and so statistically based sampling programmes have to be developed.

For long-term network studies, it has been common practice to use the high-volume gravimetric samplers that provide the mean concentration over an integrated period of say, 24 hours. In the USA, samples for the PM10 fraction are carried out every third day, and in areas where the levels exceed the specified limits, it is anticipated that the sampling frequency will be increased to daily. The use of gravimetric samplers in this programme requires the filters to be changed and weighed for each sample, which is very costly. This has been the driving force behind the development of the continuous samplers mentioned above, to enable more comprehensive coverage of aerosol concentrations to be made, provided that these systems are fully tested.

3.3.5.2.2 *Samplers designed specifically for health-related purposes.*    In the USA, measurement of an aerosol fraction in the ambient atmosphere specifically related to health effects has been carried out for the last 10 years. This US EPA definition of the thoracic aerosol fraction (EPA, 1987), which has been systematically measured with validated instruments, in the USA since 1988. It differs from the ISO definition in that it has a 50% penetration at an aerodynamic diameter of 10.6 μm and zero penetration at 16 μm, as shown in Figure 3.12. In practice, however, this difference is not significant because most instruments validated for sampling the PM10 fraction will also reliably sample the ISO thoracic fraction.

In the European Union, ambient airborne particle sampling is guided by the Black Smoke and $SO_2$ Directive (80/779/EEC). This directive makes no reference to any health-related aerosol fraction, and gives inadequate guidance on measurement methods involving either the determination of the 'blackness' of collected particles or gravimetric techniques. Following recommendations in a CEC-funded report by Wagner et al. (UK Department of Health) (1988), the directive will be soon revised to recommend the measurement of the PM10 particle fraction.

With the emphasis in the USA on measurements of the PM10 fraction, and very little development activity in Europe (including the UK) on health-related sampling, the majority of reliable instruments currently available are for the PM10 fraction. These generally make use of a validated sampling head to select the PM10 fraction of the ambient airborne particles and the collected particles are analysed in two main ways:

- Gravimetric, cumulative samplers in which the PM10 particles are deposited on a filter over a sampling period of normally 24 hours. The mass of particles collected on the filters is determined by weighing.
- Direct-reading monitors in which the selected PM10 particles are either deposited on a filter conducted with continuous assessment of the change of a property of the filter due to their presence, or conducted to an optical sensing region.

*Gravimetric, cumulative samplers.* These samplers basically comprise an omnidirectional rain-protected entry followed by a size-selective stage (normally an impactor) to select the PM10 particles which are collected on the filter. There are a number of different samplers available, ranging in flow rate from 16.7 to $1130 \, l \, min^{-1}$ and in cost from about \$7500 to \$15 000. A summary of the important features of the samplers, together with the cost (in the UK), is given in Table 3.3 and a brief description of the design features and relative performance of the sampling heads is given below.

Samplers with high flow rate such as the Graseby-Andersen PM10 Hi-vol Sampler (see Figure 3.37) have been the mainstay of routine PM10 measurements in the USA. The high flow rate of $1130 \, l \, min^{-1}$ has the advantage of providing sufficient sample both for gravimetric and chemical analysis over the specified 24-hour sampling period. They are used in studies to determine the low levels of dioxins, PCBs and PAHs found in the ambient atmosphere. The important features of the sampling head are an omnidirectional narrow slot entry, which samples particles

**Table 3.3** Examples of gravimetric, cumulative PM10 particle samplers for the ambient atmosphere currently available in UK

| Name | Flow rate ($l\,min^{-1}$) | Filter diameter (mm) | Comments |
|---|---|---|---|
| PQ167 Portable PM10 Sampling Unit | 16.7 | 47 | Uses validated PM10 inlet (US EPA protocol) connected to microprocessor-controlled pump. Battery powered – lasts for over 24 hours using quartz or glass fibre filters. |
| Partisol Model 2000 Air Sampler | 16.7 | 47 | Uses validated PM10 inlet connected to microprocessor-controlled pump. Supplied as stand-alone unit or with three additional satellites controlled by hub unit dependent upon wind speed and/or direction conditions. |
| PM10 Dichotomous Sampler | 16.7 | 2 of 37 | Uses validated SA246b PM10 inlet followed by virtual impactor to give two fractions collected on filters: 10–2.5 μm and < 2.5 μm. |
| PM10 Medium Flow Sampler | 113 | 102 | Medium flow rate sampler uses Teflon or quartz filters, specially used for X-ray fluorescence and other compositional analyses. |
| SA 1200 PM10 High volume ambient air sampler | 1200 | 200 × 250 | Standard sampler used in USA for PM10 aerosol in ambient atmosphere – high flow rate means short sampling times and large masses for gravimetric and chemical analysis. |

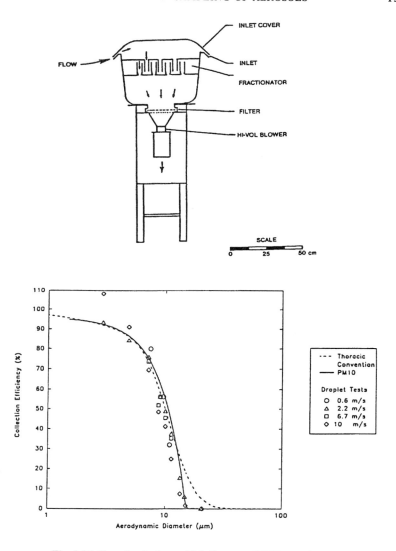

**Fig. 3.37** Graseby-Andersen high-flow-rate PM10 sampler.

independently of wind speed up to $10 \, \mathrm{m \, s^{-1}}$, and a multi-orifice single stage impactor, which allows the PM10 fraction to penetrate to a $25 \times 20 \, \mathrm{cm}$ filter.

A number of PM10 samplers with low flow rate are available. They all make use of the SA246b PM10 inlet developed and validated by Graseby Andersen. The entry (see Figure 3.38) consists of a flanged downwards-pointing circular entry with a disc rain cap held some distance above. This forms the omnidirectional entry through which particles enter, followed by a single-stage impactor which allows the PM10 fraction to penetrate to a filter. The sampling efficiency of the entry, demonstrating full agreement with the US EPA PM10 Convention, is also shown in Figure 3.38. Three devices are included in Table 3.3, each having separate features that make them to a certain extent complementary.

*Direct-reading monitors.*   For these instruments, sampling and analysis are carried out within the instrument and the concentration can be obtained almost immediately. Like the cumulative samplers with low flow rate described above, these instruments generally use the validated SA 246b PM10 inlet to select the PM10 particles which either deposit on a special filter stage or penetrate into a particle-sensing region. Instead of direct weighing, the presence of the particles either on the filter or in the sensing region gives rise to a change in some property of the zone, which can be related by calibration to the mass of particles present. A number of different instruments are available and these may be classified into the same three main categories as those instruments used in workplaces described earlier, namely: optical, resonance oscillation, and beta-particle attenuation. A summary of the main features of some examples of the direct-reading instruments for health-related purposes is given in Table 3.4, and the principles of operation are described briefly below.

- *Optical.* These instruments employ the interaction between airborne particles and visible light in a sensing region, and generally their response is dependent upon the size distribution and refractive index of the particles. They therefore require calibration to give results in terms of mass or number concentrations. *This calibration only holds provided that the nature of the particles does not change.* A number of instruments are currently available based on either single particle counter or photometry optics described above for the workplace instruments.
- *Oscillating microbalance.* The frequency of mechanical oscillation of an element such as a tapered glass tube is directly proportional to the mass of the tube. Change in effective mass of the tube, such as that due to deposition of particles on the surface of a filter at the free end of the tube, is reflected in a change in its resonant frequency. This is the principle of operation behind the Rupprecht and Patashnick Tapered Element Oscillating Microbalance (TEOM), as shown in Figure 3.39. The air is split below the standard $16.7 \, l \, min^{-1}$ PM10 inlet such that $3 \, l \, min^{-1}$ passes through a 16 mm diameter filter connected to the top of the narrow end of a hollow tapered glass tube, and $13.7 \, l \, min^{-1}$ is led away for other purposes. The inlet including the sensing system is kept at a steady 50°C to drive off any sampled water droplets. However, some concern has been expressed about the potential loss of volatile

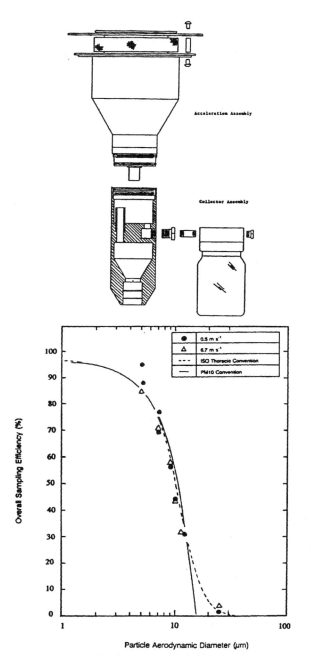

**Fig. 3.38** Graseby-Andersen low-flow-rate PM10 sampler as fitted to direct-reading instruments

**Table 3.4** Examples of direct-reading monitors for PM10 particles in the ambient atmosphere

| Name | Measurement technique | Flow rate ($l\,min^{-1}$) | Particle fraction | Concentration range[a] ($\mu g\,m^{-3}$) | Precision[a] ($\mu g\,m^{-3}$) 1 hour | Precision[a] ($\mu g\,m^{-3}$) 24 hours | Comments |
|---|---|---|---|---|---|---|---|
| TEOM Series 1400a Ambient Particulate Monitor | Tapered element oscillating microbalance | 16.7 through inlet with 3 through filter / detector | PM10 (validated head with possible 2.5 or 1.0 µm) | $0.06\text{–}1.5 \times 10^3$ | 1.5 | 0.5 | Only direct-reading monitor in which output directly related to mass. Employed as particle monitor at EUN sites. |
| W & A Beta Gauge Automated Particle Sampler | Attenuation of beta rays by particles collected on a filter | 18.9 | PM10 with EPA validated head | $4\text{–}10^4$ | 4 | 0.1 | One of a number of filter tape based beta gauges – measurement cycle 1 hour. |
| Airborne Particle Monitor APM1 | Attenuation of beta rays by particles collected on a filter | 15–30 | PM10 (non-validated) | $2\text{–}10^7$ | 56 | 2 | Cassette system with 30 filters in sequential loader. Integrity of each sample maintained for compositional analysis. |
| GRIMM Model 1.104 Dust Monitor | Light-scattering photometer | 16.7 through inlet 1.26 through detector | PM10 with EPA validated head | $1\text{–}5 \times 10^4$ indicated | Not given | Not given | Optical particle counter with in-built filter for on-site calibration, as response may be dependent upon refractive index and size of particles. |
| DataRAM Portable Real-Time Aerosol Monitor | Light-scattering photometer | 2 | PM10 or PM2.5 (non-validated) | $0.1\text{–}10^3$ indicated | 1.0 | Not given | Optical device calibrated with AC fine test dust. May need on-site calibration to give reliable mass measurements as response is dependent upon refractive index and size of particles. Entry dependent upon wind speed. |

[a] Manufacturers' figures.

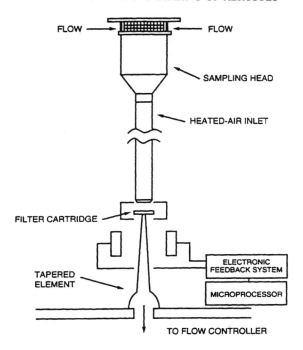

FLOW →          ← FLOW

SAMPLING HEAD

HEATED-AIR INLET

FILTER CARTRIDGE

ELECTRONIC
FEEDBACK SYSTEM

TAPERED
ELEMENT

MICROPROCESSOR

TO FLOW CONTROLLER

**Fig. 3.39** Principle of operation of Tapered Element Oscillating Microbalance.

material at this temperature. The filter is removed after a number of weeks' sampling (dependent upon the ambient particle levels) and the chemical composition of the deposited material analysed. This can provide compositional information integrated over the sampling period.

- *Beta-particle attenuation.* This involves the measurement of the reduction in intensity of beta particles passing through a dust-laden filter or collection substrate. In such instruments, the change in attenuation reflects the rate at which particles are collected on the filter and hence the concentration of the sampled particles. For most substances encountered the attenuation of the beta particles is directly dependent on the mass of particles deposited. Two main types of instrument have been developed: one using filter tape to collect the particles, and the other using a stack of conventional filters in a sequential loader. With the introduction of the TEOM instrument, the future availability of beta-particle attenuation monitors may be somewhat limited, as they have no significant advantages over the TEOM. In addition, the relationship with mass is not linear for some materials, and is also affected by the uniformity of particle deposition on the filter.

*Comparability of methods.* There is a potential, as yet unresolved, problem of the comparability of the data from the two methods. The filter of the TEOM is held at stable temperature of 50°C, while the filter stages used in the optical and beta-particle monitors and the gravimetric samplers have no specific temperature control. This means that while the TEOM can be considered to underestimate aerosol concentrations due to evaporation of the volatile components on the particles, beta-particle monitors may over-sample due to condensation of water vapour on to the filters in conditions of high humidity, and similarly optical monitors may register fine water droplets. The comparison of direct-reading methods with the gravimetric samplers is further complicated by the filter conditioning procedures necessary to obtain accurate weighing. It is not yet known to what extent this procedure gives similar filter conditions to those for the two types of direct reading instrument.

3.3.5.2.3  *Health-related samplers for coarse particles.* It is only recently that the environmental protection community has realized that particles as large as 100 µm can enter the human nose and mouth during breathing. While these large particles may be rare, they can occur close to industrial processes and during episodes of high wind speeds. Once inhaled, they will deposit in the nasopharyngeal region and if toxic (lead, radioactive particles, etc.) may enter the blood system there or in the gut. The relevant ISO health-related fraction is the *inhalable fraction*, for which there is currently no commercially available instrument. The old 'total suspended particulate' (TSP) was effectively defined by the 'high-volume' sampler used to measure this parameter. The performance of this well-known device was found to be dependent on both orientation and wind speed, with its efficiency falling well below the inhalable aerosol convention, especially for large particles (Wedding *et al.*, 1977).

As a potential European reference method, the European Community and the UK Department of the Environment provided limited funds for the development of a sampler designed specifically to match the requirements of the ISO inhalable aerosol fraction (Mark *et al.*, 1990). It was designed to mimic the essential features of humans by achieving omnidirectionality using a single orifice that rotates through 360° (see Figure 3.40). It comprises a detachable sampling head which sits on top of a rectangular cabinet which houses the head rotation mechanism (drive motor, gearbox and rotating seal), the pump, an automatic flow control system, and associated switch gear. A protective canopy is positioned over the sampling head to prevent the entry of unwanted rain and snow. The sampling efficiency of the sampler, as tested in a large wind tunnel for a range of particle sizes and types in a range of wind speeds, is also given in Figure 3.40 (Mark *et al.*, 1990). It is now available for commercial

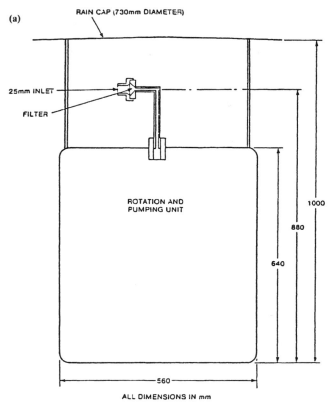

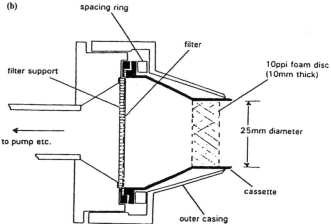

**Fig. 3.40** Prototype sampler for the measurement of inhalable aerosols in the ambient atmosphere.

exploitation, although its future use is somewhat in doubt as more emphasis for correlation with human health effects is focused on the smaller particles in the ambient atmosphere.

## 3.4    General sampling procedures

### 3.4.1    *Sampling strategy*

The sampling process involves not only the selection and operation of the appropriate samplers but also the formulation of a suitable sampling strategy. This is to ensure that: the data generated by the sampling exercise are relevant to the question posed; the inferences drawn from the data are soundly based; and the most cost-effective approach has been taken. For both occupational and environmental sampling there are two main questions to be answered that decide the main elements of the sampling strategy: why one wants to sample aerosols; and how the results will be used. Additional factors relevant to either occupational or ambient situations are then considered to produce the final sampling plan. These are given in more detail in the respective individual sections.

### 3.4.2    *Operational aspects*

Most manufacturers supply comprehensive operational instructions with the instruments that they sell, and these should be read carefully before planning the details of the sampling programme. There are some general aspects, however, that merit mention here.

#### 3.4.2.1    *Choice of filters.*    The choice of suitable filters for the gravimetric samplers mentioned above depends upon the sampling and analytical requirements. The most commonly used filters for sampling aerosols in both workplace and ambient atmospheres are glass fibre filters. These robust filters have low moisture retention, and high collection efficiencies at relatively low pressure drops. As particles are collected in the depth of the fibre bed, glass fibre filters also have the ability to collect and retain large sample masses with a low pressure build-up rate (and small change in flow rate). However, if compositional analysis or microscopic investigation of the collected particles is required, membrane filters are more suitable. In addition, glass fibre filters suffer from artefacts when sampling ambient aerosols: positive artefact mass addition due to the *in situ* conversion of sulphur dioxide to sulphate; and negative artefact mass due to volatilization of ammonium nitrate and ammonium chloride particles. Glass fibre filters also have very high chemical blank values and are not suitable for some trace element analyses such as zinc and barium. Quartz fibre filters have superior resistance to artefact effects and moisture

absorption, but they are very fragile and so require extra care when handling.

Membrane filters are made from a number of different materials, with pore sizes ranging from 0.03 to 8 μm. They provide samples better suited for trace elemental analysis studies such as INAA, ED-XRF or microscopic analysis. The membrane filters commonly used for particle sampling include polycarbonate (e.g. Nuclepore) and PTFE. Polyester filters are also available, but are becoming less widely used. However, both the polycarbonate and the PTFE filters are difficult to handle as they are very thin and susceptible to electrostatic charge build-up. This can result in the loss of coarse particles during transport. A similar problem occurs with fluorocarbon filters, which are the most suitable filters for inorganic analysis. Pure silver membrane filters are ideal for situations where impurities in the filter or high weight losses render normal membranes unsuitable. An example of this is in the sampling of coal tar pitch volatiles.

The main problem with all membrane filters is that, unlike glass fibre filters, particle collection takes place at the surface of the membrane filter. This severely limits the amount of sample that can be collected because when more than a single layer of particles is deposited on the surface, the resistance to airflow increases rapidly and there is a tendency for the deposit to be dislodged from the filter, especially during transport back to the laboratory.

Filter choice for continuous monitors is very limited. The beta-monitors can be used with either glass fibre or PTFE filters, with the 30 separate filters contained in the cassette of the Elecos APM-1 device being especially suitable for analysis.

In summary, there are a large number of filters to choose from. For workplace gravimetric studies glass fibre filters are generally used, membrane filters being used when microscopy or compositional analysis is required. For ambient gravimetric sampling, quartz fibre filters are recommended for routine particle sampling when mass concentration is determined by weighing. In general, other types of filter (Nuclepore, PTFE, fluorocarbon) are recommended if specific chemical analyses are required. However, glass filter can be used for atomic emission spectroscopy and for other chemical analyses, provided care is taken to check for impurities.

3.4.2.2 *Filter handling, conditioning and weighing*. Important aspects that are often overlooked in the methodology of sampling are the handling and conditioning of the filters. When taking new filters from their boxes it is important to inspect the state of the surfaces. For instance, glass fibre filters often have loose fibres from the cutting process that could be easily lost at any stage of the sampling process. It is essential, therefore, that these are removed prior to the first weighing. In addition, some membrane filters have been seen with a very fine powder coating when new. Such filters should be rejected and a new batch used. When laden with dust, it is essential that the filters are supported and kept upright to prevent the loss of material that could arise from either sharp impact or touching the internal walls of the containers.

There are a number of schools of thought about the conditioning of filters prior to weighing, and it is currently the subject of much debate by an ISO committee. It is recommended that all weighing is carried out under temperature and humidity control, and that the filters are allowed to condition in that atmosphere for at least 24 hours both before and after sampling. The use of desiccators is generally not recommended because when the filter is removed from the desiccator the weight will be unstable due to the absorption of moisture from the balance room atmosphere. However, desiccators may be used provided that the filters are removed at least 24 hours prior to weighing. A set of control filters to take account of changing conditions is highly recommended; a minimum of three control filters is required and these must be taken to the sampling site along with the sampling filters, but not used, so that the environmental histories of both sample and control are as similar as possible. In addition, some filters (PVC membrane, nuclepore, etc.) have excessive electrostatic charge on their surfaces, which must be neutralized with an ion gun or a radioactive source prior to weighing.

With mass concentrations of PM10 in the ambient atmosphere generally about 20 to 30 $\mu g\,m^{-3}$, care must be taken to ensure that the balance used has sufficient accuracy. This normally requires a five-place or six-place balance, available from a number of balance manufacturers. This should be placed on a solid, vibration-free bench in a room specially designated for this purpose, with temperature, and preferably humidity, control.

3.4.2.3 *Flow-rate calibration.* Besides the measurement of the mass of the collected sample, the most important parameter to determine accurately is the flow rate. Fortunately, most aerosol samplers are fitted with automatic flow rate control or compensating systems, in which the air velocity or pressure drop is continuously monitored and signals are sent via a feedback loop to either open a valve or to increase the speed of the motor when they have changed by 5%. In the environmental instruments made by Wedding and Associates, flow control is provided by means of a 'choked' flow venturi system, which limits the flow rate to a maximum value provided that sufficient vacuum potential is maintained.

While flow control devices provide major improvements in maintaining the set flow rate throughout the sampling period, it is still necessary to set the flow rate before sampling can begin. Ideally this should be carried out by measuring the flow entering the sampling inlet using either a bubble flowmeter (which is a primary airflow standard) or calibrated wet gasmeter, depending upon the flow rate of the sampler. However, this is only possible for samplers with single unidirectional entries. For those omnidirectional samplers with horizontal slit entries (mainly samplers for the ambient environment), setting the flow rate is achieved by removing the entry and using a special fitting with an orifice plate over the installed filter. Once set to the correct value, the size-selective entry is replaced and the flow control systems maintain the flow rate to within the specified limits (normally 5%).

## 3.5 Concluding remarks and future requirements

This chapter has given a brief outline of the physical processes of particle transport and has described a number of practical instruments available for the sampling of particles in the ambient atmosphere. It has been impossible to mention all samplers currently used for this purpose, only those samplers being mentioned whose sampling efficiencies are known and come close to the required specifications. This strategy is proposed with the aim of improving the accuracy (bias + precision) of the sampling of airborne particles, so that results obtained by different teams in different countries can be reliably combined and compared. With many more accurate, validated samplers available, or soon to be available, there is no excuse to use an uncharacterized or unsuitable particle sampler.

When faced with decisions as to the choice of sampler and sampler methodology for a particular application, it is worthwhile to reiterate two questions posed earlier: why does one want to sample airborne particles and how will the results be used? The answers should provide the sampling rationale relevant to the practical situation in question, and the appropriate sampling methodology can then be prescribed.

What of the future? The sampling of aerosols in all environments, including occupational and environmental, is going through an era of great change, with the emphasis in all measurements on international standardization. For aerosols, international agreement has recently been reached for a set of health-related sampling conventions. The next step, currently under way, is the establishment of agreed test protocol by which the performance of aerosol sampler can be tested for compliance with the health-related conventions. This work, which involves both laboratory and field testing, will involve the classification of samplers according to the range of conditions in which they will give reliable results. Once this has been achieved, and provided that an agreed sampling strategy is followed, there is no reason why reliable, valid measurements of individual exposures cannot be achieved. These will have the added benefit of being internationally comparable.

### 3.5.1    *Future requirements in occupational sampling*

In the occupational field, continuously recording instruments are currently used as investigational tools to determine major aerosol sources, and as educational tools to reduce aerosol exposure by altering work practices. It is expected that this will continue, but with an additional use as alarm monitors in an overall sampling strategy employing periodic personal sampling supported by continuous monitoring of all work shifts.

With an increasing incidence of allergic diseases such as asthma thought to be, in part, caused by inhaling bioaerosols, and the increasing production of process microorganisms there is a desperate need for standardized guidance on the sampling of bioaerosols. Research work is currently being carried out at a number of laboratories world-wide to develop practical, reliable standard methods. It is expected that there will not be one universally applicable method but possibly two: one for microorganisms that must be kept viable using samplers that both treat the microorganisms gently and keep them wet; and one where viability is not important and existing methodology for inorganic aerosols can be used.

### 3.5.2    *Future requirements in environmental sampling*

The main motivation for the sampling of ambient airborne particles remains for the prediction and prevention of adverse health effects in humans. The future requirements for health-related sampling of particles in the ambient atmosphere should be set by the perceived or predicted risk to human health from inhalation of the particles. Currently, it is believed that the PM10 fraction (representing those particles that penetrate below the larynx) gives the most relevant indicator of risk to health. However, with the increase in the incidence of asthma thought by some to be associated with an increase in particulate emissions from increased use of diesel vehicles, a finer fraction is being suggested. For this purpose, cyclone-based sampling heads developed in the USA for fractions with 50% penetrations at $2.5\,\mu m$ (PM2.5) and $1\,\mu m$ (PM1) are available as direct substitutes for the PM10 sampling heads. The PM10 Dichotomous Sampler, which provides two samples, 10 to $2.5\,\mu m$ and below $2.5\,\mu m$, should be a useful sampler for this purpose.

For epidemiological studies, it may be necessary to use personal samplers to provide a reliable estimate of individual exposures to ambient particles. This has been shown to be essential in the occupational field, where measurements with personal samplers can be up to 100 times those from suitable static samplers sited in the workplace. The personal samplers described in section 3.3.4 could be used to monitor individual exposures to all three health-related fractions, although studies reported so far have shown that personal exposures away from work are much lower than those at work and sampling periods may have to extend over two days.

One of the most difficult problems in relating exposure to a specific particulate pollutant to adverse health effects is the unambiguous identification of that pollutant in the soup of particles found in the ambient atmosphere. This is especially relevant for particulate emissions from diesel vehicles. One interesting new development is the Rupprecht and Patashnick Ambient Carbon Particulate Monitor. This device automatically determines hourly averages of the organic and elemental carbon particulate concentration.

Finally, Seaton et al. (1995) have proposed that the many ultrafine particles (less than 0.1 µm) that are present in the ambient atmosphere may explain the observed association between particulate air pollution and exacerbation of illness in people with respiratory disease and the rise in the numbers of deaths from cardiovascular and respiratory disease among older people. They suggest that ultrafine particles are able to penetrate readily to the interstitial tissue of the lung where they provoke a marked inflammatory response not seen for larger particles of the same chemical composition. They further suggest that the number and composition of the particles should more readily relate to disease rather than the mass of the particles deposited. In order to provide this information about the ambient particles use must be made of devices such as Condensation Nucleus Counters which give the number concentration of particles down to 0.03 µm.

## Acknowledgements

The author would like to thank Professor James H. Vincent of the University of Minnesota, USA, for many hours of fruitful discussion on aerosol sampling. Much of what is written in this chapter was discussed and clarified during these sessions.

## References

AIA (1979) Recommended technical method No. 1: Reference method for the determination of airborne asbestos fibre concentrations at workplaces by light microscopy (membrane filter method). Asbestos International Association, London.
Andersen A.A. (1958) New sampler for the collection, sizing and enumeration of viable airborne particles. J. Bacteriology, 76, 471.
Bohnet M. (1978) Particulate sampling, in W. Strauss (ed.), Air Pollution Control, Part III: Measuring and Monitoring Air Pollutants. Wiley, New York.
Buckley T.J., Waldman J.M., Freeman C.G., Lioy P.J., Marple V.A. and Turner W.A. (1991) Calibration, intersampler comparison, and field application of a new PM-10 personal air-sampling impactor. Aerosol Sci. Technol., 14, 380–387.

150    PHYSICAL AND CHEMICAL PROPERTIES OF AEROSOLS

Burton R.M. and Lundgren D.A. (1987) Wide range aerosol classifier: a size selective sampler for large particles. *Aerosol Sci. Technol.* **6**, 289–301.·

CEN (1993) *Workplace atmospheres. Size fraction definitions for measurement of airborne particles,* EN 481 Comité Européen de Normalisation.

CEN (1997) Workplace atmospheres. Assessment of performance of instruments for measurement of airborne particles. Draft CEN Prestandard. Comité Européen de Normalisation

Chatigny M.A., Macher J.M., Burge H.A. and Solomon W.A. (1989) Sampling airborne microrganisms and aeroallergens, in S.V. Hering (ed.), *Air Sampling Instruments for Evaluation of Atmospheric Contaminants,* 7th edn. American Conference of Governmental Industrial Hygienists, Cincinnati, OH.

Council of the European Communities (1980) Directive on air quality limit values and guide values for sulphur dioxide and suspended particulates (80/779/EEC). *Official Journal* L229, 30 August, pp. 30–39.

Courbon P., Wrobel R. and Fabries J-F. (1988) A new individual respirable dust sampler: the CIP10. Annals Occ. Hyg., **32**, 129–143.

Dennis R., Samples W.R., Anderson D.M. and Silverman L. (1957) Isokinetic sampling probes. *Ind. Eng. Chem.,* **49**, 294–302.

Dunmore J.H., Hamilton R.J. and Smith D.S.G. (1964) An instrument for the sampling of respirable dust for subsequent gravimetric assessment. *J. Sci. Instr.,* **41**, 669–672.

EPA (1987) Ambient monitoring reference and equivalence methods. Federal Register 40 CFR Part 53.

EPA (1987) Standards of performance for new stationary sources. Federal Register 40 CFR Part 60.

Fabries J.-F. and Wrobel R. (1987) A compact high flow rate respirable dust sampler. *Annals Occ. Hyg.,* **31**, 195–209.

Ferris B.G. and Spengler J.D. (1985) Harvard air pollution health study in six cities in the USA. *Tokai J. Exp. Clin. Med.,* **10**, 263.

Fuchs N.A. (1964) *The Mechanics of Aerosols.* Pergamon Press, Oxford.

Griffiths W.D. and DeCosemo G.A.L. (1993) The assessment of bioaerosols: a critical review. *J. Aerosol Sci.,* **25**, 195–209.

Harris G.W. and Maguire B.A. (1968) A gravimetric dust sampling instrument (SIMPEDS): preliminary results. *Annals Occ. Hyg.,* **11**, 195–201.

HSE (1997) *General Methods for the Gravimetric Determination of Respirable and Total Inhalable Dust,* MDHS 14. Health and Safety Executive, London.

HSE (1989) Control of Substances Hazardous to Health Regulations. Health and Safety Executive, London.

Hinds W.C. (1982) *Aerosol Technology.* Wiley, New York.

Hofschreuder P. and Vrins E. (1986) The aerosol tunnel sampler: a total airborne dust sampler, in *Aerosols: Formation and Reactivity.* Pergamon, Oxford, pp. 491–494.

Hollander W., Blomesath W. and Beyer A. (1990) An in-situ calibration technique for the determination of large particle aspiration efficiencies of TSP samplers: The stop distance concept. *J. Aerosol Sci.,* **21**, 41–46.

ISO (1995) *Air Quality: Particle Size Fraction Definitions for Health-Related Sampling,* ISO 7708. International Standards Organization, Geneva.

ISO (1992) *Stationary Source Emissions – Determination of Concentration and Mass Flowrate of Particulate Material in Gas-Carrying Ducts – Manual Gravimetric Method,* ISO 9096. International Standards Organization, Geneva.

Jaenicke R. (1993) Tropospheric aerosols, in P.V. Hobbs (ed.), *Aerosol–Cloud–Climate Interactions.* Academic Press, London and New York.

Kenny L.C., Aitken R.J., Chalmers C.P., Fabries J.-F., Gonzalez-Fernandez E., Kromhout H., Liden G., Mark D., Riediger G. and Prodi V. (1997) Outcome of a collaborative European study of personal inhalable sampler performance, *Annals Occ. Hyg.,* **41**, 135–153.

Lippmann M. and Harris W.B. (1962) Size-selective samplers for estimating 'respirable' dust concentrations. *Health Phys.,* **8**, 155–163.

Mark D., Vincent J.H. and Gibson H. (1985) A new static sampler for airborne total dust in workplaces. *AIHAJ,* **46**, 127–133.

Mark D. and Vincent J.H. (1986) A new personal sampler for airborne total dust in workplaces. *Annals Occ. Hyg.*, **30**, 89–102.

Mark D., Borzucki G., Lynch G and Vincent J.H. (1988) The development of a personal sampler for inspirable, thoracic and respirable aerosol. Paper presented to the Annual Conference of the Aerosol Society, Bournemouth, UK.

Mark D., Vincent J.H., Aitken R.J., Botham R.A., Lynch G., van Elzakker B.G., van der Meulen A. and Zierock K.-H. (1990) Measurement of suspended particulate matter in the ambient atmosphere. Institute of Occupational Medicine Report no. TM/90/14.

Mark D., Lyons C.P. and Upton S.L. (1993) Performance testing of the respirable dust sampler used in British coal mines. *Appl. Occ. Environ. Hyg.*, **8**(4), 370–380.

Mark D., Upton S.L., Lyons C.P., Chalmers C.P. and Kenny L.C. (1997) Wind tunnel tests of the sampling efficiency of personal inhalable samplers. *J. Aerosol Sci.*, (in press)

May K.R., Pomeroy N.P. and Hibbs S. (1976) Sampling techniques for large windborne particles. *J. Aerosol Sci.*, **7**, 53–62.

NIOSH (1975). *Criteria for a Recommended Standard – Occupational Exposure to Cotton Dust*, DHEW (NIOSH) Pub. no. 75-118. US Government Printing Office, Washington, DC.

Ott W. (1977) Development of criteria for siting of air monitoring stations. *J. Air Poll. Control Ass.*, **27**, 543.

Rupprecht E., Meyer M. and Patashnick H. (1992) The tapered element oscillating microbalance as a tool for measuring ambient particulate concentrations in real time. *J. Aerosol Sci.*, **23**, Supplement 1, S635–S638.

Seaton A., MacNee W., Donaldson K. and Godden D. (1995) Particulate air pollution and acute health effects. *Lancet*, **345**, 176–178.

Stuke J. and Emmerichs M. (1973) Das gravimetrische Staubprobenahmegerät TBF50. *Silikosebericht Nordrhein-Westfalen*, **9**, 47–51.

Upton S.L. and Barrett C.F. (1985) A wind tunnel study of the inlet efficiency of the Warren Spring Laboratory 'S' and 'Directional' samplers. Warren Spring Laboratory Report no. 526(AP)M.

Van der Meulen A. (1993) Ambient test procedure to demonstrate reference equivalency of measurement methods for fine suspended particulate matter up to 10 μm for compliance monitoring, in *Measurement of Airborne Pollutants*. Butterworth-Heinemann, Oxford, pp. 17–22.

Vincent J.H. (1989) *Aerosol Sampling: Science and Practice*. Wiley, Chichester.

Vincent J.H. and Mark D. (1990) Entry characteristics of practical workplace aerosol samplers in relation to the ISO recommendations. *Annals of Occ. Hyg.*, **34**, 249–262.

Vincent J.H., Emmett P.C. and Mark D. (1985) The effects of turbulence on the entry of airborne particles into a blunt dust sampler, *Aerosol Sci. Technol.*, **4**, 17–29.

Wagner M., von Nieding G. and Waller R.E. (1988) Health effects of suspended particulate matter, Final Report on CEC Contract No BU (84) 146 (491).

Walton W.H. (1954) Theory of size classification of airborne dust clouds by elutriation. *Brit. J. Appl. Phys.*, **5**, Supplement, S29–S40.

Wedding J.B., McFarland A.R. and Cermak J.E. (1977) Large particle collection characteristics of ambient aerosol samplers. *Environ. Sci. Technol.*, **11**(4), 387–390.

Whitby K. T. (1978) The physical characteristics of sulphur aerosols. *Atmos. Environ.*, **12**, 135–159.

Wiener R.W., Okazaki K. and Willeke K. (1988) Influence of turbulence on aerosol sampling efficiency, *Atmos. Environ.*, **22**, 917–928.

Wright B.M. (1954) A size-selecting sampler for airborne dust. *Brit. J. Indust. Med.*, **11**, 284–288.

## Further reading

Cohen B.S. and Hering S.V. (eds) (1994) *American Conference of Governmental Hygienists: Air Sampling Instruments*, 8th edn. ACGIH, Cincinnati, OH, USA.

Harrington J.M. and Gardiner K. (eds) (1995) *Occupational Hygiene*, 2nd edn., Blackwell Science, Oxford.

Willeke K. and Baron P.A. (eds) (1993) *Aerosol Measurement. Principles, Techniques and Applications*. Van Nostrand Reinhold, New York.

**Nomenclature**

| | |
|---|---|
| $A$ | aspiration efficiency, the efficiency with which particles enter directly through the plane of the sampling orifice |
| $A_I$ | inlet or entry efficiency, the efficiency with which particles enter through the plane of the sampling orifice not only directly but also after blow-off from external surfaces |
| $A_s$ | sampling efficiency, the efficiency by which particles arrive at the sensing region (e.g. filter or detection zone) after passage through orifice, transmission section and any particle size selection section |
| $B$ | aerodynamic bluffness |
| $C_c$ | Cunningham slip correction factor |
| $C_D$ | drag coefficient |
| $d$ | particle geometric diameter |
| $d_v$ | particle equivalent volume diameter |
| $D$ | characteristic dimension in a flow system |
| $D_B$ | coefficient of Brownian diffusion |
| $D_{ft}$ | coefficient of turbulent diffusion for fluid |
| $D_{pt}$ | coefficient of turbulent diffusion for particle |
| $e$ | fundamental electronic charge |
| $E$ | efficiency of particle impaction onto a bluff body |
| $E_I$ | inhalable aerosol fraction |
| $E_R$ | respirable aerosol fraction |
| $E_T$ | thoracic aerosol fraction |
| $F_D$ | drag force on a particle |
| $g$ | acceleration due to gravity |
| $k_B$ | Boltzmann constant |
| $K_{pt}$ | inertial parameter related to turbulence |
| $m$ | particle mass |
| $n$ | number of electronic charges on a particle |
| $N$ | number of particles |
| $Re$ | Reynolds number for flow system |
| $Re_p$ | Reynolds number for flow about a particle |
| $s$ | particle stop distance |
| $St$ | Stokes number |
| $U$ | freestream air velocity |
| $U_s$ | mean air velocity through sampler inlet |
| $v_x, v_y$ | mean velocity in the $x$ or $y$ direction |
| $\delta$ | dimension of sampler orifice |
| $\eta$ | air viscosity |
| $\theta$ | orientation of sampler to wind |
| $\rho_a, \rho_p$ | densities of air particle |
| $\tau$ | particle relaxation time |
| $\chi$ | particle dynamic shape factor |

# 4 Diffusion and coagulation

## S.K. FRIEDLANDER

Aerosol diffusion and coagulation both result from the Brownian motion of submicrometre particles. Particle deposition by diffusion is of fundamental importance to the functioning of gas-cleaning equipment, such as scrubbers and filters, as well as measurement instruments, such as the diffusion battery and certain types of filters. Diffusion contributes to the scavenging of small atmospheric particles by raindrops, and removal by vegetation and other surfaces, and is a significant mechanism of deposition in the lung.

The intensity of the Brownian motion increases as particle size decreases. As a result, the efficiency of collection by diffusion for particles smaller than about 0.5 µm increases with decreasing particle size; as shown in this chapter, certain gas-cleaning devices are most efficient for very small particles. For particles of diameter small compared with the characteristic length of collecting objects, rates of particle diffusion can be predicted from classical theories of molecular diffusion to surfaces. Only a few examples of point particle diffusion are given because the subject is extensively covered in the mass and heat transfer literature. For particles of finite diameter, the rate of particle diffusion to a collection surface is enhanced, compared to point particles. The particles need diffuse only to within one particle radius of the surface. This complicates the mathematical theory and requires development of a special approach described in this chapter. The finite particle diameter results in a minimum in the particle removal efficiency as a function of size for objects, such as cylinders, placed in the flow.

Particle diffusion coefficients are small compared with the kinematic viscosity of a gas (large Schmidt numbers) so the region of the gas flow near the surface from which particles are depleted is usually very narrow. This narrow region, the concentration boundary layer, is important to particle transport and is discussed in detail.

Aerosols coagulate as a result of the Brownian motion and this has a marked effect on particle size and number density. The rate of coagulation can be predicted from a theoretical analysis based on diffusion theory as discussed at the end of this chapter. Two limiting cases are considered: the coagulation of a monodisperse aerosol at short times after coagulation begins; and the asymptotic size distribution reached after very long times.

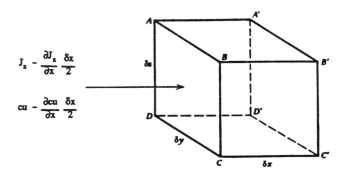

**Fig. 4.1** An elemental volume of fluid, fixed in space, in which flow and diffusion are occurring. At the centroid of the element, the particle diffusion flux in the $x$ direction is given by $J_x$. The flux across the face, $ABCD$, is shown for both diffusion and convection.

The presence of an external force field may have an important effect on diffusion and coagulation. Force fields include gravity, electrical potential gradients and thermal forces due to temperature gradients. Such phenomena are beyond the scope of this chapter. The discussion focuses on spherical particles which have been most studied, but some brief discussions for non-spherical particles are included.

## 4.1   Equation of convective diffusion

It is sometimes possible to predict rates of deposition by diffusion from flowing fluids by analysis of the equation of convective diffusion. This equation is derived by making a material balance on an elemental volume fixed in space with respect to laboratory coordinates (Fig. 4.1). Through this volume flows a gas carrying small particles in Brownian motion.

The rate at which particles are carried by the flow into the volume element across the face $ABCD$ is

$$\delta y \delta z \left[ nu - \frac{\delta x}{2} \frac{\partial nu}{\partial x} \right],$$

where $n$ is the particle concentration (number per unit volume) and $u$ is the velocity in the $x$ direction. The rate at which particles leave the volume across the opposite face is

$$\delta y \delta z \left[ nu + \frac{\delta x}{2} \frac{\partial nu}{\partial x} \right].$$

The net rate of particle accumulation for the flow in the $x$ direction is given by subtracting the rate leaving from the rate entering:

$$-\delta x \delta y \delta z \frac{\partial nu}{\partial x}.$$

Analogous expressions are obtained for the other four faces, and summing up for all three pairs, the result for the net accumulation of particles in the volume element due to the flow is

$$-\delta x \delta y \delta z \left[\frac{\partial nu}{\partial x} + \frac{\partial nv}{\partial y} + \frac{\partial nw}{\partial z}\right] = -\delta x \delta y \delta z \nabla \cdot nv. \tag{4.1}$$

Similarly, diffusional transport in the $x$ direction across the face $ABCD$ is

$$\delta y \delta z \left[J_x - \frac{\delta x}{2} \frac{\partial J_x}{\partial x}\right] \quad \text{and} \quad \delta y \delta z \left[J_x + \frac{\delta x}{2} \frac{\partial J_x}{\partial x}\right]$$

across the opposite face. The diffusion flux is given by Fick's law,

$$J_x = -D \frac{\partial n}{x}, \tag{4.2}$$

where $D$ is the coefficient of particle diffusion. After summing over all faces as before, the net diffusional accumulation in the volume element is

$$\delta x \delta y \delta z \left[\frac{\partial J_x}{\partial x} + \frac{\partial J_y}{\partial y} + \frac{\partial J_z}{\partial z}\right].$$

The rate of particle accumulation in the volume $\delta x \delta y \delta z$, taking into account flow and diffusion, is obtained by summing the two effects:

$$\frac{\partial n \delta x \delta y \delta z}{\partial t} = -\delta x \delta y \delta z \nabla \cdot nv + \delta x \delta y \delta z \nabla \cdot (D \nabla n). \tag{4.3}$$

Dividing both sides by the volume $\delta x \delta y \delta z$ and noting that $\nabla \cdot \mathbf{v} = 0$ for an incompressible fluid, the equation becomes

$$\frac{\partial n}{\partial t} + \mathbf{v} \cdot \nabla n = D \nabla^2 n \tag{4.4}$$

when the diffusion coefficient is not a function of position. This result holds both for monodisperse and polydisperse aerosols. In the polydisperse case, $n$ is the size distribution function, and $D$ depends on particle size. The continuous size distribution function $n(\mathbf{v}, \mathbf{r}, t)$ is defined by the expression

$$dN = n(\mathbf{v}, \mathbf{r}, t) \, d\mathbf{v}, \tag{4.5}$$

where $dN$ is the number of particles per unit volume of gas in the size range between the particle volume $\mathbf{v}$ and $\mathbf{v} + d\mathbf{v}$ at a point in space denoted by $\mathbf{r}$ at time $t$.

Values of $D$ can be determined as discussed in the next section. Solutions to the diffusion equation for many different boundary conditions in the absence of flow have been collected by Carslaw and Jaeger (1959) and

Crank (1975). The gas velocity distribution, **v**, can in some cases be obtained by solving the equations of fluid motion (Navier–Stokes equations) for which an extensive literature is available (Landau and Lifshitz, 1987; Rosenhead, 1963; Schlichting, 1979). In many cases, such as atmospheric transport and complex gas-cleaning devices, experimental data are necessary for the gas velocity. In this chapter, velocity distributions are introduced without derivation but with appropriate literature references. In all cases, it is assumed that particle concentration has no effect on the velocity distribution. This is true for the low aerosol concentrations usually considered, even in the case of many industrial process gases.

There is an extensive literature on solutions to equation (4.4) for various geometries and flow regimes. Many results are given by Levich (1962) and by Schlichting (1979) for boundary layer flows. Results for heat transfer, such as those discussed by Schlichting, are applicable to mass transfer or diffusion if the diffusion coefficient, $D$, is substituted for the coefficient of thermal diffusivity, $\kappa/\rho C_p$, where $\kappa$ is the thermal conductivity, $\rho$ is the gas density, and $C_p$ is the heat capacity of the gas. In some cases, the flow regimes are too complex for analytical or even numerical solutions. Dimensional analysis may then provide a useful way to correlate experimental data as discussed in this chapter.

## 4.2   Coefficient of diffusion

An expression for the coefficient of diffusion, $D$, with dimensions of square centimetres per second, can be derived as a function of particle size and gas properties. We consider diffusion in one dimension. Suppose that a cloud of Brownian particles, all the same size, is released over a narrow region around the plane corresponding to $x = 0$. The concentration everywhere else in the gas is zero. With increasing time, the particles diffuse as a result of the Brownian motion. The spread around the plane $x = 0$ is symmetrical (Figure 4.2) in the absence of an external force field acting on the particles.

The spread of the particles with time can be determined by solving the one-dimensional equation of diffusion,

$$\frac{\partial n}{\partial t} = D \frac{\partial^2 n}{\partial x^2}. \tag{4.6}$$

The solution for the concentration distribution is given by the Gaussian form (Crank, 1975)

$$n(x, t) = \frac{N_0}{2(\pi D t)^{1/2}} \exp\left(\frac{-x^2}{4Dt}\right), \tag{4.7}$$

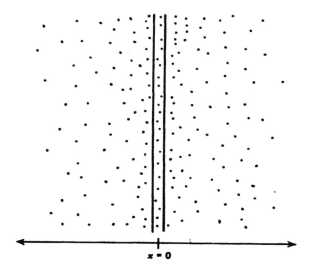

**Fig. 4.2** The spread of Brownian particles originally concentrated at the differential element around $x = 0$.

where $N_0$ is the number of particles released at $x = 0$ per unit cross-sectional area. The mean square displacement of the particles from $x = 0$ at time $t$ is

$$\overline{x^2} = \frac{1}{N_0} \int_{-\infty}^{\infty} x^2 n(x, t)\, dx. \tag{4.8}$$

Substituting (4.7) in (4.8), the result is

$$\overline{x^2} = 2Dt. \tag{4.9}$$

Thus the mean square displacement of the diffusing particles is proportional to the elapsed time.

An expression for $\overline{x^2}$ can also be derived from a force balance on a particle in Brownian motion, which for one dimension takes the form

$$m\frac{du}{dt} = -fu + F(t), \tag{4.10}$$

where $m$ is the particle mass, $u$ is the velocity, and $t$ is the time. According to equation (4.10), the force acting on the particle is divided into two parts. The first term on the right is the frictional resistance of the fluid and is assumed proportional to the particle velocity. For spherical particles much larger than the mean free path of the gas, the friction coefficient $f$ is usually based on Stokes law:

$$f = 3\pi\eta d_p, \tag{4.11}$$

where $\eta$ is the gas viscosity. Other forms for the friction coefficient are discussed below. The term $F(t)$ represents a fluctuating force resulting from the thermal motion of molecules of the ambient fluid. $F(t)$ is assumed to be independent of $u$ and its mean value, $\bar{F}(t)$, over a large number of similar but independent particles vanishes at any given time. Finally, it is assumed that $F(t)$ fluctuates much more rapidly with time than $u$. Thus over some interval, $\Delta t$, $u$ will be practically unchanged while there will be almost no correlation between the values of $F(t)$ at the beginning and end of the interval. These are rather drastic assumptions, but they have been justified by resort to models based on molecular theory. The conceptual difficulties attendant upon the use of equation (4.10) are discussed by Chandrasekhar (1943). From an analysis based on equation (4.10), it can be shown (Chandrasekhar, 1943; Friedlander, 1977) that

$$D = \frac{\overline{x^2}}{2t} = \frac{k_B T}{f}. \tag{4.12}$$

This is the Stokes–Einstein expression for the coefficient of diffusion. It relates $D$ to the properties of the fluid and the particle through the friction coefficient.

The Stokes' law form for the friction coefficient, $f = 3\pi\eta d_p$, holds for rigid spheres that move through a fluid at constant velocity with a Reynolds number, $d_p U / v$, much less than unity. Here $U$ is the velocity, and $v$ is the kinematic viscosity. The particle must be many diameters away from any surfaces and much larger than the mean free path of the gas molecules, $\lambda$, which is about 0.065 μm for air at 25°C. The range $d_p \gg \lambda$ is called the **continuum regime**.

As particle size is decreased to the point where $d_p \simeq \lambda$, the drag for a given velocity becomes less than predicted from Stokes' law and continues to decrease with particle size. In the range $d_p \ll \lambda$, the **free molecule range**, an expression for the friction coefficient, can be derived from kinetic theory (Epstein, 1924):

$$f = \frac{2}{3} d_p^2 \rho \left( \frac{2\pi k_B T}{m} \right)^{1/2} \left[ 1 + \frac{\pi\alpha}{8} \right] \qquad \frac{2\lambda}{d_p} > 1, \tag{4.13}$$

where $\rho$ is the gas density and $m$ is the molecular mass of the gas molecules.

The accommodation coefficient $\alpha$ represents the fraction of the gas molecules that leave the surface in equilibrium with the surface. The fraction $1 - \alpha$ is specularly reflected such that the velocity normal to the surface is reversed. As in the case of Stokes' law, the drag is proportional to the velocity of the spheres. However, for the free molecule range the friction coefficient is proportional to $d_p^2$, whereas in the continuum regime it is proportional to $d_p$. The coefficient $\alpha$ must, in general, be evaluated experimentally but is usually near 0.9 for momentum transfer (values differ for heat and mass transfer). The friction coefficient calculated from equation (4.13) is only 1% of that calculated from Stokes' law for a 20 Å particle.

As discussed in section 1.6, an interpolation formula is often used to cover the entire range of Knudsen numbers from the continuum to the free molecule regimes. The slip correction factor, $C_c$, is introduced as a correction to the Stokes friction coefficient as shown in equations (1.21) and (1.22). For $d_p \gg 1$, $C_c \to 1$ and $f$ approaches equation (4.11), whereas for $d_p \ll 1$, $f$ approaches the form of the kinetic theory expression given by equation (4.13).

Taking the values, given by Davies (1945), for the constants in equation (1.22) (see Table 1.2) results in $\alpha = 0.84$ in equation (4.13). Values for the diffusion coefficient and settling velocity of spherical particles, calculated over a wide particle size range, are shown in Table 4.1, together with values of the Schmidt number.

As noted above, Stokes' law is derived for the steady-state resistance to the motion of a particle. Why should it apply to the Brownian motion in which the particle is continually accelerated? The explanation is that the acceleration is always very small, so that at each instant a quasi-steady state can be assumed to exist.

For non-spherical particles, the drag depends on the orientation of the particle as it moves through the air. When $d_p \gg \lambda$, the drag can be calculated by solving the Stokes or 'creeping flow' form of the Navier–Stokes equations for bodies of various shapes. In calculating the diffusion coefficient, it is necessary to average over all possible orientations because of the stochastic nature of the Brownian motion. This calculation has been

**Table 4.1** Diffusion coefficients of spherical particles in air at 20°C, 1 atm

| $d_p$ (µm) | $C_c$ | $D \, (\mathrm{cm^2 \, s^{-1}})$ | Schmidt Number, $v/D$ |
|---|---|---|---|
| 0.001 | 216. | $5.14 \times 10^{-2}$ | 2.92 |
| 0.002 | 108. | $1.29 \times 10^{-2}$ | $1.16 \times 10^1$ |
| 0.005 | 43.6 | $2.07 \times 10^{-3}$ | $7.25 \times 10^1$ |
| 0.01 | 22.2 | $5.24 \times 10^{-4}$ | $2.87 \times 10^2$ |
| 0.02 | 11.4 | $1.34 \times 10^{-4}$ | $1.12 \times 10^3$ |
| 0.05 | 4.95 | $2.35 \times 10^{-5}$ | $6.39 \times 10^3$ |
| 0.1 | 2.85 | $6.75 \times 10^{-6}$ | $2.22 \times 10^4$ |
| 0.2 | 1.865 | $2.22 \times 10^{-6}$ | $6.76 \times 10^4$ |
| 0.5 | 1.326 | $6.32 \times 10^{-7}$ | $2.32 \times 10^5$ |
| 1.0 | 1.164 | $2.77 \times 10^{-7}$ | $5.42 \times 10^5$ |

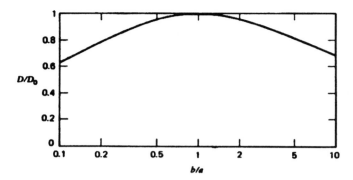

**Fig. 4.3** The coefficient of diffusion for ellipsoids of revolution as a function of the ratio of the equatorial radius $b$ to the radius of revolution $a$. $D_0$ is the coefficient of diffusion of a sphere of the same volume as the ellipsoid.

carried out by Perrin (1936) for ellipsoids of revolution. These are bodies generated by rotating an ellipse around one of its axes. We consider an ellipse with semi-axes $a$ and $b$ rotated around the $a$ axis.

For $z = b/a < 1$ (prolate or 'cigar-shaped' ellipsoid), the diffusion coefficient is

$$\frac{D}{D_0} = \frac{z^{2/3}}{(1 - z^2)^{1/2}} \ln\left(\frac{1 + (1 - z^2)^{1/2}}{z}\right), \qquad (4.14a)$$

and for $z > 1$ (oblate ellipsoid) it is

$$\frac{D}{D_0} = \frac{z^{2/3}}{(z^2 - 1)^{1/2}} \tan^{-1}(z^2 - 1)^{1/2} \qquad (4.14b)$$

where $D_0$ is the diffusion coefficient of a sphere of the same volume as the ellipsoid. For $z = 1$, $D/D_0$ is defined to be 1. If $a_0$ is the radius of the sphere, then $a_0 = az^{2/3}$.

The results of the calculation are shown in Figure 4.3. The diffusion coefficient of the ellipsoid is always less than that of a sphere of equal volume. However, over the range $10 > z > 0.1$, the coefficient for the ellipsoid is always greater than 60% of the value for the sphere. These results are not directly applicable to the diffusion of particles suspended in a shear field, because all orientations of the particle are not equally likely.

## 4.3   Similitude considerations for aerosol diffusion

Consider the flow of an incompressible gas, infinite in extent, over a body of a given shape placed at a given orientation to the flow. This is called an

**external** flow. Bodies of a given shape are said to be geometrically similar when they can be obtained from one another by changing the linear dimensions in the same ratio. Hence, it suffices to fix one characteristic length, $L$, to specify the dimensions of the body. This would most conveniently be the diameter for a cylinder or sphere, but any dimension will do for a body of arbitrary shape. Similar considerations apply for **internal** flows through pipes or ducts.

For an external flow, is assumed that the fluid has a uniform velocity, $U$, except in the region disturbed by the body. If the concentration in the mainstream of the fluid is $n_\infty$, a dimensionless concentration can be defined as follows:

$$n_1 = \frac{n}{n_\infty}. \tag{4.15}$$

Limiting consideration to the steady state, the equation of convective diffusion in the absence of an external force field can be expressed in dimensionless form as follows:

$$\mathbf{v}_1 \cdot \nabla_1 n_1 = \frac{1}{Pe} \nabla_1^2 n_1, \tag{4.16}$$

where $\mathbf{v}_1 = \mathbf{v}/U$ and $\nabla_1 = L\nabla$. The dimensionless group $LU/D$ is known as the **Peclet number** ($Pe$) for mass transfer.

In many cases, the velocity field can be assumed to be independent of the diffusional field. The steady isothermal flow of a viscous fluid, such as air, in a system of given geometry depends only on the Reynolds number when the velocity is small compared with the speed of sound.

The boundary condition for particle diffusion differs from the condition for molecular diffusion because of the finite diameter of the particle. For certain classes of problems, such as flows around cylinders and spheres, the particle concentration is assumed to vanish at one particle radius from the surface:

$$n = 0 \quad \text{at} \quad \frac{\alpha}{L} = \frac{a_p}{L} = R,$$

where $\alpha$ is a coordinate measured from the surface of the body. The dimensionless ratio $R = a_p/L$ is known as the **interception** parameter; particles within a distance $a_p$ of the surface would be intercepted even if diffusional effects were absent.

Hence the dimensionless concentration distribution can be expressed in the following way:

$$n_1 = f_1\left(\frac{\mathbf{r}}{L}, Re, Pe, R\right). \tag{4.17}$$

Two convective diffusion regimes are similar if the Reynolds, Peclet and interception numbers are the same.

The local rate of particle transfer by diffusion to the surface of the body is

$$J = -D\left(\frac{\partial n}{\partial \alpha}\right)_{\alpha=\alpha_p} = -\frac{Dn_\infty}{L}\left(\frac{\partial n_1}{\partial \alpha_t}\right)_{\alpha_1=R}. \qquad (4.18)$$

Setting the local mass transfer coefficient $k = J/n_\infty$ and rearranging, the result is

$$\frac{kL}{D} = f_2(Re, Pe, R). \qquad (4.19)$$

The particle transfer coefficient $k$ has dimensions of velocity and is often called the deposition velocity. At a given location on the collector surface, the dimensionless group $kL/D$, known as the **Sherwood number**, is a function of the Reynolds, Peclet and interception numbers. Rates of particle deposition measured in one fluid over a range of values of $Pe$, $Re$, and $R$ can be used to predict deposition rates from another fluid at the same values of the dimensionless groups. In some cases, it is convenient to work with the Schmidt number $Sc = v/D = Pe/Re$ in place of $Pe$ as one of the three groups, since $Sc$ depends only on the nature of the fluid and the suspended particles.

For $R \to 0$ ('point' particles), theories of particle and molecular diffusion are equivalent. Schmidt numbers for particle diffusion are much larger than unity, often of the same order of magnitude as for molecular diffusion in liquids. The principle of dimensional similitude tells us that the results of diffusion experiments with liquids can be used to predict rates of diffusion of point particles in gases, at the same Reynolds number.

For certain flow regimes, it is possible to reduce the number of dimensionless groups necessary to characterise a system by properly combining them. This further simplifies data collection and interpretation in several cases of considerable practical importance as shown in the sections that follow.

## 4.4 Concentration boundary layer

Flow normal to a right circular cylinder is the basic model for the theory of aerosol filtration by fibrous and cloth filters, and of particle collection by pipes and rods in a flow. The aerosol concentration at large distances from the surface is uniform; at one particle radius from the surface, the concentration vanishes.

Referring to the non-dimensional equation of convective diffusion (4.16), it is of interest to examine the conditions under which the diffusion term, on the one hand, or convection, on the other, is the controlling mode of transport. The Peclet number, $dU/D$, for flow around a cylinder of diameter $d$ is a measure of the relative importance of the two terms. For

$Pe \ll 1$, transport by the flow can be neglected, and the deposition rate can be determined approximately by solving the equation of diffusion in a non-flowing fluid with appropriate boundary conditions (Carslaw and Jaeger, 1959; Crank, 1975).

When the Peclet number is large, the physical situation is quite different. The mainstream flow then carries most of the particles past the cylinder. In the immediate neighbourhood of the cylinder, the diffusional process is important since the cylinder acts as a particle sink. Thus at high $Pe$, there are two different transport regions. Away from the immediate vicinity of the cylinder, convective transport by the bulk flow predominates. Near the surface, the concentration drops sharply from its value in the mainstream to zero (Figure 4.4). This surface region is known as the **concentration** (or **diffusion) boundary layer.** It is in many ways analogous to the velocity boundary layer that forms around the cylinder at high Reynolds numbers, with the Peclet number serving as a criterion similar to the Reynolds number. The role of the concentration boundary layer is fundamental to understanding and predicting the rate of transport of Brownian particles to surfaces. The usefulness of this concept is not limited to flows around cylinders. It applies to flows around other bodies such as spheres and wedges and to flows inside channels under certain conditions as well. Concentration boundary layers may develop in either low- or high-speed flows around collecting objects. Both cases are discussed in the sections that follow.

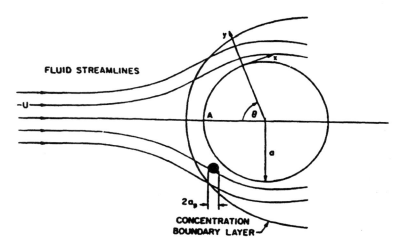

**Fig. 4.4** Schematic diagram showing concentration boundary layer surrounding a cylinder (or sphere) placed in a flow carrying diffusing small particles. Curvilinear coordinate $x$, taken parallel to the surface, is measured from the forward stagnation point $A$. Particle concentration rises from zero at $r = a + a_p$ almost to the mainstream concentration (for example, to 99% of the mainstream value) at the edge of the boundary layer.

## 4.5    Diffusion to cylinders at low Reynolds numbers

### 4.5.1    *Concentration boundary layer equation*

We consider first the case of a single cylinder set normal to an aerosol flow with low Reynolds number (Figure 4.4). This problem is of central importance to the functioning of high-efficiency fibrous filters for gas cleaning. Fibrous filters are highly porous mats of fine fibres usually containing less than 10% solid material. The spacing between the individual fibres is much greater than the diameters of the particles filtered. In the absence of electrical effects, small particles are collected by diffusion to the fibres; larger particles are removed by inertial deposition. When the fibre diameter is much larger than the mean free path of the air, continuum theory applies to the gas flow over the fibres. The equation of convective diffusion for the steady state takes the following form in cylindrical coordinates:

$$v_\theta \frac{\partial n}{r \partial \theta} + v_r \frac{\partial n}{\partial r} = D\left[\frac{\partial^2 n}{\partial r^2} + \frac{1}{r}\frac{\partial n}{\partial r} + \frac{\partial^2 n}{r^2 \partial \theta^2}\right], \tag{4.20}$$

where $v_\theta$ and $v_r$ are the tangential and radial components of the velocity. For particles of radius $a_p$ diffusing to a cylinder of radius $a$, the boundary conditions are

$$n = \begin{cases} 0 & r = a + a_p, \\ n_\infty & r = \infty. \end{cases} \tag{4.20a}$$

For fibre diameters smaller than $10\,\mu m$ and air velocities less than $10\,cm/s$, the Reynolds number is much less than unity. For isolated cylinders, the stream function for the air flow can be approximated by

$$\psi = AUa\sin\theta\left[\frac{r}{a}\left(2\ln\left(\frac{r}{a}\right) - 1\right) + \frac{a}{r}\right], \tag{4.21}$$

where $A = [2(2 - \ln Re)]^{-1}$ (Rosenhead, 1963). More approximate relations ('cell models') are usually used to take into account interactions among the fibres in developing correlations for filtration.

Even though the Reynolds number is small, there are many practical situations in which $Pe = Re \cdot Sc$ is large because the Schmidt number, $Sc$, for aerosols is very large. For $Pe \gg 1$, two important simplifications can be made in the equation of convective diffusion. First, diffusion in the tangential direction can be neglected in comparison with convective transport:

$$D\frac{\partial^2 n}{r^2 \partial \theta^2} \ll \frac{r_\theta}{r}\frac{\partial n}{\partial \theta}$$

In addition, a concentration boundary layer develops over the surface of the cylinder with its thinnest portion near the forward stagnation point.

When the thickness of the concentration boundary layer is much less than the radius of the cylinder, the equation of convective diffusion simplifies to the familiar form for rectangular coordinates (Schlichting, 1979, Chapter XII):

$$u\frac{\partial n}{\partial x} + v\frac{\partial n}{\partial y} = D\frac{\partial^2 n}{\partial y^2}, \tag{4.22}$$

where $x$ and $y$ are orthogonal curvilinear coordinates. The $x$ coordinate is taken parallel to the surface of the cylinder and measured from the forward stagnation point. The $y$ axis is perpendicular to $x$ and measured from the surface. The velocity components $u$ and $v$ correspond to the coordinates $x$ and $y$ (Figure 4.4). When the concentration boundary layer is thin, most of it falls within a region where the stream function (4.21) can be approximated by the first term in its expansion with respect to $y$:

$$\psi = 2AaU\left(\frac{y}{a}\right)^2 \sin\left(\frac{x}{a}\right). \tag{4.23}$$

The components of the velocity are related to the stream function as follows:

$$u = \frac{\partial \psi}{\partial y}, \qquad v = -\frac{\partial \psi}{\partial x}. \tag{4.24}$$

Substituting $\psi$ from equation (4.23) gives

$$u = 4AU\left(\frac{y}{a}\right)\sin\left(\frac{x}{a}\right), \tag{4.25a}$$

$$v = -2AU\left(\frac{y}{a}\right)^2\cos\left(\frac{x}{a}\right). \tag{4.25b}$$

### 4.5.2 Point particles

For the diffusion of point particles ($R \to 0$), the appropriate boundary conditions on equation (4.22) are

$$n = \begin{cases} 0 & y = 0 \\ n_\infty & y = 1. \end{cases} \tag{4.26}$$

The concentration boundary condition $n = n_\infty$ is set at $y = \infty$, even though the boundary layer form of the equation of convective diffusion (4.22) is valid only near the surface of the cylinder. This can be justified by noting that the concentration approaches $n_\infty$ very near the surface for high $Pe$.

If $x$ and $\psi$ are taken as independent variables instead of $x$ and $y$, equation (4.22) can be transformed into the following equation:

$$\left(\frac{\partial n}{\partial x}\right)_\psi = D\left[\frac{\partial}{\partial \psi}\left(u\frac{\partial n}{\partial \psi}\right)\right]_x. \qquad (4.27)$$

The $x$ component of the velocity is

$$u = \frac{\partial \psi}{\partial y} = \left(\frac{8AU}{a}\right)^{1/2}\sin^{1/2}x_1\psi^{1/2}, \qquad (4.28)$$

where $x_1 = x/a$. Substitution in equation (4.27) gives

$$\frac{\partial n}{\partial \chi} = \frac{D}{aAU}\frac{\partial}{\partial \psi_1}\left(\psi_1^{1/2}\frac{\partial n}{\partial \psi_1}\right), \qquad (4.29)$$

where

$$\chi = \int \sin^{1/2}x_1\,dx_1, \qquad \psi_1 = \frac{\psi}{2AaU}.$$

The boundary conditions in the transformed coordinates become

$$n = \begin{cases} 0 & \psi_1 = 0 \\ n_\infty & \psi_1 = \infty. \end{cases} \qquad (4.30)$$

By inspection of equation (4.29), we assume as a trial solution that $n$ is a function only of the variable $\xi = \psi_1/\chi^{2/3}$. This assumption must be checked by substitution of the expressions

$$\frac{\partial n}{\partial \chi} = -\frac{2}{3}\frac{\xi}{\chi}\frac{dn}{d\xi} \qquad (4.31)$$

and

$$\frac{\partial n}{\partial \psi_1} = \frac{1}{\chi^{2/3}}\frac{dn}{d\xi} \qquad (4.32)$$

in equation (4.29). The result of the substitution is an ordinary differential equation:

$$-\frac{APe}{3}\xi\frac{dn}{d\xi} = \frac{d}{d\xi}\left(\xi^{1/2}\frac{dn}{d\xi}\right), \qquad (4.33)$$

with the boundary condition $n = 0$ at $\xi = 0$ and $n = n_\infty$ at $\xi \to \infty$. This supports the assumption that $n$ is a function only of the variable $\xi$. Integration of equation (4.33) gives

$$n = \frac{n_\infty \int_0^{\xi^{1/2}} \exp\left(-\frac{2}{9} APez^3\right) dz}{\int_0^\infty \exp\left(-\frac{2}{9} APez^3\right) dz},$$ (4.34)

where $Pe = dU/D$, with $d$ the diameter of the cylinder. The integral in the denominator can be expressed in terms of a gamma function, $\Gamma$, as

$$\left(\frac{9}{2}\right)^{1/3} \frac{1}{3} \Gamma\left(\frac{1}{3}\right)(APe)^{-1/3} = 1.47(APe)^{-1/3}.$$ (4.35)

The rate of diffusional deposition per unit length of cylinder is

$$2D \int_0^\pi \left(\frac{\partial n}{\partial y_1}\right)_{y_1=0} dx_1 = k_{av}\pi dn_\infty.$$ (4.36)

The last expression defines the average mass transfer coefficient, $k_{av}$, for the cylinder. The concentration gradient at the surface is obtained by differentiating equation (4.34) with respect to $y$. The result is

$$\left(\frac{\partial n}{\partial y_1}\right)_{y_1=0} = \frac{(APe)^{1/3} n_\infty \sin^{1/2} x_1}{1.47\chi^{1/3}}.$$ (4.37)

Substituting in equation (4.36) and evaluating the integral with respect to $x_1$, gives (Natanson, 1957):

$$\frac{k_{av}d}{D} = 1.17(APe)^{1/3}.$$ (4.38)

The concentration gradient at the surface can be expressed in terms of an effective boundary layer thickness, $\delta_c$, as follows:

$$\left(\frac{\partial n}{\partial y}\right)_{y=0} = \frac{n_\infty}{\delta_c}.$$ (4.39)

Substituting equation (4.37), the result is

$$\frac{\delta_c}{d} \sim (APe)^{-1/3},$$ (4.40)

with a proportionality constant of order unity near the forward stagnation point. Hence the thickness of the concentration boundary layer is inversely proportional to $Pe^{1/3}$; large Peclet numbers lead to thin concentration boundary layers, as discussed in the previous section.

The theoretical expression (4.38) is in good agreement with data for diffusion in aqueous solutions over the high $Pe$ range of interest in aerosol deposition. Recalling that $Pe = Sc \cdot Re$, equation (4.38) can be rearranged to give

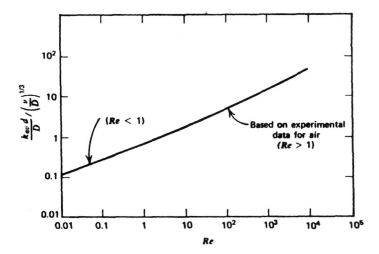

**Fig. 4.5** Diffusion to single cylinders placed normal to an air flow. The theoretical curve for low Reynolds numbers is in good agreement with experimental data for diffusion in aqueous solution (Dobry and Finn, 1956). The curve for high Reynolds numbers is based on data for heat transfer to air (data of Hilpert reported by Schlichting, 1979) corrected by dividing the Nusselt number by $(v/D)^{1/3}$. This is equivalent to assuming laminar boundary layer theory is applicable. This curve applies to the diffusion of point particles ($R \to 0$).

$$\frac{k_{av}d}{D}\bigg/\left(\frac{v}{D}\right)^{1/3} = 1.17(ARe)^{1/3} \qquad (4.41)$$

for this low Reynolds number case. At higher Reynolds numbers, a different functional form is found for the Reynolds number dependence but the general relationship

$$\frac{k_{av}d}{D}\bigg/\left(\frac{v}{D}\right)^{1/3} = f(Re) \qquad (4.42)$$

holds over a wide range of Reynolds numbers. The form of the function is shown in Figure 4.5 over both low and high Reynolds number ranges.

The efficiency of removal, $\eta_R$, is defined as the fraction of the particles collected from the fluid volume swept by the cylinder:

$$\eta_R = \frac{k_{av}\pi d n_\infty}{n_\infty U d} = 3.68A^{1/3}Pe^{-2/3}. \qquad (4.43)$$

Since $A$ is a relatively slowly varying function of Reynolds number, the efficiency varies approximately as $d^{-2/3}$, which means that fine fibres are more efficient aerosol collectors than coarse ones. Since $Pe = dU/D$, $\eta_R \sim d_p^{-2/3}$ and $d_p^{-4/3}$ for the continuum and free molecule ranges, respectively. Hence small particles are more efficiently removed by diffusion than larger particles in the range $d_p < 0.5\,\mu m$. The use of single filter fibre collection efficiencies to test this theory is discussed in a later section.

## 4.6 Diffusion at low Reynolds numbers: similitude law for particles of finite diameter

For particles of finite diameter, the interception effect becomes important. A useful similitude law that takes both diffusion and interception into account can be derived as follows (Friedlander, 1967). It is assumed that the concentration boundary layer is thin and falls within the region where the velocity distribution function is given by equations (4.25). Substituting these in equation (4.22) gives the following equation for convective diffusion:

$$4\frac{y}{a}\sin\left(\frac{x}{a}\right)\frac{\partial n}{\partial x} - 2\left(\frac{y}{a}\right)^2\cos\left(\frac{x}{a}\right)\frac{\partial n}{\partial y} = \frac{D}{AU}\frac{\partial^2 n}{\partial^2 y}. \tag{4.44}$$

We now introduce the following dimensionless variables:

$$n_1 = \frac{n}{n_\infty}, \qquad y_1 = \frac{y}{a_p}, \qquad x_1 = \frac{x}{a}, \tag{4.45}$$

where $a_p$ is the particle radius and $a$ the cylinder radius. Note that the curvilinear coordinates normal and parallel to the cylinder surface are non-dimensionalized by different characteristic lengths. Then equation (4.44) becomes

$$4y_1\sin x_1\frac{\partial n_1}{\partial x_1} - 2y_1^2\cos x_1\frac{\partial n_1}{\partial y_1} = \left(\frac{Da^2}{AUa_p^3}\right)\frac{\partial^2 n_1}{\partial y_1^2}, \tag{4.46}$$

with boundary conditions

$$n_1 = \begin{cases} 0 & y_1 = 1, \\ 1 & y_1 = \infty. \end{cases}$$

Taking $R = a_p/a$, only one dimensionless group,

$$\Pi = R^3 PeA \sim (Da^2/AUa_p^3)^{-1}, \tag{4.47}$$

appears in equation (4.46), and the boundary conditions are pure numbers. Hence the concentration distribution is

$$n_1 = f(x_1, y_1, \Pi). \tag{4.48}$$

In the boundary layer approximation, the particle deposition rate per unit length of cylinder is

$$2D\left(\frac{a}{a_p}\right)n_\infty\int_0^\pi\left(\frac{\partial n_1}{\partial y_1}\right)_{y_1=1}dx_1 = \eta_R n_\infty dU. \tag{4.49}$$

Introducing equation (4.48) in equation (4.49) leads to the following functional relationship:

$$\eta_R RPe = \frac{\pi k_{av} d_p}{D} = f_2(\Pi). \tag{4.50}$$

This is the similitude law for the diffusion of particles of finite diameter but with $R < 1$ in low-speed flows. For fixed $Re$, the group $\eta_R RPe$ should be a single-valued function of $RPe^{1/3}$ over the range in which the theory is applicable ($Pe \gg 1$, $Re < 1$, $R \ll 1$). Experimental data collected for different particle and cylinder diameters and gas velocities and viscosities should all fall on the same curve when plotted in the form of equation (4.50).

In the limiting case, $R \to 0$, $\eta_R$ is independent of the interception parameter $R$. By inspection of equation (4.50), this result is obtained if the function $f_2$ is linear in its argument $f_2 \sim \Pi$ such that

$$\eta_R = C_1 \pi A^{1/3} Pe^{-2/3}. \tag{4.51}$$

The constant $C_1 \pi = 3.68$ according to equation (4.38). In the limiting case $Pe \to \infty$, particles follow the fluid and deposit when a streamline passes within one radius of the surface. This effect is called **direct interception**. The efficiency is obtained by integrating equation (4.43) for the normal velocity component over the front half of the cylinder surface:

$$\eta_R = \frac{\int_0^{\pi/2} v_{y=a_p} \, dx_1}{Ua} = 2AR^2. \tag{4.52}$$

A result of this form can be obtained from equation (4.50) by noting that for $Pe \to \infty$, $\eta_R$ is independent of $Pe$. Then the function $f_2$ must be proportional to the cube of its argument.

Equations (4.51) and (4.52) are the limiting laws for the ranges in which diffusion and direct interception control, respectively. They show that for fixed velocity and fibre diameter, the efficiency at first decreases as $d_p$ increases because of the decrease in the diffusion coefficient (equation (4.51)); further increases in $d_p$ lead to an increase in $\eta_R$ as $R = d_p/d$ increases in equation (4.52). The result is a minimum in the plot of efficiency as a function of particle diameter. In the particle size range above the minimum, impaction and/or sedimentation usually become dominant mechanisms of particle deposition.

An analytical solution to equation (4.46) does not seem possible, but a solution can be obtained for the region near the forward stagnation point (Figure 4.4). Near the streamline in the plane of symmetry which leads to the stagnation point, $\sin x_1$ vanishes and $\cos x_1$ approaches unity so equation (4.46) becomes

$$-AR^3 Pey_1^2 \frac{dn_1}{dy_1} = \frac{d^2 n_1}{dy_1^2} \qquad (x_1 \to 0), \qquad (4.53)$$

with the boundary conditions

$$n_1 = \begin{cases} 0 & y_1 = 1 \\ 1 & y_1 = \infty. \end{cases} \qquad (4.54)$$

There is a solution to equation (4.53) with these boundary conditions for which $n_1$ is a function only of $y_1$; the following expression is obtained for the coefficient of mass transfer at the forward stagnation point:

$$k_0 = \frac{-D(\partial n/\partial y)_{y=a_p;x=0}}{n_\infty} \qquad (4.55a)$$

$$= \frac{(D/a_p)e^{-AR^3 Pe/3}}{\int_1^x \exp(-AR^3 Pez^3/3)\, dz}. \qquad (4.55b)$$

Although this result applies only at $x_1 = 0$, the deposition rate is greatest at this point and illustrates the general functional dependence on $Pe$ and $Re$ over the entire cylinder. For larger particles, inertial deposition becomes important. This effect considerably complicates the analysis and is beyond the scope of this chapter.

## 4.7   Comparison of theory with experiment

A systematic experimental test of this theory for aerosol flow around single fibres at low Reynolds numbers is difficult. However, the theory has been used to correlate data for the single fibre removal efficiency, which can be determined approximately by measuring the fraction of particles collected in a bed of fibres. The link between theory and experiment can be made as follows. In a regular array of fibres with uniform diameter, $d$, and a solids fraction, $\alpha$, let the average concentration of particles of size $d_p$ at a distance $z$ from the filter entrance be $N$ (Figure 4.6). For a single fibre, the removal efficiency is defined as

$$\eta_R = \frac{b}{d}, \qquad (4.56)$$

where $b$ is the width that corresponds to a region of flow completely cleared of all particles by the cylinder. In a differential distance, $dz$, in the flow direction, there are $\alpha dz/(\pi d^2/4)$ fibres per unit width normal to the flow direction; the removal over this distance by each fibre is

$$-dN / \frac{\alpha dz}{\pi d^2/4} = bN = (\eta_R d)N. \qquad (4.57)$$

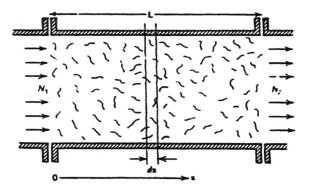

**Fig. 4.6** Schematic diagram of fibrous filter.

Rearranging and integrating from $z = 0$ to $z = L$, the thickness of the filter,

$$\eta_R = \frac{\pi d}{4\alpha L} \ln\left(\frac{N_1}{N_2}\right). \tag{4.58}$$

Since the fibre diameters are usually not all equal and the fibres are arranged in a more or less random fashion, $\eta_R$ should be interpreted as an effective fibre efficiency that can be calculated from equation (4.58) and based on an average diameter, $\bar{d}$, usually the arithmetic average. In an experimental determination of $\eta_R$, the practice is to measure $N_1$ and $N_2$, the inlet and outlet concentration of a monodisperse aerosol passed through the filter. The average fibre diameter, $\bar{d}$, can be determined by microscopic examination.

Chen (1955) and Wong *et al.* (1956) reported experimental values for single fibre efficiencies obtained in experiments with fibre mats and monodisperse liquid aerosols produced by a condensation particle generator. The filter mats used by both sets of investigators were composed of glass fibres, distributed in size. The data extrapolated to zero fraction solids have been recalculated and plotted in Figure 4.7 in the form based on the similitude analysis (equation (4.50)). The calculations are based on the average fibre diameter. Chen's data covered the ranges $62 < Pe < 2.8 \times 10^4$, $0.06 < R < 0.29$, $1.4 \times 10^{-3} < Re < 7.7 \times 10^{-2}$, and $5.2 \times 10^{-4} < Stk < 0.37$. The Stokes number, $Stk = mU/af$, where $m$ is the particle mass, is a measure of the strength of the inertial effects and must be small for the diffusion-interception theory to apply. For this data set, the analysis was satisfactory for $Stk < 0.37$.

Most of the data fell in the range $10^{-3} < Re < 10^{-1}$, and theoretical curves for the forward stagnation point (equation (4.55b)) are shown for the

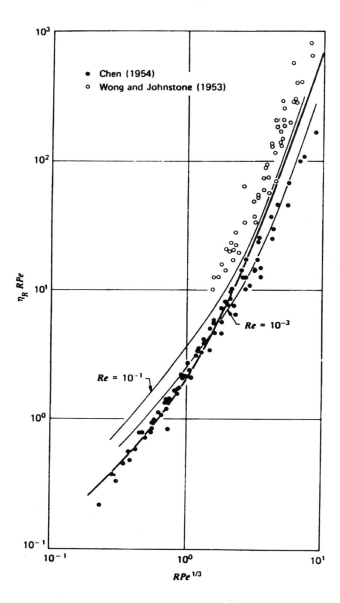

**Fig. 4.7** Comparison of experimentally observed deposition rates on glass fibre mats for dioctylphthalate (Chen) and sulphuric acid (Wong) aerosols with theory for the forward stagnation point of single cylinders (Friedlander, 1967). The theoretical curves for $Re = 10^{-1}$ and $10^{-3}$ were calculated from equation (4.55b). For all data points the Stokes number was less than 0.5. Agreement with Chen's data is particularly good. Theory for the forward stagnation point should fall higher than the experimental transfer rates, which are averaged over the fibre surface. The heavy line is an approximate best fit with the correct limiting behaviour.

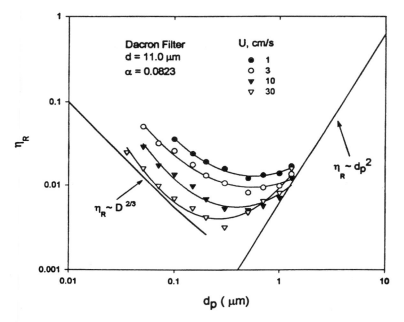

**Fig. 4.8** Efficiency minimum for single fibre removal efficiency for particles of finite diameter. For very small particles, diffusion controls according to equation (4.51) and $\eta_R \sim D^{2/3}$. The different curves result from the effects of velocity. In the interception range according to equation (4.52), $\eta_R \sim d_p^2$, and is practically independent of gas velocity (after Lee and Liu, 1982a).

limiting values of the Reynolds number. Rough agreement between experiment and theory is evident. One would expect the experimental data, based on the average deposition over the fibre surface, to fall somewhat below the theoretical curves for the forward stagnation point. This is true over the whole range for Chen's data but not for those of Wong *et al.*

In later experimental studies, Lee and Liu (1982a; 1982b) used submicrometre dioctylphthalate aerosols and dacron fibre filters with $0.035 < d_p < 1.3\,\mu m$, $1 < U < 30\,cm/s$ and fibre diameters of 11.0 and 12.9 $\mu m$. The dependence of $\eta_R$ on $\alpha$ was studied systematically. These investigators were also able to correlate their data using the result of the similarity transformation equation (4.50). As expected from theory, $\eta_R$ passes through a minimum with increasing particle diameter corresponding to the transition from the diffusional regime (equation (4.51)) to removal by direct interception (4.52) (Figure 4.8). They proposed the following correlation for the single fibre collection efficiency:

$$\eta_R = 1.6\frac{Pe^{-2/3}}{(1-\alpha)^{2/3}K^{1/3}} + 0.6\frac{R^2}{K(1+R)}, \tag{4.59}$$

where $K(\alpha) = -\frac{1}{2}\ln \alpha - \frac{3}{4} + \alpha - \frac{1}{4}\alpha^2$ and $\alpha$ is the solids fraction. This has the expected limiting forms (4.51) and (4.52).

The success of the analysis in correlating experimental data for clean filters offers convincing support for the theory of convective diffusion of particles of finite diameter to surfaces. As particles accumulate in the filter, both the efficiency of removal and the pressure drop increase, and the analysis no longer holds. Some data on this effect are available in the literature. Care must be taken in the practical application of these results because of pinhole leaks in the filters or leaks around the frames.

## 4.8 Single element particle capture by diffusion and interception at high Reynolds numbers

An analysis similar to the one for particle deposition from low-speed flows can be made for flows with high Reynolds number around blunt objects such as cylinders and spheres (Fernandez de la Mora and Friedlander, 1982). In this case, an aerodynamic boundary layer develops around the object. Within the aerodynamic boundary layer, a thin concentration boundary layer lies near the surface. The analysis takes into account both diffusion and direct interception, that is, the finite diameter of the particles. The results are important for particle deposition to cylinders much larger than those which compose high-efficiency filters, such as coarse wire filters or meshes, or heat exchanger tubes perpendicular to an aerosol flow. Other important applications are to particle deposition from the atmosphere on tree limbs, pine needles or grassy surfaces, as discussed below.

For simplicity we choose a two-dimensional geometry corresponding to the flow normal to a bluff body of arbitrary shape. The origin of co-ordinates is taken at the stagnation point, and the $y$ axis is normal to the surface at every point, $y$ being zero at the surface. Generalization to three-dimensional geometries is straightforward, and leaves the main conclusions unchanged.

The high Reynolds number velocity field in the inviscid region close to the stagnation point is given by Schlichting (1979, pp. 96–98) as

$$(u, v) = (\omega x, -\omega y); \tag{4.60}$$

within the viscous layer,

$$u = \omega x f'(\eta) \tag{4.61a}$$

$$v = -(v\omega)^{1/2}f(\eta) \tag{4.61b}$$

where $\omega$ is a constant whose value depends on the shape of the object (ribbon or cylinder), $\eta$ is the boundary layer coordinate

$$\eta = y(\omega/v)^{1/2}, \tag{4.62}$$

and the near-wall behaviour of the function $f$ is

$$f(\eta) = \tfrac{1}{2}\beta\eta^2 \tag{4.63a}$$

$$f'(\eta) = \beta\eta \tag{4.63b}$$

$(\eta \ll 1)$ with

$$\beta = f''(0) = 1.2326. \tag{4.64}$$

Therefore, sufficiently close to the stagnation point

$$u = \omega x\beta\eta, \tag{4.65a}$$

$$v = -\tfrac{1}{2}(v\omega)^{1/2}\beta\eta^2. \tag{4.65b}$$

In the region close to the wall, provided the flow has not separated and before transition to turbulence, equations (4.65) can be generalized away from the stagnation point to give

$$u = \omega a K(x_1)\eta \tag{4.66a}$$

$$v = -\tfrac{1}{2}(v\omega)^{1/2}K'(x_1)\eta^2, \tag{4.66b}$$

where we have made $x$ dimensionless using the obstacle characteristic length $a$ as in the analysis for low-speed flows

$$x_1 = x/a \tag{4.67}$$

and also

$$K' = \frac{\mathrm{d}K}{\mathrm{d}x_1}. \tag{4.68}$$

For large Reynolds numbers, the function $K$ depends on the particular shape of the obstacle, and can be calculated by standard methods of boundary layer theory (Schlichting, 1979, Chapter IX). Then, the equation of convective diffusion is, in the boundary layer approximation,

$$-D\frac{\partial^2 n}{\partial y^2} + \omega K\eta\frac{\partial n}{\partial x_1} - \tfrac{1}{2}K'(v\omega)^{1/2}\frac{\partial n}{\partial y} = 0. \tag{4.69}$$

Defining the non-dimensional distance from the surface in the same way as for low-speed flows,

$$y_1 = y/a_p, \tag{4.70}$$

equation (4.69) becomes

$$\frac{\Pi^{-3}}{3}\frac{\partial^2 n}{\partial y_1^2} - 2Ky_1\frac{\partial n}{\partial x_1} + K'y_1^2\frac{\partial n}{\partial y_1} = 0, \tag{4.71}$$

where $\Pi$ is a dimensionless group for particle deposition from high Reynolds number flows:

$$\Pi = \frac{1}{6}\frac{v}{D}a_p^3(\omega/v)^{3/2}. \tag{4.72}$$

This equation must be solved with the diffusion-interception boundary conditions

$$n = n_\infty \quad \text{for} \quad y_1 \to \infty. \tag{4.73}$$

The solution for any given obstacle ($K(x_1)$ fixed) is

$$n/n_\infty = F(x_1, y_1, \Pi), \tag{4.74}$$

and by equations (4.49) and (4.50)

$$\frac{k_{av}a_p}{D} = F_2(\Pi). \tag{4.75}$$

Also, since the Reynolds number in this case is

$$Re \sim \omega a^2/v, \tag{4.76}$$

substitution in equation (4.72) gives

$$\Pi \sim R Re^{1/2}(v/D)^{1/3}, \tag{4.77}$$

where $R = a_p/a$. The parameter $\Pi$ is related to the corresponding low Reynolds number parameter

$$\Pi_{Re<1} = R^3 Pe A \tag{4.78}$$

through the weakly varying factor $Re^{1/6}/A$

$$\Pi_{Re\gg1}/\Pi_{Re<1} \sim Re^{1/6}/A. \tag{4.79}$$

Approximate expressions for the removal efficiencies of single cylinders and spheres based on this analysis have been given by Parnas and Friedlander (1984): for cylinders

$$\eta_R = 1.88 Re^{1/6} Pe^{-2/3} + 0.80 R^2 Re^{1/2}; \tag{4.80}$$

and for spheres

$$\eta_R = 2.40 Re^{1/6} Pe^{-2/3} + 1.10 R^2 Re^{1/2}. \tag{4.81}$$

The recommended range of application is $10^2 < Re < 10^4$ and for Stokes numbers less than the critical values for impaction, $1/8$ and $1/12$, for cylinders and spheres, respectively.

## 4.9 Case of high Reynolds number: comparison with experiment

The results of this analysis have not been directly tested for flows over single cylinders. However, there is a substantial body of experimental data from wind tunnel experiments on the deposition of particles from gases on rough surfaces composed of grass blades, gravel and similar elements. The measurements, made in wind tunnels and designed to simulate atmospheric dry deposition, have been summarized by Schack *et al.* (1985). The data of Chamberlain (1966) on particle deposition from flows over artificial grass have been correlated by Fernandez de la Mora and Friedlander (1982) using equation (4.75) and plotting $k_{av}a_p/D$ against $RRe^{1/2}(v/D)^{1/3}$ in Figure 4.9. The reported particle deposition velocity $v_g$ was used in place of $k_{av}$. The characteristic length chosen for the collector was the transverse dimension of the strips forming the artificial grass elements, $a = 0.5\,cm$. As shown in Figure 4.9, the data collapse into a single curve over the wide range of experimental parameters, including six different particle sizes (32, 19, 5, 2, 1 and 0.08 µm diameter). Also shown for comparison is the theoretical curve corresponding to deposition at the stagnation point where the function $K(x_1)$ is

$$K(x_1) = \beta x_1 \tag{4.82}$$

and $x_1$ is close to zero. Then the function $F_2$ in equation (4.75) can be obtained analytically to yield

$$\frac{k_{av}a_p}{D} = F_2 = \frac{\exp(-\Pi^3)}{\displaystyle\int_1^\infty \exp(-\Pi^3\xi^3)\,d\xi}. \tag{4.83}$$

The asymptotic behaviour of $F_2$ is given by

$$\Pi \to 0, \quad F_2 \to \Pi \quad \text{(diffusion limit)}, \tag{4.84}$$

$$\Pi \to \infty, \quad F_2 \to \Pi^3 \quad \text{(interception limit)}. \tag{4.85}$$

The asymptotic behaviour for large $\Pi$ requires a slope of 3 on a log-log plot, in agreement with the data. The expected slope for lower values of $\Pi$ is unity; this is also followed reasonably well, though this is less certain because only three experimental points fall in this region, all for a single particle size, which was measured with less precision than the remaining points in the data. A quantitative comparison requires relating $\omega$ (entering into the definition of $\Pi$) with the known variables $u_*$ and $a$ appearing in Figure 4.9. For the potential flow around an obstacle one generally has

$$\omega = bU_\infty/a, \tag{4.86}$$

where the constant $b$ depends on the body geometry. In Figure 4.9, $b = 2$, corresponding to a flat strip normal to the incident (unseparated) stream, though any value of order unity would be reasonable. The correlation works beyond the expected limit of validity since significant inertial effects are likely at the larger particle sizes.

As in the case of low Reynolds number flows, the individual element removal efficiency passes through a minimum as particle size increases

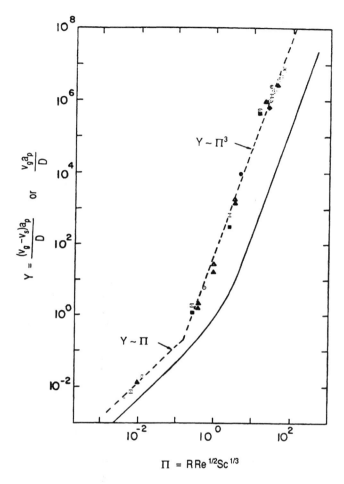

$$\Pi = R\,Re^{1/2}Sc^{1/3}$$

**Fig. 4.9** Dimensionless particle deposition velocity to artificial grass as a function of the deposition parameter. Data from Chamberlain (1966), corrected (black symbols) and uncorrected (white symbols) for gravitational settling. Two broken lines (---) with slopes 3 and 1 (corresponding to the interception and diffusion asymptotic regions) are drawn through the data. The solid line (—) shows, as a reference, the single element collection efficiency at the stagnation point of an infinite strip normal to the unseparated potential flow. The symbols correspond to various particle types and air velocities (after Fernandez de la Mora and Friedlander, 1982).

from the small submicrometre range to the micrometre sizes. However, for dry deposition from the atmosphere, collecting objects (grass blades, other vegetation, rocks, etc.) come in various sizes and shapes; this probably results in a broad minimum with respect to particle size compared to the case for uniform collectors.

### 4.10    Diffusion from a laminar pipe flow

In this section and the next, we discuss particle deposition by diffusion from laminar and turbulent flows through a smooth-walled pipe. The particle diameter is assumed to be much smaller than the tube diameter (or viscous sublayer thickness for turbulent flow) so the interception parameter which was important in the previous discussions does not play a role.

When a gas enters a smooth pipe from a large reservoir through a well-faired entry, a laminar boundary layer forms along the walls. The velocity profile in the main body of the flow remains flat. The velocity boundary layer thickens with distance downstream from the entry until it eventually fills the pipe. If the Reynolds number based on pipe diameter is less than 2100, the pipe boundary layer remains laminar. The flow is said to be fully developed when the velocity profile no longer changes with distance in the direction of flow. The profile becomes nearly fully developed after a distance from the entry of about $0.04dRe$. For example, for $Re = 1000$, the entry length extends over 40 pipe diameters from the pipe entry.

Small particles present in the gas stream diffuse to the walls as a result of their Brownian motion. Since the Schmidt number, $v/D$, is much greater than unity, the diffusion boundary layer is thinner than the velocity boundary layer and the concentration profile tends to remain flat perpendicular to the flow for much greater distances downstream from the entry than the velocity profile. As a reasonable approximation for mathematical analysis, it can be assumed that at the pipe entry, the concentration profile is flat while the velocity profile is already fully developed, that is, parabolic.

The problem of diffusion to the walls of a channel (pipe or duct) from a laminar flow is formally identical with the corresponding heat transfer (Graetz) problem when the particle size is small compared with the channel size ($R \to 0$). For a fully developed parabolic velocity profile, the steady-state equation of convective diffusion (4.4) takes the following form in cylindrical coordinates:

$$u\frac{\partial n}{\partial x} = D\left[\frac{\partial(r(\partial n/\partial r))}{r\partial r} + \frac{\partial^2 n}{\partial x^2}\right], \qquad (4.87)$$

where $u = 2U_{av}[1 - (r/a)^2]$ and $U_{av}$ is the average velocity. As boundary conditions, it is assumed that the concentration is constant across the tube inlet and vanishes at the pipe wall, $r = a$:

$$n = \begin{cases} n_1 & r < a, x = 0 \\ 0 & r = a. \end{cases} \qquad (4.88)$$

When $Pe > 100$, diffusion in the axial direction can be neglected. Solutions to this expression with these boundary conditions have been given by many investigators and the analysis will not be repeated here. For short distances from the tube inlet, a concentration boundary layer develops for the particle distribution. An analytical solution to the equation of convective diffusion gives the following expression for the fraction of the particles passing through a tube of length $L$ without depositing:

$$P = \frac{n_2}{n_1} = 1 - 2.56\Pi^{2/3} + 1.2\Pi + 0.1767\Pi^{4/3} + \ldots \qquad (4.89)$$

with $\Pi = \pi DL/Q < 0.02$, where $Q$ is the volumetric flow of air through the tube. At long distances from the tube inlet, the fraction penetrating is obtained by solving equation (4.87) using separation of variables:

$$P = \frac{n_2}{n_1} = 0.819\exp(-3.66\Pi) + 0.0975\exp(22.3\Pi)$$
$$+ 0.0325\exp(-57.0\Pi) + \ldots \qquad (4.90)$$

for $\Pi > 0.02$. Original references for these results and the corresponding expressions for flow between flat plates are given by Cheng (1993).

These results can be applied to deposition in sampling tubes and to the design of the diffusion battery, a device used to measure the particle size of submicrometre aerosols. The battery may consist of a bundle of capillary tubes, or of a set of closely spaced, parallel flat plates, through which the aerosol passes in laminar flow. The particle concentrations entering and leaving the diffusion battery are measured with a condensation particle counter. From the measured value of the reduction in concentration, the value of $\Pi$ can be determined from equation (4.89) or (4.90) or their equivalent for flat plates. The value of $D$, hence $d_p$, can be calculated since $x$, $a$ and $U$ are known for the system. For polydisperse aerosols, the usual case, the method yields an average particle diameter that depends on the particle size distribution. The theory also has application to efficiency calculations for certain classes of filters (Spurny et al., 1969) composed of a sheet of polymeric material penetrated by many small cylindrical pores.

## 4.11 Diffusion from a turbulent pipe flow

When the pipe Reynolds number is greater than about 2100, the velocity boundary layer that forms in the entry region eventually turns turbulent as the gas passes down the pipe. The velocity profile becomes fully developed, that is, the shape of the distribution ceases to change at about 25 to 50 pipe diameters from the entry.

Small particles in such a flow are transported by turbulent and Brownian diffusion to the wall. In the sampling of atmospheric air through long pipes, wall losses result from turbulent diffusion. Accumulated layers of particles will affect heat transfer between the gas and pipe walls.

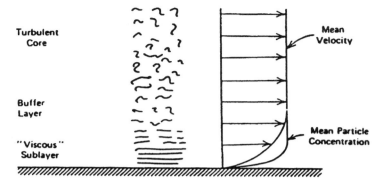

**Fig. 4.10** Schematic diagram showing the structure of turbulent pipe flow. For convenience, the flow is divided into three regions. Most of the pipe is filled with the turbulent core, with the velocity rising rapidly over the viscous sublayer. The concentration drops more sharply than the velocity because $D \ll v$ and turbulent diffusion brings the particles close to the wall before Brownian diffusion can act effectively.

Although the flow of turbulent fluids is much more complex than that of the laminar flows discussed in previous sections, semi-empirical calculations of rates of diffusional transport are actually easier to make for fully developed turbulent duct flows. The reason is that the turbulent mass transfer coefficient does not vary with distance along the pipe.

In analysing turbulent transport, it is convenient to divide the pipe flow into three different zones along a distance perpendicular to the wall (Figure 4.10). The core of the pipe is a highly turbulent region in which molecular diffusion is negligible compared with transport by the turbulent eddies. Closer to the wall there is a transition region where both molecular and eddy diffusion are important. Next to the wall itself, there is a thin sublayer in which the transfer of *momentum* is dominated by viscous forces, and the effect of weak turbulent fluctuations can be neglected. This applies also to heat and mass transfer for gases; the Schmidt and Prandtl numbers are near unity, which means that heat and mass are transported at about the same rates as momentum.

The situation is quite different for particle diffusion. In this case, $v/D \gg 1$ and even weak fluctuations in the viscous sublayer contribute significantly to transport. Consider a turbulent pipe flow. In the regions near the wall, the curvature can be neglected and the instantaneous particle flux written as follows:

$$J_y = -D\frac{\partial n}{\partial y} + nv, \qquad (4.91)$$

where $y$ is the distance measured normal to the surface and $v$ is the velocity in the $y$ direction.

In analysing turbulent pipe flows, it is assumed that the velocity and the concentration can be separated into mean and fluctuating components:

$$v = v' \quad (\text{since } \bar{v} = 0) \tag{4.92}$$

and

$$n = \bar{n} + n', \tag{4.93}$$

where the bar and prime refer to the mean and fluctuating quantities, respectively. Substituting in equation (4.91) and taking the time average, the result is

$$\bar{J} = -D\frac{\partial \bar{n}}{\partial y} + \overline{n'v'}. \tag{4.94}$$

The eddy diffusion coefficient, $\varepsilon$, is defined by

$$\overline{n'v'} = -\varepsilon\frac{\partial \bar{n}}{\partial y}. \tag{4.95}$$

Based on experimental data for diffusion controlled electrochemical reactions in aqueous solution, the following expression was proposed by Lin *et al.* (1953) for the eddy diffusion coefficient in the viscous sublayer:

$$\varepsilon = v\left(\frac{y^+}{14.5}\right)^3, \tag{4.96}$$

where $y^+ = [yU(f/2)^{1/2}]/v$, with $U$ the average velocity, $f$ the Fanning friction factor, and $v$ the kinematic viscosity. This expression for $\varepsilon$ holds when $y^+ < 5$. A similar form was found by analysing the results of a variety of measurements by other investigators (Monin and Yaglom, 1971).

Substituting equation (4.95) in (4.94), the general expression for the diffusion flux is

$$\bar{J} = -(D + \varepsilon)\frac{\partial \bar{n}}{\partial y}. \tag{4.97}$$

For particle diffusion, $v/D \gg 1$. Compared with momentum transfer, particles penetrate closer to the wall by turbulent diffusion before Brownian diffusion becomes important. (Viscous shear for momentum transfer is important at relatively large distances from the surface.) The particle concentration, which vanishes at the wall, rises rapidly practically reaching the mainstream concentration, $\bar{n}_\infty$, within the viscous sublayer in which $\varepsilon$ is given by equation (4.96). The concentration distribution and deposition flux can be obtained by integrating equation (4.97) and assuming that $\bar{J}$ is a function only of $x$ and not of $y$, the distance from the surface.

The boundary conditions are

$$\bar{n} = \begin{cases} 0 & y = 0 \\ n_\infty & y = \infty. \end{cases}$$

The result of the integration is

$$\frac{kd}{D} = 0.042 Re f^{1/2} Sc^{1/3}, \tag{4.98}$$

where the mass transfer coefficient (or particle deposition velocity) is defined by the relationship $k = -(D/\bar{n}_\infty)(\partial \bar{n}/\partial y)_{y=0}$.

## 4.12   Dynamics of coagulation

### 4.12.1   *Collision frequency function*

Particles in Brownian motion collide and adhere. The reduction in surface area that accompanies particle coalescence corresponds to a reduction in the Gibbs free energy under conditions of constant temperature and pressure. Thus aerosols are basically unstable to coagulation which causes a reduction in the total number of particles and an increase in the average size.

Brownian coagulation determines the size distribution of the sub-micrometre aerosol produced in high-temperature industrial processes such as coal combustion, titania pigment production and optical fibre fabrication. Efforts have been and continue to be made to apply basic coagulation theory to such processes. An important goal is to predict or explain particle size distributions. A coagulation limited aerosol generator has been designed for the production of particles of desired size and concentration for use in testing the efficiencies of large-scale, high-efficiency industrial filters (Koch *et al.*, 1993). Coagulation theory is also often incorporated into atmospheric aerosol models.

An expression for the time rate of change of the particle size distribution function can be derived as follows. Let $N_{ij}$ be the number of collisions occurring per unit time per unit volume between the two classes of particles of volumes $v_i$ and $v_j$. All particles are assumed to be spherical, which means that $i$ and $j$ are uniquely related to particle diameters. When two particles collide, according to this simplified model, they coalesce to form a third whose volume is equal to the sum of the original two. The collision frequency in terms of the concentrations of particles with volumes $v_i$ and $v_j$ is

$$N_{ij} = \beta(v_i, v_j) n_i n_j, \tag{4.99}$$

where $\beta(v_i, v_j)$, the collision frequency function, depends on the sizes of the colliding particles and on such properties of the system as temperature and pressure. The functional dependence on these variables is determined by the mechanisms by which the particles come into contact.

For a discrete size distribution made up of initially monodisperse particles, the rate of formation of particles of size $k$ by collision of particles of size $i$ and $j$ is $\frac{1}{2}\sum_{i+j=k} N_{ij}$, where the notation $i+j=k$ indicates that the summation is over those collisions for which $v_i + v_j = v_k$. The factor $1/2$ is introduced because each collision is counted twice in the summation. The rate of loss of particles of size $k$ by collision with all other particles is $\sum_{i=1}^{\infty} N_{ik}$. Hence the net rate of generation of particles of size $k$ is

$$\frac{dn_k}{dt} = \frac{1}{2}\sum_{i+j=k} N_{ij} - \sum_{i=1}^{\infty} N_{ik}. \qquad (4.100)$$

Substituting equation (4.99) in equation (4.100) gives

$$\frac{dn_k}{dt} = \frac{1}{2}\sum_{i+j=k} \beta(v_i, v_j)n_i n_j - n_k \sum_{i=1}^{\infty} \beta(v_i, v_k)n_i, \qquad (4.101)$$

which is the dynamic equation for the discrete size spectrum when coagulation alone is important. The solution to equation (4.101) depends on the form of $\beta(v_i, v_j)$, which is determined by the mechanism of particle collision as discussed below. The theory for the discrete spectrum, including expressions for the collision frequency function for Brownian coagulation and laminar shear, is due to Smoluchowski (1917).

### 4.12.2 Brownian coagulation

For Brownian particles much larger than the mean free path of the gas, there is experimental evidence that the collision process is diffusion limited. Consider a sphere of radius $a_i$, fixed at the origin of the coordinate system in an infinite medium containing suspended spheres of radius $a_j$. Particles of radius $a_j$ are in Brownian motion and diffuse to the surface of $a_i$, which is a perfect sink. Hence the concentration of $a_j$ particles vanishes at $r = a_i + a_j$. For the spherical symmetry, and in the absence of flow, the diffusion equation (4.4) takes the form:

$$\frac{\partial n}{\partial t} = D\frac{\partial r^2(\partial n/\partial r)}{r^2 \partial r}. \qquad (4.102)$$

For this case, the initial and boundary conditions are:

$$\text{at } r = a_i + a_j, \qquad n = 0 \text{ for all } t, \qquad (4.103a)$$

$$\text{at } r > a_i + a_j, \qquad t = 0, n = n_{\infty}. \qquad (4.103b)$$

Let

$$w = \left(\frac{n_\infty - n}{n_\infty}\right)\left(\frac{r}{a_i + a_j}\right) \qquad (4.104)$$

and

$$x = \frac{r - (a_i + a_j)}{a_i + a_j}. \qquad (4.105)$$

Then equation (4.102) can be transformed to

$$\frac{\partial w}{\partial t} = D\frac{\partial^2 w}{\partial x^2} \qquad (4.106)$$

with the boundary conditions

$$\begin{aligned} \text{at } x = 0, \quad & w = 1 \text{ for all } t, \\ \text{at } x > 0, \quad & t = 0, w = 0. \end{aligned} \qquad (4.107)$$

Equation (4.106) with these boundary conditions corresponds to one-dimensional diffusion in a semi-infinite medium for which the solution is

$$w = 1 - \text{erf}\left(\frac{x}{2(Dt)^{1/2}}\right) \qquad (4.108)$$

where erf denotes the error function. As $t \to \infty$, $w \to 1 - \text{erf}(0) = 1$, and

$$\frac{n_\infty - n}{n_\infty} \to \frac{a_i + a_j}{r}, \qquad (4.109)$$

which is the steady-state solution for the concentration distribution, also obtained by setting $\partial n/\partial t = 0$ in equation (4.102) and solving for $n$.

The rate at which particles arrive at the surface ($r = a_i + a_j$) is

$$F(t) = 4\pi D\left(r^2 \frac{\partial n}{\partial r}\right)_{r=a_i+a_j}. \qquad (4.110)$$

Differentiating equation (4.108) gives

$$= 4\pi D(a_i + a_j)n_\infty\left[1 + \frac{a_i + a_j}{(\pi Dt)^{1/2}}\right]. \qquad (4.111)$$

This is the rate per second at which particles of size $a_j$ collide with a fixed particle of size $a_j$. For sufficiently long times ($t \gg (a_i + a_j)^2/D$), this becomes

$$F = 4\pi D(a_i + a_j)n_\infty, \qquad (4.112)$$

which is equivalent to the steady-state solution (equation (4.109)).

This result holds for a fixed central particle. If the central particle is also in Brownian motion, the diffusion constant, $D$, should describe the relative motion of two particles. The relative displacement is given by $x_i - x_j$, where $x_i$ and $x_j$ are the displacements of the two particles in the $x$ direction measured from a given reference plane. The diffusion constant for the relative motion is obtained by generalizing equation (4.12):

$$D_{ij} = \frac{\overline{(x_i - x_j)^2}}{2t} \tag{4.113}$$

$$= \frac{\overline{x_i^2}}{2t} - \frac{\overline{2x_i x_j}}{2t} + \frac{\overline{x_j^2}}{2t} \tag{4.114}$$

$$= D_i + D_j. \tag{4.115}$$

The quantity $\overline{x_i x_j} = 0$ since the motion of the two particles is independent. The collision frequency function is then obtained by substitution in equation (4.112):

$$\beta(v_i, v_j) = 4\pi(D_i + D_j)(a_i + a_j) \tag{4.116}$$

For particles $0.1\,\mu m$ in radius, the characteristic time $(a_i + a_j)^2/(D_i + D_j)$ is about $10^{-3}\,s$, and the use of the steady-state solution is justified in most cases of practical interest. When the Stokes–Einstein relation holds for the diffusion coefficient and $d_p \gg \lambda$, this expression becomes

$$\beta(v_i, v_j) = \frac{2k_B T}{3\eta}\left(\frac{1}{v_i^{1/3}} + \frac{1}{v_j^{1/3}}\right)(v_i^{1/3} + v_j^{1/3}). \tag{4.117}$$

The derivation of equation (4.117) is based on the assumption that the diffusion coefficients of the colliding particles do not change as the particles approach each other. This is not correct because of the increased resistance experienced by a particle as it approaches a surface. The result is that the term $(D_i + D_j)$ tends to decrease as the particles approach each other. This effect is countered in the neighbourhood of the surface because the continuum theory on which it is based breaks down about one mean free path ($\sim 0.1\,\mu m$ at normal temperature and pressure) from the particle surface; in addition, van der Waals forces tend to enhance the collision rate. For further discussion, the reader is referred to Batchelor (1976) and Alam (1987).

For $d_p \ll \lambda$, the collision frequency is obtained from the expression derived in the kinetic theory of gases for collision among molecules that behave as rigid elastic spheres:

$$\beta(v_i, v_j) = \left(\frac{3}{4\pi}\right)^{1/6}\left(\frac{6k_B T}{\rho_p}\right)^{1/2}\left(\frac{1}{v_i} + \frac{1}{v_j}\right)^{1/2}(v_i^{1/3} + v_j^{1/3})^2, \tag{4.118}$$

where $\rho_p$ is the particle density. Fuchs (1964) has proposed a general interpolation formula for $\beta$, which takes into account the transition from the free molecule regime (equation (4.118)) to the continuum range

Table 4.2 Collision frequency function[a] in air at 23°C and 1 atm (based on Fuchs, 1964, p. 294)

| $d_{p1}(\mu m)$ | $10^{10}\beta\,(cm^3\,s^{-1})$ | | |
| | $d_{p2}(\mu m)$ | | |
| | 0.01 | 0.1 | 1 |
| --- | --- | --- | --- |
| 0.01 | 18 | | |
| 0.1 | 240 | 14.4 | |
| 1 | 3200 | 48 | 6.8 |

[a] Sometimes called the coagulation constant, although it is a function of particle size and the properties of the gas.

(equation (4.119)), and this problem is further discussed by Hidy and Brock (1970). Values of $\beta$ as a function of particle size are given in Table 4.2, which shows that the largest values of $\beta$ occur for collisions of particles of very different sizes.

4.12.2.1 *Discrete size distribution for an initially monodisperse aerosol.* A simple solution to the kinetic equation for Brownian coagulation can be obtained for nearly monodisperse systems. Setting $v_i = v_j$ in equation (4.117), the collision frequency function is

$$\beta(v_i = v_j) = \frac{8k_B T}{3\eta} = K \tag{4.119}$$

In this special case of $\beta = K$ independent of particle size, a simple analytical solution can be obtained for the discrete size distribution of an initially monodisperse aerosol. Substituting in equation (4.101) gives

$$\frac{dn_k}{dt} = \frac{K}{2} \sum_{i+j=k} n_i n_j - K n_k \sum_{i=1}^{\infty} n_i. \tag{4.120}$$

Let $\sum_{i=1}^{\infty} n_i = N_\infty$ be the total number of particles per unit volume of fluid. Summing over all values of $k$, the result is

$$\frac{dN_\infty}{dt} = \frac{K}{2} \sum_{k=1}^{\infty} \sum_{i+j=k} n_i n_j - K N_\infty^2. \tag{4.121}$$

It is not difficult to show by expanding the summation that the first term on the right-hand side is $(K/2)N_\infty^2$ so that equation (4.121) becomes

$$\frac{dN_\infty}{dt} = -\frac{K}{2} N_\infty^2. \tag{4.122}$$

Integrating once gives

$$N_\infty = \frac{N_\infty(0)}{1 + (KN_\infty(0)t/2)}, \qquad (4.123)$$

where $N_\infty(0)$ is the total number of particles at $t = 0$. For $k = 1$, substitution in equation (4.120) gives

$$\frac{dn_1}{dt} = -Kn_1N_\infty. \qquad (4.124)$$

Solving,

$$n_1 = \frac{N_\infty(0)}{(1 + t/\tau)^2}, \qquad (4.125)$$

where $\tau = 2/KN_\infty(0) = 3\eta/4k_B TN_\infty(0)$; for $k = 2$,

$$n_2 = \frac{N_\infty(0)t/\tau}{(1 + t/\tau)^3}. \qquad (4.126)$$

In general,

$$n_k = \frac{N_\infty(0)(t/\tau)^{k-1}}{(1 + t/\tau)^{k+1}}, \qquad (4.127)$$

which is the equation for the discrete size distribution, with an initially monodisperse aerosol and a collision frequency function independent of particle size. The variation in $n_k$ with time is shown in Figure 4.11. At any

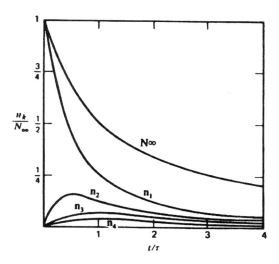

**Fig. 4.11** The variations in $N_\infty, n_1, n_2, \ldots$ with time for an initially monodisperse aerosol. The total number concentration, $N_\infty$, and the concentration of monomer, $n_1$, both decrease monotonically with increasing time. The concentrations of the doublets and larger particles pass through a maximum.

time $t$, the discrete distribution is a monotonically decreasing function of $k$. Since equation (4.127) is based on the assumption that the aerosol is monodisperse through equation (4.119), the analysis would be expected to hold best for small values of $t/\tau$.

Support for the theory of diffusion-controlled coagulation came originally from experiments with polydisperse aerosols (Whytlaw-Gray and Patterson, 1932). The measured coagulation coefficient $K$ was shown to be approximately independent of the chemical nature of the aerosol material with a value close to that predicted theoretically. Tests of the theory have also been conducted by following the coagulation of monodisperse aerosols of dioctylphthalate. Experimentally measured values of the coagulation coefficient were compared with values calculated from theory (Devir, 1963). Taking into account wall losses, good agreement between theory and experiment was obtained. Experiments of this type support the assumption that small, uncharged particles in air adhere when they collide, since the theoretical rate constants are based on this premise.

4.12.2.2  *Continuous distribution function.*  In the previous section, the evolution of the discrete size distribution of an initially monodisperse aerosol was discussed. The analysis holds best for the time period shortly after coagulation begins because of the assumption of monodispersity. In the sections that follow, it is shown that for time periods long after coagulation begins the continuous distribution function approaches an asymptotic form independent of the initial distribution function.

For the continuous distribution function, the collision rate between particles of size $v$ and $\tilde{v}$ is

$$\text{coll. rate} = \beta(v, \tilde{v})n(v)n(\tilde{v})\,dv\,d\tilde{v}, \tag{4.128}$$

where the forms of the collision frequency function discussed in previous sections are applicable. The rate of formation of particles of size $v$ by collision of smaller particles of size $v - \tilde{v}$ and $\tilde{v}$ is then given by

$$\text{formation in range } dv = \frac{1}{2}\left[\int_0^v \beta(\tilde{v}, v - \tilde{v})n(\tilde{v})n(v - \tilde{v})\,d\tilde{v}\right]dv. \tag{4.129}$$

Here we have used the result that the Jacobian for the transformation from the coordinate system $(\tilde{v}, v - \tilde{v})$ to $(\tilde{v}, v)$ is unity. The factor $1/2$ is introduced as in the discrete case because collisions are counted twice in the integral. The rate of loss of particles of size $v$ by collision with all other particles (except monomer) is

$$\text{loss in range } dv = \left[\int_0^\infty \beta(v, \tilde{v})n(v)n(\tilde{v})\,d\tilde{v}\right]dv. \tag{4.130}$$

The net rate of formation of particles of size $v$ is

$$\frac{\partial(n \, dv)}{\partial t} = \frac{1}{2} \left[ \int_0^v \beta(\tilde{v}, v - \tilde{v}) n(\tilde{v}) n(v - \tilde{v}) \, d\tilde{v} \right] dv$$
$$- \left[ \int_0^\infty \beta(v, \tilde{v}) n(\tilde{v}) n(v) \, d\tilde{v} \right] dv \qquad (4.131)$$

Dividing through by $dv$, the result is

$$\frac{\partial n}{\partial t} = \frac{1}{2} \int_0^v \beta(\tilde{v}, v - \tilde{v}) n(\tilde{v}) n(v - \tilde{v}) \, d\tilde{v}.$$
$$- \int_0^\infty \beta(v, \tilde{v}) n(\tilde{v}) n(v) \, d\tilde{v}, \qquad (4.132)$$

which is the equation of coagulation for the continuous distribution function. Methods for solving this equation are reviewed by Williams and Loyalka (1991). Analytical solutions have not been obtained for most collision frequency functions of physical interest. Numerical methods have been developed by Gelbard and Seinfeld (1978) and Landgrebe and Pratsinis (1990).

Solutions to equation (4.132) are subject to two important physical interpretations. They represent the change with time of the aerosol in a box in the absence of convection or deposition on the walls. Alternatively, they can be interpreted as the steady-state solution for an aerosol in steady 'plug' flow through a duct. In this case $\partial n / \partial t = U(\partial n / \partial x)$, where $U$ is the uniform velocity in the duct and $x$ is the distance in the direction of flow.

### 4.13 The self-preserving size distribution

A method of solving certain coagulation and condensation problems has been developed based on the use of a similarity transformation for the size distribution function (Swift and Friedlander, 1964; Friedlander and Wang, 1966). Solutions found in this way are asymptotic forms approached after long times, and are independent of the initial size distribution. Closed-form solutions for the upper and lower ends of the distribution can sometimes be obtained in this way, and numerical methods can be used to match the solutions for intermediate size particles.

The similarity transformation for the particle-size distribution is based on the assumption that the fraction of the particles in a given size range is a function only of particle volume normalized by the average particle volume:

$$\frac{n \, dv}{N_\infty} = \psi\left(\frac{v}{\bar{v}}\right) d\left(\frac{v}{\bar{v}}\right), \qquad (4.133)$$

where $\bar{v} = V/N_\infty$, the average particle volume, is a function of time and/or position. Both sides of equation (4.133) are dimensionless. Rearranging, the result is

$$n(v, t) = \frac{N_\infty^2}{V} \psi(\eta_v) \tag{4.134}$$

where $\eta_v = v/\bar{v} = N_\infty v/V$. There are also the integral relations:

$$N_\infty = \int_0^\infty n \, dv \tag{4.135}$$

and

$$V = \int_0^\infty nv \, dv. \tag{4.136}$$

It is also usually required that $n(v) \to 0$ for $v \to 0$ and $v \to \infty$. Both $N_\infty$ and $V$ are, in general, functions of time. In the simplest case, no material is added or lost from the system, and $V$ is constant. The number concentration $N_\infty$ decreases as coagulation takes place. If the size distribution corresponding to any value of $N_\infty$ and $V$ is known, the distribution for any other value of $N_\infty$ corresponding to a different time can be determined from equation (4.136) if $\psi(\eta_v)$ is known. The shapes of the distribution at different times are similar when reduced by a scale factor. For this reason, the distribution is said to be 'self-preserving'.

The determination of the form of $\psi$ is carried out in two steps. First, the special form of the distribution function is tested by substitution in the equation of coagulation for the continuous distribution function with the appropriate collision frequency function. If the transformation is consistent with the equation, the partial integrodifferential equation (4.132) is reduced to an ordinary integrodifferential equation for $\psi$ as a function of $\eta_v$. The next step is to find a solution of this equation subject to the integral constraints (4.135) and (4.136), and the limits on $n(v)$. Solutions have been given by Friedlander and Wang (1966) and more recently by Vemury et al. (1994). The results of the numerical solution are shown in Figure 4.12. For some collision kernels, it may not be possible to find solutions for $\psi(\eta_v)$ that satisfy the usual boundary conditions.

The change in the particle size distribution function with time for coagulating cigarette smoke was measured by Keith and Derrick (1960). Smoke issuing from a cigarette was rapidly mixed with clean air, and the mixture introduced into a 12 litre flask where coagulation took place. The dilution ratio was 294 volumes of air to 1 volume of raw smoke.

To follow the coagulation process, samples of the smoke were taken at intervals over a period of 4 min from the flask and passed into a conifuge, a centrifugal aerosol collector and classifier. Size distribution curves were measured, together with values for the total number of particles per unit volume, obtained by the graphical integration of the size distribution curves. The volume fraction of aerosol material was $V = 1.11 \times 10^{-7}$,

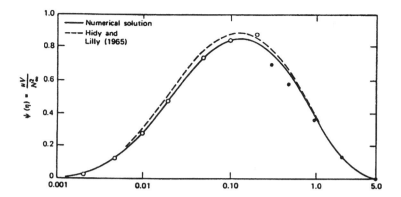

**Fig. 4.12** Self-preserving particle size distribution for Brownian coagulation. The form is approximately lognormal. The result obtained by solution of the ordinary integrodifferential equation for the continuous spectrum is compared with the limiting solution of Hidy and Lilly (1965) for the discrete spectrum, calculated from the discrete form of the coagulation equation. Shown also are points calculated from analytical solutions for the lower and upper ends of the distribution (Friedlander and Wang, 1966). See also Vemury *et al.* (1994).

based on measured values for the mass concentrations of the tobacco smoke. This value was checked by numerical integration of the size distribution corresponding to a nominal ageing time of 30 s.

In Figure 4.13, experimental data for the change in the distribution function with time are compared with calculations based on self-preserving size distribution theory. Agreement is fair; the experimental results fall significantly higher than theory at the upper end of the distribution (large particle sizes).

The time to reach the self-preserving form depends on the shape of the initial distribution – the closer the initial distribution to the asymptotic form, the faster the approach. For initially monodisperse aerosols, Vemury *et al.* (1994) found that the time lag to reach the self-preserving distribution was

$$\tau_{\text{SP}} = 5\left\{ \left(\frac{3}{4\pi}\right)^{1/6} \left(\frac{6k_{\text{B}}T}{\rho_{\text{p}}}\right)^{1/2} v_0^{1/6} N_0 \right\}^{-1} \tag{4.137}$$

for the free molecule regime and

$$\tau_{\text{SP}} = 13\left[\frac{2k_{\text{B}}T N_0}{3\eta}\right]^{-1} \tag{4.138}$$

for the continuum regime. The criterion for attainment of the self-preserving form was that the geometric standard deviation (GSD) of the distribution function should be within ±1% of the asymptotic (self-

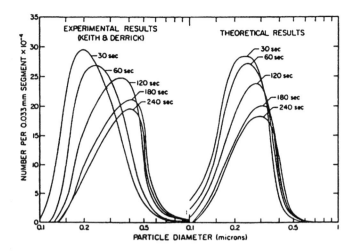

**Fig. 4.13** Experimental measurements and self-preserving size distribution theory compared for an ageing tobacco smoke aerosol. Calculation based on $V = 1.11 \times 10^{-7}$ and experimental values of $N_\infty$ (Friedlander and Hidy, 1969).

preserving) GSD. The self-preserving distribution can be approximated by a log-normal distribution function with GSDs of about 1.44 and 1.46 for the continuum and free molecule regimes, respectively.

When the initial size distribution is lognormal, the time to reach the self-preserving distribution depends strongly on how far the initial GSD is from the value for the asymptotic form. In practical applications, each parcel of gas may have a different time/temperature history. At high particle concentrations, the size distribution in any gas parcel may be self-preserving but different from distributions in the other gas parcels. That is, the distribution may be locally self-preserving. Sampling of many gas parcels or collection of the particles in the entire gas stream will produce a composite of locally self-preserving distributions. The spread of the composite distribution will be larger than that of any individual self-preserving distribution, which represents a minimum. Such distributions can also be multimodal if there are several significantly different residence times. Of course, if many self-preserving distributions are completely mixed and aged, a new self-preserving distribution results.

Finally, we note that for aerosols composed of fractal-like agglomerates, there is a considerable enhancement in the rate of coagulation, compared with spherical particles of the same mass. This enhancement is clearly demonstrated in calculations of the corresponding self-preserving distributions (Wu and Friedlander, 1993). The results of many simulations of the agglomeration of fractal-like particles are given in the collection of papers edited by Family and Landau (1984).

## Acknowledgement

The work on this chapter was supported in part by US NSF grant CTS-9527999.

## References

Alam M.K. (1987) The effect of van der Waals and viscous forces on aerosol coagulation. *Aerosol Sci. Technol.*, **6**, 41–52.

Batchelor G.K. (1976) Brownian diffusion of particles with hydrodynamic interaction. *J. Fluid Mech.*, **74**, 1–29.

Carslaw H.S. and Jaeger J.C. (1959) *Conduction of Heat in Solids*. Oxford University Press, London.

Chamberlain A.C. (1966) Transport of lycopodium spores and other small particles to rough surfaces. *Proc. Roy. Soc. A*, **296**, 45–70.

Chandrasekhar S. (1943) Stochastic problems in physics and astronomy, *Reviews of Modern Physics*, **15**, 1. (Reprinted in N. Wax (ed.) (1954) *Noise and Stochastic Processes*. Dover, New York.)

Chen C.Y. (1954) Filtration of aerosols by fibrous media. *Annual Report Eng. Exp. Station, University of Illinois*, 30 January.

Chen C.Y. (1955) Filtration of aerosols by fibrous media. *Chem Rev.*, **55**, 595–623.

Cheng Y.S. (1993) Condensation detection and diffusion size separation techniques, in *Aerosol Measurement* (eds K. Willeke and P.A. Baron). Van Nostrand Reinhold, New York.

Crank J. (1975) *The Mathematics of Diffusion*. Oxford University Press, Oxford.

Davies C.N. (1945) Definitive equations for the fluid resistance of spheres. *Proc. Phys. Soc.*, **57**, 259–270.

Devir S.F. (1963) Coagulation of aerosols. *J. Colloid Sci.*, **18**, 744–756.

Dobry R. and Finn R.K. (1956) Mass transfer to a cylinder at low Reynolds numbers. *Ind. Eng. Chem.*, **48**, 1540–1543.

Epstein P.S. (1924) Resistance experienced by spheres in their motion through gases. *Phys. Rev.*, **23**, 710–733.

Family F. and Landau D.P. (1984) *Kinetics of Aggregation and Gelation*. Elsevier, New York.

Fernandez de la Mora and Friedlander S.K. (1982) Aerosol and gas deposition to fully rough surfaces – filtration model for blade-shaped elements. *Int. J. Heat Mass Transfer*, **2S**, 1725–1735.

Friedlander S.K. (1967) Particle diffusion in low speed flows. *J. Colloid Interface Sci.*, **23**, 157–164.

Friedlander S.K. (1977) *Smoke, Dust and Haze*, Wiley-Interscience, New York.

Friedlander S.K. and Hidy G.M. (1969) New concepts in aerosol size spectrum theory, in *Proceedings of the 7th International Conference on Condensation and Ice Nuclei* (ed. J. Podzimek). Academia, Prague.

Friedlander S.K. and Wang C.S. (1966) Self-preserving particle size distribution for coagulation by Brownian motion. *J. Colloid Interface Sci.*, **22**, 126–132.

Fuchs N.A. (1964) *The Mechanics of Aerosols*. Pergamon Press, New York.

Gelbard F.M. and Seinfeld J.H. (1978) Numerical solution of the dynamic equation for particulate systems. *J. Comput. Phys.*, **28**, 357–376.

Hidy G.M. and Brock J.R. (1970) *Dynamics of Aerocolloidal Systems*. Pergamon, New York.

Hidy G.M. and Lilly D.K. (1965) Solutions to the equations for the kinetics of coagulation. *J. Colloid Sci.*, **20**, 867–874.

Keith C.H. and Derrick J.E. (1960) Measurement of particle size distribution and concentration of cigarette smoke by the conifuge. *J. Colloid Sci.*, **15**, 340–356.

Koch W., Windt H. and Karfich N. (1993) Modeling and experimental evaluation of an aerosol generator for very high number currents based on a free turbulent jet. *J. Aerosol Sci.*, **24**, 909–918.

Landau L.D. and Lifshitz E.M. (1987) *Fluid Mechanics*, 2nd edn. Pergamon Press, Oxford.

Landgrebe J.L. and Pratsinis S.E. (1990) A discrete-sectional model for particulate production by gas-phase chemical-reaction and aerosol coagulation in the free-molecular regime. *J. Colloid Interface Sci.*, **139**, 63–86.

Lee K.W. and Liu B.Y.H. (1982a) Experimental study of aerosol filtration by fibrous filters. *Aerosol Sci. Technol.*, **1**, 35–46.

Lee K.W. and Liu B.Y.H. (1982b) Theoretical study of aerosol filtration by fibrous filters. *Aerosol Sci. Technol.*, **1**, 147–161.

Levich V.G. (1962) *Physicochemical Hydrodynamics*. Prentice Hall, Englewood Cliffs, N.J.

Lin C.S., Moulton R.W. and Putnam G.L. (1953) Mass transfer between solid wall and fluid streams. *Ind. Eng. Chem.*, **45**, 636–646.

Monin A.S. and Yaglom A.M. (1971) *Statistical Fluid Mechanics: Mechanics of Turbulence*, Vol. 1. MIT Press, Cambridge, MA.

Natanson G.L. (1957) Deposition of aerosol particles by electrostatic attraction upon a cylinder around which they are flowing. *Dokl Akad. Nauk SSSR*, **112**, 696–699.

Parnas R. and Friedlander S.K. (1984) Particle deposition by diffusion and interception from boundary-layer flows. *Aerosol Sci. Technol.*, **3**, 3–8.

Perrin F. (1936) Brownian motion of an ellipsoid. Part II Free rotation and depolarisation of fluorescence: translation and diffusion of ellipsoidal molecules. *J. Phys. Radium*, **7**, 1–11.

Rosenhead L. (ed.) (1963) *Laminar Boundary Layers*. Oxford University Press, Oxford.

Schack C.J., Pratsinis S.E. and Friedlander S.K. (1985) A general correlation for deposition of suspended particles from turbulent gases to completely rough surfaces. *Atmos. Environ.*, **19**, 953–960.

Schlichting H. (1979) *Boundary Layer Theory*, 7th edn. McGraw-Hill, New York.

Smoluchowski M. (1917) Mathematical theory of the kinetics of coagulation of colloidal solutions. *Z. Physik Chem.*, **92**, 129–168.

Spurny K.R., Lodge J.P., Frank E.R. and Sheesley D.C. (1969) Aerosol filtration by means of nuclepore filters: structural and filtration properties. *Environ. Sci. Tech.*, **3**, 453–464.

Swift D.L. and Friedlander S.K. (1964) The coagulation of hydrosols by Brownian motion. *J. Colloid Sci.*, **19**, 621–647.

Vemury S., Kusters K.A. and Pratsinis S. (1994) Time-lag for attainment of the self-preserving particle size distribution by coagulation. *J. Colloid Interface Sci.*, **165**, 53–59.

Whytlaw-Gray R. and Patterson H.S. (1932) *Smoke: A Study of Aerial Disperse Systems*. Edward Arnold, London.

Williams M.M.R. and Loyalka S.K. (1991) *Aerosol Science Theory and Practice: With Special Applications to the Nuclear Industry*. Pergamon Press, Oxford.

Wong J.B. and Johnstone H.F. (1953) *Collection of Aerosols by Fiber Mats*, Technical Report 11. Eng. Exp. Station, University of Illinois.

Wong J.B., Ranz W.E. and Johnstone H.F. (1956) Collection efficiency of aerosol particles and resistance to flow through fiber mats. *J. Appl. Phys.*, **27**, 161–169.

Wu M.K. and Friedlander S.K. (1993) Enhanced power-law agglomerate growth in the free-molecule regime. *J. Aerosol Sci.*, **24**, 273–282.

## Nomenclature

| | |
|---|---|
| $a$ | radius of cylinder |
| $a_p$ | radius of particle |
| $C_p$ | heat capacity |
| $D$ | particle diffusion coefficient |
| $d_p$ | particle diameter |

| | |
|---|---|
| $f$ | friction coefficient |
| $F(t)$ | force resulting from thermal motion of molecules |
| $J$ | diffusion flux |
| $K$ | coagulation coefficient |
| $k$ | local mass transfer coefficient (deposition velocity) |
| $k_{av}$ | average mass transfer coefficient |
| $k_B$ | Boltzmann constant |
| $L$ | characteristic length |
| $m$ | molecular mass, mass of particle |
| $n$ | number concentration |
| $N_{i,j}$ | number of collisions per unit time per unit volume |
| $Pe$ | Peclet number |
| $R$ | interception parameter |
| $Re$ | Reynolds number |
| $Sc$ | Schmidt number |
| $Stk$ | Stokes number |
| $t$ | time |
| $U$ | velocity |
| $v_\theta$ | tangential component of velocity |
| $v_i, v_j$ | volume of particles $i$ and $j$ |
| $v_r$ | radial component of velocity |
| $\bar{x}^2$ | mean square displacement |
| | |
| $\alpha$ | accommodation coefficient |
| $\beta(v_i, v_j)$ | collision frequency |
| $\delta_c$ | effective boundary layer thickness |
| $\varepsilon$ | eddy diffusion coefficient |
| $\eta$ | viscosity |
| $\eta_R$ | removal efficiency |
| $\kappa$ | thermal conductivity |
| $\lambda$ | mean free path of gas molecules |
| $v$ | kinematic viscosity |
| $\rho$ | density |
| $\tau_{SP}$ | time to reach self-preserving distribution |
| $\psi$ | stream function |

# 5 Electrical and thermodynamic properties

C.F. CLEMENT

## 5.1 Introduction

Aerosol particles or droplets are surrounded by the gaseous medium which supports them, and the subject of this chapter is the relationship between the two phases. Their relation is governed by kinetics as well as thermodynamics. Thermodynamics, in differences between temperatures and chemical potentials of the phases, drives heat and mass exchange between the particles and the gas, but the rates of exchange are controlled by kinetics. In contrast to the exchange between two bulk phases with a single interface, no true thermodynamic equilibrium may be reached because the particles and droplets have different sizes. For an uncharged aerosol, the most thermodynamically stable state is a single giant particle or droplet. This means that observed aerosols will, as a result of exchanges, generally only reach quasi-equilibria which then slowly change with time.

Electrical properties of aerosols arise from their charging and the exchange of charge with the surrounding gas. At normal temperatures gases remain neutral in equilibrium because their energy is much too low for ionic dissociation to occur. The principal mechanisms for the appearance of charges and charge carriers in the atmosphere are non-equilibrium, high-energy ion-formation processes induced by radioactivity and cosmic rays. These charges are collected by aerosol particles, and a steady-state charge distribution will generally result on an aerosol from the competing kinetics of ionic charging by positive and negative charge addition to the aerosol. This is not a thermodynamic equilibrium, as the reverse process in which an aerosol particle spontaneously emits a charge does not occur. Other aerosol charging mechanisms include contact charging resulting from collisions with surfaces and photoelectric charging resulting from the emission of electrons following the absorption of high-energy electromagnetic radiation. Steady-state ion kinetics generally governs aerosol charge distributions and is described in section 5.3, together with some electrical effects of charging.

Almost exclusively, current descriptions of aerosols refer to a dilute or very dilute aerosol phase in a gas. This certainly applies to theoretical treatments of exchange processes which treat individual particles interacting with an average gaseous medium, and which neglect correlations between the particles. We adopt this dilute description here, although the topic is slightly elaborated upon in section 5.2. As the aerosol concentration increases, the aerosol–gas mixture will transform into a two-fluid mixture which will have very different properties. Two-phase flows are important in chemical engineering, and their behaviour and description are complex (Butterworth and Hewitt, 1977). An exception to particles behaving independently in a dilute aerosol in a gas arises if there is a net space charge on the aerosol. The long-range electrostatic forces between particles then result in a repulsive electric field which acts to cause the expansion of the particle cloud in the gas.

A general description of the aerosol and gas phases is given in section 5.2, including the distinction between continuum and molecular treatments of the media. Because of the small sizes of aerosol particles, molecular descriptions are often needed to describe exchanges with the medium. In section 5.3 we describe the acquisition of charge by aerosol from ions in gas and related topics.

Growth and evaporation of aerosols occur when a condensable vapour in the medium gets out of thermodynamic equilibrium. This generally implies both heat and mass transfer between the aerosol and a surrounding vapour–gas mixture, and this subject is described in section 5.4, including a description of droplets containing soluble species. Condensation is an important process for forming aerosols, and the initial stage of nucleation, in which viable growing nuclei are formed, is described in section 5.5. A short description is included of an experimental technique which provides some of the best information on aerosol nucleation and growth rates. Finally, in section 5.6, the importance of chemical reactions involving aerosol is discussed, together with concepts necessary to extend the theory of condensation given here to surface reactions with gas molecules.

## 5.2  Aerosol–vapour–gas mixtures

The most common examples of aerosols occur in the atmosphere; this may be regarded as a mixture of gas molecules constituting dry air, water vapour molecules, and aerosol particles and droplets. While a gas or vapour has identical molecular constituents, this is never true of an aerosol, and we allow for the differences here by assuming a range of spherical particles whose number concentration (number per unit volume) at a point $\mathbf{r}$ and whose radius lies between $R$ and $R + \mathrm{d}R$ is $n(\mathbf{r}, R)\,\mathrm{d}R$. Important physical properties of the aerosol are then obtained by summing or averaging over this size distribution in terms of its moments:

$$M_n(\mathbf{r}) = \int R^n n(\mathbf{r}, R)\,\mathrm{d}R. \qquad (5.1)$$

The most important moments are: *total number concentration,*

$$N_a(\mathbf{r}) = M_0; \tag{5.2}$$

*aerosol surface concentration,*

$$A(\mathbf{r}) = 4\pi M_2; \tag{5.3}$$

and *aerosol mass concentration,*

$$\rho_a(\mathbf{r}) = (4\pi/3)\rho_p M_3. \tag{5.4}$$

For aerosols consisting of other simple basic shapes, such as rods or ellipsoids, these moments are easy to modify, and are still integral moments, but for fractal aerosols the moments would generally be non-integral in terms of a 'radius' for the aerosol.

For a vapour–gas–aerosol mixture, basic thermodynamic quantities will be expressed in terms of the moments. For a gas, the pressure is given by $p = N k_B T$, where $N$ is the number concentration, $T$ the temperature, and $k_B$ is Boltzmann's constant. Aerosol particles do not exert the same pressure as gas molecules on walls as they will often stick to them rather than bounce off them, altering the momentum transfer. However, since $N_a$ will always be many orders of magnitude less than gas number concentrations, any contribution of an aerosol to the pressure can always be neglected. We are concerned with dilute aerosols, which means that the aerosol volume fraction, $M_3 \ll 1$, but we cannot always neglect the contribution of the aerosol to the total mass concentration or density of the mixture,

$$\rho = \rho_g + \rho_v + \rho_a, \tag{5.5}$$

where $\rho_g$ and $\rho_v$ are the gas and vapour densities, respectively. Since $\rho_g$ is typically about $1\,\mathrm{kg\,m^{-3}}$ and $\rho_p$ is of the order of $10^3\,\mathrm{kg\,m^{-3}}$, $\rho_g$ and $\rho_a$ are comparable for $M_3 = 10^{-3}$. Aerosol mass concentrations are commonly greater than $1\,\mathrm{g\,m^{-3}}$, for which $M_3$ is about $10^{-6}$. This means that aerosols, often in combination with condensible vapours, are capable of significantly changing the density and convective properties of gases, a subject which has been considered in relation to severe accidents (Clement, 1988). Two lethal mixtures denser than air occur naturally which roll down mountains destroying everything in their paths: powder snow avalanches, where tiny snow particles become airborne; and a glowing cloud of incandescent ash in superheated gas from volcanoes, such as was emitted in the Mt Pelée eruption in 1902.

The internal energy of the mixture is also additive in its constituents. In particular, this applies to the concentration of total enthalpy or heat content,

$$\rho h = \rho_g h_g + \rho_v h_v + \rho_a h_a, \tag{5.6}$$

where $h$ is the enthalpy per unit mass and specific heats at constant pressure are

$$(\partial h/\partial T)_p = C_p. \tag{5.7}$$

Values of $C_p$ are comparable for gases and condensed matter, so that if $\rho_a \ll \rho_g$, the heat content of the aerosol can be neglected in (5.6) when considering temperature changes. Neglect of the aerosol does not apply when there are phase changes such as condensation from vapour into aerosol, when the latent heat is given by

$$L(T) = h_v(T) - h_a(T). \tag{5.8}$$

When an amount, $\Delta\rho_v$, of the vapour condenses into aerosol, the temperature rise $\Delta T$ is given by energy conservation as

$$\rho C_p \Delta T = L\Delta\rho_v. \tag{5.9}$$

As the vapour density moves along its equilibrium value, $\rho_{ve}(T)$, the importance of the heat release is specified by the dimensionless ratio

$$\Delta_b = \rho C_p/(L d\rho_{ve}(T)/dT). \tag{5.10}$$

When $\Delta_b$ is small, the rise in temperature from latent heat release limits condensation into aerosol. For large $\Delta_b$, the heat capacity of the medium is relatively large, and most of the vapour can easily be condensed into aerosol by lowering the temperature. For water vapour in air, the transition temperature at which $\Delta_b = 1$ is about 4°C, and it is no coincidence that water mists are easily formed by heat transfer at this temperature. A fuller discussion of the thermodynamics and kinetics of aerosol formation at constant pressure was given by Clement (1985).

## 5.2.1 Concentrations and uniformity

It is important to discuss briefly the spatial uniformity of aerosols in relation to that of the gaseous medium. There are fundamental differences in their properties stemming from differences in diffusion. For gases, diffusivities are about $10^{-5}\,\mathrm{m^2\,s^{-1}}$ at normal temperature and pressure, whereas for a $1\,\mu m$ aerosol particle the value is $10^6$ times smaller. Although the aerosol diffusivity is inversely proportional to the radius, which makes the factor $10^4$ at $0.01\,\mu m$, most aerosols will diffuse very much more slowly than gases, and as a consequence will not have uniform spatial concentrations. The non-uniformity can persist for long times and large distances and is readily observable in plumes and clouds. Fluctuations in aerosol concentrations are thus an important topic. Random fluctuations in thermodynamic quantities are formally related to dissipative processes such as diffusion by the **fluctuation–dissipation**

**theorem** (Landau and Lifshitz, 1980), but, because the diffusivity is so small, this is practically irrelevant for aerosols whose fluctuations reflect on their sources and physical instabilities, such as convection, which occur in the atmosphere.

Aerosol mixing by turbulence is commonly described by turbulent diffusion which introduces a term, $-D_T \nabla^2 n$, where $D_T$ is the turbulent diffusion coefficient, into the evolution equation for the size distribution in space, but it is important to realize that this is an approximation which has two limitations. First, mixing is only efficient down to the smallest-scale length of the turbulence, which may not be small (metres) in the atmosphere, so that large fluctuations exist in concentration measurements over smaller lengths. Second, turbulence, unlike diffusion, does not always act to transfer concentrations from regions where it is high to regions where it is low. Because of the inertia of aerosol particles, they can be ejected from vortices, and also from regions of high turbulence into regions of low turbulence.

### 5.2.2  Molecular and continuum descriptions

For aerosol particles of small enough size, a continuum description of the surrounding gas is not sufficient and must be replaced by a molecular description. A measure for the appropriate description is given by the *Knudsen number*,

$$Kn = \lambda_m / R, \tag{5.11}$$

where $\lambda_m$ is the molecular mean free path in the gas, which may be expressed in terms of the molecular mass, $\mu_v$, and diffusivity, $D$, by $\lambda_m = 2D/(\mu_v/2R_G T)^{1/2}$, where $R_G$ is the gas constant.

At normal temperature and pressure, the mean free path is about $0.1\,\mu m$, so that usually molecular effects are small for $R \gg 0.1\,\mu m$ where $Kn \ll 1$, whereas they dominate for $R \ll 0.1\,\mu m$ where $Kn \gg 1$. In the **continuum regime** ($Kn \ll 1$) and in the **kinetic regime** ($Kn \gg 1$), the interactions and exchanges between an aerosol particle and the surrounding gas are easier to understand and calculate than in the **transition regime** where $Kn = O(1)$. In the simpler regimes we can use continuum theories and associated empirical constants, e.g. vapour diffusivity in the gas, or molecular gas theory based on the Boltzmann equation (Chapman and Cowling, 1970) to describe the problem. While solutions of the Maxwell–Boltzmann equation in principle describe the transition and continuum regimes, they are often very complicated and the problem arises of reconciling basic molecular collision integrals with the empirical continuum constants.

### 5.2.3 *Exchange processes*

In this chapter we are concerned with the exchange of mass and heat between an aerosol particle or droplet and the surrounding gas. Such exchanges occur in condensation on and evaporation from aerosols, and also in chemical reactions between aerosol and the surrounding gaseous medium. The basic exchange processes which occur specifically for a droplet growing by vapour condensation are illustrated in Figure 5.1, and would be identical for a gaseous constituent reacting chemically with molecules on the droplet surface. In both cases any heat released, the latent heat (5.8) in the case of condensation, must be transferred outwards to the gas or by radiation because the droplet has a negligible heat capacity. The physical processes which govern the exchange rates shown in Figure 5.1 and the parameters specifying their magnitudes are given in Table 5.1. We now discuss some of their general properties: the resulting aerosol growth rates are given in section 5.4.3.

First, the heat and mass transfer processes occur in series, as do the two pairs of continuum and kinetic processes, MD and MK, HC and HK.

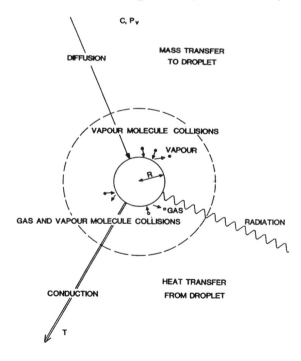

**Fig. 5.1** Local heat and mass transfer processes to and from a surrounding gas at temperature $T$ and vapour mass concentration $c$ at vapour pressure $P_v$ for a droplet growing by vapour condensation.

**Table 5.1** Physical exchange processes between an aerosol droplet and the medium

| Transfer process | Magnitude parameter | |
|---|---|---|
| Mass transfer MD | Diffusion through gas layer | $D$ (vapour–gas diffusivity) |
| Mass transfer MK | Molecular surface collision with capture or surface reaction | $S_p$ (sticking probability) or $\sigma_c$ (condensation coefficient) $\gamma_r$ (reaction coefficient) |
| Heat transfer HC | Heat conduction through layer | $k$ (thermal conductivity) |
| Heat transfer HK | Molecular surface collision with rebound | $\alpha$ (thermal accommodation coefficient) |
| Radiative heat transfer RAD | Radiation from droplet surface | $\varepsilon$ (surface emissivity) |

Radiative heat transfer occurs in parallel with the two conductive heat transfer processes.

Second, at normal temperatures, radiative transfer is often (but not always) negligible in comparison to conductive heat transfer. However, as radiative emission across a large temperature drop rises as $T^4$, radiative heat transfer is important for high-temperature aerosols.

Third, neglecting radiative heat transfer, droplet growth or evaporation is controlled by the slowest of the four processes, MD, MK, HC and HK. For $Kn \gg 1$ we expect this to be MK or HK, and for $Kn \ll 1$ MD or HC, but the actual slowest process is strongly influenced by the sizes of the kinetic constants. The relative magnitudes of the heat and mass transfer processes depend on the amount of heat removal necessary, and thus on the latent heat $L$. Their relative importance can be expressed in terms of a dimensionless **surface condensation number** (Clement, 1985) which in the continuum region is

$$Cn_s = \text{rate of latent heat removal by conduction/rate of mass transport}$$
$$= k(1 - c_e(T))/[LD_v\rho \, dc_e(T)/dT],$$

$$(5.12)$$

where $k$ is the thermal conductivity, $D_v$ the vapour diffusivity, and $c_e(T) = \rho_{ve}(T)/\rho$ is the equilibrium vapour concentration. An analogous quantity, $Cn_k$, can be defined in the kinetic region (Barrett and Clement, 1988), where it is inversely proportional to the vapour molecule sticking probability, $S_p$. Two regions for the condensation and evaporation of aerosols can always be defined:

$$Cn_s \gg 1, \quad \text{aerosol growth rate controlled by mass transfer}$$
$$Cn_s \ll 1, \quad \text{aerosol growth rate controlled by heat transfer.}$$
$$(5.13)$$

For water droplets of continuum sizes ($> 1\ \mu m$) at atmospheric pressure, $Cn_s = 1$ at about 4°C and, if $S_p = 1$, the analogous kinetic $Cn_k = 1$ at about 1°C in the kinetic regime. As both values of $Cn$ rapidly decrease as $T$ increases, this means that heat transfer controls the rate of water mist and cloud growth at moderate and high temperatures.

Finally, the continuum rate constants, $k$ and $D$, in Table 5.1 can be measured or estimated fairly accurately (less than 10% error) for most molecules, but there has been considerable uncertainty over the values of the kinetic constants, $S_p$ and $\sigma_c$, although values of $\alpha$ for gases except the lightest (e.g. He) are believed to be close to unity. Values of $S_p$ for important atmospheric constituents such as water and sulphuric acid have often been quoted as low as 0.02 (Pruppacher and Klett, 1978; Van Dingenen and Raes, 1991). However, direct measurements of aerosol growth of water, $n$-propanol and binary mixtures (Wagner, 1982; Rudolph et al., 1991) using accurate techniques (see section 5.5.3) have found that $S_p$ cannot be much less than unity. Combining these results with direct experimental evidence from surface science and theoretical considerations, Clement et al. (1996) have deduced that $S_p$ should generally be close to unity at normal temperatures. In many experiments where low values were found, condensation or evaporation was clearly limited by heat transfer not considered in their interpretation. There is good evidence that surface layers of molecules different from the condensing or evaporating species (e.g. hydrophobic molecules in the case of water) can form on droplet surfaces which greatly inhibit the condensing species entering the bulk. Thus barriers to aerosol condensation and evaporation can arise additional to those given in Table 5.1. Mass transfer across such barriers, as well as those across non-uniform layers relevant to growth of multi-component droplets or particles, must be considered separately from basic condensation or evaporation of a single substance. Extra barriers are more likely to arise in evaporating droplets than in condensing droplets, but it is unlikely that they should exist in all droplets throughout the atmosphere.

## 5.3   Aerosol charging and electrical effects

The charging of aerosol is a widespread phenomenon even though temperatures are far too low in the atmosphere for spontaneous ion emission to take place. Aerosol behaviour is affected by charging once it exceeds a certain level, and we describe some of these effects in section 5.3.7. The deliberate charging of aerosols to make use of these effects is also possible, and is used in some of the applications summarized in section 5.3.8.

Charging takes place naturally by the capture of ions by aerosol particles or droplets. Radioactive aerosol, considered in section 5.3.5, is different in that particles can charge themselves by the emission of high-energy elementary particles (mainly electrons) in the radioactive decay process. A charging mechanism particularly suited to ultrafine particles is photoelectric emission, and this can be used for material analysis (see, for example, Fendel and Schmidt-Ott, 1993), but it will not be described here. In the atmosphere in thunderstorms large charge transfer processes take place between ice particles. For a discussion of this subject, and of how cloud microphysics is related to electricity, see Pruppacher and Klett (1978).

To describe an aerosol containing particles with $je$ charges, we introduce concentrations $n_j(\mathbf{r}, R)$, which sum to total concentrations $N_j(\mathbf{r})$, so that the total concentration is now

$$N_a(\mathbf{r}) = \sum_j N_j(\mathbf{r}) = \sum_j \int n_j(\mathbf{r}, R)\, dR. \tag{5.14}$$

Aerosols will then have a distribution both in size and charge, which extends to negative charges, but, where size is irrelevant, $N_j(\mathbf{r})$ alone will be used to denote the size distribution. We begin, in section 5.3.1, by briefly describing the ions responsible for charging, and then, in sections 5.3.2 and 5.3.3 the processes of diffusion charging and field charging, respectively. The limits to charging are given in section 5.3.4. Following a discussion of radioactive aerosols in section 5.3.5, we briefly examine the subject of time-scales and aerosol neutralization in section 5.3.6. Finally, as mentioned above, the topics of electrical effects on aerosols and their applications are covered in sections 5.3.7 and 5.3.8.

### 5.3.1   Ions and mobilities

Charged ions are introduced into the atmosphere by high-energy non-equilibrium processes. High-energy cosmic rays from space, secondary particles produced by their collisions, and high energy $\alpha$ and $\beta$ particles from radioactive decays by nuclei in the atmosphere and on the ground, collide with atmospheric molecules and generally knock out electrons leaving positive ions behind. The electrons rapidly attach to molecules, usually $O_2$, and both types of ion transform into relatively stable *small ions*, which consist of a singly charged ion surrounded by a cluster of water molecules (Mohnen, 1974). Typical clusters are $H_3O^+(H_2O)_n$, $H^+(H_2O)_n$ and $O_2^-(H_2O)_n$ and have lifetimes of the order of $100\,s$. Apart from radioactivity itself, these small ions are the source of charge for aerosol particles to which they can become attached. They also recombine with each other, often in the presence of a third body. In the Earth's electric field, or any other field, they move with a velocity, $\mathbf{v} = \mu\mathbf{E}$, where $\mu$ is the *electrical mobility* which is related to the ion's diffusivity, $D$, by Einstein's relation,

**Table 5.2** Ion parameters in air at 1 atm pressure and $T = 20°C$

| Ion property | Origin | Value |
|---|---|---|
| Production rate, $q$ | Soil and air radioactivity and cosmic rays | $10^7 \, \text{m}^{-3} \, \text{s}^{-1}$ |
| Ion recombination rate, $\alpha$ | Positive and negative ion collision | $1.6 \times 10^{-12} \, \text{m}^3 \, \text{s}^{-1}$ |
| Ion concentrations, $n_+, n_-$ | Maximum from recombination, $(q/\alpha)^{1/2}$ | $2.5 \times 10^9 \, \text{m}^{-3} \, \text{s}^{-1}$ |
|  | Typical observed range | $5 \times 10^8 – 4 \times 10^9 \, \text{m}^{-3} \, \text{s}^{-1}$ |
| Positive ion mobility, $\mu_+$ | Ion motion | $1.14 \times 10^{-4} \, \text{m}^2 \, \text{V}^{-1} \, \text{s}^{-1}$ |
| Negative ion mobility, $\mu_-$ | Ion motion | $1.25 \times 10^{-4} \, \text{m}^2 \, \text{V}^{-1} \, \text{s}^{-1}$ |

$$\mu = De/k_B T. \tag{5.15}$$

Properties of the small ions are summarized in Table 5.2. The fact that the negative ions are slightly more mobile than the positive ions means that the *ion asymmetry parameter*,

$$x = n_+\mu_+/(n_-\mu_-), \tag{5.16}$$

differs from unity even if the ion concentrations, $n_+$ and $n_-$, are equal. The loss rate of ions to an aerosol particle with a charge $je$ can be expressed in terms of *attachment coefficients*, $\beta_{1j}(R)$ for positive ions and $\beta_{-1j}(R)$ for negative ions, so that the positive ion concentration, $n_+$, obeys the equation,

$$\partial n_+/\partial t + \nabla \cdot \mathbf{i}_+ = q_+ - \alpha n_+ n_- - n_+ \sum_j \int \beta_{1j}(R) n_j(\mathbf{r}, R) \, dR, \tag{5.17}$$

where $q$ is the ion production rate and the ion current is

$$\mathbf{i}_+ = n_+\mathbf{v} + \mu_+ n_+\mathbf{E} - D_+\nabla n_+, \tag{5.18}$$

where $\mathbf{v}$ is the gas velocity.

A similar equation applies to the negative ion concentration with the ion current containing the electric field contribution, $-\mu_- n_-\mathbf{E}$. In a steady-state situation with recombination dominating ion removal, the first two terms on the RHS of (5.17) remain, resulting in the maximum ion concentration given in Table 5.2. At atmospheric ion production rates and aerosol concentrations the aerosol loss term in (5.17) is often significant.

A final point is worth making regarding the ions. If electrons are being produced and the gas contains no oxygen or other molecule to which electrons can easily attach, electron attachment would predominate on an aerosol because of the electron's vastly greater mobility than that of positive ions. The parameter $x$ will be small, and, in spite of an overall charge neutrality, the aerosol will become strongly negatively charged.

### 5.3.2   *Diffusion charging*

To determine the ion attachment coefficients to spherical particles of charge $je$ in the continuum regime, a steady-state version of (5.17) and (5.18) must be solved. In a volume around the particle, whose radius is large compared with the particle radius but small compared with the radius of a spherical volume per particle in the aerosol, neglect of ion production and recombination is justified (for calculations which test these approximations and consider spatially non-uniform ion production, see Clement *et al.*, 1994), and the equation $\nabla \cdot \mathbf{i} = 0$, with $\mathbf{v} = 0$ and a radial electric field corresponding to the charge, was solved by Gunn (1954) and Pluvinage (1946) to obtain coefficients applying to uniform ion concentrations, $n_+$ and $n_-$, at infinity:

$$\beta_{1j}(R) = je\mu_+/\varepsilon_0[\exp(2\lambda j) - 1], \tag{5.19}$$

$$\beta_{-1j}(R) = je\mu_-/\varepsilon_0[1 - \exp(-2\lambda j)], \tag{5.20}$$

where $\varepsilon_0$ is the permittivity of the vacuum and the dimensionless aerosol size parameter is

$$\lambda = e^2/(8\pi\varepsilon_0 R k_B T) = 8.351/[R(\mu m)T(K)]. \tag{5.21}$$

When an aerosol is surrounded by bipolar ions of concentrations $n_+$, $n_-$, the aerosol charging equations for the concentrations $N_j$ are, for all integral $j$ (Hoppel and Frick, 1986; Clement and Harrison, 1992),

$$dN_j/dt = \beta_{1,j-1}n_+N_{j-1} - \beta_{1,j}n_+N_j + \beta_{-1,j+1}n_-N_{j+1} - \beta_{-1,j}n_-N_j. \tag{5.22}$$

In a steady state, these equations have an exact solution which has been obtained by several authors (see Pruppacher and Klett, 1978, p. 587; Clement and Harrison, 1991; Poluektov *et al.*, 1991):

$$N_j = N_0 x^j[(\sinh \lambda j)/\lambda j]\exp(-\lambda j^2) \tag{5.23}$$

$$\approx N_0 \exp[-j^2 e^2/(8\pi\varepsilon_0 R k_B T)]. \tag{5.24}$$

The final approximate Boltzmann form has often been used in aerosol science, and has led to the belief that the charge distribution has a thermodynamic origin (e.g. Keefe *et al.*, 1959; Matsoukas, 1994). However, aerosol charging at normal temperatures is the consequence of steady-state kinetics, and the exponential energy dependence arises from ion penetration through potential barriers (Clement *et al.*, 1995a). Filippov (1994) has generalized the derivation of the exact distribution (5.23) to apply to non-spherical particles with the particle radius, $R$, in $\lambda$ replaced by the particle capacitance, $C$. The mean charge predicted by the distribution (5.23) is

$$J = \sum_j jN_j \Big/ \sum_j N_j \approx (\ln x)/2\lambda, \tag{5.25}$$

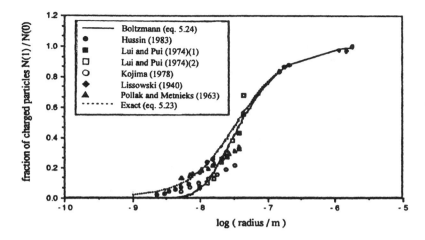

**Fig. 5.2** Comparison between theory and experiment for the fraction of singly charged particles as a function of radius (adapted from Clement and Harrison, 1991).

the final approximation, obtained by Gunn (1954), having been shown to be very accurate (Clement and Harrison, 1992).

For particles of the size of the ionic mean free path or smaller, difficulties arise in calculating attachment coefficients: there is no pure molecular limit or kinetic regime in which Boltzmann distributions can be assumed as in section 5.2.2 because of the long-range Coulomb potential from charged particles. Numerical calculations generally have to be performed. Fuchs (1963; 1964a) introduced a theory based on a 'limiting sphere', a radius of the order of the mean free path larger than the particle radius. The image force is important at small radii (Brock, 1970), and Hoppel (1977) also included three-body trapping. Marlow and Brock (1975) derived analytic first-order corrections at large $Kn$ with allowance for the image force. Hoppel and Frick (1986) have performed extensive calculations, resulting in tabulated attachment coefficients, using corrected Fuchs theory, and Fillipov (1993) has recently developed a numerical Monte Carlo algorithm to apply to the intermediate Knudsen regime. Comparison between theory and experiment for aerosol charging is generally reasonable, but it is interesting to note that the simple exact continuum theory expression (5.23) improves the fit to the fractional ratio $N_1/N_0$ shown in Figure 5.2 over that of the Boltzmann distribution (5.24), extending into the very small-size region. Comparisons with experiment for whole-charge distributions, $N_j/N_a$ for aerosols of diameters, $0.5\,\mu m$, $1.0\,\mu m$ and $2.0\,\mu m$ made by Emets *et al.* (1993) show considerably improved fits for the distribution (5.23) over those for (5.24).

### 5.3.3   *Field charging*

Ions will move much faster in an electric field, $\mathbf{E}$, than aerosol particles with much smaller mobilities, so that high fields charge aerosol directly. The equations $\nabla \cdot \mathbf{i} = 0$, with $\mathbf{i}_+$ given by (5.18) neglecting the diffusion term and $\mathbf{v} = 0$, and a similar form for $\mathbf{i}_-$, are then solved for the ion concentrations with uniform fluxes in the field $\mathbf{E}$ at infinity, and the combined particle and imposed field outside the particle. In terms of the ion asymmetry parameter, $x$, and the dielectric constant, $\varepsilon_p$, for a particle, its steady-state charge is

$$
\begin{aligned}
J &= (4\pi\varepsilon_0/e)[3\varepsilon_p/(\varepsilon_p + 2)]R^2 E(x^{1/2} - 1)/(x^{1/2} + 1) \\
&= 6.9 \times 10^{-4} E(\text{V m}^{-1})R^2(\mu\text{m})[3\varepsilon_p/(\varepsilon_p + 2)](x^{1/2} - 1)/(x^{1/2} + 1);
\end{aligned}
\tag{5.26}
$$

see Jantunen and Reist (1983), who give the result in terms of a parameter different than, but related to, $x$. The ion asymmetry factor varies between $-1$ and $+1$, and no charging occurs in the symmetric case $x = 1$. The dielectric factor has the order of unity as usually $\varepsilon_p < 10$. For micrometre-sized particles, the charging is clearly small unless the field reaches the order of $E = 10^4 \text{ V m}^{-1}$, a factor of 100 larger than the Earth's normal fair-weather field. Field charging can be neglected in the Earth's field.

Combined field and diffusion charging has no simple analytic solutions, but has been investigated numerically by Liu and Kapadia (1978) and Filippov (1990).

### 5.3.4   *Limits to charging*

Indefinite amounts of charge cannot be put on a small particle or droplet as it becomes unstable against losing charge or against breaking up. The *Rayleigh limit* to the charge on a droplet (Rayleigh, 1882) arises from the onset of an unstable quadrupole oscillation of the droplet surface which causes the droplet to break up. Roughly speaking, the electrostatic repulsive force between different parts of the droplet overcomes the surface tension restoring a spherical surface. The limit can be expressed as being reached when the electric stress at the surface reaches the stress from surface tension:

$$
E^2/8\pi = 2\gamma_L/R,
\tag{5.27}
$$

where $\gamma_L$ is the liquid surface tension and the electric field at the surface in terms of the droplet charge, $je$, is

$$
E = je/4\pi\varepsilon_0 R.
\tag{5.28}
$$

The charge limit becomes

$$j = (4\pi\varepsilon_0/e)(2\gamma_L R^3)^{1/2} = 46\,700 R^{3/2}(\mu\text{m}), \qquad (5.29)$$

for water droplets at $0°C$ ($\gamma_L = 0.0761\,\text{J m}^{-2}$).

Droplet fission can occur before the Rayleigh limit is reached, as is shown by recent experiments of Davis and Bridges (1994) where water droplets containing dodecyl sulphate surfactant and 1-dodecanol droplets were observed to fission at about 90% of the limit. For solid particles the maximum charge is limited by positive ion emission and electron emission for positively and negatively charged particles, respectively, the ion limit being an order of magnitude or more higher than the electron emission limit and the Rayleigh limit (Hinds, 1982, pp. 301–302).

When a droplet evaporates, its size will decrease but its charge will remain, thus approaching the stability limit. The fate of the droplet is of considerable interest in electrosprays and in atmospheric cloud physics (Pruppacher and Klett, 1978). The droplet discharge rate by ions depends on charge and size (see equations (5.19) and (5.20)) and it is possible they can be discharged fast enough so that fission does not occur. A model for evaporation and discharge which examines the criterion for droplet disintegration has been developed by Lin et al. (1994).

### 5.3.5  Radioactive aerosols

Natural radioactive aerosols are formed when the involatile decay products of gaseous radon ($^{222}$Rn) and thoron ($^{220}$Rn) form small clusters with other molecules, known as the 'unattached' fraction of radionuclides, which subsequently become attached to atmospheric aerosol particles, making the 'attached' fraction. A comprehensive review of the processes involved has been given by Porstendörfer (1994). In nuclear reactor accidents, large-sized aerosol could be explosively formed, and volatile fission product vapours could be emitted as a result of overheating and condense to form nuclear aerosols. At Chernobyl, both types of aerosol were formed, and the small-sized Cs fission product aerosol was widely dispersed in the atmosphere. The study of nuclear aerosols and their containment plays a major role in nuclear safety analyses (Schikarski et al., 1988).

Radioactive aerosols differ from other aerosols in that they are capable of charging themselves directly by the emission of charged particles, one electron for $\beta^-$ decay and several electrons usually knocked out during $\alpha$ decay. The positive charge left on the aerosol particle may be completely or partially neutralized by negative ions produced; a large number $I$ of ion pairs ($10^3$–$10^5$) are produced per decay. The processes involved are

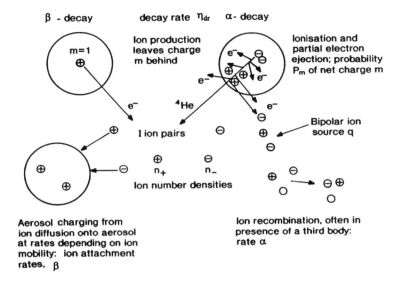

**Fig 5.3** Processes leading to the charging of radioactive aerosols.

illustrated in Figure 5.3, and the still limited experimental evidence for charging is summarized in Table 5.3. A charging theory has been constructed (Clement and Harrison, 1992) based on rate equations which, for $\beta^-$ decay, are obtained by augmenting (5.22) with an additional term, $\eta_{dr}(N_{j-1} - N_j)$, where $\eta_{dr}$ is the decay rate for the particle which will usually depend on $R$. The resulting charge distribution depends on the parameters $\lambda$ and $x$ as before, and an additional charging parameter,

$$y(R) = \varepsilon_0 \eta_{dr}(R)/e\mu_- n_-. \tag{5.30}$$

**Table 5.3** Experimental evidence for the charging of radioactive aerosols

| Author | Decay | Experiment | Charge $j$ observed |
|--------|-------|------------|---------------------|
| Ehrenhaft (1925) | $\alpha$ | Hg or Se spheres with Ra decay products attached | 3–5 per decay |
| Ivanov *et al.* (1969) | $\alpha$ | Monodisperse dibutylphthalate aerosol with Ra | 9–40 |
| Ivanov and Kirichenko (1970) | $\beta$ | $^{198}$Au | 100–300 |
| Yeh *et al.* (1976) | $\beta$ | Monodisperse $^{198}$Au | 2–9 |
| Yeh *et al.* (1978) | $\alpha$ | $^{238}$Pu O$_2$ | about 3 |

The distribution can be expressed exactly in terms of products (Clement and Harrison, 1992; Emets *et al.*, 1993), but is very well approximated by the Gaussian form (Clement *et al.*, 1995b),

$$n_j(R) = [n(R)/(2\pi\sigma^2)^{1/2}]\exp[-(j-J)^2/(2\sigma^2)],\tag{5.31}$$

where

$$J = y/\{1 - (x-1)/[\exp(2\lambda y) - 1]\}, \qquad \lambda y > 0.22,\tag{5.32a}$$

$$J = y + (x-1)/(2\lambda), \qquad \lambda y < 0.22,\tag{5.32b}$$

$$\sigma = [y + 1/(2\lambda)]^{1/2}.\tag{5.33}$$

These approximate expressions have been shown to reproduce well exact results for charging modifications to Brownian coagulation rates for two radioactive aerosols (Clement *et al.*, 1995b). The general properties of the charging depend on the ion concentration $n_-$ appearing in (5.30) which, when recombination dominates ion removal, may be estimated as $(I\eta_{av}N_{av}/\alpha)^{1/2}$, where $I$ is the ion pair production rate per decay, and $\eta_{av}N_{av}$ is the average spatial decay rate from the aerosol. The value of $y$ and the charging are only considerable when recombination does dominate, and its properties are then as follows: large radioactive particles become strongly positively charged; small slightly active particles and non-radioactive particles become slightly negatively charged if, as expected, $x < 1$. In regions of very high ion production, such as a reactor containment, only very active particles will be charged significantly, but in the environment, isolated 'hot' particles could reach charging levels of up to $j = 10^5$ (Clement and Harrison, 1995).

### 5.3.6 Time-scales for neutralizing aerosols

To obtain an aerosol not subject to electrical effects, an initial aerosol is often neutralized by a radioactive source which produces bipolar ions. Because $x$ is close to unity, ion attachment then produces a mean charge, $J$, close to zero. It is important to distinguish such an aerosol, which has a charge distribution whose mean is zero, from one where all charged particles are removed by electric fields, a subsequent procedure which is often also carried out. The time-scale for neutralization is implicit in the set of rate equations (5.22), and is therefore always inversely proportional to the ion concentration, $n$, and an attachment coefficient, $\beta_0$. In the continuum regime where the attachment coefficients are given by (5.19) and (5.20), an exact time-scale can be derived for $J$ to reach its steady-state value in the case when the parameter $x = 1$ ($n_+ = n_-$ and $\mu_+ = \mu_-$). By multiplying equations (5.22) by $j$ and summing over $j$, and then substituting explicitly for the $\beta_j$, the following explicit equation is obtained for $J$:

$$dJ/dt + \beta_0 nJ = \eta_{dr},\tag{5.34}$$

where the extra terms introduced into the equations for a radioactive aerosol add the $\eta_{dr}$ term to the RHS. Here, $\beta_0 = e\mu/\varepsilon_0$, so that the neutralization time-scale is

$$\tau_s = 1/\beta_0 n = \varepsilon_0/e\mu n. \tag{5.35}$$

This result was obtained in this way by Clement and Harrison (1992), but is identical to that of Liu and Pui (1974). Mayya and Sapra (1996) have studied charge neutralization over the entire particle size range and have obtained an analytic correction to (5.35) which applies for unsymmetrical ion situations. For small particles in the molecular regime, they used the attachment coefficients of Hoppel and Frick (1986) to show that, in going from continuum to molecular regime in aerosol size, the neutralization rate constant $\beta_0$ approximately halves with a corresponding doubling of the time-scale.

### 5.3.7  Electrical effects of charging

Many aspects of aerosol behaviour are affected by charging. The simplest is the direct action of an electric field which, for a micrometre-sized particle charge $je$ in the continuum regime and subject to Stokes' drag, acquires a velocity of

$$v_a = jeE/(8\pi\eta R) = 3.51 \times 10^{-10}jE(\text{V m}^{-1})/R(\mu\text{m}) \text{ m s}^{-1} \tag{5.36}$$

where $\eta$ is the gas viscosity (taken as $1.815 \times 10^{-5}$ for air at 20°C). Both $j$ and $E$ must be large for a sizeable velocity to be induced. In effect, this result specifies the electrical mobility of an aerosol particle, which is very much smaller than that of a nanometer-sized particle or cluster.

At a conducting wall, deposition may be enhanced because the presence of an image charge, $-je$, leads to an attractive electric field at a distance $x$ from the wall of

$$E = je/16\pi\varepsilon_0 x^2. \tag{5.37}$$

For the Brownian coagulation of spherical aerosol particles $(j_1, R_1)$ and $(j_2, R_2)$, the effect of charging was found by Fuchs (1964b) to modify the coagulation rate by a factor

$$f(Y) = Y/(\exp Y - 1), \tag{5.38}$$

where

$$Y = j_1 j_2 e^2/[4\pi\varepsilon_0 k_B T(R_1 + R_2)] = 16.7 j_1 j_2/[T(\text{K})(R_1 + R_2)(\mu\text{m})]. \tag{5.39}$$

For particles of $R = 1\,\mu\text{m}$ or less, these simple formulae are subject to corrections (Loyalka, 1976).

Some experimentally observed and theoretically proposed effects are summarized in Table 5.4. The results of the experiments of McMurry and

**Table 5.4** Electrical effects on aerosol transport

| Transport process | Effect | Reference |
|---|---|---|
| Wall deposition of aerosol | Experiments and theory show that electrostatic effects can dominate for 0.05–1 µm size range. | McMurry and Rader (1985) |
| Coagulation and scavenging | Experiments with gold aerosols show the presence of radioactivity and charging has large effects on the rates of the processes. | Rosinski *et al.* (1962) |
| Scavenging of aerosol by raindrops | Strongly enhanced by electrical forces in thunderclouds. Enhancement possibly reduced by high ion concentrations produced by nuclear debris. | Rosenkilde and Serduke (1982) |
| Earth's *electrode effect* | Earth's vertical electric field changed near surface by effects of ion production and motion towards electrode. | Hoppel (1967) |
| Charge-induced diffusion | The randomness of charging of particles in static and oscillating electric fields induces diffusive drift along field lines | Mayya and Malvankar (1993), Mayya (1994) |

Rader (1985), shown in Figure 5.4, illustrate a classic general result in aerosol science in that electrostatics effects can fill in the minimum in deposition rates expected in the size range 0.05–1 µm if only gravitational settling and diffusion are considered. Electrostatic effects on coagulation are presently an under-researched area, but the experiments of Rosinski *et al.* (1962) show that large effects can be expected. The theoretical

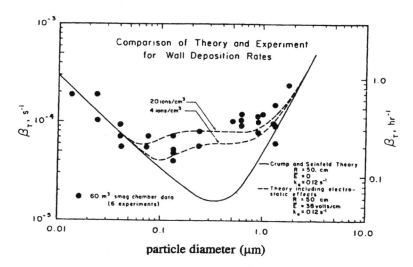

**Fig. 5.4** Comparison between experimental wall deposition rates in smog chambers, $\beta_T$, and theoretical predictions with and without electrostatic effects (McMurry and Rader, 1985). (Reproduced with permission of Elsevier Science. © 1985 by American Association of Aerosol Research.)

investigation of Hoppel (1967) into the electrode effect reveals the nonlinear and complex nature of the likely behaviour of ions and aerosol in electric fields. Ion motion near surfaces can alter the net space charge and thus the electric field. The changes in relative positive and negative ion concentrations then affect the aerosol charging and motion. The recent work of Mayya and Malvankar (1993) and Mayya (1994) points to a new type of diffusion produced by the stochastic nature of the aerosol charging process by ions, whose magnitude can exceed that of normal diffusion.


### 5.3.8  *Applications of electrical effects*

There is a vast array of applications of electrical effects to aerosols ranging from their use in aerosol measurement to industrial applications for air cleaning and spraying. A summary of some of the most important applications is given in Table 5.5. Electrostatic precipitators, employing high electric fields, are widely used in industry but are inefficient for submicrometre particles, so that combinations of electrostatics and filtration, which can be very efficient, are mentioned in the table. In most of the applications, the aerosol physics needed has been described here, involving the charging of particles and their motion in electric fields. Details can, however, be very sophisticated, such as the design of the fields to levitate particles in a stable way. The exception to the physics coverage is for electrosprays, where the physics involved is very complicated and

**Table 5.5** Applications of electrical effects on aerosols

| Application | Physics Principles | Recent References |
| --- | --- | --- |
| Differential mobility analysers (DMA), Differential mobility particle sizers (DMPS) | Charge fine and ultrafine particles. Motion in an electric field then measures the mobility and enables size/mass to be deduced. | Winklmayr et al. (1991) Wiedensohler et al. (1994) |
| Electret filters | Polymer sheets with large permanent electric polarization so that there are opposite charges on the two faces. | Brown et al. (1994) |
| Electrostatics in granular filters | Aerosol deposition enhancement in filters from charging and electric fields. | Shapiro et al. (1988) |
| Electrodynamic balance (EDB) | Suspension or levitation of particles by electric fields. Geometry (usually hyperbolic) and inclusion of a.c. field ensure stability. | Davis (1992) Jacko and Reed (1994) |
| Electrosprays | Atomization of a liquid as the result of high charging and interaction with an applied electric field. | Bailey (1988) Grace and Marijnissen (1994) |

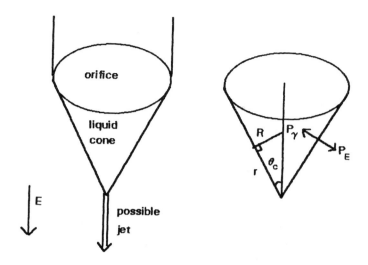

**Fig. 5.5** Taylor cone of a charged liquid emerging from an orifice in an electric field, also showing the geometry and surface tension and electrical stresses at the wall.

not yet completely understood. An issue of the *Journal of Aerosol Science* has been devoted to the subject of electrosprays (see, among others, Grace and Marijnissen, 1994).

The discussion here is limited to the topic of **Taylor cones** (Taylor, 1964), which can occur as a meniscus on a liquid at a nozzle subject to a high electric field, as shown in Figure 5.5. Under some conditions, including high flow rates, a narrow jet of micrometre size can be emitted from the tip of the cone which then breaks up into droplets. Taylor (1964) considered the stress balance at the surface of a liquid cone between the inward stress from the surface tension, $\gamma_L$, and the outward electrical stress, $P_E$,

$$P_\gamma = \gamma_L/R = \gamma_L \cot \theta_c/r = P_E = \frac{1}{2}\varepsilon_0 E^2, \qquad (5.40)$$

where the coordinates $r$ and $R$ are shown in Figure 5.5.

Shtern and Barrero (1994) show that the $1/r$ dependence gives a fluid velocity with the same dependence corresponding to a wide class of conical similarity flows with varying polar and azimuthal dependence. They can correspond to a variety of physical origins in the Taylor cone, which itself is just one of many functioning modes of spray production (Cloupeau and Prunet-Foch, 1994). As well as the surface tension, droplet production is affected by the conductivity, electrical permittivity, viscosity and density of the liquid in addition to its flow rate and variables describing the electric field. There is only a limited understanding at present of the influence of all these variables on spray production (Grace and Marijnissen, 1994).

## 5.4   Condensation and evaporation

The formation and disappearance of water mists and clouds are the commonest examples of condensation and evaporation of aerosols occurring in the atmosphere. They arise because of temperature changes and corresponding changes in the water vapour saturation,

$$S = p_v/p_{ve}(T) \qquad (5.41)$$

where $p_{ve}(T)$ is the equilibrium vapour pressure over a plane liquid surface, which varies with temperature according to the Clausius–Clapeyron equation (Landau and Lifshitz, 1980),

$$(d/dT)\ln p_{ve}(T) = \mu_v L/(R_G T^2) = \beta/T^2. \qquad (5.42)$$

This means that $p_{ve}(T)$ has an exponential dependence on temperature, behaving over an interval where $L$ is approximately constant as

$$p_{ve}(T) = p_{ve}(T_0)\exp[-\beta(1/T - 1/T_0)]. \qquad (5.43)$$

The nonlinear dependence of $p_{ve}$ on temperature has a profound effect on where aerosol can form in vapour–gas mixtures, which is generally a region where there is supersaturation, i.e. $S > 1$. We illustrate this in Figure 5.6, which describes the vapour pressure variation across a boundary layer between an unsaturated or saturated flow ($S \leq 1$) and a colder wall where there is condensation and $S = 1$. Across a laminar flow or in a static situation, and with no aerosol condensation, both temperature and vapour pressure (or concentration) are linear with position and therefore with each other. This linear relation with each other persists across turbulent boundary layers if they are the same size for heat and mass transfer, a condition which is satisfied for the Lewis number, $Le$, of the vapour–gas mixture equal to unity:

$$Le = \text{rate of heat transport by conduction/}$$
$$\text{rate of mass transport by diffusion} \qquad (5.44)$$
$$= k/(D_v C_p\rho).$$

Lewis numbers are close to unity for simple molecules in gases, e.g. 0.85 for water vapour in air at normal temperature and pressure, but can become large for big complex molecules for which $D$ is small.

In the situation shown in Figure 5.6, the vapour pressure reaches its equilibrium value, $p_{ve}(T_w)$, at the wall temperature, $T_w$, but is below

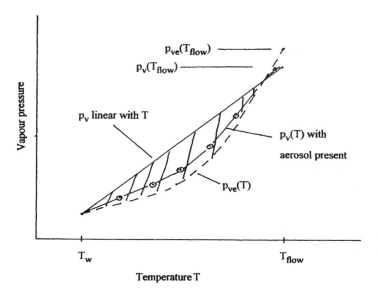

**Fig. 5.6** Actual ($p_v$) and equilibrium ($p_{ve}$) vapour pressures as functions of temperature through the boundary layer of a cooled unsaturated flow. The hatched area indicates the supersaturated region.

saturation at $T_{flow}$. The nonlinearity of $p_{ve}(T)$ results in the supersaturated hatched region shown. With an aerosol present or nucleated, $p_v(T)$ moves nearer to its equilibrium value from condensation. If the aerosol moves into the unsaturated tube interior flow region, by turbulent transport for example, it will start to evaporate and $p_v(T)$ will rise above the linear value. This behaviour can be seen in water vapour–air mixtures in the 'steaming' of cold wet surfaces after rain, and mists above cold water and snow. Before discussing other mechanisms for aerosol formation, we need to describe vapour saturation above droplet surfaces.

### 5.4.1 Vapour saturation and the Kelvin effect

Aerosol droplets have spherical curved surfaces, and the Kelvin effect states that the equilibrium vapour pressure over a curved liquid surface, radius $R$, is increased over that for a plane surface so that

$$p_{ve}(R, T)/p_{ve}(T) = \exp(R_\gamma/R) \approx 1 + R_\gamma/R, \qquad (5.45)$$

where $R_\gamma$ is defined in terms of the liquid surface tension ($\gamma_L$) and liquid density ($\rho_L$) by

$$R_\gamma = 2\mu_v\gamma_L/(\rho_L R_G T). \qquad (5.46)$$

For water droplets, $R_\gamma$ is 1.071 nm at 20°C decreasing to 0.696 nm at 100°C so that the expanded form is generally adequate in (5.45). Because the vapour pressure varies with $R$, a cloud of water droplets is not in thermodynamic equilibrium and water will evaporate from smaller-size droplets to condense on larger sizes, the *Ostwald ripening process* discussed in section 5.4.7.

### 5.4.2 *Aerosol formation mechanisms*

Aerosols may form by vapour condensation in supersaturated vapour–gas mixtures depending on the presence of existing aerosol or nuclei. In their presence, only a small supersaturation with $S - 1 > R_\gamma/R$ is necessary to induce growth. For nuclei of a different species, the process may involve growing a liquid drop on a particle surface, a process known as **heterogeneous nucleation** (section 5.5.2). If little or no aerosol exists, vapour can spontaneously produce growing droplets by **homogeneous nucleation** (section 5.5.1) which usually requires large supersaturations to produce measurable aerosol number concentrations.

To form an aerosol requires the production of vapour supersaturation in the first place, and equations describing changes in saturation have been given by Clement (1987; 1991) which can represent the condensation processes shown in Figure 5.7. In adiabatic expansion of a vapour–gas mixture, the temperature changes as

$$T/T_0 = (p/p_0)^{(\gamma-1)/\gamma}, \tag{5.47}$$

where $\gamma = C_p/C_v$ (=1.4 for air) is the ratio of the specific heats. Once $p_v$ exceeds $p_{ve}(T)$ and aerosol condensation begins, heat is released and the expansion is no longer adiabatic in that the expansion is no longer isentropic. The heat release changes the local temperature according to equation (5.9). Such expansions occur in rising columns of gas in the atmosphere where the pressure falls with height according to $\partial p/\partial z = -\rho g$, where $g$ is gravitational acceleration, and the full set of equations which represent cloud growth in such columns is described by Manton (1983).

The response to pressure changes is an example of a process where thermodynamics alone specifies the ultimate aerosol mass concentration, $\rho_a$. For all the processes, the criterion for determining $\rho_a$ is given in Table 5.6, and thermodynamics alone with the conservation of mass and energy, including (5.9), is also sufficient in the case of mixing. Where heat and mass transfer to or from walls is involved, the mass of vapour condensed is controlled by their rates, expressed in terms of $Le$ (5.44) and a condensation number $Cn_s$, defined similarly to (5.12), but with the temperature derivative of the actual vapour concentration, $c_v$, at the wall, which can be obtained from the slope of the curve for $p_v(T)$ in Figure 5.6. In particular, the likelihood of aerosol formation depends on the

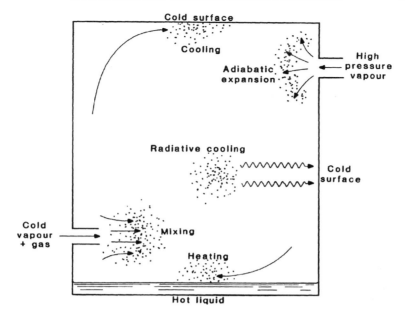

**Fig 5.7** Processes in vapour–gas mixtures leading to aerosol formation and growth.

difference, $Le - 1$, between the Lewis number and unity, and several cases have been considered by Clement (1985). If $Cn_s$ reaches its limiting equilibrium value given by (5.12), this corresponds to *maximum* aerosol formation, and explicit calculations were made by Clement and Ford (1989) for maximum aerosol densities which could be produced in tube flows and flows over heated pools.

In the final three cases shown in Table 5.6, the pressure is assumed to be maintained as constant. In all cases, whether or not $\rho_a$ is determined by thermodynamics, other properties of the aerosol, including the number density $N_a$, are determined by the dynamics of nucleation and growth, subjects discussed later in more detail.

**Table 5.6** Formation of aerosols by condensation mechanisms

| Cooling mechanism | Aerosol mass concentration specified by |
|---|---|
| Adiabatic expansion of vapour–gas mixture | Thermodynamics alone |
| Mixing of specified volumes of two vapour–gas mixtures | Thermodynamics alone |
| Cooling of a mixture at a surface | Heat and mass transfer to surface |
| Heating of a mixture by a hotter liquid condensate at a surface | Heat and mass transfer from surface |
| Radiative cooling of a vapour–gas mixture | Total heat radiated |

### 5.4.3  *Single-droplet growth rates*

An imbalance in vapour concentration in the medium and its equilibrium value at a droplet surface leads to net mass and heat exchange between the droplet and the medium by the exchange processes considered in section 5.2.1. To evaluate the molecular fluxes at the surface we need to consider the molecular momentum distribution which reduces to the Maxwell form in the kinetic region and where it gives the simple Hertz–Knudsen formula for the mass flux from the droplet,

$$F_M = (\mu_v/2\pi R_G)^{1/2}(\sigma_e p_{ve}(T_d)/T_d^{1/2} - \sigma_c p_{v\infty}/T_\infty^{1/2})$$
$$\approx S_p[v_v(T_d) - Sv_v(T_\infty)], \tag{5.48}$$

where $T_d$ and $T_\infty$ are the temperatures on the droplet surface and in the surrounding gas, $v_v$ is the vapour molecular flux, and the evaporation coefficient, $\sigma_e$, and condensation coefficient, $\sigma_c$ (or sticking probability, $S_p$), are only strictly equal at equilibrium. It is important to note that this formula does not apply to condensation on a *plane* surface (Barrett and Clement, 1992) where many experiments to measure $S_p$ have been made. The corresponding heat flux is given by Barrett and Clement (1988) in terms of a heat transfer coefficient,

$$H = \alpha_g c_{bg} v_g(T_\infty) + \alpha_v c_{bv} Sv_v(T_\infty), \tag{5.49}$$

where $c_b = h_v(T) - \frac{1}{2}R_G T/\mu_v$ and $\alpha$ is an effective thermal accommodation coefficient (in practice close to unity).

The continuum diffusive mass flux and conductive heat flux are based on molecular gas theory (Hirschfelder *et al.*, 1954) and there are small additional fluxes from Stefan flow, thermal diffusion, and the Dufour effect whose contributions to droplet growth in the continuum regime have been explored in detail by Kulmala and Vesala (1991).

Steady-state currents always apply across boundary layers in the continuum region as diffusive transit times are of the order of $R^2/D_v < 10^{-5}$ s. Growth rates are obtained by balancing the mass flux with the heat flux to remove latent heat. In the continuum regime when the supersaturation $S - 1$ is not small, corrections are needed (Williams, 1995), but otherwise the droplet growth rate has the form, omitting the Kelvin correction,

$$\dot{R} = \frac{1}{\rho_L} \frac{(S-1) + [RAD]}{[MASS] + [HEAT]}, \tag{5.50}$$

where [RAD] is a driving force proportional to the radiative heat flux, $Q_R$, from a droplet, and [MASS] and [HEAT] represent the resistances in series to growth from mass and heat transfer as described in section 5.2.3. In this form, the growth rate can be generalized to include additional resistances in series by adding them into the denominator. MacKenzie and

Haynes (1992) have applied it to the growth of stratospheric ice crystals where it is necessary to add an additional resistance representing surface kinetics.

In the continuum region (5.50) becomes (Barrett and Clement, 1988) a more accurate version of the Mason equation (Mason, 1971):

$$\dot{R} = \frac{\rho_{ve}(T_\infty)}{\rho_L R} \frac{D_v p}{p - S p_{ve}(T_\infty)} \frac{C n_s(T_\infty)}{C n_s(T_\infty) + 1} (S - 1 + [\text{RAD}]), \tag{5.51}$$

which clearly shows the role of $Cn_s(T_\infty)$ in determining whether mass or heat transfer controls the growth rate. A similar expression can be defined in terms of $Cn_k(T_\infty)$ for the growth rate in the kinetic regime, and Barrett and Clement (1988) suggested the use of a simple formula based on interpolation in the transition regime:

$$[\text{MASS}] = R(p - S p_{ve}(T_\infty))/(p D_v \rho_{ve}(T_\infty)) + 1/(S_p v_v(T_\infty)) \tag{5.52}$$

$$[\text{HEAT}] = (\beta L/T_\infty^2)(R/k + 1/H), \tag{5.53}$$

$$[\text{RAD}] = (\beta Q_R/4\pi R^2 T_\infty^2)(R/k + 1/H). \tag{5.54}$$

Calculations have been performed to see where the four processes, MD and HC in the continuum regime and MK and HK in the kinetic regime, limit water droplet growth (Barrett and Clement, 1988). The results depend strongly on the value chosen for $S_p$ and the regions in the temperature–droplet size plane are reproduced in Figure 5.8 for $S_p = 1$ and $p = 5$ bar. As the pressure reduces the $Cn = 1$ lines move down to lower

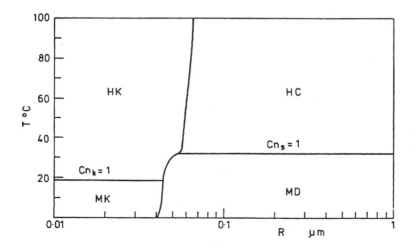

**Fig. 5.8** Regions in temperature ($T$) – radius ($R$) space where different physical processes control water droplet growth in air at $p = 5$ bar for a water molecule sticking probability of $S_p = 1$ (Barrett and Clement, 1988).

$T$, and the lines between the kinetic and continuum regimes move to the right to larger $R$. If $S_p$ is sharply reduced to 0.02–0.05, the MK region expands upwards and to the right to reach $R > 1\ \mu m$ at low temperatures.

For mass transfer alone, more elaborate theories have been constructed which modify the continuum growth rate according to the value of $Kn$. Fuchs and Sutugin (1970) introduced a correction factor $F$ to the MD limit which, in a modified form due to Hegg (1990) and Kreidenweis et al. (1991), is

$$F = F(Kn)/\{1 + 1.33KnF(Kn)[1/S_p - 1]\}, \qquad (5.55)$$

$$F(Kn) = (1 + Kn)/(1 + 1.71Kn + 1.33Kn^2). \qquad (5.56)$$

### 5.4.4 Multi-component droplets and binary growth

The most important examples of multi-component droplets are water droplets containing soluble material which may (sulphuric acid, ammonium nitrate) or may not (inorganic salts) have a significant vapour pressure. In the latter case just water vapour transport to the droplet occurs as before, and the theory only needs modification of the equilibrium water vapour pressure over the droplet. According to a modified Raoult's law, the vapour pressure is proportional to the mole fraction, $X_L = M_L\mu_s/(M_L\mu_s + M_s\mu_L)$ of the solvent, where $M_L$ and $M_s$ are the liquid and solute masses (Seinfeld, 1986), so that (5.45) is replaced by

$$p_{ve}(R, T, X_L)/p_{ve}(T) = \Gamma_L X_L \exp(R_\gamma/R), \qquad (5.57)$$

where $\Gamma_L$ is the *activity coefficient*. Such a description applies in general to droplets which are mixtures of liquids, and activities have generally to be found from experiment. For dilute solutions ($M_s \ll M_L$), the ratio may be rewritten as (Manton, 1983)

$$S_e(R_s, R) = 1 + R_\gamma/R - b(R_s/R)^3, \qquad (5.58)$$

where the exponential has been expanded as in (5.45), $R_s$ is the radius of a sphere corresponding to the mass, $M_s$, in the droplet at solid density, $\rho_s$, and

$$b = i\mu_L\rho_s/(\mu_s\rho_L), \qquad (5.59)$$

where $i$ is the van't Hoff factor to allow for dissociation in the liquid, a quantity which is formally related to the activity coefficient.

In expressions (5.50) and (5.51) for droplet growth, the factor $S - 1$ should be replaced by $S - S_e(R_s, R)$ for droplets containing soluble material. This implies that droplets grow if $S > S_e(R_s, R)$, but evaporate if $S < S_e(R_s, R)$. Plots of $S_e(R_s, R)$ showing the equilibrium saturation corresponding to $R_s$ and $R$ are known as **Köhler curves**. In Figure 5.9 the

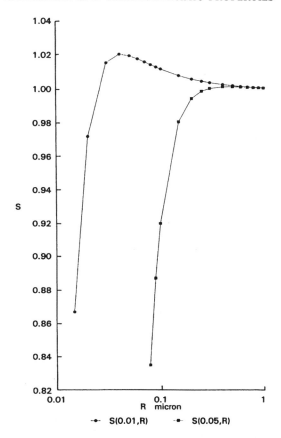

**Fig. 5.9** Köhler curves or equilibrium saturations, $S(R_s, R)$, for water droplets at 10°C containing an equivalent radius, $R_s$, of ammonium sulphate, as functions of droplet radius, $R$.

curves for ammonium sulphate $(b = 0.74)$ in water at 10°C $(R_\gamma = 1.13\,\text{nm})$ are shown for $R_s = 0.01\,\mu\text{m}$ and $0.05\,\mu\text{m}$. They have maxima at a critical radius, $R_c$, and saturation, $S_c$,

$$R_c = (3bR_s^3/R_\gamma^3)^{1/2}, \tag{5.60}$$

$$S_c = 1 + [2/(27b)^{1/2}](R_\gamma/R_s)^{1/2}. \tag{5.61}$$

For $S < S_c$ and $R < R_c$, the droplets have a stable equilibrium value on the curves, a fact responsible for haze droplets in the atmosphere. For $R > R_c$, a droplet grows indefinitely if $S > S(R)$, but evaporates to reach its stable value if $S < S(R)$. A nucleus is said to be **activated** once $R > R_c$.

For droplets containing two or more vaporizable materials, e.g. ammonia and water or sulphuric acid and water, a theoretical description becomes much more complicated, even if, as is usually the case in the atmosphere, concentrations of the vapours are both small so that they have uncoupled diffusive mass currents to a droplet. Considerable progress has been made in obtaining useful expressions for binary growth and evaporation (Kalkkinen *et al.*, 1991; Kulmala *et al.*, 1993a; 1993b). Such binary and ternary aerosols (e.g. including nitric acid as well as water and sulphuric acid) are particularly important in the stratosphere (Meilinger *et al.*, 1995), where they play a large role in the chemical reactions leading to holes in the ozone layer.

### 5.4.5    *Clouds and changes in size distribution*

A cloud is specified by a size distribution, $n(\mathbf{r}, R)$, which is taken here to cover the size range greater than a nucleation size, $R_N$. Ignoring all except the growth term, the general dynamic equation for the distribution reduces to

$$\partial n(R)/\partial t + \partial(n\dot{R})/\partial R = 0, \qquad (5.62)$$

where the growth rate, $\dot{R}$, given by (5.50), has a general $R$-dependence of the form,

$$\dot{R} = \mu_G(S(t) - S_e(R, R_s))/(1 + \alpha_G R), \qquad (5.63)$$

where $\mu_G$ and $\alpha_G$ depend on thermodynamic and kinetic quantities.

With $S$ independent of $R$, general techniques are available (Clement, 1978) to obtain explicit solutions for the evolving distribution (see, for example, Barrett and Clement, 1991). Here, we evaluate the increase in the aerosol mass concentration, $\rho_a(t)$, of droplets over a nucleated size, $R_N$, resulting from growth:

$$
\begin{aligned}
d\rho_a/dt &= (4\pi/3)\rho_p \int (\partial n/\partial t)R^3 \, dR \\
&= (4\pi/3)\rho_p R_N^3 n(R_N)\dot{R}(R_N) + 4\pi\rho_p \int R^2 \dot{R} n \, dR,
\end{aligned}
\qquad (5.64)
$$

where the integrals run from $R_N$ to $\infty$, and the contribution of the time derivative to the integral has been evaluated using (5.62) and partial integration. The first term in (5.64), which is the additional mass from freshly nucleated droplets, is usually negligible, but the corresponding contribution, $n(R_N)\dot{R}(R_N)$, to the total number concentration derivative, $dN_a/dt$, obtained by performing the same procedure with (5.3), gives the entire result for the increase in number concentration. From the form of $\dot{R}$ given by (5.63) with $S_e(R, R_s)$ approximated by unity, which is in the kinetic regime when $\alpha_G R \ll 1$ and the continuum regime when $\alpha_G R \gg 1$, the limiting mass transfer rates from vapour to the aerosol cloud are:

$$d\rho_a/dt = 4\pi\rho_p\mu_G(S(t) - 1)M_2, \qquad \text{kinetic regime} \qquad (5.65)$$

$$d\rho_a/dt = 4\pi\rho_p(\mu_G/\alpha_G)(S(t) - 1)M_1, \qquad \text{continuum regime}. \qquad (5.66)$$

For tiny particles, the surface area of the aerosol controls the mass transfer rate, but the rate becomes relatively smaller for large particles where diffusion and conduction through boundary layers result in $M_1$ as the controlling moment. Since $S = \rho_v/\rho_{ve}(T)$, and mass conservation means $\rho_v + \rho_a$ is constant, these equations are essentially differential equations for $\rho_v$. In the continuum case, taking $\mu_G/\alpha_G$ from (5.51) and neglecting small terms, (5.66) becomes

$$d\rho_v/dt + 4\pi D_v M_1[Cn_s/(Cn_s + 1)](\rho_v - \rho_{ve}) = 0. \qquad (5.67)$$

If mass transfer controls aerosol growth, this immediately gives the equilibration time-scale between a continuum-sized aerosol and its surrounding mixture,

$$t_e = (4\pi D_v M_1)^{-1}, \qquad Cn_s \gg 1. \qquad (5.68)$$

In the opposite limit where heat transfer is controlling growth, the temperature is changing so that $Cn_s$ is a function of time, and it is necessary to consider the equation for $T$ in addition. A derivation in this case (Barrett et al., 1992) gives, neglecting small terms,

$$t_e = (4\pi D_v M_1)^{-1}(Cn_s + 1)(Cn_s + Le)^{-1},$$
$$\approx [4\pi M_1 k/(\rho C_p)]^{-1}, \qquad Cn_s \ll 1. \qquad (5.69)$$

Similar time-scales depending on the molecular rate constants, and proportional to $M_2^{-1}$, could be derived for aerosols in the kinetic regime. The time-scales are short, since with a typical value of $D_v = 10^{-5} \, \text{m}^2\,\text{s}^{-1}$, and a number concentration of $10^9 \, \text{m}^{-3}$, (5.68) gives $t_e \approx 8/R(\mu\text{m})\,$s, where $R$ is the mean radius. Unless aerosol concentrations are much smaller than typical atmospheric urban values, overall equilibrium between the aerosol and condensible vapours will be rapidly reached and maintained.

## 5.4.6   Redistributive processes

We have just seen that aerosols are likely to equilibrate rapidly with surrounding condensible vapours in the sense that no further overall mass transfer takes place and $d\rho_v/dt = 0$. This does not mean, however, that the diverse individual droplets and particles are all in equilibrium, and in many cases mass will continue to be redistributed over an aerosol size distribution. A similar situation arises when an aerosol is cooling or heating up in a radiation field; this arises because the radiation driving term [RAD] in (5.51) is an addition to the exact driving term, $S - S_e(R, R_s)$, from supersaturation, and the form of [RAD] as it appears in (5.54) is dependent on $R$. Some results obtained in studies of the subject (Barrett and Clement, 1990; Barrett *et al.*, 1992) are summarized here.

When the form (5.58) is used for $S_e(R, R_s)$ in (5.63) and [RAD], now denoted by $G(R)$, is added, the growth rate becomes,

$$\dot{R} = \mu_G(S - 1 - R_y/R + b(R_s/R)^3 + G(R))/(1 + \alpha_G R). \tag{5.70}$$

Starting from an arbitrary value of $S$, there will be a rapid equilibration, as described above, followed by slower change to the size distribution. This may or may not involve further total mass change with a finite $d\rho_a/dt$ depending on whether heat is being transferred into or out of the vapour–gas–aerosol mixture, by radiation for example. In either case, the growth rate can be re-expressed in terms of $d\rho_a/dt$ and additional terms using the integral, $4\pi\rho_p \int R^2 \dot{R} n \, dR$, in (5.64). The result appears in terms of moments of the distribution and in the continuum regime takes the relatively simple form

$$
\begin{aligned}
\dot{R} = (d\rho_a/dt)/4\pi\rho_p M_1 R && \text{overall growth} \\
+ (\mu_G/\alpha_G)(R_y/R)(N_a/M_1 - 1/R) && \text{Ostwald ripening} \\
- (\mu_G/\alpha_G)(bR_s^3/R)(M_{-2}/M_1 - 1/R^3) && \text{solvent redistribution} \\
+ (\mu_G/\alpha_G)(1/R)[G(R) - \langle RG(R)\rangle/M_1], && \text{radiative redistribution}
\end{aligned}
$$
$$\tag{5.71}$$

where $\langle RG(R)\rangle$ represents the moment of the size distribution and is proportional to the moment of $Q_R$.

The three final terms involve mass redistribution because they change sign across the size distribution. For Ostwald ripening the sign is positive for large $R$, so that large droplets grow at the expense of small ones, making a cloud of pure drops basically unstable. The phenomenon also occurs in colloids and has an extensive literature, including the existence of an asymptotic theory for $R \gg R_y$ (Lifshitz and Slezov, 1959; 1961) with an asymptotic distribution and time-scale (Barrett *et al.*, 1992):

$$t_{OR} = 2.25\alpha_G R_{av}^3/(\mu_G R_y) = 17.4 R_{av}^3(\mu m) \text{ s}, \tag{5.72}$$

for water droplets at 20°C, where $R_{av} = M_1/N_a$ is the mean aerosol radius. In practice, small water droplets in the atmosphere would not disappear, but become stabilized according to their solute content (section 5.4.4).

Solvent redistribution can occur following rapid condensation at high $S$. Each droplet with a given $R_s$ tends to a stable size as large ones evaporate and small ones grow. Some calculations indicating relatively short time-scales were performed by Barrett *et al.* (1992).

For radiative redistribution, the case of pure radiative cooling, when the temperature falls, and the situation with $d\rho_a/dt = 0$, corresponding to the radiative interaction of a cloud whose temperature is maintained as

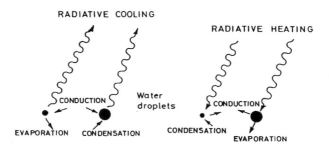

**Fig. 5.10** Expected physical behaviour of large and very small water droplets in a cloud interacting with radiation alone (Barrett and Clement, 1990).

constant, were considered by Barrett and Clement (1990) and Barrett *et al.* (1992). In the cooling case, the conditions for redistribution to occur are that $Q_R$ is not proportional to $R$, which is satisfied for water droplets for which $G(R) = $ constant $R\{1 - \exp[-0.3R(\mu m)]\}$, and $Le < 1$ (0.85 for water vapour in air). For a water droplet cloud interacting with radiation alone, we can have the interesting situations illustrated in Figure 5.10. While the bulk of the size distribution is either growing or evaporating, very small droplets may be doing the opposite. The effect arises because rates of heat and mass transfer through the boundary layers differ ($Le < 1$) and the relative amounts of heat radiated at different sizes do not correspond to the same relative amounts conducted through boundary layers.

## 5.5 Nucleation

Nucleation is the formation of a new stable embryonic particle or droplet which is capable of growth by condensation. Molecules in a gas are continuously forming transitory clusters, but generally the clusters are unstable in the sense that their probability of gaining a molecule is less than their probability of losing a molecule. The situation for clusters in a supersaturated vapour is shown in Figure 5.11, where $\beta(n)$ is the growth probability per unit time for a cluster of $n$ molecules and $\gamma(n)$ is its decay probability. Large clusters will grow because $S$ is greater than the equilibrium saturation outside the surface and $\beta(n) > \gamma(n)$, whereas small clusters are unstable because of the Kelvin effect. There exists a critical cluster size where $\beta(n_c) \approx \gamma(n_c)$, and, as $S$ approaches unity for equilibrium over a plane surface, this point moves to the right in the diagram, reaching $n_c = \infty$ at $S = 1$. For $n_c > 2$ a **nucleation barrier** exists between the monomers and the stable clusters which must be penetrated to produce

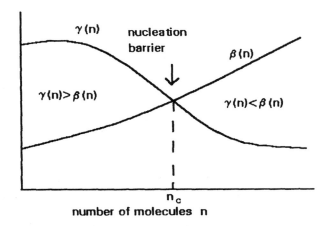

**Fig. 5.11** Growth ($\beta(n)$) and decay ($\gamma(n)$) rates of clusters of $n$ molecules which have to penetrate through a nucleation barrier to be nucleated at the unstable equilibrium critical cluster at $n = n_c$.

stable growing clusters, and the nucleation rate may be defined as

$$J_N = (d/dt) \sum_{n > n_c} c_n, \qquad (5.73)$$

where $c_n$ is the concentration of clusters of $n$ molecules (number per unit volume). The barrier may be expressed in terms of a free energy difference between the monomer and the critical cluster and, like the penetration rate of other energy barriers in physics, the nucleation rate falls off exponentially with the size of the barrier. If the barrier is too large, practically no nuclei are formed, and results of nucleation are often expressed in terms of a minimum detectable nucleation rate of $J_N = 1 \, \mathrm{cm}^{-3} \, \mathrm{s}^{-1}$.

Important exceptions to nucleation through a barrier exist in cases when all clusters down to the dimers ($n = 2$) are stable. It is clear that in these cases the rates of creation of stable clusters depend solely on collision rates and the monomer concentration or production rate, and thus on kinetics rather than on energies and thermodynamics. The situation is realized in practice by gaseous chemical reactions which produce very involatile products. An important example in the atmosphere may be the creation of ammonium sulphate particles from ammonia and sulphuric acid molecules. The study of barrierless nucleation with a source, 'source-enhanced nucleation', has quite a long history, and it has recently been shown that the ultimate number of particles produced with a constant monomer production rate varies according to how the growth coefficients $\beta(n)$ depend on $n$ (Lushkinov and Kulmala, 1995). Physical growth rates have power dependencies $\beta(n) \propto n^s$, with $s$ between 0 and 1 (Clement and

Wood, 1979). Only a finite number of particles are produced with fast growth rates, $s > \frac{1}{2}$, an infinite number are created if $s < \frac{1}{2}$, and the result depends on the monomer production rate for $s = \frac{1}{2}$.

When a nucleation barrier exists, we have to distinguish between two main cases. *Homogeneous nucleation* takes place in a pure vapour, although it may involve two or more vapour species, as in binary nucleation etc., and the critical cluster consists of just the nucleating species. *Heterogeneous nucleation* occurs when the cluster forms on a substrate consisting of an existing aerosol particle or nucleus. In the atmosphere, homogeneous nucleation of water vapour does not occur and condensation takes place on *cloud condensation nuclei* (CCN) which generally consist of activated aerosol containing soluble species. At all times, condensation can occur on any nucleus which is already activated so that its growth probability is greater than the decay probability for molecules. This brings us back to equations (5.64)–(5.67) for the aerosol and vapour mass concentrations and their generalizations, in which physical processes are increasing the vapour saturation. To discover whether nucleation will occur or not, the equation for the vapour concentration must be examined to see whether the loss term for condensation on existing, or just nucleated, aerosol is large enough for the saturation to reach, or remain at, a value where the nucleation rate is significant. In nucleation through a barrier, we may conjecture that growth on existing nuclei will always eventually cut off further nucleation, so that it will occur in bursts. In barrierless nucleation a similar cut-off only occurs in the case of fast growth rates (Lushkinov and Kulmala, 1995). Models to calculate the numbers of particles produced in a burst have been constructed by Warren and Seinfeld (1985) and Barrett and Clement (1991), and the response to saturation changes has been further considered by Barrett (1992).

In general, a much lower supersaturation is needed for heterogeneous nucleation than for homogeneous nucleation, and this extends to the special case of *ion-induced nucleation*, in which the heterogeneous nucleus is a charged ion or cluster. The propensity to nucleate on ions provides the physical basis of the Wilson cloud chamber used to see the motion of high-energy particles which produce ions along their tracks. The presence of charge can greatly modify the free energy of molecular clusters (see, for example, Budd *et al.*, 1985), enhancing nucleation, and experimentally the introduction of ions into vapour–air jets by corona discharge has been shown to augment vapour condensation (Vatazhin *et al.*, 1995).

### 5.5.1  *Homogeneous nucleation*

The simplest approach to homogeneous nucleation stems from thermodynamics and makes the **capillarity approximation**. In this picture a critical nucleus is a tiny liquid drop which has a macroscopic value for its surface tension, $\gamma_L$, and is in unstable equilibrium with the surrounding vapour. The nucleation rate is then given by the probability of forming the critical nucleus times a kinetic growth factor, $K_N$, i.e.

$$J_N = K_N \exp[-\Delta G_c/(k_B T)], \tag{5.74}$$

where the change in free energy or minimum work required to create the droplet has volume and surface contributions proportional to its volume, $V = 4\pi R_c^3/3 = n_c v_L$, where $v_L$ is the volume of a liquid molecule, and $A = 4\pi R_c^2$:

$$\Delta G_c = \Delta G_v + \Delta G_s = -n_c \Delta \mu_c + \gamma_L A = -(p_L - p_v)V + \gamma_L A, \tag{5.75}$$

where the chemical potential change is that from equilibrium over a plane surface for the vapour,

$$\Delta \mu_c = k_B T \ln(p_v/p_{ve}) = k_B T \ln S. \tag{5.76}$$

The use of this expression and the pressure difference across the droplet surface, $p_L - p_v = 2\gamma_L/R_c$, shows that $\Delta G_v = -(2/3)\Delta G_s$, and gives the classical nucleation theory results,

$$R_c = 2v_L \gamma_L/(k_B T \ln S), \tag{5.77}$$

$$\Delta G_c = -(16\pi/3)\gamma_L^3[v_L/(k_B T \ln S)]^2. \tag{5.78}$$

Further details of this derivation may be found in Chapter XV of Landau and Lifshitz (1980). Becker and Döring (1935) showed that the same result could be obtained as an approximation to a rate theory based on the equations for the cluster concentrations,

$$dc_n/dt = J(n - 1) - J(n), \tag{5.79}$$

where the currents which add to or subtract from the concentration are

$$J(n) = \beta(n)c_n - \gamma(n)c_{n+1}. \tag{5.80}$$

Such equations apply to the barrierless nucleation case, and form a basis for nucleation theory in general independent of whether thermodynamics is applicable or not (Clement, 1992). The principal assumptions needed are that clusters exist as physical entities and that significant growth only takes place by monomer addition so that the growth rates, $\beta(n)$, are proportional to $c_1$, the monomer concentration, and contain a flux to the droplet surface, given by $v_v(T)$ as in (5.48) times the surface area. The principal difficulty is the direct lack of knowledge of the decay rates, $\gamma(n)$, in molecular physics, so that they have to be determined from equilibrium cluster concentrations when $J(n) = 0$ and (5.80) becomes the detailed balance equation. The considerable application of statistical mechanics to nucleation theory is a response to the need to calculate such concentrations; see, for example, Reiss et al. (1990).

The nucleation rate was obtained by Becker and Döring (1935) by taking a steady state constant current, $J(n) = J_N$ up to a value, $m$, of $n$ above the barrier, and neglecting the decay rate at this point, to give the formal result:

$$J_N = \beta(1)c_1 / \left[ 1 + \sum_{n=2, m-1} \prod_{j=2, n} \gamma(j)/\beta(j) \right]. \tag{5.81}$$

The products are sharply peaked at the barrier height ($\gamma(n) > \beta(n)$, $n < n_c$) which enables the sum, replaced by an integral, to be evaluated by the method of steepest descent. The form for $\gamma(n)$ can be specified in terms of free energy differences, known, for example, in the capillarity approximation. The result contains the Zeldovich correction to the simplest kinetic form for $K_N$, and in the capillarity approximation gives the form (5.74) with $\Delta G_c$ specified by (5.78) and

$$K_N = [2\gamma_L/\pi m]^{1/2} [Sp_{ve}(T)/k_B T]^2, \tag{5.82}$$

where $m$ is the molecular mass. The derivation of similar results in a case relevant to solids is treated in more detail by Clement and Wood (1980).

A nucleation theorem which follows from thermodynamics, but may have a more general validity (Kaschiev, 1982), relates the number of molecules, $n_c$, in a critical cluster to the isothermal dependence of $J_N$ on $S$:

$$[\partial \ln J_N / \partial \ln S]_T = n_c. \tag{5.83}$$

This is a useful relation with which to examine experimental results. Classical theory accounts reasonably in some cases for the isothermal behaviour of $J_N$ and deduced values of $n_c$, but, together with alternative theories, fails to account for the temperature dependence of $J_N$ as is illustrated in Figure 5.12. This agreement with variation of $S$, but lack of it for variation with $T$, has been noted and a corresponding empirical form has been proposed for $\Delta G_c$ (McGraw and Laaksonen, 1996). There is no necessity for nucleation to be an isothermal process, since latent heat must be transported away from a growing droplet and thus nucleation might be viewed as a temperature fluctuation as well as a number fluctuation. A model including temperature changes was proposed by Feder et al. (1966), and a model based on rate equations which removes some deficiencies in previous work has been proposed (Barrett et al., 1993). Calculations suggest relatively small changes to isothermal theory, including a small dependence of $J_N$ on pressure. For binary nucleation, not covered here, disagreement with classical theory is even worse than in the unary case. Classical nucleation theory is widely used in aerosol science, for lack of anything better, but major improvements need to be made before its results can be trusted. Empirical and scaling approaches to nucleation data (Hale, 1986) are necessary to make predictions.

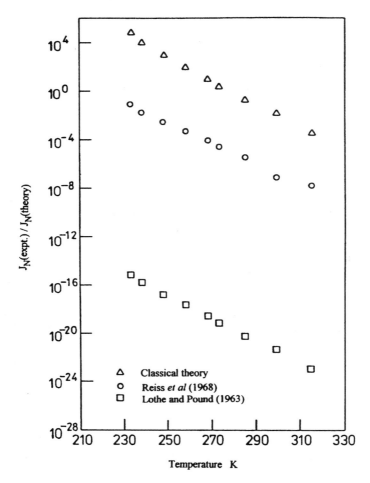

**Fig. 5.12** Comparison of experimental results, $J_N$(expt.) for the homogeneous nucleation of *n*-nonane as a function of temperature with various theoretical predictions taken from Katz *et al.* (1988).

### 5.5.2 *Heterogeneous nucleation*

Heterogeneous nucleation is the predominant process in the atmosphere, although there is now good experimental evidence that binary homogeneous nucleation of sulphuric acid–water droplets does occur (e.g. Covert *et al.*, 1992). Since the nucleation occurs on a substrate which can have a variety of different compositions and geometries, the subject of heterogeneous nucleation is very complex. In the atmospheric case, there is extensive coverage in Pruppacher and Klett (1978). To illustrate

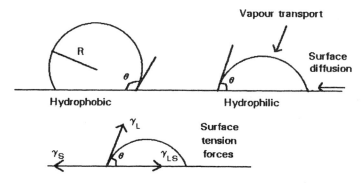

**Fig. 5.13** Critical water droplets for heterogeneous nucleation on a plane solid surface with the balancing surface tension forces at a point of contact of the air, liquid (L) and solid (S) phases.

the principles involved in extending homogeneous nucleation theory, we shall consider here just the simple case of a water droplet forming on a plane solid surface, as illustrated in Figure 5.13. The angle $\theta$ between the liquid and solid at the point of contact is known as the **contact angle** or **wetting angle** which, with water, makes the solid hydrophobic or hydrophilic according as to whether the angle is greater or less than 90°. Mechanical equilibrium at the point of contact means that the component along the surface of the surface tension forces shown in Figure 5.13 must balance:

$$\gamma_S - \gamma_{LS} = \gamma_L \cos\theta, \tag{5.84}$$

where $\gamma_L$ and $\gamma_S$ refer to the liquid and solid interfaces with air and $\gamma_{LS}$ to the liquid–solid interface. For a droplet to form, their values are restricted to make $-1 \leq \cos\theta \leq 1$.

To apply thermodynamic nucleation theory, the free energy change, $\Delta G_c$, needed to create a critical nucleus must be calculated as before by altering (5.75). For the droplet shapes shown in Figure 5.13, the volume $V = (1/3)\pi R^3 (1 - \cos\theta)^2 (2 + \cos\theta)$, and the surface areas of the base and liquid cap are $\pi R^2 \sin^2\theta$ and $2\pi R^2 (1 - \cos\theta)$, respectively. The change in surface energy is then

$$\begin{aligned}
\Delta G_s &= (\gamma_{LS} - \gamma_S)\pi R^2 \sin^2\theta + \gamma_L 2\pi R^2 (1 - \cos\theta) \\
&= \gamma_L \pi R^2 (1 - \cos\theta)^2 (2 + \cos\theta),
\end{aligned} \tag{5.85}$$

where (5.84) has been used. We see that the results for the homogeneous nucleation of a liquid droplet should apply with the surface tension multiplied by the factor $(1 - \cos\theta)^2 (2 + \cos\theta)$. The factor reduces to zero as $\theta$ tends to zero and the nucleation barrier vanishes.

Such simple theories have had very mixed success in explaining atmospheric nucleation, as real geometries and mechanisms are more complicated. In addition to vapour transport to nucleating droplets, surface diffusion of adsorbed molecules may play a role. For water, the existence of soluble salts and sulphuric acid in the atmosphere which can attach to aerosol particles greatly enhances their affinity to water. Even black carbon particles representative of engine exhausts, which might be thought to be hydrophobic, have been shown experimentally (Carleton *et al.*, 1995) to condense sulphuric acid irreversibly into pores (up to 14% by mass as the humidity increases) to become good CCN. For ice nucleation, simple heterogeneous theories (Fletcher 1962; 1969) fail to account for experimental results (Gorbunov *et al.*, 1980), and the theory has been extended to allow for ice embryo growth in three stages (Gorbunov and Kakutkina, 1982). Even without any deficiencies in classical nucleation theory, its application to heterogeneous situations requires a good understanding of local geometries and surfaces to be successful.

### 5.5.3    *Measurements of nucleation and growth*

Starting with the Wilson cloud chamber (Wilson, 1897) in which there was adiabatic expansion of a vapour–gas mixture, a variety of methods have been used to produce supersaturated vapours and consequent nucleation and growth. The methods include static or steady-state flow diffusion chambers, turbulent mixing and supersonic nozzles, all of which suffer from difficulties in interpreting the results because of spatial non-uniformity in temperatures and concentrations. Developments of the adiabatic expansion approach by Wagner and Strey (for a recent summary of progress, see Strey *et al.*, 1994) have probably provided the most reliable method for measuring aerosol nucleation and growth rates, and we give here a brief description of the physical principles used in their nucleation pulse measurements.

To produce a well-defined nucleation pulse, an adiabatic expansion in a chamber is followed by a slight recompression after a specified time during which steady-state nucleation takes place. The recompression is such as to reduce the supersaturation so that further nucleation is negligible, but the remaining supersaturation allows the nuclei produced to grow. Drops are illuminated by a HeNe laser beam and the transmitted light flux and light scattered at a fixed angle (15°) are measured. This constant-angle Mie scattering (CAMS) can be used to measure growth accurately as maxima in scattering cross-sections repeat themselves as droplets grow. The technique simultaneously measures number concentrations and size and has been described in detail by Wagner (1985). Initially, a two-piston expansion chamber was used for the expansion and recompression, but this gave only non-isothermal curves of

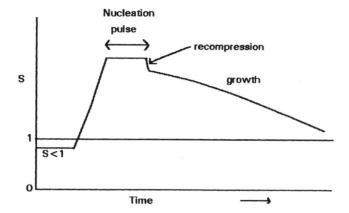

**Fig. 5.14** Vapour saturation as a function of time during nucleation pulse experiments. After Strey *et al.* (1994).

$J_N$ against $S$. To be able to measure isothermal $J_N$–$S$ curves, the chamber has been modified to contain two rapidly operating valves (Strey *et al.*, 1994).

Saturation changes during an experiment are illustrated in Figure 5.14. The duration of the nucleation pulse can be varied, and the basic hypothesis of steady-state nucleation theory that the rate is independent of time at constant saturation has been verified. Results obtained from pulsed nucleation include the following:

- Classical nucleation theory does not predict the correct temperature dependence of nucleation rates.
- Any dependence of nucleation rates on the pressure of the gas present is small.
- For binary nucleation, rates are many orders of magnitude different from predicted rates, and some cases have been found where molecules which are immiscible on a macroscopic scale apparently conucleate at rates greater than those of the single species.
- Molecular sticking probabilities or condensation coefficients are close to unity (section 5.2.2).

### 5.5.4  *Aerosol phase changes and hysteresis*

In response to temperature changes, aerosol droplets and particles can freeze and melt, respectively, and other types of changes may take place in multi-component aerosol. For example, crystallization can occur in water solutions containing salts in response to decreasing humidity. All such phase changes frequently exhibit the property of **hysteresis** which is particularly common in the small volumes of aerosol droplets. In other words, the phase change which occurs with heat release, i.e. freezing or crystallization, does not happen at the normal temperature so that the droplet becomes highly supercooled or supersaturated, respectively.

The subjects of freezing and hysteresis are discussed in Pruppacher and Klett (1978) as they particularly affect atmospheric droplets, and two properties are depicted in Figure 5.15 in simplified forms. Rising water droplets occur in convective clouds and often reach very low temperatures, down to about −40°C, before they freeze. In addition to the dependence on droplet size shown in Figure 5.15a, the freezing temperature also falls with increase in cooling rate. Both the effects can be explained as nucleation phenomena in that a given nucleation rate actually realizes nucleation more rapidly in a larger volume or in a longer time. The effect of seeding water droplets with atmospheric dust on the nucleation process in freezing has been investigated by Stoyanova *et al.* (1994) who find that ice nucleation is controlled by only one type of centre in unseeded droplets, but by three new types characterized by activity types and 'wetting angles' in seeded droplets. Their results are consistent with interpretations based on classical homogeneous nucleation theory.

Ice particles start to melt as soon as the temperature rises above 0°C and so freezing and melting has a large hysteresis loop in temperature. For relative humidity changes and crystallization, a hysteresis loop shows up in the droplet radius, as shown in Figure 5.15b; as the relative humidity decreases, a crystal nucleus appears at large supersaturation, and the droplet quickly loses water and size. According to recent ideas, the ozone hole aerosol consisting of hydrated sulphuric acid and eventually nitric

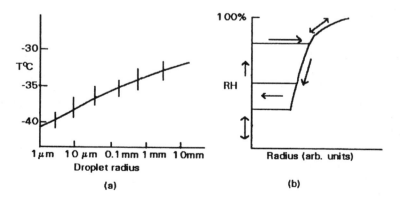

**Fig. 5.15** Non-equilibrium effects observed for phase changes in small droplets: (a) observed freezing temperatures for droplets of different radii; (b) hysteresis loop in the size of droplets containing NaCl for relative humidity (RH) changes leading to and from crystallization. Adapted from experimental results quoted by Pruppacher and Klett (1978).

acid, also shows hysteresis effects in its freezing and melting. Long times are needed for small aerosol particles and droplets to reach internal equilibrium, and it is not safe to assume that aerosol particles or droplets are always in their equilibrium thermodynamic state.

## 5.6   Chemistry with aerosols

The presence of an aerosol can alter or induce chemical reactions in a gas. Reactions may take place between two gaseous molecules on the aerosol surface, between a gaseous molecule and an aerosol molecule on the surface, or in the interior, of an aerosol droplet or particle. Examples of all these processes occur in the atmosphere where aerosols play a major role in the physics and chemistry of pollution (Seinfeld, 1986). Specific examples are the following:

- *Heterogeneous surface reactions* (Storozhev, 1995). Molecules become adsorbed onto an aerosol surface, interact with the surface and each other, possibly through multi-stage processes, and finally desorption of reaction products may take place. Storozhev (1995) evaluated possible maximum contributions of such processes to reactions relevant to atmospheric chemistry and found many examples where these contributions could be larger than that of pure gas-phase reactions.
- *The ozone hole.* Without the presence of an aerosol in the stratosphere, studies involving solely gas phase reactions would not predict the destruction of significant numbers of ozone molecules by Cl or halogens released by the interaction of sunlight on such chemicals as chloro-fluorocarbons (CFCs). The observed presence of stratospheric ozone holes in the Antarctic and Arctic springs is attributed to the presence of sulphuric acid aerosols on which nitric acid and water may condense and chemical reactions take place. The sizes of the observed holes have been correlated with the amount of such aerosol produced from $SO_2$ emissions from volcanoes. Reactions on aerosol surfaces are also likely to play a significant role in ozone photochemistry in the troposphere (Crutzen, 1996).
- *Aerosols and the greenhouse effect.* Emissions of $SO_2$ into the atmosphere are believed to lead to the production of sulphuric acid or sulphate aerosol whose radiative interaction produces a net cooling to offset some of the warming from increased levels of $CO_2$ in the atmosphere (see, for example, Jones *et al.*, 1994). Much of the chemical conversion of $SO_2$ to sulphuric acid by oxidation takes place by chemical reactions inside cloud droplets (see, for example, Langner and Rodhe, 1991), and only aqueous phase oxidation growth in cloud droplets can explain observed atmospheric aerosol growth in short time-scales (10–30 min) (Smith *et al.*, 1996).

The description of molecular processes inside droplets or particles is outside the scope of material covered in this chapter, and indeed the processes may be essentially time-dependent if molecules have to diffuse into solid particles. However, description of molecular transport onto aerosol surfaces and interactions on them is mainly a straightforward modification or extension of the theory given above in connection with condensation. For example, the process of adsorption onto an aerosol surface involves the two mass transfer processes of diffusion and molecular capture given in Table 5.1, except that now the molecular sticking coefficient must be replaced by an **attachment coefficient** (Baltensperger *et al.*, 1996; Porstendörfer *et al.*, 1979) which may be much less than unity to allow for the reversal of the attachment process. If chemical reactions and desorption of reaction products follow, an effective adsorption or reaction coefficient for each product, $r$, may be defined as advocated by Baltensperger *et al.* (1996), e.g.

$$\gamma_r = \frac{\text{no. of molecular surface collisions resulting in emission of } r}{\text{total no. of molecular surface collisions}} \quad (5.86)$$

The sum of such $\gamma_r$ and a similar adsorption coefficient for the molecule resulting in a surface bound state would give the overall interaction coefficient removing the molecule from the gas phase at the surface. The various molecular reactions occur in parallel once the molecule reaches the surface, and it is this total coefficient which must be used in determining whether diffusion or molecular transport governs their magnitude.

Provided that heat transfer does not play a role, it would then be straightforward to modify the results given in section 5.4.5 for molecular concentrations and equilibration time-scales for molecule–aerosol surface chemical reactions and they will be specified by the moments $M_1$ or $M_2$ of the aerosol size distribution. If large amounts of heat are emitted by the chemical reactions involved and these occur at sufficient frequency, heat transfer may control processes when reaction rates are sufficiently temperature-dependent. Experimental techniques with which to examine gas–aerosol chemical reactions are still being developed and now include optical resonance spectroscopy (Taflin and Davis, 1990) and labelling with radioactive isotopes (Ammann *et al.*, 1995).

## References

Ammann M., Baltensperger U., Bochert U.K., Eichler B., Gäggeler H.W., Jost D.T., Türler A. and Weber A.P. (1995) Study of HI, HBr, and NO₂ adsorption on graphite and silver aerosol particles using short-lived isotopes. *J. Aerosol Sci.*, **26**, 61–70.
Bailey A.I. (1988) *Electrostatic Spraying of Liquids*. Wiley, New York.
Baltensperger U., Amman M., Kalberer M. and Gäggeler H.W. (1996) Chemical reactions on aerosol particles: concept and methods. *J. Aerosol Sci.*, **27**, S651–S652.

Barrett J.C. (1992) Nucleation with changing saturation. *J. Aerosol Sci.*, **23**, S141–S144.

Barrett J.C. and Clement C.F. (1988) Growth rates for liquid drops. *J. Aerosol Sci.*, **19**, 223–242.

Barrett J.C. and Clement C.F. (1990) Growth and redistribution in a droplet cloud interacting with radiation. *J. Aerosol Sci.*, **21**, 761–776.

Barrett J.C. and Clement C.F. (1991) Aerosol concentrations from a burst of nucleation. *J. Aerosol Sci.*, **22**, 327–335.

Barrett J.C. and Clement C.F. (1992) Kinetic evaporation and condensation rates and their coefficients. *J. Colloid Interface. Sci.*, **150**, 352–364.

Barrett J.C., Clement C.F. and Ford I.J. (1992) The effect of redistribution on aerosol removal rates. *J. Aerosol Sci.*, **23**, 639–656.

Barrett J.C., Clement C.F. and Ford, I.J. (1993) Energy fluctuations in homogeneous nucleation theory for aerosols. *J. Phys. A.*, **26**, 529–548.

Becker R. and Döring W. (1935) Kinetische Behandlung der Keimbildung in übersättigten Dämpfen. *Ann. Physik.*, **24**, 719–752.

Brock J.R. (1970) Aerosol charging: the role of the image force. *J. Applied Phys.*, **41**, 843–844.

Brown R.C., Wake D., Thorpe A., Hemingway M.A and Roff M.W. (1994) Theory and measurement of the capture of charged dust particles by electrets. *J Aerosol Sci.*, **25**, 149–163.

Budd T., Marshall M. and Kwok C.S. (1985) Droplet growth and diffusion in a low-pressure cloud chamber. *J. Aerosol Sci.*, **16**, 145–155.

Butterworth D. and Hewitt G.F. (eds) (1977) *Two Phase Flow and Heat Transfer*. Oxford University Press, Oxford.

Carleton K.L., Sonnenfroh D.M., Rawlins W.T., Wyslouzil B.E. and Arnold S. (1995) Single aerosol experimental studies of chemical modification of background and engine exhaust plume aerosols, in *AAAR'95, Abstracts of 14th Annual Meeting*, Pittsburgh, PA, p. 310.

Chapman S. and Cowling T.G. (1970) *The Mathematical Theory of Non-uniform Gases*. Cambridge University Press, Cambridge.

Clement C.F. (1978) Solutions of the continuity equation. *Proc. Roy. Soc. London A*, **364**, 107–119.

Clement C.F. (1985) Aerosol formation from heat and mass transfer in vapour–gas mixtures. *Proc. Roy. Soc. London A*, **398**, 307–339.

Clement C.F. (1987) *The Supersaturation in Vapour–Gas mixtures Condensing into Aerosols*. UKAEA Harwell Report AERE-TP 1223.

Clement C.F. (1988) Convective behaviour in severe accidents, in *Proc. Int. Symposium on Severe Accidents in Nuclear Power Plants*, IAEA, Vienna, Vol. 2, pp. 125–134.

Clement C.F. (1991) Condensation and evaporation in clouds, in M. Kulmala and K. Hämeri (eds), *Workshop on Condensation*, Helsinki, 1991, *Report Series in Aerosol Science* No. 17. Finnish Association for Aerosol Research, Helsinki.

Clement C.F. (1992) Aerosol growth and nucleation from a molecular viewpoint, in N. Fukata and P.E. Wagner (eds), *Nucleation and Atmospheric Aerosols: Proc. Thirteenth Int. Conf.* A. Deepak, Hampton, VA, pp. 327–336.

Clement C.F. and Ford I.J. (1989) Maximum aerosol densities from evaporation and condensation processes. *J. Aerosol Sci.*, **20**, 293–302.

Clement C.F. and Harrison R.G. (1991) Charge distributions on aerosols, in B.C. O'Neill (ed.), *Electrostatics 1991: Proc. Eighth Int. Conf.*, Univ. Oxford, Institute of Physics Conf. Series no. 118. IOP, Bristol, pp. 275–280.

Clement C.F. and Harrison R.G. (1992) The charging of radioactive aerosols. *J. Aerosol Sci.*, **23**, 481–504.

Clement C.F. and Harrison R.G. (1995) Electrical behaviour of radioactive aerosols in the environment. *J. Aerosol Sci.*, **26**, S571–S572.

Clement C.F. and Wood M.H. (1979) Equations for the growth of a distribution of small physical objects. *Proc. Roy. Soc. London A*, **368**, 521–546.

Clement C.F. and Wood M.H. (1980) The principles of nucleation theory applied to the void swelling problem. *J. Nuclear Mat.*, **89**, 1–8.

Clement C.F., Barrett J.C. and Harrison R.G. (1995a) The diffusive penetrability of particles into energy barriers. *J. Aerosol Sci.*, **26**, 735–743.

Clement C.F., Calderbank D.M.J. and Harrison R.G. (1994) Radioactive aerosol charging with spatially varying ion concentrations. *J. Aerosol Sci.*, **25**, 623–637.

Clement C.F., Clement R.A. and Harrison R.G. (1995b) Charge distributions and coagulation of radioactive aerosols. *J. Aerosol Sci.*, **26**, 1207–1225.

Clement C.F., Kulmala M. and Vesala T. (1996) Theoretical consideration on sticking probabilities. *J. Aerosol Sci.*, **27**, 869–882.

Cloupeau M. and Prunet-Foch B. (1994) Electrohydrodynamic spraying functioning modes: a critical review. *J. Aerosol Sci.*, **25**, 1021–1036.

Covert D., Kapustin V.N., Quinn P.K. and Bates T.S. (1992) New particle formation in the marine boundary layer. *J. Geophys. Res.-Atmospheres*, **97**, 20 581–20 589.

Crump J.G. and Seinfeld J.H. (1981) Turbulent deposition and gravitational sedimentation in a vessel of arbitrary shape. *J. Aerosol Sci.*, **12**, 405–415.

Crutzen P. (1996) The role of particulate matter in ozone photochemistry (stratospheric and tropospheric), in *Nucleation and Atmospheric Aerosols 1996* (eds M. Kulmala and P.E. Wagner). Pergamon/Elsevier Science, Oxford, pp. 268–270.

Davis E.J. (1992) Principles and applications of the electrodynamic balance for aerosol studies, in I.L. Anderson (ed.), *Advances in Chemical Engineering*, Vol. 18, pp. 1–94. Academic Press, New York.

Davis E.J. and Bridges M.A. (1994) The Rayleigh limit of charge revisited: light scattering from exploding droplets. *J. Aerosol Sci.*, **25**, 1179–1199.

Ehrenhaft F. (1925) The electrical behavior of radioactive colloidal particles of the order of $10^{-5}$ cm as observed separately in a gas. *Phil. Mag.*, **49**, 633–648.

Emets E.P., Kascheev V.A. and Poluektov P.P. (1993) Statistics of aerosol charging. *J. Aerosol Sci.*, **24**, 867–877.

Feder J., Russell K.C., Lothe J. and Pound G.M. (1966) Homogeneous nucleation and the growth of droplets in vapours. *Adv. Phys.*, **15**, 111–178.

Fendel W. and Schmidt-Ott A. (1993) Material separation of nanoparticles via multiple photoelectric charging. *J. Aerosol Sci.*, **24**, S73–S74.

Filippov A.V. (1990) A mixed aerosol-particle charge. The asymptote and interpolation formulas for the electrification current. *J. Appl. Mech. Tech. Phys. (USA)*, **30**, 851–857.

Filippov A.V. (1993) Charging of aerosol in the transition region. *J. Aerosol Sci.*, **24**, 423–436.

Filippov A.V. (1994) Charge distribution among non-spherical particles in a bipolar ion environment. *J. Aerosol Sci.*, **25**, 611–615.

Fletcher N. (1962) *The Physics of Rainclouds.* Cambridge University Press, Cambridge.

Fletcher N. (1969) Active sites and ice crystal nucleation. *J. Atmos. Sci.*, **26**, 1266–1271.

Fuchs N.A. (1963) On the stationary charge distribution on aerosol particles in a bipolar ionic atmosphere. *Geophys. Pura. Appl.*, **56**, 185–193.

Fuchs N.A. (1964a) On the steady-state distribution of the charges of aerosol particles in a bipolar ionized atmosphere. *Izv. Geophys. Ser.*, **4**, 579–586.

Fuchs N.A. (1964b) *The Mechanics of Aerosols.* Pergamon Press, New York, pp. 291–294.

Fuchs N.A. and Sutugin A.G. (1970) *Highly Dispersed Aerosols.* Ann Arbor Science, Ann Arbor, MI.

Gorbunov B.Z. and Kakutkina N.A. (1982) Ice crystal formation on aerosol particles with a non uniform surface. *J. Aerosol Sci.*, **13**, 21–28.

Gorbunov B.Z., Kakutkina N.A. and Koutzenogii K.P. (1980) Studies of silver iodide ice-forming activity: verification of theory. *J. Appl. Meteor.*, **19**, 71–77.

Grace J.M. and Marijnissen J.C.M. (1994) A review of liquid atomization by electrical means. *J. Aerosol Sci.*, Special Issue: Electrosprays: Theory and Application, **25**, 1005–1019.

Gunn R. (1954) Diffusion charging of atmospheric droplets by ions, and the resulting combination coefficients. *J. Meteorol.*, **11**, 329–347.

Hale B.N. (1986) Application of a scaled homogeneous nucleation-rate formalism to experimental data at $T \leq T_c$. *Phys. Rev. A*, **33**, 4156–4163.

Hegg D.A. (1990) Heterogeneous production of cloud condensation nuclei in the marine atmosphere. *Geophys. Res. Lett.*, **17**, 2165–2168.

Hinds W.C. (1982) *Aerosol Technology. Properties, Behavior and Measurement of Airborne Particles.* Wiley, New York.

Hirschfelder J.O., Curtiss C.F. and Bird R.B. (1954) *Molecular Theory of Gases and Liquids.* Wiley, New York.

Hoppel W.A. (1967) Theory of the electrode effect. *J. Atmos. Terr. Phys.*, **29**, 709–721.

Hoppel W.A. (1977) Ion attachment coefficients and the diffusional charging of aerosols, in H. Dolezalek and R. Reiter (eds), *Electrical Processes in Atmospheres*. Dietrich Steinkopf Verlag, Darmstadt, pp. 60–69.

Hoppel W.A and Frick G.M. (1986) Ion-attachment coefficients and the steady-state charge distribution on aerosol in a bipolar ion environment. *Aerosol Sci. Technol.*, **5**, 1–21.

Hussin A., Scheibel H.G., Becker K.H. and Postendorfer J. (1983) Bipolar diffusion charging of aerosol particles I: Experimental results within the diameter range 4–30 nm. *J. Aerosol Sci.*, **14**, 671–677.

Ivanov V.D. and Kirichenko V.N. (1970) Spontaneous unipolar charging of beta-active 'hot' aerosol particles. *Sov. Phys. Dokl.*, **14**, 859–862.

Ivanov V.D., Kirichenko V.N. and Petryanov I.V. (1969) Charging of alpha-active aerosols by secondary electron emission. *Sov. Phys. Dokl.*, **13**, 902–904.

Jacko R.B. and Reed D.A. (1994) A banded double ring electrodynamic balance for the suspension of submillimeter sized particles. *J. Aerosol Sci.*, **25**, 289–294.

Jantunen M.J. and Reist P.C. (1983) General field charging theory for aerosol particle charging and neutralizing in unipolar and bipolar ion fields. *J. Aerosol Sci.*, **14**, 127–133.

Jones A., Roberts D.L. and Slingo A. (1994) A climate model study of indirect radiative forcing by anthropogenic sulphate aerosols. *Nature*, **370**, 450–453.

Kalkinen J., Vesala T. and Kulmala M. (1991) Binary droplet evaporation in the presence of an inert gas: an exact solution of the Maxwell–Stefan equations. *Int. Comm. Heat and Mass Transfer*, **18**, 117–126.

Kaschiev D. (1982) On the relation between nucleation work, nucleation size, and nucleation rate. *J. Chem. Phys.*, **76**, 5098–5102.

Katz J.L., Hung C.H. and Krasnopoler M. (1988) The homogeneous nucleation of nonane, in P.E. Wagner and G. Vali (eds), *Atmospheric Aerosols and Nucleation*. Springer-Verlag, Berlin, pp. 356–359.

Keefe D., Nolan P.J. and Rich T.A. (1959) Charge equilibrium in aerosols according to the Boltzmann law. *Proc. R. Irish Acad. A*, **60**, 27–45.

Kojima H. (1978) Measurements of equilibrium charge distributions in bipolar ionic atmosphere. *Atmos. Envir.*, **12**, 2363–2368.

Kreidenweis S.M., Yin F., Wang S.C., Grosjean D., Flagan R.C. and Seinfeld J.H. (1991) Aerosol formation during photooxidation of organosulfur species. *Atmos. Environ. A*, **25**, 2491–2500.

Kulmala M. and Vesala T. (1991) Condensation in the continuum regime. *J. Aerosol Sci.*, **22**, 337–346.

Kulmala M., Vesala T. and Wagner P.E. (1993a) An analytical expression for the rate of binary condensational particle growth. *Proc. Roy. Soc. London A*, **441**, 589–605.

Kulmala M., Laaksonen A., Korhonen P., Vesala T., Ahonen T. and Barrett J.C. (1993b) The effect of atmospheric nitric acid vapour on cloud condensation nucleus activation. *J. Geophys. Res.*, **98** (D12), 22 949–22 958.

Landau L.D. and Lifshitz E.M. (1980) *Statistical Physics*, 3rd edn. Pergamon Press, Oxford.

Langner J. and Rodhe H. (1991) A global three-dimensional model of the tropospheric sulfur cycle. *J. Atmos. Chem.*, **13**, 225–263.

Lifshitz I.M. and Slezov V.V. (1959) Kinetics of diffusive decomposition of supersaturated solid solutions. *Sov. Phys. JETP*, **8** (35), 331–339.

Lifshitz I.M. and Slezov V.V. (1961) The kinetics of precipitation from supersaturated solid solutions. *J. Phys. Chem. Solids*, **19**, 35–50.

Lin J.-C., Chang Y.-C., Gentry J.W. and Ranade M.B. (1994) Models for discharge and evaporation in electrospray pyrolysis. *J. Aerosol Sci.*, **25**, S217–S218.

Lissowski P. (1940) Das Laden von Aerosolteilchen in einer bipolaren Ionenatmosphäre. *Acta Phisicochimica URSS*, **13**, 157–192.

Liu B.Y.H. and Kapadia A. (1978) Combined field and diffusion charging of aerosol particles in the continuum regime. *J. Aerosol Sci.*, **9**, 227–242.

Liu B.Y.H. and Pui D.Y.H. (1974) Electrical neutralization of aerosols. *J. Aerosol Sci.*, **5**, 465–472.

Lothe J. and Pound G.M. (1963) *Condensation and Evaporation*. Pergamon, Oxford.

Loyalka S.K. (1976) Brownian coagulation of aerosols. *J. Colloid Interface Sci.*, **57**, 578–579.

Lushkinov A.A. and Kulmala M. (1995) Source-enhanced condensation in monocomponent disperse systems. *Phys. Rev. E*, **52**, 1658–1668.

MacKenzie A.R. and Haynes P.H. (1992) The influence of surface kinetics on the growth of stratospheric ice crystals. *J. Geophys. Res.*, **97**, 8057–8064.

Manton M.J. (1983) The physics of clouds in the atmosphere, *Rep. Progr. Phys.*, **46**, 1393–1444.

Marlow W.H. and Brock J.R. (1975) Unipolar charging of small aerosol particles. *J. Colloid Interface Sci.*, **50**, 32–38.

Mason B.J. (1971) *The Physics of Clouds*. Oxford University Press, Oxford.

Matsoukas T. (1994) Charge distributions in bipolar particle charging. *J. Aerosol Sci.*, **25**, 599–609.

Mayya Y.S. (1994) Charging-induced drift and diffusion of aerosol particles in oscillating electric fields. *J. Aerosol Sci.*, **25**, 277–288.

Mayya Y.S. and Malvankar S.V. (1993) Charge-induced diffusion of aerosol particles moving under external electric fields in ionised gases. *J. Colloid Interface Sci.*, **156**, 78–84.

Mayya Y.S. and Sapra B.K. (1996) Variation of the charge neutralization coefficient in the entire particle size range. *J. Aerosol Sci.*, **27**, 1169–1178.

McGraw R. and Laaksonen A. (1996) Scaling properties of the critical nucleus in classical and molecular-based theories of vapor–liquid nucleation. *Phys. Rev. Lett.*, **76**, 2754–2757.

McMurry P.H. and Rader D.J. (1985) Aerosol wall losses in electrically charged chambers. *Aerosol Sci. Technol.*, **4**, 249–268.

Meilinger S.K., Koop T., Luo B.P., Huthwelker T., Carslaw K.S., Krieger U., Crutzen P.J. and Peter T. (1995) Size-dependent stratospheric droplet composition in lee wave temperature fluctuations and their potential role in PSC freezing. *Geophys Res. Lett.*, **22**, 3031–3034.

Mohnen V.A. (1974) Formation, nature and mobility of ions of atmospheric importance, in H. Dolezalek and R. Recter (eds), *Proc. 5th Int. Conf. on Atmospheric Electricity*, Garmisch-Partenkirchen Steinkopff, Darmstadt, pp. 1–17.

Pluvinage P. (1946) Etude théorique et experimentale de la conductibilité électrique dans les nuages non orageux. *Ann. Geophys.*, **2**, 31–54.

Pollak L.W. and Metneiks A.L. (1963) On the validity of Bolzmann's distribution law for the charges of aerosol particles in electrical equilibrium. *Geofisica*, **53**, 111–132.

Poluektov P.P., Emets E.P. and Kascheev V.A. (1991) On steady-state distribution of aerosol particle electric charges. *J. Aerosol Sci.*, **22**, S237–S240.

Porstendörfer J. (1994) Properties and behaviour of radon and thoron and their decay products in the air. *J. Aerosol Sci.*, **25**, 219–263.

Porstendörfer J., Röbig G. and Ahmed A. (1979) Experimental determination of the attachment coefficients of atoms and ions on monodisperse aerosols. *J. Aerosol Sci.*, **10**, 21–28.

Pruppacher H.R. and Klett J.D. (1978) *Microphysics of Clouds and Precipitation*. D. Reidel, Dordrecht, Netherlands.

Rayleigh, Lord (1882) On the equilibrium of liquid conducting masses charges with electricity. *Phil. Mag.*, **14**, 184–186.

Reiss H., Katz J.L. and Cohen E.R. (1968) Translation-rotation paradox in the theory of nucleation. *J. Chem. Phys.*, **48**, 5553–5560.

Reiss H., Tabazadeh A. and Talbot J. (1990) Molecular theory of vapour phase nucleation: the physically consistent cluster. *J. Chem. Phys.*, **92**, 1266–1274.

Rosenkilde C.E. and Serduke F.J.D. (1982) Electrical aspects of rainout, in H.R. Pruppacher, R.G. Semonin and W.G.N. Slinn (eds), *Precipitation Scavenging, Dry Deposition and Resuspension, Proc. 4th Int. Conf.*, Santa Monica, CA, Vol. 1. Elsevier, New York, pp. 573–581.

Rosinski J., Werle D. and Naganoto C.T. (1962) Coagulation and scavenging of radioactive aerosols. *J. Colloid Sci.*, **17**, 703–716.

Rudolph R., Majerowicz A., Kulmala M., Vesala T., Viisanen Y. and Wagner P.E. (1991) Kinetics of particle growth in supersaturated binary vapor mixtures. *J. Aerosol Sci.*, **22**, S51–S54.

Schikarski W.O. *et al.* (1988) Nuclear aerosol science. *Nuclear Technol.*, **81**, 137–306.

Seinfeld J.H. (1986) *Atmospheric Chemistry and Physics of Air Pollution*. Wiley Interscience, New York.

Shtern V. and Barrero A. (1994) Striking features of fluid flows in Taylor cones related to electrosprays. *J. Aerosol Sci.*, **25**, 1049–1063.

Shapiro M., Gutfinger C. and Laufer G. (1988) Electrostatic mechanisms of aerosol collection by granular filters: a review. *J. Aerosol Sci.*, **19**, 651–677.

Smith M.H., O'Dowd C.D. and Lowe J.A. (1996) Observations of cloud-induced aerosol growth, in M. Kulmala and P.E. Wagner (eds), *Nucleation and Atmospheric Aerosols 1996*. Pergamon/Elsevier Science, Oxford, pp. 937–940.

Storozhev V.B. (1995) Estimation of the contribution of heterogeneous reactions on the surface of aerosol particles to the chemistry of the atmosphere. *J. Aerosol Sci.*, **26**, 1179–1187.

Stoyanova V., Kaschiev D. and Kupenova T. (1994) Freezing of water droplets seeded with atmospheric aerosols and ice nucleation activity of the aerosols. *J. Aerosol Sci.*, **25**, 867–877.

Strey R., Wagner P.E. and Viisanen Y. (1994) The problem of measuring homogeneous nucleation rates and the molecular contents of nuclei: progress in the form of nucleation pulse measurements. *J. Phys. Chem.*, **98**, 7748–7758.

Taflin D.C. and Davis E.J. (1990) A study of aerosol chemical reactions by optical resonance spectroscopy. *J. Aerosol Sci.*, **21**, 73–86.

Taylor G.I. (1964) Disintegration of water drops in an electric field. *Proc. Roy. Soc. London A*, **280**, 383–397.

Van Dingenen R. and Raes F. (1991) Determination of the condensation accommodation coefficient of sulfuric acid on water-sulfuric acid aerosol. *Aerosol Sci. Technol.*, **15**, 93–106.

Vatazhin A., Lebedev A., Likhter V., Shulgin V. and Sorokin A. (1995) Turbulent air-stream jets with a condensed dispersed phase: theory, experiment, numerical modeling. *J. Aerosol Sci.*, **26**, 71–93.

Wagner P.E. (1982) Aerosol growth by condensation, in *Aerosol Microphysics II* (ed. W.H. Marlow). Springer-Verlag, Berlin, pp. 129–178.

Wagner P.E. (1985) A constant-angle Mie scattering method (CAMS) for investigation of particle formation processes. *J. Colloid Interface Sci.*, **105**, 456–467.

Warren D.R. and Seinfeld J.H. (1985) Prediction of aerosol concentrations resulting from a burst of nucleation. *J. Colloid Interface Sci.*, **105**, 136–142.

Wiedensohler A., Büscher P., Hansson H.-C., Martinsson B.G., Stratman S., Ferron G. and Busch B. (1994) A novel particle charger for ultrafine aerosol particles with minimal particle losses. *J. Aerosol Sci.*, **25**, 639–649.

Williams M.M.R. (1995) Growth rates of liquid drops for large saturation ratios. *J. Aerosol Sci.*, **26**, 477–487.

Williams M.M.R and Loyalka S.K. (1991) *Aerosol Science Theory and Practice*. Pergamon Press, Oxford.

Wilson C.T.R. (1897) Condensation of water vapour in the presence of dust-free air and other gases. *Phil. Trans. Roy. Soc. London A*, **189**, 265–274.

Winklmayr W., Reischl G.P., Lindner A.O. and Berner A. (1991) A new electromobility spectrometer for the measurement of aerosol size distributions in the size range from 1 to 1000 nm. *J. Aerosol Sci.*, **22**, 289–296.

Yeh H.C., Newton D.J., Raabe O.G. and Boor D.B. (1976) Self-charging of Au[198] labelled monodisperse gold aerosols studied with a miniature electrical spectrometer. *J. Aerosol Sci.*, **7**, 245–253.

Yeh H.C., Newton D.J. and Teague S.V. (1978) Charge distribution on plutonium-containing aerosols produced in mixed-oxide reactor fuel fabrication and the laboratory. *Health Phys.*, **35**, 500–503.

**Nomenclature**

| | |
|---|---|
| $A$ | area |
| $b$ | dimensionless coefficient (5.59) |
| $c_b$ | modified enthalpy for heat transfer ($= h_v - R_G T/\mu_v$) |
| $c_e$ | equilibrium vapour concentration ($= \rho_{ve}/\rho$) |
| $c_n$ | concentration of clusters of $n$ molecules |
| $C_p$ | specific heat at constant pressure |
| $C_v$ | specific heat at constant volume |
| $Cn_k$ | kinetic condensation number |
| $Cn_s$ | surface condensation number (5.12) |
| $D$ | molecular diffusivity |
| $D_T$ | turbulent diffusivity |
| $e$ | unit of charge |
| $\mathbf{E}$ | electric field |
| $f$ | function (5.38) |
| $F$ | mass flux from droplet (5.48) |
| $g$ | acceleration due to gravity |
| $G(R)$ | radiative driving term [RAD] as a function of $R$ |
| $G_c$ | Gibbs free energy |
| $h$ | enthalpy or heat content per unit mass |
| $H$ | heat transfer coefficient (5.49) |
| $i$ | van't Hoff factor (5.59), ion current ($+/-$ subscripts) |
| $I$ | ion pair production per decay |
| $j$ | integral number of unit charges |
| $J$ | mean charge (5.25) |
| $J_N$ | nucleation rate (5.73) |
| $J(n)$ | cluster concentration growth current (5.80) |
| $k$ | thermal conductivity |
| $k_B$ | Boltzmann constant |
| $K_N$ | kinetic growth factor for nucleation (5.74) |
| $Kn$ | Knudsen number (5.11) |
| $L$ | latent heat of vaporization |
| $Le$ | Lewis number (5.44) |
| $m$ | molecular mass |
| $M_L$ | liquid mass in droplet |
| $M_n$ | moment of aerosol size distribution (5.1) |
| $M_s$ | solute mass in droplet |
| $n_j(R)$ | aerosol charge and size distribution |
| $N$ | number concentration |
| $N_j$ | aerosol charge distribution |
| $p$ | pressure |
| $P_E$ | electric stress (5.40) |
| $P_\gamma$ | surface stress (5.40) |

| | |
|---|---|
| $q$ | ion production rate |
| $Q_R$ | radiative heat flux from droplet |
| $r$ | space coordinate |
| $R$ | aerosol radius; radial distance (in 5.40, see Figure 5.5) |
| $R_{av}$ | mean aerosol radius |
| $R_c$ | critical radius for nucleation |
| $R_G$ | gas constant |
| $R_s$ | radius of solute spherical mass in droplet |
| $R_\gamma$ | Kelvin radius (5.46) |
| $\dot{R}$ | aerosol radius growth rate ($dR/dt$) |
| $S$ | saturation |
| $S_c$ | critical saturation (5.61) |
| $S_e$ | equilibrium saturation for droplet containing solute (5.58) |
| $S_p$ | molecular sticking probability |
| $t$ | time |
| $t_{OR}$ | Ostwald ripening time-scale (5.72) |
| $T$ | temperature |
| $\mathbf{v}$ | velocity |
| $V$ | volume |
| $x$ | ion asymmetry parameter (5.16); coordinate (5.37) |
| $X_L$ | solvent mole fraction |
| $y$ | radioactive aerosol charging parameter (5.30) |
| $Y$ | function of variables for two particles (5.39) |
| $z$ | vertical coordinate |
| | |
| $\alpha$ | ion recombination coefficient; thermal accommodation coefficients (subscripted) |
| $\alpha_G$ | coefficient in aerosol growth rate (5.63) |
| $\beta$ | coefficient in Clausius–Clapeyron equation (5.42) |
| $\beta_0$ | basic ion attachment coefficient |
| $\beta_{+1j}, \beta_{-1j}$ | ion attachment coefficients |
| $\beta(n)$ | $n$-cluster growth probability |
| $\gamma$ | specific heat ratio; surface tension (subscripted L or S) |
| $\gamma_r$ | surface chemical reaction rate coefficient (5.86) |
| $\gamma(n)$ | $n$-cluster decay probability |
| $\Gamma_L$ | activity coefficient |
| $\Delta_b$ | dimensionless ratio (5.10) |
| $\varepsilon_0$ | permittivity of the vacuum |
| $\varepsilon_p$ | dielectric constant |
| $\eta$ | gas viscosity |
| $\eta_{dr}$ | radioactive decay rate |
| $\theta$ | angle |
| $\lambda$ | aerosol size parameter (5.21) |
| $\lambda_m$ | molecular mean free path |

| | |
|---|---|
| $\mu$ | electrical mobility (5.15); molecular mass (subscripted v or L) |
| $\mu_c$ | chemical potential |
| $\mu_G$ | growth rate coefficient (5.63) |
| $v_v(T)$ | molecular kinetic mass flux at temperature $T$ |
| $\rho$ | density |
| $\sigma$ | standard deviation of Gaussian distribution |
| $\sigma_c$ | condensation coefficient |
| $\sigma_e$ | evaporation coefficient |
| $\tau_S$ | neutralization time-scale (5.35) |

## Subscripts

| | |
|---|---|
| a | pertaining to aerosol |
| av | pertaining to average or mean value |
| c | pertaining to critical value |
| e | pertaining to equilibrium |
| g | pertaining to gas |
| $j$ | pertaining to charge $je$ |
| L | pertaining to liquid |
| $n$ | pertaining to number $n$ |
| p | pertaining to particle or pressure |
| s | pertaining to surface |
| S | pertaining to solid |
| v | pertaining to vapour or volume |
| w | pertaining to wall |
| $+, -$ | pertaining to sign of charge |
| $\infty$ | pertaining to value at infinity |

# 6 Fibrous filtration

J.I.T. STENHOUSE

## 6.1 Introduction

Fibrous filters are used to remove aerosol particles from gas flows. A mat of fine fibres which are arranged more or less normal to the direction of flow but in random orientation in that plane is used to collect the particles. A scanning electron micrograph of a section of such a mat, or medium, is shown in Figure 6.1. The spaces between fibres are quite large relative to the particles. As the gas passes through the mat the particles are collected on the surface of the fibres by a range of mechanisms. If a particle is not captured by the first fibre it sees, as the gas in which it is suspended passes into and through the mat, it will have further opportunities of capture as

**Fig. 6.1** Scanning electron micrograph of the surface of a fibrous filter.

it penetrates deeper into it. In this way the process is probabilistic – the deeper the mat the less the chance of penetration. Particles accumulate within the mat and the pressure drop increases until it reaches an unacceptable level when the filter has to be replaced with a new unit. For this reason these depth filters are used to clean gases in situations where the particulate concentration is low.

Typical applications include dust removal in air-conditioning systems, clean room installations, engine intakes, face masks, process gas filtration and other applications where high efficiency of particulate removal is necessary.

Many types of filter are available to clean a gas to the required standard. Prefilters, or roughing filters, may be in the form of panels, pads, rolls, etc. The fibres are normally of large diameter, up to about 200 μm, and the face velocity is often high, sometimes up to about 2.5 m s$^{-1}$. An adhesion assisting fluid is usually applied to the fibres to prevent particle bounce and ensure retention. Rolls, pleated media and envelopes of media are also used. These are high-velocity, low-pressure-drop units which remove coarser particles.

Panel filters are supplied as both prefilters and secondary filters. In this latter category very high efficiencies can be achieved with extremely small particles in the submicrometre range. These include high-efficiency particulate air (HEPA) and ultrahigh-efficiency particulate air (UHEPA) filters. The fibres are fine – frequently in the submicrometre diameter region, the medium face velocity may be as low as 25 mm s$^{-1}$, and the clean filter pressure drop as low as 150 Pa. Fine glass fibres are very often used. A large number of unit designs are available. Probably the most popular is the $609 \times 609$ mm by 298 mm panel. In the deep pleat version of this the nominal medium surface area is 20 m$^2$ and the maximum rated capacity about 1700 m$^3$ h$^{-1}$. In the mini pleat version, in which the spacing between pleats is reduced, the medium area is increased to 35 m$^2$. Typical applications are in clean rooms and in nuclear installations.

Face masks contain pads of fibrous media, and here it is important to achieve the required efficiency with minimum pressure drop as well as ensuring a good fit. Fibrous filters are also used in engine intakes to remove potentially abrasive material. Cartridge filters are widely used to clean process gases. Again submicrometre glass fibres are used in cartridges to clean, for example, compressed air. Stainless steel fibres down to about 1.5 μm diameter make extremely robust media and ceramic fibrous media are available for very high-temperature work.

A vast amount of experimental and theoretical work has been carried out on the subject and a number of comprehensive texts have been written. Among the most important of these are the books by Davies (1973) and Brown (1993).

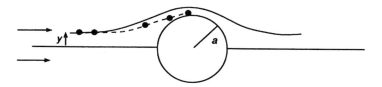

**Fig. 6.2** Single-fibre efficiency. The particle deviates from the gas streamline on its critical trajectory – a particle starting further from the centre line would miss the fibre. The single fibre efficiency is $y/a$.

## 6.2   Single-fibre efficiency

It is common practice to consider the behaviour of particles with respect to a single ideal fibre, normal to the direction of flow, within the depth of a filter. This is then related to the overall behaviour of the medium. The single-fibre efficiency is the ratio of the volume of air cleaned by the fibre to the fibre swept volume. This is illustrated in Figure 6.2, in which the critical particle trajectories are shown – particles inside the space represented by $2y$ are collected and the swept volume is represented by the space $2a$. The single-fibre efficiency here is simply

$$\eta_{si} = \frac{y}{a}.$$

Consider a differential depth, $dl$, in a filter of unit cross-sectional area. The length of fibre in this cross section is

$$\frac{\alpha dl}{\pi a^2},$$

where $\alpha$ is the packing density (volume of fibre/volume of filter).

If the volumetric gas rate is $V$ then the interstitial velocity in the filter is $V/(1 - \alpha)$. The volume of gas cleaned in this differential volume is thus

$$dV = \frac{2\eta_{si}\alpha V}{\pi a(1 - \alpha)dl}.$$

Since the total volumetric rate of gas passing through this element is $V$, the fraction of gas cleaned is

$$\frac{dV}{V} = \frac{2\eta_{si}\alpha dl}{\pi a(1 - \alpha)} = -\frac{dc}{c},$$

where $c$ is the aerosol particle concentration. Integrating over the depth of the filter,

$$\frac{c}{c_0} = \exp\left(\frac{-2\alpha L\eta_{si}}{\pi a(1 - \alpha)}\right).$$

This is the filter penetration, and since

$$\text{efficiency} = 1 - \text{penetration}$$

the overall efficiency of the filter, $\eta_0$, is given by

$$\eta_0 = 1 - \left(\frac{c}{c_0}\right) = 1 - \exp\left(\frac{-2\alpha L\eta_{\text{si}}}{\pi a(1 - \alpha)}\right).$$

This relates the overall filtration efficiency to the single-fibre efficiency for a filter of depth $L$. In the following sections the prediction of the single-fibre efficiency will be discussed.

## 6.3   Flow fields in fibrous media

A prerequisite of any particle–fibre collision theory is a description of the flow field within a filter. The true flow field is extremely complex due to the randomness of the structure and defies analytical description. Consequently, a simplified mathematical model must be used. The objective is a description of the velocity resolutes in the proximity of the fibre, preferably in dimensionless terms. Particles are captured more easily if the streamlines have a high tortuosity and pass close to the fibre surface. The tortuosity increases with fibre Reynolds number and medium packing density.

To obtain a solution to this problem it is necessary to solve the Navier–Stokes equation with the appropriate boundary conditions:

$$\mathbf{u} \cdot \nabla\mathbf{u} = \frac{\nabla p}{\rho} + v \cdot \nabla^2\mathbf{u}.$$

Analytical solutions have been derived for fairly simple models of the system and more recent attempts have involved iterative numerical techniques.

### 6.3.1   *Description of fields close to single fibres*

The simplest model for the flow field is that of an isolated fibre in an infinite medium. Even with this an analytical solution for the full equation is not available. Lamb (1911) solved the Oseen approximation of the Navier–Stokes equation:

$$U_0 \cdot \nabla\mathbf{u} = \frac{\nabla p}{\rho} + v \cdot \nabla^2\mathbf{u}.$$

In his solution the influence of fibre Reynolds number is included but the effect of neighbouring fibres is ignored. This has some application in the case of relatively open filters which are operated at high face velocities

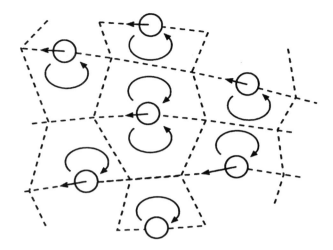

**Fig. 6.3** Happel–Kuwabara flow net. The bank of parallel cylinders is considered to be moving from right to left through the fluid.

although only for a very limited range of Reynolds numbers (in excess of 0.2). The great majority of filters are operated with very low Reynolds numbers and medium packing densities in the range 0.04–0.10. In these conditions the influence of packing density on the field is much more important than Reynolds number.

The flow through a filter can be more realistically described using a cellular model. The filter is considered to consist of a number of cells, each comprising a single fibre surrounded by a concentric envelope of fluid. The diameter of the cell is determined by the proximity of neighbouring fibres and is thus related to the packing density of the medium. In this way each cell may be considered as an independent entity and the flow field in the envelope can be considered in isolation from the rest of the medium. A cellular model of this type was published by Happel (1959) and Kuwabara (1959) to describe the flow through banks of parallel cylinders. The network of cylinders is illustrated in Figure 6.3 and the individual cell in Figure 6.4. The cell radius is related to the packing density, or solidity, of the medium by

$$\alpha = \frac{a^2}{b^2}.$$

The equation of creeping motion, in which the Reynolds number is assumed to be zero, was solved for the field within the envelope:

$$0 = \frac{\nabla p}{\rho} + v \cdot \nabla^2 \mathbf{u}.$$

MATHEMATICAL MODEL

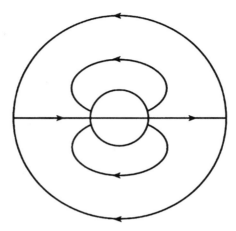

**Fig. 6.4** The mathematical model is a cylindrical envelope of fluid surrounding the cylinder. The cylinder is again moving across the cell. For the stationary cylinder and moving fluid case the main stream velocity is added to the field.

Happel assumed the boundary condition of zero slip at the fibre surface and zero shear stress at the surface of the imaginary cell. Kuwabara assumed zero vorticity at the cell surface. The general solution for the stream function for this model published by both authors is given by

$$\psi = \left( Ar + \frac{B}{r} + Cr \ln\left(\frac{r}{a}\right) + Dr^3 \right) \sin\theta.$$

The values of the constants are shown in Table 6.1. The velocity resolutes in cylindrical polar coordinates are then obtained from

$$u_r = \frac{1}{r}\frac{\partial\psi}{\partial\theta}, \qquad u_\theta = -\frac{\partial\psi}{\partial r}.$$

Following flow visualization work by Kirsh and Fuchs (1967) it is generally accepted that the Kuwabara solution is more appropriate in fibrous filtration studies.

A major problem with these fields is that there is a step change in the velocity vector at the outside boundary which cannot be correct. Spielman and Goren (1968) proposed a field based on the introduction of a permeability term on the left-hand side of the creeping motion equation. The problem with the change in velocity at the boundary is obviated.

A simplified version of the stream function for each of these fields, which is applicable close to the fibre surface, can be expressed as

$$\psi = \frac{aU_0}{2\xi}\left(\frac{a}{r} - \frac{r}{a} + 2\frac{r}{a}\ln\left(\frac{r}{a}\right)\right)\sin\theta,$$

where the hydrodynamic factor, $\xi$, is given in Table 6.2.

All these analytical solutions suffer from the disadvantage that they include either the effect of Reynolds number or packing density but not both. They are derived from simplistic pictures of a very complex system but are relatively easy to use. Predictions obtained using these models must inevitably be approximate.

In the models described above the boundary condition of zero velocity at the fibre surface is assumed. Unfortunately, with fibres smaller than about $2\,\mu m$ diameter there is some slip at the surface and this has to be taken into account by modifying the boundary condition. This was achieved by Pich (1966) who modified the Kuwabara model to apply for finite Knudsen numbers. Although the methods of describing the flow fields discussed here result from idealized models and are thus approximations, they have the benefit of simplicity.

**Table 6.1** Values of constants for the general solution of the stream function for the models of Happel and Kuwabara

| | |
|---|---|
| $A$ | $\frac{(\alpha - 1)}{2}J$ |
| $B$ | $\frac{a^2}{2}\left(1 - \frac{\alpha}{2}\right)J$ |
| $C$ | $J$ |
| $D$ | $\frac{-\alpha}{4a^2}J$ |
| $J$ | $V_0\left(-\frac{1}{2}\ln\alpha - \frac{3}{4} + \alpha - \frac{\alpha^2}{4}\right)^{-1}$ |

**Table 6.2** Expressions for $\xi$ in the simplified version of the stream function

| Field | $\xi$ | $\xi$ for very low $\alpha$ |
|---|---|---|
| Lamb | $2 - \ln N_{\mathrm{Re}}$ | |
| Happel | $-\frac{1}{2}\ln\alpha - \frac{1}{2} + \frac{\alpha^2}{2(1 + \alpha^2)}$ | $-\frac{1}{2}\ln\alpha - \frac{1}{2}$ |
| Kuwabara | $-\frac{1}{2}\ln\alpha - \frac{3}{4} + \alpha - \frac{\alpha^2}{4}$ | $-\frac{1}{2}\ln\alpha - \frac{3}{4}$ |
| Spielman and Goren | $K_0(k_1^{1/2}a)/k_1^{1/2}aK_1(k_1^{1/2}a)$ | $-\ln(\frac{1}{2}k_1^{1/2}a) - 0.5772$ |

$K_0$ and $K_1$ are Bessel functions of zero and first order and $k_1$ is the filter permeability. The terms in the right-hand column are applicable for very low values of $\alpha$.

### 6.3.2  *Description of flow through arrays of fibres*

Finite-element methods have been applied to the flow through arrays (Fardi and Liu, 1992; Liu and Wang, 1996); however, it is necessary to set up a network or mesh throughout the field. A more satisfactory technique would be to determine continuous stream functions throughout the field. A useful advance was presented by Brown (1984; 1986) who obtained the stream function for flow through a two-dimensional array, such that the dissipation of energy due to viscous drag is a minimum, subject to the boundary conditions of the system. Both packing density and Reynolds number effects are included and the entrance boundary condition difficulties, as with the Kuwabara model, are eliminated. The model can be adapted to non-cylindrical fibre systems and the formation of standing eddies is illustrated. An alternative method of solution is the boundary-element method numerical technique (Hildyard *et al.*, 1985). The extension of this from cylinders to fibres of arbitrary cross-section is immediate without significant increases in complexity or computing time.

It should be made clear that all the above are gross simplifications of the flow through a real filter. The fibres are not parallel and equidistant. Figure 6.1 shows this clearly. The structure is very complex and the flow through it has defied description. The models are such that the pressure drop and collection efficiency calculated using them are almost twice the experimental values. However, their use leads to qualitative estimates of the behaviour of filters with respect to the various particle collection mechanisms which, with some empirical correction, can be applied predictively.

## 6.4   Mechanisms of particle collision

A number of mechanisms play their part in determining the single-fibre efficiency. These are summarized in Figure 6.5. The behaviour of the smallest particles is described first.

### 6.4.1  *Diffusional collection*

Fine particles, usually less that about $0.3\,\mu m$ diameter, undergo significant Brownian diffusion so that they deviate from the gas streamlines as in Figure 6.5a. Because of this there is a finite probability of particle–fibre collision. The closer the streamline passes to the fibre surface the greater is this probability. It is also clear that the smaller the particle the greater the Brownian movement and thus the higher the collection efficiency. Lower gas velocities will increase the particle residence time close to the surface and increase collection. Quantitative expressions for this have been obtained from approximate boundary-layer type analyses. More rigorous

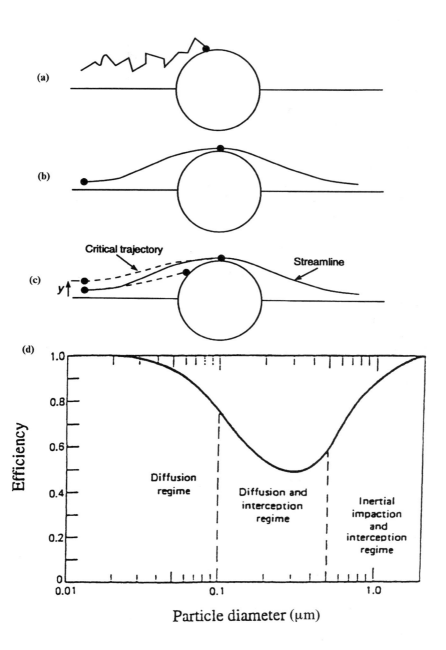

**Fig. 6.5** Particle capture mechanisms: (a) diffusional collection; (b) interception; (c) inertial collection. Effect of particle diameter on collection efficiency according to each mechanism (d).

but still approximate solutions of the diffuso-convective equation, and computer finite-element and trajectory models are available. The behaviour of a diffusing aerosol is described in dimensionless form by

$$\frac{\partial c'}{\partial \tau} + \bar{u}\nabla c' = \frac{2}{Pe}\nabla^2 c',$$

where

$$c' = \frac{c}{c_i}, \qquad \tau = \frac{tU_0}{a}, \qquad Pe = \frac{2aU_0}{D_{AB}}.$$

$Pe$ is the Peclet number, which is the ratio of the convective to diffusive mechanisms. This appears as the major dimensionless parameter in all expressions for collection by this mechanism. The hydrodynamic parameter must also have some effect on the process, albeit a minor one since the concentration boundary layer is quite thin. Since the particle density does not appear in the Stokes–Einstein equation the mechanism is insensitive to this parameter. One of the most widely accepted expressions is (Kirsch and Fuchs, 1968):

$$\eta_D = \frac{2.9}{\xi^{1/3}}Pe^{-2/3}.$$

This has been reasonably well verified experimentally, especially with regard to the value of the exponent.

### 6.4.2   Interception

In this mechanism the particle travels along the streamline and impacts with the fibre simply because of its size (Figure 6.5b). The critical group is the interception parameter,

$$N_R = \frac{d_p}{d_f}.$$

It is determined theoretically by equating the stream function at 90° to the fibre at a distance $1 + N_R$ from the axis with that upstream in the planar flow a distance $y$ from the axis. The efficiency is then

$$\eta_R = \frac{y}{a}.$$

It is then simple to show that where the Kuwabara field is applied, for example, the single-fibre efficiency is given by

$$\eta_R = \frac{1}{2\xi}\left\{2(1 + N_R)\ln(1 + N_R) - (1 + N_R)(1 + \alpha)\right.$$
$$\left. + (1 + N_R)^{-1}\left(1 - \frac{\alpha}{2}\right) - \frac{\alpha}{2}(1 + N_R)^3\right\}.$$

In the regime where aerodynamic slip becomes important, for fibre diameters less than about 2 μm, Pich (1966) has shown that for relatively low packing densities the above can be modified to give

$$\eta_R = \frac{(1 + N_R)^{-1} - (1 + N_R) + 2(1 + 1.996Kn_f)(1 + N_R)\ln(1 + N_R)}{2(-0.75 - 0.5\ln\alpha) + 1.996Kn_f(-0.5 - \ln\alpha)}.$$

### 6.4.3  Inertial impaction

This mechanism is shown in Figure 6.5c, which depicts the critical trajectory of the particle which just grazes the side of the fibre. Clearly those particles which start their trajectories inside this are captured and those outside escape, so the efficiency is given by the ratio of the starting point of the critical trajectory to the fibre radius.

The trajectories are determined by the Stokes number, conventionally defined in the case of fibrous filtration as the ratio of the particle stop distance to the fibre diameter, i.e.

$$Stk = \frac{\rho_p U_0 d_p^2}{18\eta d_f}.$$

It should be noted that in some work the inertia parameter, which has double this value, is used. These trajectories are of course strongly affected by the tortuosity of the flow field which is controlled by both the filter packing density, $\alpha$, and the fibre Reynolds number. The effect of both of these is illustrated in Figure 6.6. Unfortunately, as explained earlier, there

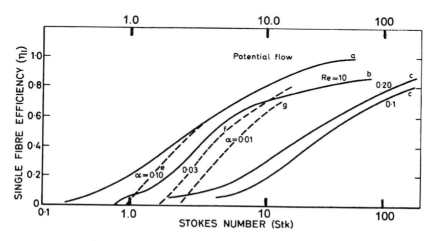

**Fig. 6.6** Efficiency of inertial collision – comparison of theories. Adapted from Stenhouse and Freshwater (1976, Figure 2).

is as yet no theoretical model of the field which takes the effect of both $\alpha$ and $N_R$ into account simultaneously. Further, interception must be included in any calculation since it controls the capture radius.

Using an approximate calculation method which is applicable only at low $Stk$, Stechkina $et$ $al.$ (1969) introduced the following based on the Kuwabara flow field:

$$\eta_{IR} = \eta_R + \frac{J}{4\xi^2} Stk,$$

where $J = (29.6 - 28\alpha^{0.62})N_R - 27.5N_R^{2.8}$. It is applicable for $N_R < 0.4$ and $\alpha < 0.11$. Based on experimental data from real filters Nguyen and Beekmans (1975) included both $\alpha$ and $N_R$ in an empirical expression:

$$\eta_I = \frac{Stk^3 f^3}{Stk^3 f^3 + 0.77\left(1 + \frac{4}{Re^{0.5}} + \frac{65}{Re}\right)Stk^2 f^2 + 0.58};$$

$f$ is related to the packing density in real filters by:

$$f = 1 + 4\alpha + 2250\alpha^2.$$

### 6.4.4  Gravitational settling

The gravitational settling parameter is the ratio of the gravitational settling speed to the main stream velocity:

$$G = \frac{\rho_p d_p^2 C_H g}{18\eta U_0}.$$

For upward and downward flow the single-fibre efficiency, $\eta_G$, is approximated by $1 - G(1 + N_R)$ and $G(1 + N_R)$, respectively. If the flow is horizontal then $\eta_G$ is of order $G^2$. Clearly this is only important when the face velocity is low and the particles have a high mass.

### 6.4.5  Electrostatic collection

This can be extremely important and a range of media are now available which specifically invoke the mechanism; however, it requires either the particle or the fibre or both to be charged before it can have any effect. The same degree of penetration can be achieved but the pressure drop is much less. If the filter is not charged, then the effect is usually ignored since in the practical situation the particle charge is not known and the most pessimistic case of zero particle charge is assumed.

The electrostatic force is relatively long-ranged and the mechanism will be most significant when the face velocity is low and the particle residence time in the proximity of the fibre is high. The single-fibre efficiency may be determined by including the electrostatic force resolutes, in dimensionless form, in the trajectory equations. This requires the introduction of dimensionless groups for each of the three simple symmetrical cases (Kraemer and Johnstone, 1955):

- *Fibre and particle charged.* This is a simple coulombic force given in dimensionless form by

$$\frac{N_{Qq}}{r'}; \quad \text{where} \quad N_{Qq} = \frac{Qq}{3\pi^2 \varepsilon_0 \eta d_p d_f U_0} \quad \text{and} \quad r' = \frac{r}{a}.$$

Kramer and Johnstone (1955) have suggested that the single-fibre efficiency due to this is given by $\eta_{Ec} = \pi N_{Qq}$.

- *Fibre neutral, particle charged.* The particle produces an image force as it approaches the fibre, which again in its dimensionless form and in the radial direction is given by

$$\frac{N_{0q}}{r' - 1}; \quad \text{where} \quad N_{0q} = \frac{q^2}{12\pi^2 \eta U_0 \varepsilon_0 d_p d_f^2} \left(\frac{\varepsilon_2 - 1}{\varepsilon_2 + 1}\right).$$

Lundgren and Whitby (1965) showed experimentally that $\eta_{Ei} = 1.5 N_{0q}$.

- *Fibre charged, particle neutral.* In this case charges are induced on the particle such that if, for example, the fibre is negative then the side of the particle closest to the fibre will become positively charged and that on the reverse side negatively charged. The net effect is an attraction force. Again, in its dimensionless form, this is approximated by

$$\frac{N_{Q0}}{(r')^3}; \quad \text{where} \quad N_{Q0} = \frac{Q^2 d_p^2}{3\pi^2 \varepsilon_0 \eta d_f^3 U_0} \left(\frac{\varepsilon_1 - 1}{\varepsilon_1 + 2}\right).$$

Values of $\eta_{Ei}$ have been given as $\pi N_{Q0}$, $(1.5\pi N_{Q0})^{1/3}$ and $0.84 N_{Q0}^{0.75}$ (for $\alpha = 0.03$) (Kraemer and Johnstone, 1955; Natanson, 1957; Stenhouse, 1974).

The effect of charging is shown in Figure 6.7. Electrostatically enhanced media have been classified by Brown (1993) according to the method of charging. Triboelectric charging takes place when two materials such as ebony and fur are rubbed together. An early example of media charged in this manner is the resin-wool filter developed by Hansen. Charging takes place during the carding process and the surface of the wool fibre becomes coated with negatively charged resin particles as shown in Figure 6.8. Another more recent example is the mixed fibre material in which two types of synthetic fibre are oppositely charged during the carding process. Corona charging is used to form electret fibres which are in the form of narrow ribbons with a positive charge on one side and negative on the other. These are widely used and have been well studied. The fibres in the

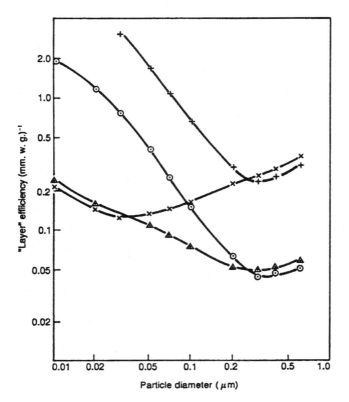

**Fig. 6.7** Layer efficiencies of fibrous filters tested with submicrometre aerosols. From Trottier and Brown (1990).

above materials are relatively coarse, in the region of 20 μm diameter. If induction charging during melt extrusion is used then finer charged fibres can be produced.

A major problem with all electrostatically enhanced media is charge retention. High humidities and temperatures have been shown to accelerate charge leakage (Brown, 1993) and filters deteriorate in use because of charge shielding and some neutralization by deposited particles.

### 6.4.6  *Particle adhesion and rebound*

In the above it has been assumed that once a particle collides with a fibre it will be retained. This is not necessarily true, especially in dry systems operated in conditions of relatively high inertia. A typical case would be a primary filter with a high face velocity intended to collect fairly coarse dust. An example of this is shown in Figure 6.9. The problem may be obviated in

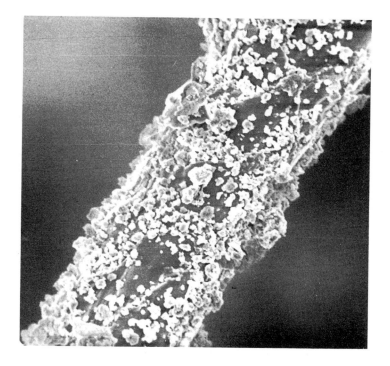

**Fig. 6.8** Fibre (22 μm diameter) from a resin-wool filter.

practice by the application of an adhesion assisting fluid to the filter. However, the question remains concerning the critical conditions above which
adhesion is no longer certain, and the probability of adhesion thereafter.

It can be easily illustrated that the major mechanism for particle non-
adhesion is bounce or rebound. Most of the kinetic energy of the incoming
particle will be available as rebound energy to overcome the adhesion
energy. This bounce process is very rapid as calculation shows that the
time to maximum compression of a 5 μm diameter elastic particle
travelling at $0.5\,\mathrm{m\,s^{-1}}$ is of the order of $10^{-7}\,\mathrm{s}$.

The three forces responsible for adhesion in fibrous filtration are van
der Waals, electrostatic and capillary forces. Of these the London van der
Waals force is the most important in dry filters. For an undeformed
spherical particle on a flat smooth surface this force, $\overset{\circ}{F}$, and energy, $\overset{\circ}{E}$, are
expressed as

$$\overset{\circ}{F} = \frac{A_H R}{6z^2}, \qquad \overset{\circ}{E} = \frac{-A_H R}{3z}.$$

$R$ is the radius of the particle in contact with the flat surface – it could

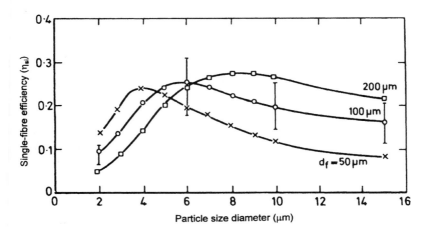

**Fig. 6.9** Experimental single-fibre efficiency for collection of feldspar particles on model filters of grids of parallel stainless steel fibres. Adapted from Stenhouse and Freshwater (1976).

in fact be the radius of a surface protrusion on a particle of dust. $A_H$ is the Hamaker van der Waals constant which is normally in the range $10^{-18}$–$10^{-20}$ J, and $z$ the minimum separation distance, which is usually taken as 0.3 nm. Particle or surface deformation during impact could significantly increase this force (Freshwater and Stenhouse, 1972) and adsorbed layers – for example, of water molecules – will have a strong effect.

In normal circumstances in air filtration electrostatic adhesion forces are not considered to be as strong as the van der Waals force. There will be an attraction force if the particle or substrate is charged, and this is usually the case; however, the charge is not likely to be large enough for the mechanism to be important. Contact charging may, on the other hand, result in large electrostatic forces. Krupp (1967) showed that in the case of surfaces with very high contact potentials the electrostatic force could be greater than the van der Waals force for particles with surface asperities.

Capillary forces may be very large and operate where the fibre is coated with a thin layer of adhesion assisting fluid or in cases of very high relative humidity. The force, between a sphere and a flat surface, is simply given by

$$F = 4\pi\gamma R,$$

where $\gamma$ is the surface tension. Although the force is strong, it is doubtful if there is sufficient time for the meniscus to form during impact; yet there is no doubt that coated filters are much more efficient in the high inertia regime. It is likely that the fluid dissipates much of the kinetic energy of the incoming particle. Following capture the mechanism will assure retention.

The rebound phenomenon was observed directly by Dahneke (1975), who measured particle velocities before and after bouncing from a surface by measuring the transit time between two laser beams. He also defined the capture limit velocity, $v^*$, as the incident velocity at the limit between particle bounce of capture. The total adhesion energy is the sum of the incident adhesion energy, $E_i$, prior to any deformation or dissipation, and an added term which may, for example, be due to deformation, $\Delta E$. The energy available for removal, or rebound, is related to the total incident energy $(E_{kinetic} + E_i)$ through the coefficient of restitution, $e$:

$$E_{adhesion} = E_i + \Delta E, \qquad E_{removal} = e^2(E_{kinetic} + E_i).$$

Equating and taking $\Delta E$ as the difference between the adhesion energies seen by incoming and rebounding particles, the capture limit is found to be (Dahneke, 1971):

$$v^* = \left\{ \frac{-2[\Delta E + (1 - e^2)E_i]}{me^2} \right\}^{1/2}.$$

In some work $\Delta E$ is taken as zero, but this may be unrealistic. Further more detailed analyses have been presented more recently (Xu et al., 1993; Andres, 1995; Dahneke, 1995). The trajectories of small particles close to the surface have been followed using high-speed cine photography (Broom, 1979). This shows the effect of impact angle and surface roughness. Hiller and Loeffler (1978), in another energy model, point out the importance of plastic deformation. In air filtration the velocity of the particle is reduced as it approaches the fibre so the kinetic energy of impact, which is the major energy made available through compression for rebound, is also reduced. There is thus a rapid increase in this energy as inertial impaction becomes effective so it is almost axiomatic that high levels of inertia lead to low adhesion probability.

## 6.5 Pressure drop

Almost all filters are operated with low fibre Reynolds numbers in the viscous flow regime, so the pressure drop is directly proportional to the face velocity. At higher Reynolds numbers, above about 1.0, more energy is dissipated and the relationship becomes nonlinear. In principle the pressure drop can be calculated by summing the drag forces on all the fibres in a unit area of the medium. The drag force per unit length of fibre, $F$, is given by

$$F = \frac{4\pi U_i \eta}{\xi},$$

where $U_i$ is the interstitial velocity, or $U_0/(1 - \alpha)$, and $\xi$ the hydrodynamic factor as in Table 6.2. The pressure drop across the filter is then

$$\Delta p = \frac{4 \alpha l U_i \eta}{\xi d_f^2}.$$

This of course applies to a perfectly homogeneous filter or one in which the fibres are equally spaced and far apart. This is a gross simplification and significantly over-predicts the pressure drop. It is necessary therefore to add a homogeneity factor, $\varepsilon$, which is the ratio of the theoretical pressure drop using the above, with the Kuwabara value for $\xi$, and the experimental pressure drop. $\varepsilon$ is found to lie in the range 1.13–2.25. Where very small fibres are used, comparable to the mean free path of the gas, slip flow at the fibre surface has to be taken into account. A modified Kuwabara hydrodynamic factor applicable up to a fibre Knudsen number, $Kn_f$, of 0.25 can then be used (Pich, 1966):

$$\xi_{Kn} = \frac{\xi + Kn_f \left( \dfrac{\alpha^2}{2} - \dfrac{1}{2} - \ln \alpha \right)}{1 + 2 Kn_f}.$$

From a large number of experimental observations using widely differing filter materials and values of $\alpha$ covering the range 0.006–0.3, Davies (1952) determined the following empirical expression:

$$\Delta p = \frac{64 \eta l U_0 \alpha^{1.5} (1 + 56 \alpha^3)}{d_f^2}.$$

### 6.6    Calculation of filter performance

The overall efficiency of a filter can be found by solving the equation for the overall efficiency of the filter, at the end of section 6.2, if the single-fibre efficiency due to the appropriate mechanisms is known. To this end it has been accepted practice simply to use the sum of the individual contributions:

$$\eta_{si} = \eta_D + \eta_R + \eta_{DR} + \eta_I + \eta_G + \eta_E.$$

This is theoretically incorrect, but it is justified on the grounds that only one or two of the mechanisms will be significant at any time. Generally diffusion is important for particles less than 0.3 μm in diameter and inertia for those above about 0.5 μm. Where operation is in the diffusional and interception regimes it is necessary to add an interference term, $\eta_{DR}$. Similarly, inertia and interception will occur together and particle inertia can diminish electrostatic effects.

Two methods of calculating $\eta_{si}$ have been developed for the case where interception and diffusion occur together.

The fan model is based on the use of a model filter, consisting of planes of structured fibre grids placed at random orientations. This acts as an intermediate between theory and observations of real filter behaviour. It was introduced by Fuchs *et al.* (1973), described in some detail later (Kirsch and Stechkina, 1978) and was modified by Kirsch and Chechuev (1985). The single-fibre efficiency of the fan model filter, $\eta_{si}^f$, is calculated first, and this is then modified to give $\eta_{si}$ by incorporating an inhomogeneity factor $\varepsilon_f$ which is determined from the media pressure drop:

$$\eta_{si}^f = \eta_D^f + \eta_R^f + \eta_{DR}^f$$

$$\eta_D^f = 2.7Pe^{-2/3}\{1 + 0.39(\xi^f)^{-1/3}Pe^{-1/3}Kn\}$$

$$\eta_{DR}^f = 1.24(\xi^f)^{-1/2}Pe^{-1/2}N_R^{2/3}$$

$$\eta_R^f = (2\xi^f)^{-1}\{(1 + N_R)^{-1} - (1 + N_R) + 2(1 + N_R)\ln(1 + N_R)$$
$$+ 2.86Kn_f(2 + N_R)R(1 + N_R)^{-1}\}$$

$$\xi_f = -0.5\ln\alpha - 0.52 + 0.64\alpha + 1.43(1 - \alpha)Kn_f.$$

The single-fibre efficiency for real filters is calculated from:

$$\eta_{si} = \frac{\eta_{si}^f}{\varepsilon_f}, \quad \text{where} \quad \varepsilon_f = \frac{F_0^f}{F_0},$$

$$\Delta p = \frac{4F\eta U_0\alpha l}{\pi d_f^2}, \quad \frac{1}{F} = \frac{1}{F_0} + \frac{1.43(1 - \alpha)\varepsilon_f^{1/2}Kn_f}{4\pi} \quad \text{and}$$

$$F_0^f = 4\pi\{-0.5\ln\alpha - 0.52 + 0.64\alpha\}^{-1}.$$

The model developed by Lee and Liu (1982) and later modified (Liu and Rubow, 1990) is also semi-empirical but the complication of having the fan model as an intermediate is eliminated. The experimental data used in its development are excellent. A modified Peclet number, $Pe^*$, and interception number, $N_R^*$, are used:

$$C_D = 1 + 0.388Kn_f\left\{\frac{(1 - \alpha)Pe}{\xi_K}\right\}^{1/3}; \quad C_R = 1 + \frac{1.996Kn_f}{N_R}$$

$$Pe^* = PeC_D^{-3/2}; \quad \frac{N_R^{*2}}{1 + N_R^*} = \frac{N_R^2}{1 + N_R}C_R.$$

Then

$$Y = 1.6X + 0.6X^3,$$

where

$$Y = \frac{\eta_{si}Pe^*N_R^*}{(1 + N_R^*)^{1/2}}, \quad X = \left\{\frac{(1 - \alpha)Pe^*}{\xi_K}\right\}^{1/3}\frac{N_R^*}{(1 + N_R^*)^{1/2}}.$$

In the above $\eta_{si}$ is the single-fibre efficiency due to both diffusion and interception. It is based on a comparison between theory using the Kuwabara flow field and experimental data from real filter tests. An inhomogeneity factor of 1.6 is included since this gives the best fit in the non-slip regime. For values of $X$ less than about 1.0 ($d_p < 0.05\,\mu m$) the method leads to an overestimate of efficiency.

## 6.7    Filter uniformity

In theoretical models of fibrous filters it is normally assumed that the medium is perfectly homogenous and comprised of an assemblage of parallel equidistant fibres. Examination of Figure 6.1 shows that this is nowhere near the truth. On a micro scale there are wide variations in fibre spacing. If the scale of scrutiny is increased to larger, though still small, volumes then there will be a distribution of packing densities and thus permeabilities between these. This will cause substantial variations in local face velocities. We could then picture the medium as a three-dimensional matrix of such small volumes with a range of properties about a mean – this has been referred to as the macrostructure.

It has already been explained that filter inhomogeneity leads to a reduction in pressure drop and efficiency. It is unavoidable in real filters, and the difference between theory and experiment is taken care of by the introduction of an empirical inhomogeneity factor. Although convenient, this method is not entirely satisfactory and more thorough analyses of uniformity effects have recently taken place. Schweers and Loffler (1994) measured the effect of macrostructure by mapping the local velocity as a function of position on a 1 mm square grid. The measured local velocities were described by a normal distribution with a standard deviation, in their case, of 30% of the average. The variation of fibre face velocities within the filter is presumably much higher. They carried out a theoretical investigation of the effects based on a three-dimensional model of the medium made up from cubes representing different local conditions. Their results showed that the effects are strongly dependent on the particle collection mechanisms and that the use of non-selective corrections, such as $\varepsilon$, is not universally valid. Basically, non-uniformity causes the efficiency curve to be smoothed out, with a reduction in diffusional and inertial collection. An alternative theoretical approach is to model the medium as a series of parallel fibres randomly located but having the correct average packing density. The flow is modelled in this geometry and large numbers of trajectory calculations carried out (Trottier, 1996). As would be expected, there is a similar smoothing of the efficiency curve.

The microscopic structure of fibrous filters has been analysed by embedding them in a thermosetting resin and then analysing sections by microscopy and image analysis techniques (Schmidt and Loffler, 1989; Vaughan and Brown, 1996). The fibre arrangement was found to be slightly more 'clumped' rather than random, but this must depend on the specific medium tested. Other studies of nearest-neighbour distances have been carried out with similar results (Molter and Fissan, 1995). This kind of information is essential if the medium's performance is ever to be predicted from single-fibre 'micro' models. The volume averaging process is extremely difficult, but progress has been made in that area (Quintard and Whitaker, 1995). An alternative and extremely interesting approach to particle collection in non-homogenous media has been developed by Shapiro (1996), who has derived a general equation governing the change of aerosol filtration length along curved air streamlines.

Fibres of more than one size can be mixed to form multi-component filters. If a filter is made from two separate layers of fibres of different sizes but the same packing density, then the total pressure drop should be the sum of the two individual pressure drops. Werner and Clarenburg (1965) found that when the fibres were mixed to give a multi-component filter the pressure drop was reduced. This was considered to be caused by interference effects. Unfortunately, the penetration is also affected. The practice of mixing grades of fibres to give filters of the required properties is, however, useful and is practised by some manufacturers. The properties of a mixture can be estimated for engineering purposes as simple linear functions of the weight fractions, $w_i$, of the individual components (Trottier, 1996):

$$\Delta p = \sum_i w_i \Delta p_i, \qquad \ln(pen) = \sum_i w_i \ln(pen_i).$$

## 6.8 Loading

The vast bulk of work on fibrous filtration has been aimed at determining the behaviour of clean filters. Since penetration falls with loading for most filters (electrostatic media are an important exception) it is perfectly reasonable that we should consider the worst case. However, the life expectancy of a filter is of considerable economic importance since it will have to be replaced when the resistance across it becomes excessive. As particles accumulate in the medium they act as very effective targets and both efficiency and pressure drop increase. Dendrites grow on the fibres and the structure of the medium becomes finer. An example of these dendritic structures is shown in Figure 6.10. The effect on pressure drop and penetration when filters, which are the same in all but fibre packing density, are loaded with monodisperse particles of stearic acid is shown in Figures 6.11 and 6.12 (Graef et al., 1995). Eventually dendrites bridge

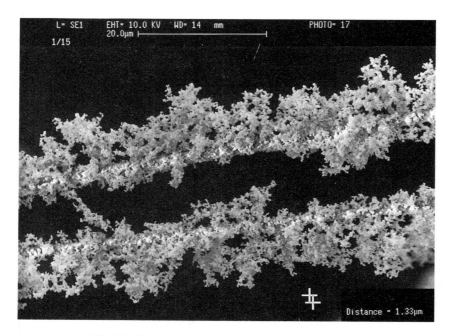

**Fig. 6.10** Dendritic structure of particle deposits in a filter.

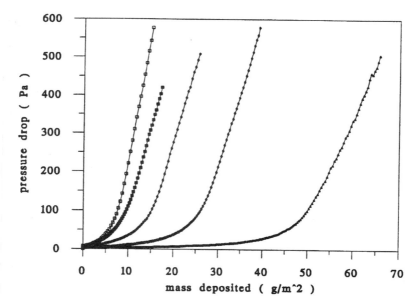

**Fig. 6.11** Pressure drop as a function of loading in a fibrous filter ($d_p = 1.35\,\mu\mathrm{m}$, $U_0 = 0.08\,\mathrm{m\,s^{-1}}$) using stearic acid aerosol.

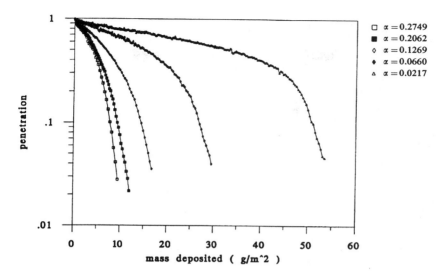

**Fig. 6.12** Penetration as a function of loading in a fibrous filter ($d_p = 1.35\,\mu\text{m}$, $U_0 = 0.08\,\text{m s}^{-1}$) using stearic acid aerosol.

the gaps between fibres and the filter becomes clogged. When this occurs the filter is extremely efficient and a surface layer grows. It is easiest to describe the characteristics by reference to the pressure drop curves. In the initial stages there is almost a linear increase in pressure drop, in the later stages there is again an almost linear response as surface filtration proceeds and finally the clogging point has to be defined. This is a very complicated process, especially given the random fibre arrangement and stochastic path of the particles, yet it represents the simplest case of filter loading. The introduction of particle size distributions, incomplete adhesion, electrostatic phenomena and liquid particles all add to the complexity.

The collection efficiency of a single fibre is obviously strongly related to the load of particles already collected on it. Since this varies with both position in the filter and time, the equation for the overall efficiency of the filter, at the end of section 6.2, is invalid. It is therefore necessary to consider the filter as a series of thin layers. The collection efficiency in each layer, $\eta_\sigma$, is a function of both the initial efficiency, $\eta_{si}$, and local loading, $\sigma$. If this function were known the behaviour of the whole filter could be found by numerical integration and by application of a simple mass balance its loading characterisics calculated.

In the initial stages of loading a simple linear function has been used:

$$\eta_\sigma = 1 + \lambda\sigma,$$

where $\lambda$ is a collection efficiency raising factor and $\sigma$ the local loading which is expressed in load per unit volume. The first term describes collection on the fibres and the second collection on deposited particles. By considering each deposited particle as a target, independent of other deposits, Radushkevich and Velicho (1962) derived a theoretical expression for $\lambda$, and an experimental value of $0.05\,\mathrm{m^3\,kg^{-1}}$ for operation over a limited range of variables in the inertial interception regime ($0.007 <$ $\eta_{si} < 0.2$) has been measured (Japuntich et al., 1992). More frequently $\lambda$ is taken to be directly proportional to $\eta_{si}$ (Yoshioka et al., 1969; Myojo et al., 1984). A simple and attractive method has been to carry out trajectory analyses using, for example, Monte Carlo simulations of particle build-up in the Kuwabara cell.

As particle build-up proceeds the simple linear equation is no longer adequate and a more complex picture of the process is required. A mixture of statistical and trajectory analyses has been carried out in which rigid simple dendrites were assumed. Monte Carlo simulations have also been extended again using the Kuwabara field to describe the flow round the fibre. Studies of dendrite formation for monodisperse particles (Kanaoka et al., 1980; 1983), polydisperse particles (Banes and Schollmeyer, 1987), particles in electrostatically charged filters (Baumgartner and Loeffler, 1987a) and particles with a probability of rebound (El-Shobokshy et al., 1994) have been carried out. These take three-dimensional loading into account and involve the use of considerable computer time. The effect of the dendritic structures themselves on the flow field is ignored, so these methods can only be applicable to the early stages of loading. Clearly the mechanism of filtration will influence the shape of the dendrites. In the extreme, pure diffusion will lead to all-round coverage and pure inertial impaction to front loading, as is shown in Figure 6.13 (Kanaoka et al., 1986). The inclusion of crossed fibres and the influence of the deposit on the hydrodynamic field are taken into account in the numerical simulation of Filippova and Hanel (1996), but this work will require vast computer time.

A number of other modelling approaches have been taken. The simplest is the fibre thickening model in which the fibre diameter is taken to increase with particle accumulation (Juda and Chorsciel, 1970). This leads to an increase in packing density and has to include a number of empirical parameters. Since it would be difficult to predict either the shape of the dendritic structure or its density from theory, the inclusion of two empirical constants in such a model seems entirely reasonable. The method has had limited success and with further development, perhaps taking into account some results from the more ambitious computer simulations, could lead to a useful predictive technique.

The clogging point is a strong function of fibre packing density in the filter – at low packing density there is simply a large volume to fill with

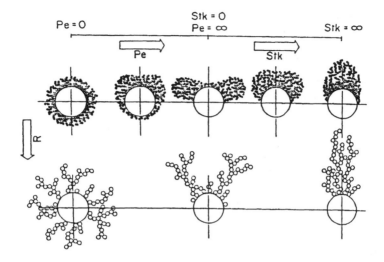

**Fig. 6.13** Effect of collection mechanism on shape of deposits built up during loading. *R* refers to the interception parameter. From Kanaoka *et al.* (1986).

deposit before clogging occurs. This is clearly shown in Figure 6.11, and Japuntich *et al.* (1997) found experimentally that the load to coarse clogging is directly proportional to filter pore size in the inertial interception regime. The clogging point is not of course a simple function of filter packing density but will also be related to filter uniformity. Depending on the regime the structure may become more or less uniform with particle collection. Lean zones may be self-perpetuating due to preferential collection in dense areas, or the structure may even out.

Surface filtration, which follows clogging, was studied by Kirsch and Lahtin (1975), who measured the voidage of the cake for a range of types of dust. This can vary from about 0.7 to 0.93, depending on dust properties and regime of collection. Computer simulation has been applied to surface growth and to the penetration of fines into the cake, and it has been shown experimentally that, where particle–particle adhesion is uncertain, compression can occur (Schmidt, 1995; Christ and Renz, 1996).

In low-adhesion systems it is possible for the efficiency to increase initially and then fall off almost to zero. Particles may be collected on fibre surfaces until a pseudo-equilibrium is reached where further accumulation is prevented by a mixture of air drag on the deposit and the inertial impaction of incoming 'large' particles. A cascade, or avalanche, may then take place in which the dislodged agglomerate will trigger a series of further dislodgements through the filter. When this stage is reached the system is acting merely as an agglomerator.

The deterioration of electrostatically charged media with loading is well known (Brown *et al.*, 1988). Figure 6.14 shows the increase in penetration through a two-fibre medium for a range of particle sizes. Experimental work comparing the behaviour of filters exposed to neutral particles and aerosol charged to its equilibrium Boltzmann distribution shows that shielding is the primary mechanism responsible for this. As particles are collected the charged sites become shielded so that the electrostatic mechanism diminishes. At the same time mechanical filtration is increased by dendrite formation and eventually the medium clogs. Unfortunately, in this case the penetration goes through a maximum (Baumgartner and Loeffler, 1987b; Walsh and Stenhouse, 1997).

The description of filter behaviour during loading obviously presents a formidable challenge. The structure changes continuously and the manner of the changes is clearly a function of a very wide range of variables. It is only in very recent years that significant progress has been made in the study and description of the structure of virgin filter media. The description of partially loaded media presents a much stronger challenge. Perhaps a start has been made in the introduction of fractal analysis. Scanning electron micrographs have been taken of partially loaded fibres from within the depth of media and these have been subsequently analysed to determine the contour fractal dimensions of the deposits. It has been shown (Trottier, 1996) that the average dimensions are 1.33 and 1.19 for

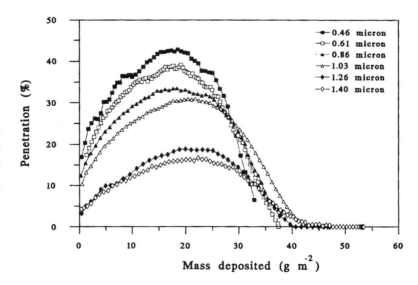

**Fig. 6.14** Penetration through filter samples loaded with stearic acid particles of Boltzmann equilibrium charge distribution, over a size range 0.46–1.40 μm at face velocity 0.10 m s$^{-1}$. From Walsh and Stenhouse (1997).

capture in the diffusional and inertial interception regimes, respectively. This at least gives a description of the structure which may be utilized in further theoretical development.

## References

Andres R.P. (1995) Inelastic energy transfer in particle/surface collisions. *Aerosol Sci. Technol.*, **23**, 40–50.

Banes T. and Schollmeyer E. (1987) Computer simulation of the filtration process in a fibrous filter collecting polydisperse dust. *J. Aerosol Sci.*, **17**, 191–200.

Baumgartner H. and Loeffler F. (1987a) Three-dimensional numerical simulation of the deposition of polydisperse aerosol particles on filter fibres – extended concept and preliminary results. *J. Aerosol Sci.*, **18**, 885–888.

Baumgartner H. and Loeffler F. (1987b) Particle collection in electret fibres filters – a basic theoretical and experimental study. *Filtr. Sep.*, **24**, 346–351.

Broom G.P. (1979) Adhesion of particles in fibrous air filters. *Filtr. Sep.*, **16**, 661–669.

Brown R.C. (1984) A many fibre model of airflow through a fibrous filter. *J. Aerosol Sci.*, **15**(5), 583–593.

Brown R.C. (1986) Many fibre theory of airflow through a fibrous filter. *J. Aerosol Sci.*, **17**(4), 685–697.

Brown R.C. (1993) *Air Filtration, An Integrated Approach to the Theory and Applications of Fibrous Filters.* Pergamon Press, Oxford.

Brown R.C., Wake D., Gray D.B., Blackford D.B. and Bostock G.J. (1988) Effect of industrial aerosol on electrically charged filter materials. *Ann. Occup. Hyg.*, **32**, 271–294.

Christ A. and Renz U. (1996) Numerical simulation of filter cake build-up on surface filters, in E. Schmidt, P. Gang, T. Pilz and A. Dittler (eds), *High Temperature Gas Cleaning.* Univ. Karlsruhe MVM, pp. 169–182.

Dahneke B. (1971) The capture of particles by surfaces. *J. Colloid. Int. Sci.*, **37**, 342–353.

Dahneke B. (1975) Further measurements of the bouncing of small latex spheres. *J. Colloid. Int. Sci.*, **51**, 58–65.

Dahneke B. (1995) Particle bounce or capture-search for an adequate theory: I. Conservation-of-energy model for a simple collision process. *Aerosol Sci. Technol.*, **23**, 25–39.

Davies C.N. (1952) The separation of airborne dust and particles. *Proc. Inst. Mech. Eng.*, **1B**(5), 185–213.

Davies C.N. (1973) *Air Filtration.* Academic Press, London.

Davies C.N. and Peetz V. (1956) Impingement of particles on a transverse cylinder. *Proc. Roy. Soc. A*, **234**, 269–295.

El-Shobokshy M.S., Al-Sanea S.A. and Adnan A.M. (1994) Computer simulation of monodisperse aerosol collection in fibrous filters. *Aerosol Sci. Technol.*, **20**, 149–160.

Fardi B. and Liu B.Y.H. (1992) Flow field and pressure drop of filters with rectangular fibres. *Aerosol Sci. Technol.*, **17**, 36–44.

Filippova O. and Hanel D. (1996) Numerical simulation of particle deposition in filters. *J. Aerosol Sci.*, **27**, 627–628.

Freshwater D.C. and Stenhouse J.I.T. (1972) The retention of large particles in fibrous filters. *AIChemE J.*, **18**, 786–791.

Fuchs N.A., Kirsch A.A. and Stechkina I.B. (1973) A contribution to the theory of fibrous aerosol filters. *Faraday Symposium.* Chem. Soc. No. 7, 143–156.

Graef A., Stenhouse J.I.T. and Walsh D.C. (1995) The effect of solid aerosol on prefilter material performance. *J. Aerosol Sci.*, **26**, S741–S742.

Happel J. (1959) Viscous flow relative to arrays of cylinders. *AIChemE J.*, **5**, 174–177.

Hildyard M.L., Ingham D.B., Heggs P.J. and Kelsmanson M.A. (1985) in C.A. Brebbia and G. Maier (eds). *Boundary Elements.* Springer-Verlag, New York, pp. 9–81.

Hiller R. and Loeffler F. (1978) The influence of bouncing of solid particles and oil drops on the filtration efficiency in fibre filters. *Deposition and filtration of particles from gases and liquids.* Soc. Chem. Ind., London, pp. 81–94.

Japuntich D. A., Stenhouse J.I.T. and Liu B.Y.H. (1992) The behaviour of fibrous filters in the initial stages of loading. *J. Aerosol Sci.*, **23**, S761–S764.

Japuntich D.A., Stenhouse J.I.T. and Liu B.Y.H. (1997) Effective pore diameter and monodisperse particle clogging of fibrous filters. *J. Aerosol Sci.*, **28**, 147–158.

Juda J. and Chorsciel S. (1970) Comparison co-efficient of filter materials. *Staub-Reinhaltung der Luft*, **30**, 196–198.

Kanaoka C., Emi H. and Myojo T. (1980) Simulation of the growth process of a particle dendrite and evaluation of a single fibre collection efficiency with dust load. *J. Aerosol Sci.*, **11**, 377–389.

Kanaoka C., Emi H. and Tanthapanichakoon W. (1983) Convective diffusional deposition and collection efficiency of aerosol on a dust loaded fibre. *AIChemE J.*, **29**, 895–902.

Kanaoka C., Emi I., Hiragi S. and Myojo T. (1986) Morphology of particulate agglomerates on a cylindrical fibre and collection efficiency of a dust-loaded filter, in *Aerosols, Formation and Reactivity, Second Int. Conf., Berlin.* Pergamon, Oxford, pp. 674–677.

Kirsch A.A. and Chechuev P.V. (1985) Diffusion deposition of aerosol in fibrous filters at intermediate Peclet numbers. *Aerosol Sci. Technol.*, **4**, 11–16.

Kirsch A.A. and Fuchs N.A. (1967) The fluid flow in a system of parallel cylinders perpendicular to the flow direction at small Reynolds numbers. *J. Phys. Soc. Japan*, **22**(5), 1251–1255.

Kirsch A.A. and Fuchs N.A. (1968) Studies on fibrous filters. III. Diffusional deposition of aerosols in fibrous filters. *Ann. Occup. Hyg.*, **11**, 299–304.

Kirsch A.A. and Lahtin U.B. (1975) Gas flow in high-porous layers of high-dispersed particles. *J. Colloid. Int. Sci.*, **52**, 270.

Kirsch A.A. and Stechkina I.B. (1978) The theory of aerosol filtration with fibrous filters, in D.T. Shaw (ed.), *Fundamentals of Aerosol Science.* Wiley-Interscience, New York, pp. 165–256.

Kraemer H.F. and Johnstone H.F. (1955) Collection of aerosol particles in the presence of electrostatic fields. *Ind. Eng. Chem.*, **47**, 2426–2434.

Krupp H. (1967) Particle adhesion. Theory and experiment. *Adv. Colloid Int. Sci.*, **1**, 111–239.

Kuwabara S. (1959) The forces experienced by randomly distributed circular cylinders or spheres in viscous flow at small Reynolds numbers. *J. Phys. Soc. Japan*, **14**, 527–532.

Lamb H. (1911) Motion of a sphere through a viscous fluid. *Philos Mag.*, **21**, 112–121.

Langmuir I. (1949) *U.S.O.S. Report No. 865. Office of Technical Services, Washington.*

Lee K.W. and Liu B.Y.H. (1982) Theoretical study of aerosol filtration in fibrous filters. *Aerosol Sci. Technol.*, **1**, 146–161.

Liu B.Y.H. and Rubow K.L. (1990) Efficiency, pressure drop and figure of merit of high efficiency fibrous and membrane filter media, in *Proceedings of Fifth World Filtration Congress, Nice.* Société Française de Filtration, Cachan, France..

Liu Z.G. and Wang P.K. (1996) Numerical investigation of viscous flow fields around multifibre filters. *Aerosol Sci. Technol.*, **25**, 375–391.

Lucke T., Knosche C., Adam R. and Tittel R. (1993) Calculation of fluid flow and particle trajectories in a system of randomly placed parallel cylinders – a new model for aerosol filtration. *J. Aerosol Sci.*, **24**, S555–S556.

Lundgren D.A. and Whitby K.T. (1965) Effect of particle electrostatic charge in filtration by fibrous filters. *Ind. Eng. Chem., Process Design & Development*, **4**, 345–349.

Molter W. and Fissan H. (1995) Gewinnung innerer Strukuraten von HEPA-Glasfaserfiltern, Teil 2. *Staub-Reinhaltung der Luft*, **55**, 379–382.

Myojo T., Kanaoka C. and Emi H. (1984) Experimental observation of collection of a dust-loaded filter. *J. Aerosol Sci.*, **15**, 483–489.

Natanson G.L. (1957) Deposition of aerosol particles by electrostatic attraction upon a cylinder around which they are flowing. *Dokl. Akad. Nauk USSR*, **112**, 696–699.

Nguyen X. and Beekmans J.M. (1975) Single fibre capture efficiency of aerosol particles in real and model filters in the inertial interception domain. *J. Aerosol Sci.*, **6**, 205–212.

Pich J. (1966) Theory of aerosol filtration by fibrous filters and membrane filters, in C.N. Davies (ed.), *Aerosol Science.* Academic Press, London, pp. 223–306.

Quintard M. and Whitaker S. (1995) Aerosol filtration: An analysis using the method of volume averaging. *J. Aerosol Sci.*, **26**, 1227–1255.

Raduschkevich L.V. and Velicho M.V. (1962) Theory of precipitating highly disperse aerosols from a stream onto an ultrafine cylinder. *Dokl. Akad. Nauk.*, **146**(2), 406–408.

Schmidt E. (1995) Experimental investigations into the compression of dust cakes deposited on filter media. *Filtr. Sep.*, **32**, 789–793.

Schmidt E. and Loeffler F. (1989) Preparation von Staubkuchen. *Staub-Reinhaltung der Luft*, **49**, 429–432.

Schweers E. and Loeffler F. (1994) Realistic modelling of the behaviour of fibrous filters through consideration of filter structure. *Powder Technol.*, **80**, 191–206.

Shapiro M. (1996) An analytical model for aerosol filtration by nonuniform filter media. *J. Aerosol Sci.*, **27**, 263–280.

Spielman L. and Goren S. (1968) Model for predicting pressure drop and filtration efficiency in fibrous media. *Env. Sci. Technol.*, **2**, 279–287.

Stechkina I.B., Kirsch A.A. and Fuchs N.A. (1969) Studies on fibrous filters. IV. Calculation of aerosol deposition in model filters in the range of maximum penetration. *Ann. Occup. Hyg.*, **12**, 1–8.

Stenhouse J.I.T. (1974) The influence of electrostatic forces in fibrous filtration. *Filtr. Sep.*, **11**, 25–26.

Stenhouse J.I.T. and Freshwater D.C. (1976) Particle adhesion in fibrous air filters. *Trans. I. Chem. E.*, **54**, 95–99.

Stenhouse J.I.T. and Harrop J.A. (1970) The theoretical prediction of inertial impaction efficiencies in fibrous filters. *Chem. Eng. Sci.*, **25**, 1113–1115.

Trottier R.A. (1996) Enhancement of the collection efficiency of fibrous filtration in the region of maximum penetration. *PhD thesis, Loughborough University*.

Trottier R.A. and Brown R.C. (1990) The effect of aerosol charge and filter charge on the filtration efficiency of submicrometer particles. *J. Aerosol Sci.*, **21**, S689–S692.

Vaughan N.P. and Brown R.C. (1996) Observation of the microscopic structure of fibrous filters. *Filtr. Sep.*, **33**, 741–748.

Walsh, D.C. and Stenhouse, J.I.T. (1997) The effect of particle size, charge, and composition on the loading characterisics of an electrically active fibrous filter material. *J. Aerosol Sci.*, **28**, 307–321.

Werner R.M. and Clarenburg L.A. (1965) Pressure drop across single-component glass fibre filters. *Ind. Eng. Chem. Process Design & Development*, **4**(3), 289–283.

Xu M., Willeke K., Biswas P. and Pratsinis S.E. (1993) Impaction and rebound of particles at acute incident angles. *Aerosol Sci. Technol.*, **18**, 143–155.

Yoshioka N., Emi H., Yasunami I.M. and Sato H. (1969) Filtration of aerosols through fibrous packed bed with dust loading. *Chem. Eng. Tokyo*, **33**, 1013–1018.

# Nomenclature

| | |
|---|---|
| $a$ | fibre radius (m) |
| $b$ | cell radius (m) |
| $c$ | aerosol particle concentration ($kg\,m^{-3}$) |
| $c_0$ | aerosol particle concentration at filter inlet ($kg\,m^{-3}$) |
| $c_i$ | aerosol particle concentration at large distance from fibre ($kg\,m^{-3}$) |
| $c'$ | dimensionless concentration, $c/c_i$ |
| $d_f$ | fibre diameter (m) |
| $d_p$ | particle diameter (m) |
| $e$ | coefficient of restitution |
| $f$ | packing density function |
| $g$ | gravitational acceleration ($m\,s^{-2}$) |
| $k_1$ | filter permeability |

| | |
|---|---|
| $l$ | filter depth (m) |
| $m$ | particle mass (kg) |
| $p$ | pressure ($N\,m^{-2}$) |
| $q$ | particle charge (C) |
| $r$ | position in cylindrical polar coordinates (m) |
| $u$ | gas velocity ($m\,s^{-1}$) |
| $u_r, u_\theta$ | cylindrical polar velocity resolutes ($m\,s^{-1}$) |
| $w_i$ | weight fraction of component in medium |
| $z$ | minimum surface separation distance (m) |
| | |
| $A$ | coefficient |
| $A_H$ | Hamaker constant (J) |
| $B$ | coefficient |
| $C$ | coefficient |
| $C_H$ | slip correction factor |
| $D$ | coefficient |
| $D_{AB}$ | particle diffusion coefficient ($m^2\,s^{-1}$) |
| $\overset{\circ}{E}$ | van der Waals adhesion energy (J) |
| $E_i$ | incident adhesion energy (J) |
| $E_r$ | rebound adhesion energy (J) |
| $\Delta E$ | $E_r - E_i$ (J) |
| $\overset{\circ}{F}$ | van der Waals adhesion force (N) |
| $F$ | drag force per unit length of fibre ($N\,m^{-1}$) |
| $G$ | gravity settling parameter |
| $J$ | packing parameter |
| $K_0, K_1$ | Bessel functions of zero and first order |
| $Kn_f$ | fibre Knudsen number, $a/$(mean free path) |
| $N_{Qq}$ | electrostatic parameter |
| $N_{0q}$ | electrostatic parameter |
| $N_{Q0}$ | electrostatic parameter |
| $N_R$ | interception parameter |
| $Pe$ | Peclet number |
| $Q$ | fibre charge per unit length ($C\,m^{-1}$) |
| $R$ | radius of curvature at contact point (m) |
| $Re$ | fibre Reynolds number |
| $Stk$ | Stokes number |
| $U_0, U_i$ | mainstream velocity, interstitial velocity ($m\,s^{-1}$) |
| $V$ | superficial velocity (volumetric gas flow per unit area) ($m\,s^{-1}$) |
| | |
| $\alpha$ | filter packing density |
| $\varepsilon$ | inhomogeneity factor |
| $\varepsilon_f$ | inhomogeneity factor used in fan model |
| $\varepsilon_0$ | permittivity of free space ($F\,m^{-1}$) |
| $\varepsilon_1$ | dielectric constant of particle |

| | |
|---|---|
| $\varepsilon_2$ | dielectric constant of fibre |
| $\eta$ | gas viscosity $(N\,s\,m^{-2})$ |
| $\eta_{si}$ | single fibre efficiency |
| $\eta_D$ | single fibre efficiency by diffusion |
| $\eta_R$ | single fibre efficiency by interception |
| $\eta_{DR}$ | single fibre efficiency by diffusion and interception – added term |
| $\eta_I$ | single fibre efficiency by inertia |
| $\eta_G$ | single fibre efficiency by gravitational deposition |
| $\eta_E$ | single fibre efficiency by electrostatics |
| $\eta^f_{si,D,R,DR}$ | single fibre efficiency according to fan model |
| $\eta_\sigma$ | single fibre efficiency at load $\sigma$ |
| $\theta$ | cylindrical polar coordinate |
| $\lambda$ | efficiency raising factor $(m^3\,kg^{-3})$ |
| $\nu$ | kinematic viscosity, $(\eta/\rho)$ $(m^2\,s^{-1})$ |
| $\xi$ | hydrodynamic factor |
| $\xi_{Kn}$ | hydrodynamic factor with slip correction |
| $\xi^f$ | hydrodynamic factor with slip correction as in fan model |
| $\rho$ | gas density $(kg\,m^{-3})$ |
| $\rho_p$ | particle density $(kg\,m^{-3})$ |
| $\sigma$ | local particle loading $(kg\,m^{-3})$ |
| $\tau$ | dimensionless time |
| $\psi$ | stream function |

# 7 Atmospheric aerosols

U. BALTENSPERGER and S. NYEKI

## 7.1 Introduction

Atmospheric aerosols originate from either naturally occurring processes or anthropogenic activity. Major natural aerosol sources include volcanic emissions, sea spray and mineral dust emissions, while anthropogenic sources include emissions from industry and combustion processes. Within both categories further distinction of so-called primary and secondary sources may be made. Direct emissions of aerosols into the atmosphere constitute primary sources, while secondary sources arise from the gas-to-particle conversion (GPC) of gaseous precursor compounds such as nitric oxide and nitrogen dioxide (collectively known as $NO_x$), sulphur dioxide ($SO_2$) and hydrocarbons.

A useful overview of the complex life cycle of atmospheric aerosols has already been given in Figure 1.3, which summarizes aerosol sources, transformation mechanisms while resident in the atmosphere, and subsequent sink processes. The size fraction with diameter $d > 2\,\mu m$ is usually referred to as the coarse mode, while the fraction below this size is the fine mode. The latter mode can be further divided into the accumulation mode ($d \sim 0.1–2\,\mu m$) and the nucleation mode ($d < 0.1\,\mu m$). Due to the $d^3$ dependence of aerosol mass, the coarse mode is typified by the largest mass concentration; similarly, the accumulation mode is typified by the surface area concentration and the nucleation mode by the number concentration.

As aerosol size is one of the most important parameters in describing aerosol properties and their interaction with the atmosphere, its determination and use are of fundamental importance. Various instruments measure different equivalent aerosol diameters, defined in many of the aerosol texts listed at the end of this introduction. Aerosol size is commonly presented as an aerodynamic diameter, which normalizes for density and shape.

Aerosols in the nucleation mode arise from GPC via either heterogeneous or homogeneous nucleation. The former refers to condensation growth on existing nuclei, while the latter refers to the formation of new nuclei through condensation. Heterogeneous nucleation preferentially occurs on nuclei with a large surface area, typically the accumulation mode, and hence only low supersaturations are required. For instance, water can condense on nuclei at supersaturations below 1–2%, in contrast to homogeneous nucleation which requires values in excess of 300% in the absence of impurities. Examples of GPC are combustion processes and the ambient formation of nuclei from gaseous organic emissions. High initial number concentrations during formation having $d < 0.1\,\mu m$ are reduced rapidly through coagulation, resulting in aerosol lifetimes of the order of minutes.

Particles in the accumulation range arise typically from the condensation of low-volatility vapours and from coagulation of smaller particles in the nucleation mode with themselves or with the larger accumulation mode particles. Particles tend to accumulate in this mode as there is a minimum efficiency in sink processes. Of these processes wet deposition (in-cloud and below-cloud scavenging) is the major sink process.

Particles in the coarse mode are usually produced by weathering and wind erosion processes. Dry deposition (primarily sedimentation) is the dominant removal process, followed by wet deposition. Chemically their composition reflects their sources, and hence inorganic compounds such as mineral dust and sea-salt are found in addition to organic compounds such as biological (spores, pollens and bacteria) and biogenic particles resulting from direct emission of hydrocarbons into the atmosphere. As the sources and sinks of the coarse and fine mode are different, there is only a weak association of particles in the two modes.

The above brief summary on atmospheric aerosols serves to illustrate the complex processes involved in modelling their behaviour and assessing their influence on climate. It is mainly for the latter reason that interest in aerosols has grown of late. The present discussion is intended as an intermediate introduction to the topic of atmospheric aerosols, where it is presumed that the reader already has a knowledge of aerosol basics. General aspects of aerosol science may be found in Hinds (1982), Hidy (1984), Finlayson-Pitts and Pitts (1986), Seinfeld (1986), Lodge (1989) and Baron and Willeke (1993). For aspects of atmospheric composition and climate change, see Hobbs and McCormick (1988), Rowland and Isaksen (1988), Hobbs (1993), Jennings (1993), Charlson and Heintzenberg (1995), Kouimtzis and Samara (1995) and Singh (1995). More specialized topics concerning atmospheric aerosols include: desert aerosols (Leinen and Sarnthein, 1989), biomass burning (Levine, 1991), radioactive aerosols (Chamberlain, 1991), Arctic pollution (Sturges, 1991) and atmospheric acidity (Radojevic and Harrison, 1992).

## 7.2 Aerosol sources and transformations

### 7.2.1 Size distribution, composition and concentration

Knowledge of the different moments of an aerosol distribution is important for several different reasons. Optical effects are largely

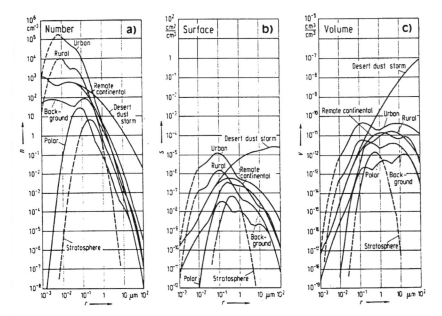

**Fig. 7.1** Number, surface and volume concentrations for various atmospheric aerosols; $n = dN/d\log r$, $s = dS/d\log r$ and $v = dV/d\log r$. Aerosol types are self-explanatory, apart from background aerosol which refers to the tropospheric aerosol 5 km above the continents and 3 km above the oceans. After Jaenicke (1988).

influenced by the aerosol diameter, chemical reactions by the surface area and health effects by the number, surface area and mass. Of these moments, the mass concentration has been commonly preferred by regulatory authorities in the past, mainly due to the availability of instrumental techniques. Modern techniques and mass-production, however, allow greater inter-comparison and wider coverage of atmospheric aerosol parameters.

It will be shown that the atmospheric aerosol may be typified by several categories, according to source and geographical location. Figure 7.1 illustrates typical size distributions for number, surface and volume concentration for various aerosol types, where the bimodal form in each is generally evident. The graphs indicate that the mode diameter increases in going from a number through to a volume distribution, again emphasizing the way in which submicrometre aerosols dominate the number distribution and supermicrometre aerosols the volume distribution.

The number size distribution of the atmospheric aerosol may be approximated by an empirical power-law equation for radii $r > 0.1\,\mu m$ (Junge, 1963):

$$dN/d\log r = cr^{-\nu}, \tag{7.1}$$

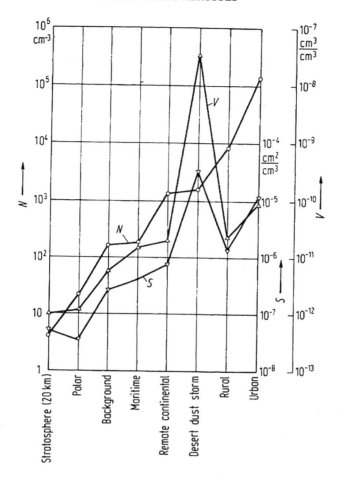

**Fig. 7.2** Integrals of number $N$, surface $S$ and volume $V$ concentration for the aerosol types given in Figure 7.1. After Jaenicke (1988).

where $N$ is the aerosol concentration, $c$ is a constant and $v$ depends on the aerosol type. The relative abundance of each aerosol type in Figure 7.1 is presented in Figure 7.2 as number, surface and mass concentrations. Aerosols at the Earth's surface vary greatly in number concentration from below $20\,cm^{-3}$ for the polar regions to over $10^5\,cm^{-3}$ for an urban aerosol. Apart from the variation in the geographical extent of aerosols, the vertical extent also varies substantially. For instance, a background aerosol at 3 km above the oceans or 5 km above the continents exhibits a number concentration $\sim 150\,cm^{-3}$, and the stratospheric aerosol at 20 km exhibits $\sim 10\,cm^{-3}$. Urban, remote continental and remote maritime

**Table 7.1** Typical mass composition ($\mu g \, m^{-3}$) of various chemical species in urban, remote continental and remote maritime aerosol types (after Pueschel, 1995)

| Element or compound | Urban aerosol – photochemical smog | Remote continental aerosol | Remote maritime aerosol |
|---|---|---|---|
| $SO_4^{2-}$ | 16.5 | 0.5–5 | 2.6 |
| $NO_3^-$ | 10 | 0.4–1.4 | 0.05 |
| $Cl^-$ | 0.7 | 0.08–0.14 | 4.6 |
| $Br^-$ | 0.5 | – | 0.02 |
| $NH_4^+$ | 6.9 | 0.4–2.0 | 0.16 |
| $Na^+$ | 3.1 | 0.02–0.08 | 2.9 |
| $K^+$ | 0.9 | 0.03–0.01 | 0.1 |
| $Ca^{2+}$ | 1.9 | 0.04–0.3 | 0.2 |
| $Mg^{2+}$ | 1.4 | – | 0.4 |
| $Al_2O_3$ | 6.4 | 0.08–0.4 | – |
| $SiO_2$ | 21.1 | 0.2–1.3 | – |
| $Fe_2O_3$ | 3.8 | 0.04–0.4 | 0.07 |
| $CaO$ | – | 0.06–0.18 | – |
| Organics | 30.4 | 1.1 | 0.9 |
| Total | 103.6 | 2.95–12.4 | 12.0 |
| Selected Mass Fractions and Molar Ratios | | | |
| $SO_4^{2-}$ (%) | 15.9 | 30.2–45.7 | 22.6 |
| $NO_3^-$ (%) | 9.6 | 13.3–22.7 | 0.44 |
| $NH_4^+/SO_4^{2-}$ | 2.2 | 2.1–3.4 | 0.47 |

Note: Soot values not reported. For typical values see section 7.2.2.3.

aerosols in Figure 7.2 may as a first approximation be described by typical chemical compositions, appearing in Table 7.1. Sulphate is seen to be a major component of urban and remote continental regions and sodium chloride of remote maritime regions. The aerosol chemical composition and geographical type are thus recognized as being fairly specific to an aerosol source and are discussed below in greater detail.

### 7.2.2  Sources

The distinction between natural and anthropogenic and between primary and secondary aerosols has already been outlined; however, certain sources cannot be so clearly defined. Smoke aerosol, arising from natural wild fires, is often either categorized as anthropogenic in origin due to the dominance of the latter or not at all. On the other hand, mineral dust entrained into the atmosphere from agriculturally eroded regions has been considered as a natural rather than an anthropogenic source.

Estimates of annual emissions to the atmosphere for major aerosol sources in the troposphere are given in Table 7.2. The range of values highlights the uncertainties involved in estimating emissions, and stems from the large spatial and temporal variation of sources and sinks. Further

**Table 7.2** Global emission source strengths for atmospheric aerosols in teragrams per year ($1\,\mathrm{Tg} = 1 \times 10^{12}\,\mathrm{g}$)

| Aerosol component | d'Almeida et al. (1991) | Pueschel (1995) | IPCC (1995)[a] | Particle size mode |
|---|---|---|---|---|
| **Natural** | | | | |
| Primary | | | | |
| Sea salt | 1000–10 000 | 300–2000 | 1300 | coarse |
| Mineral dust | 500–2000 | 100–500 | 1500 | mainly coarse |
| Primary organic aerosols/ | | | | |
| biological debris | 80 | 3–150 | 50 | coarse |
| Volcanic ash | 25–250 | 25–300 | 33 | coarse |
| Secondary | | | | |
| Sulphate from biogenic gases | 345–1100 | 121–452 | 90 | fine |
| Sulphate from volcanic $SO_2$ | | 9 | 12 | fine |
| Nitrate from $NO_x$ | | 75–700 | 22 | fine/coarse[b] |
| Organics from biogenic VOC[c] | | 15–200 | 55 | fine |
| Natural Total | 1950–13 430 | 648–4311 | 3062 | |
| **Anthropogenic** | | | | |
| Primary | | | | |
| Industrial dust | 10–90 | 167 | 100 | coarse/fine |
| Biomass burning | 3–150 | 29–72 | 80 | fine |
| Soot (all sources) | | 24 | 10 | mainly fine |
| Secondary | | | | |
| Sulphate from $SO_2$ | 175–325 | 70–220 | 140 | fine |
| Nitrate from $NO_x$ | | 23–40 | 40 | fine/coarse |
| Ammonium from $NH_3$ | | 269 | | |
| Organics from VOC[c] | | 15–90 | 10 | fine |
| Anthropogenic Total | 188–565 | 597–882 | 380 | |
| Overall Total | 2138–13 995 | 1245–5193 | 3442 | |

[a] 'best' estimate.
[b] Relative fractions uncertain.
[c] VOC = volatile organic compounds.

uncertainty stems from the use of different terminology and source categorization in the literature. The values from the IPCC (1995) report will be used to describe source emissions, mainly to retain consistency in the following discussion. Table 7.2 suggests that the anthropogenic fraction currently represents about 11% of total emissions. This is not insignificant, as the major fraction lies in the accumulation mode. The recent estimates of $SO_4^{2-}$ anthropogenic emissions are seen to exceed those from natural sources, evidence of increasing industrialization since about 1850. In considering these globally averaged figures it must be remembered that they are not indicative of local burdens, owing to the short tropospheric lifetime of aerosol particles and their non-uniform geographical distribution. On a global scale natural sources will dominate due to their emission from large-area sources such as the deserts and the oceans, although aerosol lifetimes for this coarse mode will be lower than for the fine mode, dominated by anthropogenic emissions. In contrast, anthropogenic emissions over relatively smaller areas, such as the industrialized regions of Europe, the USA and Japan, are likely to exceed the contributions from natural sources. Each emission source in Table 7.2 is described next.

### 7.2.2.1    Natural sources – primary emissions

7.2.2.1.1  *Sea-salt aerosol.* This results from the bursting of bubbles, formed by wave and wind action at the ocean surface. The ejected sea-spray droplets then either return to the surface or evaporate to form inorganic/organic aerosols, which may then be entrained into the boundary layer by wind turbulence. Wind speed is a major controlling factor in sea-salt aerosol concentration, roughly exhibiting a linear dependence. The atmospheric emission, estimated at $1300 \, \text{Tg yr}^{-1}$, is seen to be a major component of the natural tropospheric aerosol. Aerosol chemical composition depends on sea-water composition, which is fairly uniform globally and contains the following major dissolved species with mass mixing ratios in brackets: $Na^+$ (31%), $Mg^{2+}$ (3.7%), $K^+$ (1.1%), $Ca^{2+}$ (1.2%), $Cl^-$ (55%) and $SO_4^{2-}$ (7.7%). These ions exist mainly as the salts $NaCl$, $KCl$, $CaSO_4$ and $Na_2SO_4$. Soluble and insoluble organic compounds may also be an important component of marine aerosols, the fraction depending on a number of parameters such as location and season. The above concentrations may not necessarily reflect the composition of maritime aerosols, due to the enrichment of particular species during formation or subsequent chemical transformation in the atmosphere. Typical aerosol mass concentrations for a maritime aerosol are illustrated in Table 7.1. As a result of the uniform aerosol chemical composition over the oceans, source regions cannot be easily identified.

The mass median diameter of sea-salt aerosol near the sea surface is about $8 \, \mu\text{m}$ and, because of their short lifetime and inefficient light scattering, the largest particles exhibit a minimal interaction with the atmosphere. Hence, $1300 \, \text{Tg yr}^{-1}$ is considered as representative for the fraction transported in the lower marine boundary layer (MBL). Sea-salt aerosol concentrations, given in Figure 7.2, are generally not representative over continental interiors due to atmospheric removal processes, and rapidly decrease in concentration with distance inland.

7.2.2.1.2  *Mineral dust.* Wind-blown or aeolian mineral dust from desert and semi-arid regions is an important source of tropospheric aerosols (d'Almeida, 1989; Pye, 1987) and of particular interest in long-range transport and palaeoclimatological studies (Leinen and Sarnthein, 1989). Sources of mineral dust arise from the physical and chemical weathering of rock and soils. Wind speed is the main controlling factor in

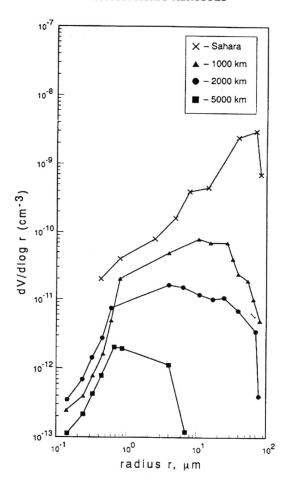

**Fig. 7.3** Mineral dust size distribution from a Saharan source as it evolves with downwind distance across the Atlantic Ocean. After Duce (1995). Reproduced by permission of John Wiley & Sons.

entraining particles into the atmosphere, among other factors such as soil moisture and surface composition. The atmospheric size distribution of soils has a bimodal structure close to the source, in which the range $d \sim 10$–$200 \, \mu m$ consists mainly of quartz grains and for $d < 10 \, \mu m$ of clay particles. Quartz grains will preferentially sediment close to their source, resulting in a fractionation process from a quartz/clay to a clay aerosol with increasing distance. An example of an evolving size distribution from a Saharan source is given in Figure 7.3, as a result of dry deposition only. Measurements at several downwind distances illustrate the way in which the mass median radius and total number concentration decrease. The maximum radius $r$ in the volume distribution at $r \sim 30\text{--}50\,\mu m$ shifts to about $1\,\mu m$ at a $5000\,km$ distance, where it appears to stabilize at $r \sim 0.5\text{--}1.5\,\mu m$ (Prospero *et al.*, 1989). If the chemical composition of long-range transported mineral dust is considered, then a close relation to the average crustal composition is found. Principal elemental constituents are oxides and carbonates of silicon, aluminium, calcium and iron. Due to the inert nature of mineral dust, chemical transformation processes in the atmosphere are thus considered minor, although surface chemical reactions may be important.

Mineral dust is estimated to contribute $1500\,Tg\,yr^{-1}$ to global atmospheric emissions, originating from areas totalling about 10% of the Earth's surface and comparing to a similar sea-salt emission from oceans covering an area larger than 70% of the surface. The principal source regions of mineral dust include: the Saudi Arabian peninsula, the US Southwest, and the Sahara and Gobi deserts. Only the latter two regions are considered as significant sources of long-range transported dust, occurring mainly westward over the tropical North Atlantic and eastward over the North Pacific, respectively (Prospero *et al.*, 1989). While direct anthropogenic emissions of mineral dust from agricultural activity are considered minor, the indirect effects from increased land erosion and desertification need to be considered. Aeolian dust from such regions, of which the Sahel in Africa is an example, may be significant additional sources.

7.2.2.1.3 *Primary organic aerosols/biological debris.* Natural emissions from the biosphere, and to a smaller extent from combustion processes, are included in this category. Continental sources from vegetation include plant waxes and fragments, pollen, spores, fungi and decaying material. Maritime sources consist of marine surfactants from bubble bursting processes. The total global emission of $50\,Tg\,yr^{-1}$ consists of a slightly higher maritime than continental contribution, where coarse mode aerosol sizes predominate.

7.2.2.1.4 *Volcanic emissions.* The spectacular eruptions of Mount St Helens (USA, 1980), El Chichón (Mexico, 1982) and Mount Pinatubo (Philippines, 1991) have highlighted the importance of considering emissions from volcanoes and fumaroles (i.e. vents) to the atmosphere (see special 1992 edition of *Geophysical Research Letters*, **19**, 149–218). Volcanic activity occurs on a sporadic basis and is mainly located in the Northern Hemisphere. Apart from volcanic ash emissions (principally composed of $SiO_2$, $Al_2O_3$ and $Fe_2O_3$), major emissions of the gaseous species $SO_2$, $H_2S$, $CO_2$, $HCl$, $HF$ and water vapour also occur. An estimated $33\,Tg\,yr^{-1}$ of ash particles are emitted into the coarse mode and therefore have a greater regional than global impact. Of greater long-term importance is the emission of $SO_2$ and subsequent conversion to an

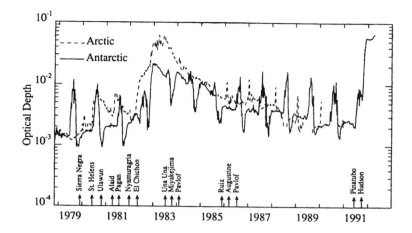

**Fig. 7.4** Polar aerosol optical depth from SAM II and SAGE satellite solar extinction measurements. Measurements at 1.0 μm and column-integrated from 2 km above the tropopause upwards. Injections of aerosols associated with El Chichón and Mt Pinatubo are seen to be superimposed on normal seasonal variations. After McCormick *et al.* (1993).

estimated 12 Tg yr$^{-1}$ of sulphate aerosol. These figures suggest that highly explosive eruptions may contribute up to 10–20% of the total natural sulphur emission.

Explosive eruptions such as Mount Pinatubo have been observed to inject large amounts of SO$_2$ buoyantly into the stratosphere. Mount Pinatubo was estimated to emit 9 Tg of sulphur (S) compared to 3.5 Tg(S) for El Chichón. A non-volcanic background aerosol appears to be maintained in the stratosphere through the upward flux of carbonyl sulphide (COS), emitted from the oceans, and its subsequent chemical transformation to H$_2$SO$_4$ droplets. The emission of SO$_2$ from volcanoes may significantly enhance background levels, as illustrated in Figure 7.4, where the aerosol optical depth (see section 7.5.1.1 for a definition) in the stratosphere is given. Peak optical depths occur about 6 months after an eruption, during which H$_2$SO$_4$ droplets of a sufficient size to interact with radiation are formed. Stratospheric aerosols generally exhibit a bimodal structure where the range $d \sim 0.001$–1 μm is composed of H$_2$SO$_4$ droplets ($\sim 75\%$ H$_2$SO$_4$ and $\sim 25\%$ water) and sizes $d > 1$ μm may be attributed to ash particles. The residence time of stratospheric aerosols has been estimated at 6–9 months and depends on altitude and latitude. Removal mechanisms are primarily sedimentation, subsidence and exchange through tropopause folds which may result in enhanced tropospheric concentrations of sulphate aerosols in remote regions.

7.2.2.2 *Natural sources – secondary emissions*.   Secondary natural aerosols may be formed from a number of natural precursor gas sources, containing sulphur, nitrogen and hydrocarbons, of which the main source of natural sulphate aerosol is the release of the gas dimethyl sulphide (DMS, $CH_3SCH_3$) from the oceans. DMS is formed from the biological activity of phytoplankton and eventually forms aerosol sulphate via the photooxidation to methanesulphonic acid and $SO_2$. The contribution to sulphate from DMS and other sources except sea water is otherwise known as non-sea-salt (nss) sulphate, to differentiate it from sea water as a source. Emissions of nss sulphate are estimated at $90 \, Tg \, yr^{-1}$. Other maritime sulphur sources, such as $H_2S$, $CS_2$ and COS, contribute less than 10% of total sulphur emissions, of which COS has the longest lifetime (about 40 years) and may thus participate in stratospheric chemistry. The seasonal variation of DMS follows the productivity cycle of the oceans and may be an order of magnitude higher in the Northern Hemisphere summer season than in winter. Globally averaged concentrations of 100 ppt are typical in the MBL and decline rapidly with altitude to several parts per thousand in the free troposphere (FT). These values compare typically with 20 ppt or less over continental surfaces. In general, less than 50% of the sulphate in the MBL is of maritime origin, the rest being attributed to soil dust and anthropogenic sulphate. The fraction of nss sulphate in moles per cubic metre is given by:

$$[nss \, SO_4^{2-}] = [SO_4^{2-}(total)] - 0.0605[Na^+],$$

where the value 0.0605 represents the average $[SO_4^{2-}]/[Na^+]$ molar ratio in sea water. Concentrations of nss $SO_4^{2-}$ in the remote MBL range from 5–200 ppt for the Southern Hemisphere oceans to 100–700 ppt for the North Atlantic and illustrate the enhanced anthropogenic contribution from the long-range transport of continental air in the Northern Hemisphere.

Nitrate aerosols are less characterized than those from sulphate, in part due to the complexity of nitrogen chemistry in the atmosphere. Their formation from nitrogen precursor gases has two main natural sources: $NO_x$ from lightning and soils; and $N_2O$ from bacterial activity in soils and the oceans. The emission rate of nitrate aerosols is estimated at $22 \, Tg \, yr^{-1}$.

The release of volatile organic compounds (VOCs) from vegetation forms condensable organics after partial oxidation. Major sources include terpenes (conifers) and isoprenes (broad-leaf trees). Emissions of secondary organic aerosols are estimated at $55 \, Tg \, yr^{-1}$. The mechanisms of formation are described further in section 7.2.2.4.

7.2.2.3 *Anthropogenic sources – primary emissions*

7.2.2.3.1 *Biomass burning*.   The burning of vegetation is collectively termed biomass burning, which includes natural wild fires and anthropogenic prescribed fires. The latter has been by far the larger source in the past two decades, where the annual clearing of forest and savannah for agricultural purposes is estimated to account for 95% of all biomass emissions. A combined emission of $80 \, \text{Tg yr}^{-1}$ is categorized as anthropogenic in Table 7.2 and consists of incomplete combustion products containing soot, sulphate, nitrate and hydrocarbon compounds to varying degrees. Additional biomass sources include the use of fuel-wood and the burning of agricultural waste. It is estimated that up to 87% of biomass burning occurs in the tropics during the dry seasons, i.e. December to March in the Northern Hemisphere and June to September in the Southern Hemisphere (Andreae, 1991). Besides particle emissions, gaseous emissions of $CO_2$, $CO$, $CH_4$ and VOCs are also important.

7.2.2.3.2 *Industrial aerosols.* Aerosol emissions from industrial processes are estimated at $100 \, \text{Tg yr}^{-1}$ and have a diverse number of sources. Current major sources in industrialized countries include aerosols formed from incombustible inorganic compounds in oil and coal fuels, coal and mineral dust from mining, stone-crushing, cement manufacture, metal foundries and grain elevators. The more recent implementation of abatement strategies – particle and gas scrubbers, clean fuel technology, etc. – has meant that emissions are approaching one-tenth the level of several decades ago in the older industrialized nations. In contrast, emissions are increasing in emerging nations where the implementation of modern technologies is not keeping apace with rapid economic development.

7.2.2.3.3 *Soot aerosols.* Anthropogenic soot aerosols from the combustion of oil and coal fuels in power generation, heating and vehicular transport, have a source strength of $\sim 10 \, \text{Tg yr}^{-1}$. Included in this figure are contributions from biomass sources. While current estimates of biomass soot emissions to the fine mode, at $12 \, \text{Tg yr}^{-1}$ (Penner, 1995), and fossil fuel, at $6 \, \text{Tg yr}^{-1}$, are somewhat larger than the combined IPCC figure of $10 \, \text{Tg yr}^{-1}$, the relative contributions of each source are illustrated. Emission inventories are hampered by the exact definition of soot content, which is somewhat dependent on the measurement technique, and hence this component is also otherwise known as black, graphitic or elemental carbon. Depending on the source and burning conditions, the soot fraction is highly variable. For instance, biomass soot and soot from spark-ignition engines have a high organic carbon content, in contrast to diesel soot which has a high elemental carbon content. For average urban conditions, 10–20% of total aerosol emissions are composed of carbonaceous material, of which 60–80% exists in organic form and the rest as elemental carbon (QUARG, 1993). Average rural conditions exhibit similar fractions, although the absolute mass concentrations may be several magnitudes lower over the continents and oceans, as indicated in Table 7.1. Soot is a ubiquitous component of the atmospheric aerosol and may be used as an anthropogenic tracer for a number of reasons: it is chemically inert, and poor hygroscopic properties prevent rapid scavenging through wet deposition.

### 7.2.2.4   *Anthropogenic sources – secondary emissions*

**7.2.2.4.1   *Sulphate and nitrate aerosols*.**   The main source of secondary particles in the atmosphere is the oxidation of $SO_2$ and $NO_x$. It is estimated that about 50% of $SO_2$ and $NO_x$ are oxidized before being deposited (Langner and Rodhe, 1991; Dentener and Crutzen, 1993). Oxidation may either occur in the gas or condensed phase. Gas-phase oxidation to both $H_2SO_4$ and $HNO_3$ is dominated by $OH^-$ and other atmospheric radicals. Of the above remaining 50%, about 50% of $NO_x$ and 90% or more of $SO_2$ are oxidized to $HNO_3$ and $H_2SO_4$ respectively, by heterogeneous reactions. Due to the low vapour pressure of $H_2SO_4$, oxidation of $SO_2$ always results in formation of aerosol mass, in contrast to $HNO_3$, which is distributed between the gas and aerosol phases. Although the chemical transformation of gases into particles depends on many factors – chemical kinetics (e.g. molecular rates) and physical kinetics (e.g. plume mixing and dispersion, oxidant concentration, sunlight, catalytic aerosol surfaces) – conversion rates of $SO_2$ are generally around 1–2% per hour and about 10–20 times higher for nitrate. The oxidation of non-$SO_2$ gases depends largely on their photochemical activity. Aerosol number and mass concentrations lie in the fine mode. Table 7.2 indicates that current sulphate emissions from anthropogenic sources exceed natural sources, while different estimates exist in the case of nitrate.

The largest source of secondary anthropogenic aerosol comes from fossil fuel emissions of $SO_2$ and subsequent conversion to $H_2SO_4$. Over continental surfaces, where gaseous ammonia is present, $H_2SO_4$ forms the components $NH_4HSO_4$ (acidic) and $(NH_4)_2SO_4$ (salt). These salts may exist simultaneously and are evidenced by varying molar ratios of $NH_4^+/SO_4^{2-}$ according to the atmospheric aerosol type, as in Table 7.1. In contrast, $H_2SO_4$ is the major component in the upper troposphere and the stratosphere.

Atmospheric emissions from fossil fuel combustion have been increasing steadily since about 1850. The current industrial $SO_2$ emission of about 70–90 Tg(S) yr$^{-1}$ accounts for about 80–85% of the $SO_2$ annual flux in the Northern and 30% in the Southern Hemispheres (see Möller, 1995; Berresheim *et al.*, 1995). Of these emissions, 90% are estimated to arise in the industrialized regions of the Northern Hemisphere, of which little is transported to the Southern Hemisphere. The long inter-hemispheric mixing time, of the order of a year, compares to aerosol lifetimes of around a week in the lower troposphere, hence resulting in minimal mixing. Concentrations in the clean continental planetary boundary layer

(PBL) range from about 20 ppt to 1 ppb and between 20 and 50 ppt in the MBL.

Current estimates of $NO_x$ emissions, from biomass burning and $NH_3$ oxidation, at $\sim 17\,Tg(N)\,yr^{-1}$, are lower than the anthropogenic value of $\sim 32\,Tg(N)\,yr^{-1}$ from fossil fuel combustion. The formation of $HNO_3$ from $NO_x$ is a major removal mechanism for $NO_x$ in the troposphere as most $HNO_3$ is subsequently lost through wet and dry deposition (Dentener and Crutzen, 1993).

Sulphate aerosols are stable to the concentration of atmospheric $H_2SO_4$, and temperature and humidity conditions encountered in the atmosphere, in contrast to ammonium nitrate and chloride aerosols. For these aerosols, the following reversible reactions occur to form the parent gaseous components under conditions of low atmospheric ammonia concentration, high temperature and low humidity:

$$NH_4NO_3\,(s) \leftrightarrow NH_3\,(g) + HNO_3\,(g)$$

$$NH_4Cl\,(s) \leftrightarrow NH_3\,(g) + HCl\,(g),$$

where (g) and (s) denote gaseous and solid components, respectively. The main source of atmospheric HCl is from refuse incineration and coal combustion.

Ammonia, being the commonest alkaline gas in the atmosphere, plays an important role in the neutralization of acid species. Conversion to ammonium salts is a function of not only altitude, due to the major sources being at ground level, but also temperature and humidity, and undergoes a diurnal and seasonal cycle. Major natural and anthropogenic sources respectively include: soils and organic decomposition; and fertilizers and animal farming. Emission are estimated at $269\,Tg\,yr^{-1}$.

The aqueous phase production of aerosol material on existing cloud condensation nuclei (CCN; aerosols acting as nuclei for the formation of cloud droplets), is an important mechanism in non-precipitating clouds. Cloud droplets may undergo on average ten evaporation/condensation cycles (Pruppacher and Jaenicke, 1995) before precipitable droplets are formed. During this process, gaseous species are scavenged and undergo chemical transformation, while aerosols and other droplets are scavenged by coagulation and phoresis mechanisms. As a result, the aerosol mass and hygroscopicity increases, in turn increasing the CCN activity. The conversion of $SO_2$ and $NH_3$ dissolved in droplets appears to be an efficient process for the production of $NH_4HSO_4$ and $(NH_4)_2SO_4$. Measurements of production rates in clouds indicate that $\sim 0.9\text{–}2.8\,\mu g\,m^{-3}$ sulphate is formed, possibly a 10–15 times larger sink for $SO_2$ than gas-phase oxidation. Such processes in cloud droplets have been postulated to be responsible for the rather uniform composition of the background aerosol in the troposphere (Jaenicke, 1993).

7.2.2.4.2    *Organics from VOCs.* Anthropogenic aerosol sources from VOCs, with an emission of $10\,\text{Tg yr}^{-1}$, include hydrocarbons from petroleum fumes and industrial solvents. A recent estimate places anthropogenic sources at $15–90\,\text{Tg yr}^{-1}$, of which $3\,\text{Tg yr}^{-1}$ is attributed to biomass burning (Pueschel, 1995). For natural and anthropogenic VOCs to act as aerosol precursors, they must first undergo oxidation. Compounds with low supersaturation vapour pressures preferentially undergo GPC by nucleation and condensation. A broad analysis of ambient organic aerosols indicates the following common functional groups: $-COOH$, $-CHO$, $-NO_3$ and $-OH$, with empirical formulae $X-(CH_2)_m-Y$, where X and Y are functional groups.

### 7.3    Heterogeneous chemistry

Heterogeneous reactions are reactions of gaseous molecules with solid or liquid surfaces. Atmospheric aerosol particles offer surfaces for such reactions, which otherwise would be kinetically unfavourable. Heterogeneous reactions have two different implications: first, they modify the aerosol composition by formation of condensable material; and second, they significantly influence gas-phase chemistry. After being neglected for a long time, heterogeneous reactions have recently gained great interest in tropospheric and even more in stratospheric chemistry, the latter being exemplified by the Antarctic ozone depletion. An example of a heterogeneous reaction is the following adsorption reaction:

$$A\,(g) + \{B\} \rightarrow \{AB\},$$

where species in { } denote surface-bound compounds. The rate with which gaseous molecules stick to aerosol particles is described by the following equation:

$$J = N_{\text{mol}} \sum_i N_i(d)B(d), \qquad (7.2)$$

where $J$ is the flux, $N_{\text{mol}}$ the number concentration of the molecule of interest, $d$ the aerosol diameter, $N_i(d)$ the particle number concentration with particle diameter $d$ and $B(d)$ the attachment coefficient. While there are a variety of different equations in the literature, a simplified formula (Porstendörfer *et al.*, 1979) gives $B(d)$ as

$$B(d) = \frac{2\pi Dd}{\dfrac{8D}{\bar{c}d\gamma} + \dfrac{1}{1 + 2\lambda_{\text{MFP}}/d}}, \qquad (7.3)$$

where $D$, $\bar{c}$ and $\lambda_{\text{MFP}}$ are the diffusion coefficient, the mean thermal velocity and the mean free path of the diffusing molecule, respectively, and $\gamma$ a dimensionless sticking coefficient. The latter determines whether a heterogeneous reaction is of importance in atmospheric chemistry. Even though the occurrence of these reactions is widely accepted today, large uncertainties remain over the actual value of $\gamma$, and further research is needed.

In the case of unit sticking ($\gamma = 1$) it can be seen from equation (7.3) that for small particles ($d \ll \lambda_{MFP}$) $B(d)$ is proportional to $d^2$, while for large particles ($d \gg \lambda_{MFP}$) $B(d)$ is proportional to $d$. However, for $\gamma \leq 0.01$, which is the case for most tropospheric reactions, equation (7.3) can be simplified to

$$B(d) = \frac{1}{4}\pi \bar{c} d^2 \gamma \tag{7.4}$$

for all particle diameters of interest ($d \leq 10\,\mu m$).

### 7.3.1  Tropospheric heterogeneous chemistry

Heterogeneous reactions can take place either on dry aerosol particles, haze particles (i.e. wet particles at a relative humidity (RH) above the deliquescence point) or cloud particles (i.e. particles that were activated due to RH $> S_C$, the critical supersaturation).

7.3.1.1  *Dry aerosols.*  It has been known for quite some time that the reaction between sea-salt and nitric acid results in liberation of HCl:

$$HNO_3\,(g) + NaCl\,(s) \rightarrow HCl\,(g) + NaNO_3\,(s).$$

Another example is ozone destruction on aerosol particle surfaces:

$$O_3 + \{C\} \rightarrow \{C(O_3)\}$$
$$\{C(O_3)\} \rightarrow O_2 + \{C(O)\}$$
$$\{C(O)\} \rightarrow CO, CO_2.$$

A fraction of the oxygen remains on the surface, which reduces the initial sticking coefficient from around $10^{-4}$ to $2 \times 10^{-5}$ with reaction time.

7.3.1.2  *Haze.*  As mentioned above, haze is composed of particles above their deliquescence point, which thus offer wet surfaces for heterogeneous reactions to occur. An important example is the nighttime formation of nitric acid:

$$NO_2 + O_3 \rightarrow O_2 + NO_3$$
$$NO_3 + NO_2 + M \rightarrow N_2O_5 + M$$
$$N_2O_5 + H_2O \xrightarrow{\text{aerosol}} 2HNO_3.$$

During the day, these reactions are of minor importance due to the rapid photolysis of $NO_3$. Model calculations have shown that with these reactions, the yearly average global $NO_x$ burden decreases by 50%, due to a decreased residence time in the atmosphere. Including these aerosol reactions, the observed nitrate wet deposition patterns in North America and Europe are better simulated.

High $HONO/NO_2$ ratios have been found in urban areas, which cannot be explained by gas-phase reactions. Thus, heterogeneous reactions are probably responsible for formation of high HONO concentrations, which again occur during the night:

$$2NO_2 + H_2O \xrightarrow{\text{aerosol}} HONO + HNO_3.$$

Nitric acid is retained on the particle surface, while nitrous acid is desorbed to the gas phase. In the presence of light, HONO is rapidly photolysed:

$$HONO + hv \ (< 400\,\text{nm}) \rightarrow NO + OH^{\cdot}$$

and is thus an efficient source of radicals. All these reactions in haze are characterized by high solute concentrations, and the chemistry in these non-ideal solutions needs to be further investigated.

7.3.1.3  *Clouds.* Equations (7.2) to (7.4) also apply in this case; however, in the context of cloud droplets $\gamma$ is usually called the accommodation coefficient. The most important example of heterogeneous reactions in cloud droplets is the aqueous oxidation of $SO_2$. The man oxidation paths here are the reactions with aqueous $H_2O_2$ and ozone:

$$SO_2\,(aq) + H_2O_2\,(aq) \rightarrow H_2SO_4$$
$$SO_2\,(aq) + O_3\,(aq) + H_2O \rightarrow H_2SO_4 + O_2.$$

These reactions are not expected to play an important role in haze, because the liquid water content (LWC) of the latter is about three orders of magnitude smaller than in a typical cloud or fog.

### 7.3.2  *Stratospheric heterogeneous chemistry*

It is now well established that heterogeneous chlorine activation reactions on polar stratospheric clouds (PSCs) play a central role in polar ozone depletion. In contrast, the chemical composition and formation mechanism of the polar clouds is still a topic of considerable debate. Above the ice frost point (189 K), PSCs exist as either nitric acid trihydrate (NAT) around a frozen sulphuric acid tetrahydrate (SAT) core, or ternary solutions of $HNO_3–H_2O–H_2SO_4$. Both are called type I PSCs, the former being solid and the latter being liquid (Tolbert, 1996). When the temperature falls below the ice frost point, water ice particles form on

SAT particles to make type II PSCs. On both types of PSCs, hetero-geneous reactions convert inert hydrogen chloride (HCl) and chlorine nitrate ($ClONO_2$) to reactive molecular chlorine and hypochlorous acid (HOCl):

$$ClONO_2\,(g) + HCl\,(ads) \rightarrow Cl_2\,(g) + HNO_3\,(ads)$$
$$ClONO_2\,(g) + H_2O\,(ads) \rightarrow HOCl\,(g) + HNO_3\,(ads).$$

Nitric acid remains in the condensed phase and therefore does not react with active chlorine to re-form inert $ClONO_2$. When the sun emerges at the start of the Antarctic spring, its visible rays dissociate the molecular chlorine and hypochlorous acid to form chlorine radicals. These radicals initiate a number of reactions which destroy ozone (Pueschel, 1995):

$$Cl + O_3 \rightarrow ClO + O_2$$
$$ClO + ClO \rightarrow Cl_2O_2$$
$$Cl_2O_2 + h\nu \rightarrow Cl + ClOO$$
$$ClOO \rightarrow Cl + O_2.$$

Similar reactions occur with gaseous $BrONO_2$. The validity of this ozone depletion scheme has been verified by a variety of independent methods, including satellite observations, lidar measurements and balloon observations. The major source of chlorine compounds in the stratosphere are industrial chlorine compounds such as chlorofluorocarbons (CFCs) used as refrigerants and aerosol propellants. CFC production has increased steadily since the 1950s; however, the 1990 Montreal Protocol forced severe restrictions on it, resulting in a recent decline of tropospheric concentrations (Montzka et al., 1996). These results suggest that the amount of reactive chlorine and bromine will probably reach a maximum in the stratosphere between 1997 and 1999 and will decline thereafter.

## 7.4   Aerosol sink processes

The time spent in the atmosphere by an aerosol particle is a complex function of its physical and chemical characteristics as well as the time and location of release. As already mentioned, actual sink mechanisms to the Earth's surface may be divided into two categories: dry and wet deposition.

Removal by dry deposition occurs by retention to the Earth's surfaces – water, vegetation, soil, buildings, etc. (Nicholson, 1988). A measure of the dry deposition velocity $v_{DEP}$ (cm s$^{-1}$) of a particular aerosol or chemical species to a particular surface is defined as:

$$v_{DEP} = \frac{\text{aerosol flux to the surface}}{\text{aerosol mass concentration}}. \qquad (7.5)$$

Defined in this way, $v_{\text{DEP}}$ represents the deposition velocity from a number of different mechanisms, such as diffusion and sedimentation. Figure 7.5 shows a typical graph of dry deposition velocity as a function of aerosol size. The removal of small particles by diffusion ($d < 0.1\,\mu\text{m}$) is mainly influenced by surface characteristics, such as the roughness length of surface features $z_0$, and the stability of the atmospheric surface layer, while the removal of large particles by sedimentation ($d > 1\,\mu\text{m}$) is mainly influenced by aerosol diameter and density $\rho$. The deposition velocity due only to gravitational sedimentation, $v_{\text{D}}$, is defined as:

$$v_{\text{D}} = \frac{\rho g d^2 C_{\text{c}}}{18\eta}, \qquad (7.6)$$

where $g$ is the acceleration due to gravity, $C_{\text{c}}$ the Cunningham slip correction factor and $\eta$ the viscosity of air. The curves for sedimentation in Figure 7.5 illustrate the minor importance of this sink mechanism below $d \sim 1\,\mu\text{m}$.

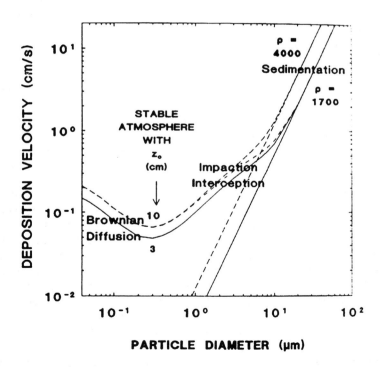

**Fig. 7.5** Dry deposition velocity versus aerosol diameter for two different particle densities ($\text{kg}\,\text{m}^{-3}$) and roughness lengths ($z_0$, $\text{cm}^{-3}$) at a $5\,\text{m}\,\text{s}^{-1}$ wind velocity. After Ruijgrok *et al.* (1995).

A minimum in $v_{DEP}$ for the intermediate size range (0.1–1 μm) or accumulation mode, coincides with a minimum in diffusion and sedimentation efficiency and explains the longer atmospheric lifetime of this mode. While impaction and interception mechanisms gain in relative importance, the dominant sink for the accumulation mode is wet deposition. Overall, removal by dry deposition is considered to be minor compared to wet deposition, despite the former occurring continuously at all times.

Wet deposition mechanisms consist of in-cloud and below-cloud scavenging. In-cloud scavenging refers to the incorporation of aerosols into cloud droplets, where a mass scavenging coefficient $C_{SCAV}$ can be described as (Hobbs, 1993):

$$C_{SCAV} = \frac{aerosol_{CD}}{aerosol_{Total}} = \frac{aerosol_{Total} - aerosol_{Interstitial}}{aerosol_{Total}}, \qquad (7.7)$$

in which $aerosol_{Total}$ is the total aerosol mass concentration entering the cloud, $aerosol_{Interstitial}$ is the aerosol that remains interstitial to the cloud droplets and $aerosol_{CD}$ is the aerosol in cloud droplets. Such a definition includes the contribution from nucleation scavenging and the aqueous formation of aerosols in clouds. Values of $C_{SCAV} \sim 0.2$–1 have been found with $C_{SCAV}$ decreasing with increasing aerosol concentration (Gillani et al., 1995), while values $C_{SCAV} > 1$ indicate the aqueous production of sulphate in the droplets. In below-cloud scavenging, falling droplets collect aerosols by impaction for $d > 1$ μm, and by diffusion and phoresis for $d < 1$ μm. Below-cloud scavenging mechanisms are considered less efficient than in-cloud mechanisms.

Through the overall removal of aerosols by wet and dry deposition, a residence time in the atmosphere may be considered. The residence time $\tau$ as a function of the aerosol radius appears graphically in Figure 7.6. The delineations indicate the different mechanisms responsible for the removal of aerosols from the atmosphere and are seen to be limited to specific size ranges. As already mentioned, coarse-mode aerosols are removed efficiently by sedimentation, while the nucleation mode is depleted by thermal diffusion or coagulation. The latter mechanism causes a rapid reduction in the aerosol number concentration, but does not constitute an actual sink as particle mass is only transferred into the accumulation mode. A maximum in $\tau$, corresponding to the accumulation mode, varies with altitude due to the decreasing frequency of wet deposition: less than 8 days for the lower troposphere (below about 1.5 km), $\sim 3$ weeks for the upper troposphere and $\sim 200$ days for the tropopause and above. It is worth noting that the same size range interacts most effectively with visible radiation, as shown in section 7.5.

The above typical residence times are of course temporally and spatially dependent on weather patterns as well as on the physicochemical properties of the aerosol. An average lifetime for sulphate particles in the

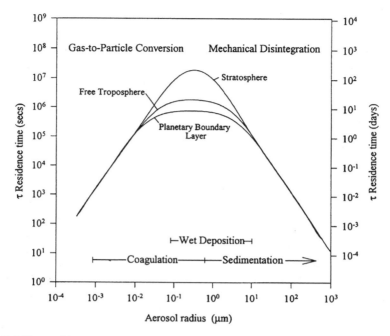

**Fig. 7.6** The residence time of atmospheric aerosols according to various removal mechanisms and altitude versus aerosol radius. Major removal mechanisms of interest here include coagulation, wet deposition and sedimentation removal. Modified after Jaenicke (1988).

range $d \sim 0.01$–$1.0\,\mu m$, released or formed in the mid-latitude boundary layer, is around 4–5 days and is mainly influenced by the frequency of precipitation. Similar residence times have been estimated for soot and biomass smoke aerosols. Radioactive species have been used as continental air mass tracers and to estimate aerosol lifetimes. For instance, the aerosol-bound radionuclide $^{137}Cs$ from the Chernobyl nuclear power station accident was estimated to have a lifetime of 9 days. Radon is, however, the commonest radioactive tracer used in long-range transport studies (Heimann *et al.*, 1990), for a number of reasons: first, radon is an inert gas; second, the main source is land surfaces; and third, a 3.8 day half-life allows sufficient resolution of the age of air masses.

## 7.5 Climate forcing

Natural as well as anthropogenic aerosols have the potential to change the global radiation balance, an aspect more recently highlighted in studies of global climate change (Penner *et al.*, 1993; IPCC, 1992; 1995). The increased emission of anthropogenic aerosols to the atmosphere may explain the discrepancy between observed temperature increases and the larger increases expected from greenhouse gases. Aerosols are considered to be responsible for a negative forcing (forcing is defined as a perturbation of energy input to the climate, measured in watts per square metre) or cooling of the Earth–atmosphere system, in contrast to a positive forcing (i.e. warming from greenhouse gases). It is important to note that in this definition of climate forcing, complex feedback systems, such as the hydrological cycle, are not considered as they are in the discussion of climate change. This makes the consideration of aerosols more straight-forward, but does not fully consider their overall impact.

Aerosols may influence the atmosphere in two important ways: direct and indirect effects. Direct effects refer to the scattering and absorption of radiation with a subsequent influence on the climatic system and planetary albedo. Indirect effects refer to the increase in available CCN due to an increase in anthropogenic aerosol concentration. This is suspected to increase the cloud droplet number concentration for a constant cloud LWC. As a result, the increase in cloud albedo is predicted to influence the Earth's radiation budget (Twomey, 1977). Cloud lifetimes and precipitation frequencies are also thought to be affected.

The effect of aerosols on the climate is, however, poorly quantified and it is at present still unclear what magnitudes are involved, let alone the sign of the forcing for the indirect aerosol effect. Figure 7.7 illustrates recent estimates of the mean annual radiative forcing for various climate change mechanisms, averaged globally for the period from 1850 to 1990 (IPCC, 1995). Identified mechanisms are at present thought to be dominated by a positive greenhouse forcing from the anthropogenic emissions of $CO_2$, $CH_4$, $N_2O$ and halocarbons. Additional mechanisms, such as the depletion of stratospheric ozone, the formation of tropospheric ozone by photochemical smog and variation in the solar irradiance, are believed to contribute smaller forcings. Tropospheric aerosols most probably contribute a negative forcing, although the large uncertainty in aerosol radiative forcing compared to that for greenhouse gases does not allow a meaningful net value from all mechanisms to be defined. As mentioned above, aerosols may be responsible for the lower than predicted temperature increase in the Northern Hemisphere. While this may apply on a global basis, such a simple argument cannot be applied on a regional basis for a number of reasons. First, in the comparison of greenhouse and aerosol forcing, it must be emphasized that while greenhouse gases have lifetimes measured in decades to centuries and are globally well mixed, aerosols have lifetimes measured in days and have a very variable geographical extent. Second, aerosol forcing will respond far more rapidly to changes in aerosol emissions than greenhouse forcing, which is still influenced by past accumulated emissions. Third, aerosol radiative forcing

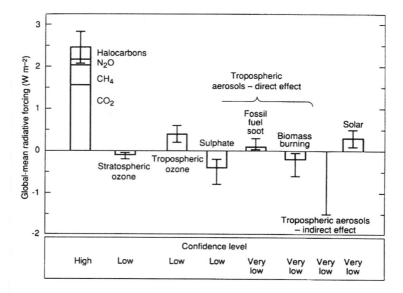

**Fig. 7.7** Radiative forcing estimates of the indicated climate change mechanisms. Confidence levels (high to very low) in each estimate are reflected in the uncertainty range of the error bars. Due to the episodic and sporadic nature of volcanoes, they are omitted from the above estimates, although their effects on time-scales of the order of decades may be significant. After IPCC (1995).

is more restricted to source and downwind regions, which are then more likely to experience an overall negative forcing. Lastly, greenhouse gas forcing is not as diurnally and seasonally variable as aerosol forcing, which has a greater influence during daylight and in the summer. These difficulties greatly hamper the collection of representative data and the modelling of aerosol direct and indirect effects.

### 7.5.1 Direct effects – optical properties of aerosols

The interaction of aerosols with radiation is greatest when the aerosol dimensions are similar to the radiation wavelength ($\lambda$), in other words when the aerosol size parameter $\pi d/\lambda \sim 1$. Hence, the longer-lived accumulation mode (0.1–1.0 μm) influences the short-wave solar radiation spectrum to a greater extent than the long-wave infrared spectrum emitted by the Earth's atmosphere and surface. At these longer infrared wavelengths, peaking at $\lambda \sim 7\,\mu$m, the low residence time of coarse-mode aerosols is considered to result in a smaller radiative interaction.

In order to assess the effect of tropospheric aerosols on the radiation balance, the following aspects are considered: the radiative properties of aerosols; and estimates of aerosol direct radiative forcing.

7.5.1.1 *Radiative properties of aerosols*. The interaction of radiation with aerosols may be described by a number of parameters. Incident radiation scattered or absorbed from an incident beam is defined by the total scattering ($\sigma_{SP}$) and absorption coefficients ($\sigma_{AP}$), which are a measure of the fractional loss in beam intensity per metre due to scattering and absorption, respectively. Incident radiation scattered into the backward hemisphere is described by the backwards hemispheric scattering coefficient ($\sigma_{BSP}$). The sum of $\sigma_{SP}$ and $\sigma_{AP}$ is the extinction coefficient $\sigma_{EXT}$, which when vertically integrated over the beam path length $dl$ gives the aerosol optical depth $\Omega$:

$$\sigma_{SP} + \sigma_{AP} = \sigma_{EXT} \tag{7.8}$$

$$\Omega = \int \sigma_{EXT}\, dl. \tag{7.9}$$

A measure of the relative importance of the scattering to the extinction coefficient is the single-scattering albedo $\bar{\omega}$:

$$\bar{\omega} = \sigma_{SP}/\sigma_{EXT}. \tag{7.10}$$

The extinction coefficient and its components are often approximated as being proportional to $\lambda^{-\mathring{a}}$

$$\sigma_{EXT} = K\lambda^{-\mathring{a}}, \tag{7.11}$$

where $\mathring{a}$ is the Ångström exponent and $K$ a constant. Furthermore, $\mathring{a}$ may be related to the Junge parameter $v$ by the relation

$$\mathring{a} = v - 2. \tag{7.12}$$

Typical values of $\mathring{a}$ for gases are $\mathring{a} \sim 4$, urban aerosols $\mathring{a} \sim 2$, rural haze $\mathring{a} \sim 1\text{–}2$ and coarse aerosols $\mathring{a} \sim 0$. Hence if $\mathring{a}$ can be measured then information on the number size distribution is gained and vice versa.

The asymmetry factor $g_{ASY}$ is defined as the cosine weighted mean of the angular scattering phase function $\beta(\phi)$, where the latter describes the amount of light scattered through an angle $\phi$:

$$g_{ASY} = \frac{\int \beta(\phi) \cos \phi \, d(\cos \phi)}{\int \beta(\phi) \, d(\cos \phi)}. \tag{7.13}$$

The value of $g_{ASY}$, ranging from $-1$ to $1$ for complete backscattering and forward scattering respectively, is important in radiative models of the atmosphere to take the angular scattering of radiation into account. As direct measurement is not possible, it may be parametrized by the ratio $\sigma_{BSP}/\sigma_{SP}$.

Present methods to assess global direct radiative forcing consider both local and column-integrated aerosol optical properties. These two different approaches are necessary to determine the vertical structure of the atmosphere and to validate ground-based integral observations with those of aircraft and satellites. Both approaches adopt a number of assumptions and are ideally carried out simultaneously to verify the internal consistency of current physicochemical optical models. However, few measurements of all necessary optical and chemical aerosol parameters exist at present.

Local measurements involve directly measuring the above aerosol parameters, some aspects of which are described further in section 7.6.3. Aerosol optical properties may be derived from their direct and *in situ* measurement or by calculating their properties through knowledge of the aerosol chemical composition, size distribution and refractive index. As many general circulation models (GCMs) calculate global aerosol mass concentration distributions, the ability simultaneously to derive distributions of scattering and absorption by aerosols is advantageous. Parameters which relate aerosol scattering and absorption coefficients to measures of the aerosol concentration and composition are the specific scattering, $\alpha_{SP}$, and specific absorption, $\alpha_{AP}$, efficiencies given for either a particular chemical species or the fine-mode mass:

$$\alpha_{SP} = \sigma_{SP}/M, \tag{7.14}$$

$$\alpha_{AP} = \sigma_{AP}/M, \tag{7.15}$$

where $M$ is the appropriate aerosol mass concentration. Typical values are given in Table 7.3, which shows a mean value of $\alpha_{SP} \sim 5\,\mathrm{m^2\,g^{-1}}$ for sulphate and $\alpha_{AP} \sim 10\,\mathrm{m^2\,g^{-1}}$ for soot. The latter high value emphasizes the important role of soot aerosols in atmospheric extinction despite a small emission term in comparison to other anthropogenic emissions. As mentioned previously, the accumulation mode $(0.1 \le d \le 1.0\,\mu\mathrm{m})$ is predominantly responsible for optical interaction and hence values are usually reported for this range, although values for the entire size range are also commonly reported. Specific efficiencies are dependent on the particle size distribution, the wavelength of the incident light, as well as the ambient RH. The latter is of importance when measuring both $\sigma_{SP}$ and

**Table 7.3** Specific scattering ($\alpha_{SP}$) and absorption ($\alpha_{AP}$) coefficients for the indicated aerosol species. Also indicated is $f(\mathrm{RH} = 80\%)$, the fractional increase for $\alpha_{SP}$ at an 80% relative humidity. All measurements $\lambda = 525\,\mathrm{nm}$. Figures in parentheses represent approximate standard deviations. Adapted from IPCC (1995)

| Optical property | Sulphate | Organic carbon | Mineral dust | Soot | Fine aerosol |
|---|---|---|---|---|---|
| $\alpha_{SP}$ (m² g⁻¹) | 5 (3.6–7) | 5 (3.0–7) | $\alpha_{EXT} \sim 0.7$ | 3 | 3 (2–4) |
| $\alpha_{AP}$ (m² g⁻¹) | 0 | 0 | | 10 (8–12) | (0–10) |
| $f(\mathrm{RH} = 80\%)$ | 1.7 (1.4–4) | 1.7 (1.4–4) | < 0.05 | (0–1.7) | 1.7 (0–4) |

$\sigma_{\text{BSP}}$ in the PBL. An increase in RH results in condensation of water on aerosol surfaces. For atmospheric aerosols, no effect is generally observed under 60% RH. At larger values the aerosol may grow abruptly at the deliquescence point, due to the condensation of water, resulting in an increase in $\sigma_{\text{SP}}$. The effect is important and is demonstrated in Table 7.3, where $f(\text{RH} = 80)$ denotes the relative increase of the $\sigma_{\text{SP}}$ at RH = 80%. Values indicate a factor 1.7 increase in most cases, although the standard deviations in parentheses demonstrate the dependence on chemical composition. Soot does not always exhibit an increase below 80% RH and a value is therefore only presented as a standard deviation. The influence of high RH and low supersaturations on aerosols is considered in section 7.5.2.

Column-integrated properties refer to aerosol properties averaged over altitude, from the surface to essentially the top of the atmosphere. Such measurements have been commonly conducted using sun photometers by subtracting the contribution from atmospheric gases to give the aerosol optical depth. Despite numerous *in situ* sun photometer measurements, satellites are gaining in importance for several reasons: global coverage of most of the Earth's surface is possible; continuous measurements allow secular trends to be identified; and calibration can be verified over the long term. While measurement of the aerosol optical depth over ocean surfaces is essentially straightforward, the varying albedo of land surfaces is problematic (i.e. seasonal variations of biosphere, agricultural land use). Figure 7.8 illustrates the aerosol optical depth at 0.58–0.68 µm over the oceans from the NOAA 11 polar orbiting environmental satellite. Notable features are the enhanced aerosol burden throughout the Northern Hemisphere and the westward emission of Saharan mineral dust, off the

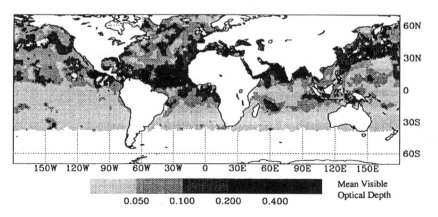

**Fig. 7.8** Mean visible aerosol optical depth over ocean areas for 23–29 May 1991, retrieved from the NOAA 11 polar orbiting environmental satellite. After Harshvardhan (1993).

**Table 7.4** Representative values of observed aerosol optical properties in the lower troposphere for 500–550 nm wavelength and relative humidity less than 60% (adapted from IPCC, 1995)

| Parameter | Polluted continental | Clean continental | Clean maritime |
|---|---|---|---|
| Optical depth ($\Omega$) | 0.2–0.8 | 0.02–0.1 | 0.05–0.1 |
| Single-scattering albedo ($\bar{\omega}$) | 0.8–0.95 | 0.9–0.95 | $\sim 1$ |
| Back/total scattering ratio ($R$) | 0.1–0.2 | 0.13–0.21 | 0.15 |
| Total scattering coefficient ($\sigma_{SP}$; m$^{-1}$) | $50$–$300 \times 10^{-6}$ | $5$–$30 \times 10^{-6}$ | $5$–$20 \times 10^{-6}$ |
| Absorption coefficient ($\sigma_{AP}$; m$^{-1}$) | $5$–$50 \times 10^{-6}$ | $1$–$10 \times 10^{-6}$ | $0.01$–$0.05 \times 10^{-6}$ |
| Fine mass concentration ($\mu$g m$^{-3}$) | 5–50 | 1–10 | 1–5 |
| CN number concentration (cm$^{-3}$) | $10^3$–$10^5$ | $10^2$–$10^3$ | $< 10^2$ |
| CCN number concentration (cm$^{-3}$; 0.7–1% supersaturation) | 1000–5000 | 100–1000 | 10–200 |
| Ångström exponent ($\mathring{a}$) | 1–2 | 1–2 | 1.5–2.1 |

coast of West Africa. Although the aerosol optical depth over land surfaces is at present difficult to deduce from satellite data, geochemical mass balances estimate that the 11% anthropogenic contribution to the total annual emission (see Table 7.2) represents 48% of the total aerosol optical depth (Andreae, 1995). Aerosol species primarily responsible for this concerning result are sulphates, biomass smoke and soot.

Table 7.4 summarizes the range of optical properties representative of polluted continental, clean continental and clean maritime aerosol types. Polluted continental aerosol number and mass concentrations are almost 10 times larger than for remote regions. The optical depth also exhibits similar behaviour, while a lower single-scattering albedo indicates the higher proportion of aerosol absorbing species in polluted continental air. A detailed representative database of the optical parameters in Table 7.4 for various representative locations around the globe remains to be gathered, along with chemical and microphysical properties.

7.5.1.2 *Estimates of aerosol direct forcing.* The application of GCMs in assessing the influence of atmospheric aerosols is hampered by the large spatial and temporal variation in their properties. Present assessments of the role of aerosols have come from models as opposed to observations, which further hampers the validation of these models. Most models predict a regional offset of greenhouse forcing by sulphate aerosols, mainly confined to the industrialized regions of the eastern USA, central Europe and eastern China. However, as mentioned previously, the regional forcing is not expected to be indicative of regional climate response, as the atmospheric circulation results in a non-local response to a local forcing.

Direct aerosol effects in Figure 7.7 were estimated to contribute a forcing of between $-0.25$ and $-1.5\,\mathrm{W\,m^{-2}}$ to the global radiation budget. Current estimates of the globally averaged radiative forcing for direct aerosol effects range from $-0.2$ to $-0.9\,\mathrm{W\,m^{-2}}$ for sulphate since industrial times (see IPCC, 1995). Consideration of the Northern Hemisphere forcing indicates a slightly higher value in the range $-0.43$ to $-1.1\,\mathrm{W\,m^{-2}}$. Estimates for biomass aerosols range from $-0.05$ to $-0.6\,\mathrm{W\,m^{-2}}$ (Penner *et al.*, 1992; Andreae, 1995) and are difficult to assess as sources are seasonal and mainly restricted to the tropics. Variation in the above figures is mainly due to the differing sophistication of GCMs, while the uncertainty depends on assumed optical properties and the global modelled distribution of aerosol species. Hence, the importance of obtaining more widespread geographical data on aerosol physicochemical properties is emphasized.

Additional sources that may also exhibit a significant effect on the radiation budget include: mineral dust; sulphate aerosols from volcanoes injected into the stratosphere; and other aerosol categories, such as nitrates and anthropogenic hydrocarbons.

Estimates have so far been mainly confined to the first two categories. The climatic impact of volcanoes is difficult to assess, but the recent El Chichón and Mount Pinatubo eruptions have allowed the effect of a large transient forcing on the climate system to be studied. Current GCMs have yielded temperature predictions over short time-scales that are in reasonable agreement with observations. Forcing from mineral dust was until recently assumed to be of minor importance. Recent estimates suggest that a mineral dust direct forcing of $-0.75\,\mathrm{W\,m^{-2}}$ contributes about one-third of the forcing from *natural* aerosols (Andreae, 1995), while the contribution from agriculturally eroded regions may be a significant additional source of aeolian dust (Tegen *et al.*, 1996). Volcanic emissions to the stratosphere influence the climate by warming the lower stratosphere through aerosol absorption and by reducing the net radiation transmitted to the troposphere and surface. However, as stratospheric aerosol residence times are of the order of 1 year, the radiative influence is also restricted to similar time-scales. Mount Pinatubo is estimated to have contributed a maximum forcing of $-4\,\mathrm{W\,m^{-2}}$ and about $-1\,\mathrm{W\,m^{-2}}$ up to 2 years later (Hansen *et al.*, 1992), which illustrates the large transient cooling effect when compared to the other radiative forcing mechanisms in Figure 7.7.

## 7.5.2 Indirect effects – aerosol effects on clouds

The indirect effect of increased aerosol emissions to the atmosphere refers to the possible modification of cloud properties, through the alteration of CCN physicochemical properties. The presence of CCN in the atmosphere allows cloud droplet formation to occur at supersaturations below 1–2%, whereas without CCN, supersaturations of several hundred per cent would be required. The physical mechanisms of these processes have been investigated and established, but atmospheric observations and secular studies remain sparse due to the complexity of the overall interaction and the difficulty in assessing the impact on climate. Not only are aerosol–cloud interactions difficult to model, but also cloud parametrizations themselves are still an outstanding issue.

The critical supersaturation at which CCN formation occurs depends mainly on aerosol composition, size and age and may be described by the Köhler theory. Figure 7.9 illustrates the influence of increasing RH on an NaCl aerosol. For any initial dry aerosol size, increase of the RH will not cause condensation to occur on the aerosol until the deliquescence point is reached, when an abrupt increase in size occurs due to the formation of an aqueous droplet. Pure salts exhibit an abrupt phase transition, characteristic of the chemical species, in contrast to mixed compositions where the transition is poorly defined. For example, $(NH_4)_2SO_4$ deliquesces at 80% RH, NaCl at 75% and $NH_4HSO_4$ at 39%. In addition, pure salts may be characterized by a hysteresis cycle (see Figure 7.9) observed on decrease of the RH. Recrystallization is found to occur at a lower RH than

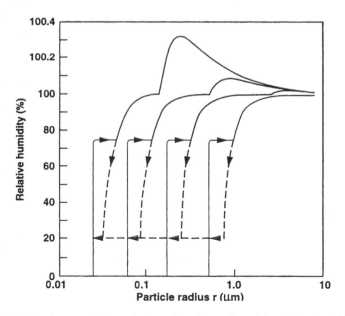

**Fig. 7.9** Köhler diagram of the variation of particle radius with relative humidity for CCN of various sizes. The curves depict a deliquescence point at 75%, typical for NaCl, and hysteresis effects upon decreasing humidity. After Pueschel (1995).

the deliquescence point, as the droplet remains in a supersaturated state. With increasing RH and supersaturation a critical supersaturation $S_C$ is reached, corresponding to a critical diameter $d_{CRIT}$ at which the aerosol becomes 'activated'. At this point droplets are in an unstable equilibrium with their environment and may grow uncontrollably or return to a stable equilibrium, to the left of the $S_C$ point on the curve, where they will exist as unactivated droplets or haze.

The value of $S_C$ depends on the soluble ion content and dry diameter of the aerosol. The larger both parameters are, the lower the critical supersaturation required to activate the aerosol. As a result the larger, hygroscopic aerosols tend to form CCN first. Typical supersaturations in marine stratus clouds, estimated at 0.1%, imply that dry aerosol diameters with $d > 0.1\,\mu m$ will be active to produce cloud droplets, whereas for $d < 0.1\,\mu m$ the aerosol will remain interstitial to cloud droplets.

Effective CCN sources are in general secondary aerosols of either natural or anthropogenic origin. Sulphate (Charlson *et al.*, 1992) and biomass/organic (Penner, 1995) aerosols are estimated to be the major anthropogenic CCN sources, and nss sulphate the major natural source. While smoke particles from biomass burning may exhibit CCN activation of up to 100%, only 1–2% of smoke from petroleum fuels is active, probably due to the lower concentration of soluble species. A similar low activity is observed for mineral dust, which also renders it a minor CCN source. The principal aerosol–cloud interactions are schematically summarized in Figure 7.10 for marine and continental air. Many of the life-cycle physical and chemical aspects have already been considered in sections 7.1–7.4 and large uncertainties still exist, especially in the further interaction of clouds on the radiative budget of the Earth–atmosphere system.

In order to assess the indirect effects of aerosols, the following aspects are further discussed: cloud properties and estimates of aerosol indirect radiative forcing.

### 7.5.2.1 *Cloud properties.*

Cloud physical properties over continental regions differ from those over oceans. The larger CCN concentration over land surfaces results in increased cloud droplet concentrations ($N_{CLOUD}$) and since for both types of cloud the LWC is similar, continental clouds will have a smaller average droplet size than marine clouds. These considerations are borne out by observations, where marine cumulus clouds have a median $N_{CLOUD} \sim 45\,cm^{-3}$ and a broad droplet size spectrum with median $\sim 30\,\mu m$ and continental cumuli have a median $N_{CLOUD} \sim 230\,cm^{-3}$ and a narrow size spectrum with median $\sim 10\,\mu m$.

Cloud optical thickness, $\delta$, may be parametrized in a simple way by

$$\delta \propto L/r_e, \tag{7.16}$$

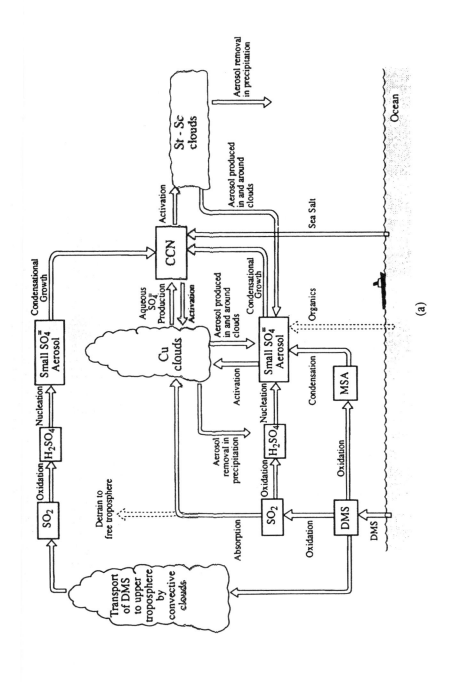

(a)

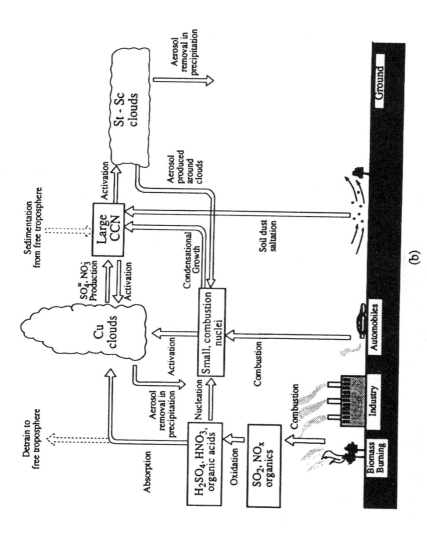

**Fig. 7.10** Schematic of aerosol–cloud interactions for (a) marine air and (b) continental air. After Hobbs (1993). Cloud types: Cu, cumulonimbus; St, Stratus; and Sc, stratocumulus.

where $L$ is the cloud LWC and $r_e$ is the effective radius of the cloud droplets. In addition, $L$ may be approximated by

$$L \propto \frac{4}{3} \pi r_e^3 N_{\text{CLOUD}},$$    (7.17)

and solving for $r_e$ and differentiating gives

$$\frac{\Delta \delta}{\delta} = \frac{1}{3} \frac{\Delta N}{N}.$$    (7.18)

Hence for a constant LWC, an increase in $N$ through anthropogenic sources will result in an increase in $\delta$. The situation is, however, complicated by many other factors apart from the inherent assumptions in the above derivation. The albedo $A_0$ of a cloud may be approximated by

$$A_0 = \frac{(1 - g_{\text{ASY}})\delta}{1 + (1 - g_{\text{ASY}})\delta}$$    (7.19)

(see Hobbs, 1993). For scattering of solar radiation by clouds, $g_{\text{ASY}} \sim 0.85$. Assuming that the cloud LWC and depth remain constant then combining both of the above equations yields:

$$\frac{\Delta A_0}{\Delta N} = \frac{A_0(1 - A_0)}{3N}.$$    (7.20)

Analysis of equation (7.20) reveals that $\Delta A_0 / \Delta N$, otherwise known as the susceptibility, is most sensitive when $A_0$ lies between $\sim 0.25$–$0.75$ and $N$ is small, values which are typical of marine clouds. This indicates that a small increase in the anthropogenic emission of CCN is likely to have a greater influence in maritime than continental regions (i.e. in the Southern Hemisphere). As clouds are already optically thick to long-wave radiation, only short-wave radiation is expected to be influenced by cloud properties. Hence, enhancement of the short-wave albedo should result in increased reflection of solar radiation back to space and a cooling of the Earth's surface.

The susceptibility of marine stratiform clouds to an increase in CCN concentration has been observed. Aerosol emissions from ship stack exhausts result in an increased cloud albedo, seen as trails on satellite images (Coakley et al., 1987). The increase in droplet concentration and reduction in size have been confirmed simultaneously by in-situ and remote sensing measurements. Observational evidence for an increase in cloud droplet concentration indicates that for a 10-fold increase in the aerosol size range $d \sim 0.1$–$0.3\,\mu\text{m}$, a two- to five-fold increase in droplet concentration results. An additional effect, as a result of an increased droplet concentration and reduction in mean droplet size, is a reduction in precipitation efficiency and increased cloud lifetime (Albrecht, 1989). Such observations are, however, at present difficult to quantify.

The processes described above suggest that the increase in industrial and biomass burning from anthropogenic activity is likely to have an influence on the global CCN distribution and population. Further global measurements are necessary to substantiate these observations, before the influence on climate can even be accurately considered.

### 7.5.2.2 *Estimates of aerosol indirect forcing.*

As a consequence of the large uncertainties in the interaction of aerosols with clouds, an estimate of the sign and magnitude of indirect forcing is fraught with assumptions. However, it appears that the radiative balance of the Earth may be influenced by the backscattering of incident solar radiation as a result of a cloud albedo increase. Present estimates of aerosol indirect forcing, as illustrated in Figure 7.7, indicate a similar sign and magnitude to direct forcing, at $-0.9\,\mathrm{W\,m^{-2}}$ with a range of uncertainty of 0 to $-1.5\,\mathrm{W\,m^{-2}}$. Many studies are presently conducting research in this area as the confidence in the forcing is designated as 'very low' in Figure 7.7.

Apart from uncertainties in many aspects concerning the indirect effect of aerosols, a recent study has suggested that the incorporation of soot aerosols from biomass burning may be influencing cloud albedo. A study of cloud cumulus and stratocumulus clouds over the Amazon basin during the burning season showed that the average droplet size reduced from 15 to $9\,\mu\mathrm{m}$ when smoke aerosols were present. A reduction in cloud reflectance was observed and attributed to absorption by soot in the smoke aerosols. While this remains an unresolved problem, in part due to the small number of observations, recent studies have suggested that the absorption of short-wave radiation by clouds may warrant inclusion in radiative transfer models (Cess *et al.*, 1995; Ramanathan *et al.*, 1995).

The formation of nss sulphate via the emission of DMS has been proposed as a cloud-climate feedback mechanism and essentially reasons along the following lines. The present greenhouse warming of the Earth's surface/atmosphere should also warm the ocean surface waters. As a result, an increase in phytoplankton activity and hence DMS emissions is then postulated to increase the CCN population and the albedo of marine stratiform clouds, thereby offsetting a global temperature rise (Charlson *et al.*, 1987). While the mechanism and different pathways are complex, the large areal extent of stratiform clouds, covering 25% of the oceans, renders such a feedback mechanism of potential importance.

It is clear that a great deal of further research is required. To end this section, two interesting aspects that may limit the indirect effect are mentioned (IPCC, 1995). The first concerns the availability of water vapour, which may only be able to activate a specific CCN fraction, despite the anthropogenic emission of additional CCN. The second aspect concerns the effect of increase in CCN on cloud albedo, which may level off at high CCN concentrations, thereby limiting the indirect aerosol effect.

## 7.6    Background aerosols – the European Alps as a case study

The study of aerosol properties in remote regions is important as the study of secular trends allows the impact of anthropogenic emissions to the atmosphere to be assessed. This section discusses one such remote region, the central European Alps, and will consider current knowledge of high-alpine background aerosols. High-elevation sites are important monitoring sites for a variety of reasons. First, they are often considered to represent the free troposphere (FT) and have therefore been selected for aerosol measurements by a variety of monitoring programmes, such as the Global Atmosphere Watch (GAW) programme, run under the auspices of the World Meteorological Society (WMO, 1991). It is, however, of great importance to know how representative these sites are of the FT. Second, important archives such as glaciers are located at high-elevation areas. These archives can only be interpreted in terms of past pollution if the transport processes to high-alpine sites and accumulation mechanisms in glaciers are reasonably well understood.

The ubiquitous nature of anthropogenic aerosols has been demonstrated by the detection of soot, as a tracer of combustion processes, in such remote environments as the Arctic (Heintzenberg, 1982), the FT above Mauna Loa in Hawaii (Clarke and Charlson, 1985), the South Pole (Hansen *et al.*, 1988) and over the oceans (Clarke, 1989). The Arctic late winter/early spring haze, formed as a result of the long-range transport of polluted air from mid-latitude sources in Eurasia, is perhaps the most documented case (see the special editions of *Atmospheric Environment*, Vol. 23, No. 11, 1989; Vol. 27A, No. 17/18, 1993). The presence of long-range transported pollution in remote regions highlights the importance of monitoring background aerosol properties.

An introduction to high-alpine background aerosols will begin with a discussion on the meteorology and geography of the European Alps, followed by current knowledge on background aerosols, and conclude with a discussion on Alpine glaciers as historical records of past pollution. General aspects of central European climatology are considered by Wallén (1977) and mountain climatology by Barry (1992).

### 7.6.1    *Meteorology and geographical extent of the European Alps*

The central European Alps run from France in the west, through Switzerland and into Austria in the east and border Germany in the north and Italy in the south. The large extent of the Alpine Massif (about 700 km in length and 100–250 km wide) and its elevation (up to 4810 m for Mont Blanc) cause a significant modification of the general atmospheric circulation. Southwesterly winds predominate in the western Alps and advect moist, humid air to the region, in contrast to the eastern Alps where continental polluted air is generally advected by northwesterly winds. Precipitation systems are mainly cyclonic and large-scale, resulting in decreasing precipitation eastwards through the Alps. At about 3000 m, snow cover typically occurs on 350 days of the year and 80% of the precipitation may be attributed to snow.

Mountainous terrain has a complex interaction with wind fields mainly due to three phenomena: the channelling effect of wind in valleys; the lifting effect of ridges and the subsequent development of turbulence; and slope and valley winds as a result of solar irradiance.

The warming of the ground just after sunrise leads to convective upslope winds along south-facing valley slopes. A compensating downslope wind in the centre of the valley enhances mixing and therefore warming of the air mass, which results in a valley wind. This results in the transportation of pollutants from the populated plains to the valleys, where entrainment into the atmosphere above the mountain ridges may occur. During the night the opposite occurs, where aged or non-polluted air masses drain down the valleys and onto the plains.

### 7.6.2 Background aerosol measurements

The observation of aerosol parameters at elevated sites in the Alpine region allows measurements to be conducted in the FT away from the influence of the PBL. While aerosol measurements have been conducted at a number of NOAA baseline stations for several decades, measurements of climatically important parameters in the FT and at a central European site are scarce. These NOAA sites include: Barrow in Alaska (8 m), Mauna Loa in Hawaii (3397 m), Samoa (77 m) and the South Pole (2841 m) (Bodhaine, 1983; NOAA, 1994).

Meteorological observations in the Alps have a long history and hence mountain climatology has been more thoroughly investigated here than elsewhere. The main observatories conducting atmospheric research are the high-alpine stations at the Zugspitze in Germany (2962 m), the Sonnblick in Austria (3106 m) and the Jungfraujoch in Switzerland (3454 m). Early aerosol measurements largely focused on Saharan dust episodes, which are historically documented and occur more frequently in the spring and summer seasons in the Alps. Other measurements have been made, albeit for varying lengths of time and intermittently during the last several decades. The Zugspitze high-alpine research station has monitored aerosol number concentration, radioactivity and associated meteorological parameters over extended periods. Reiter and co-workers (Reiter *et al.*, 1985, and references therein) performed extensive analyses on aerosol chemical composition, size distributions and transport pathways, based on measurements from 1972 to 1982. Only recently have aerosols regained interest in the establishment of the GAW programme. Lidar measurements, which give vertical profiles of the integrated backscatter coefficient, have also been conducted over several decades in the French and German Alps. The eruptions of El Chichón and Mount Pinatubo were events of particular interest and were monitored at both sites (Jäger, 1992; Chazette et al., 1995).

A number of aerosol, gaseous and cloud chemistry campaigns have been conducted at the Sonnblick. Summer to winter ratios of aerosol $NO_3^-$, $SO_4^{2-}$ and $NH_4^+$ occurred in the range 8–17. Concentrations of the major soluble ions in gaseous and aerosol samples were found to be similar to those at other background sites and illustrated the suitability of the Sonnblick for such measurements (Grasserbauer et al., 1994; Kasper and Puxbaum, 1994).

A number of research groups in Italy have also measured aerosol mass concentrations and conducted elemental analysis (Braga Marcazzan et al., 1993) and measured the aerosol extinction coefficient, resolved vertically in an alpine valley using sun photometers (Tomasi and Vitale, 1984).

The high-alpine station at the Jungfraujoch has been monitoring the aerosol optical depth (Blumthaler and Ambach, 1994) and column abundance of trace gas species (van Roozendael et al., 1994) for a number of years. The further consideration of alpine background aerosols will focus more closely on recent measurements at this site (Baltensperger et al., 1997), mainly due to the overall scarcity of continuous aerosol measurements at other locations. The large areal extent of the Alps emphasizes the need for further international research on climatically important aerosol parameters. This aspect has been recognized by the GAW programme, in which recommended aerosol parameters for measurement at regional stations include (i) optical depth, (ii) condensation nuclei (CN) concentration, i.e. the total number concentration as measured by a CN counter, and (iii) the chemical composition of size-fractionated aerosol; and at global baseline stations, (i) aerosol scattering and absorption coefficients, (ii) CCN concentration, (iii) lidar vertical aerosol profile, (iv) aerosol mass and number concentration, (v) diffuse and global solar radiation, and (vi) parameters as for regional stations (WMO, 1991).

Some of these parameters have been measured over varying periods in the Alps, and instruments for their measurement are listed in Table 7.5, along with some experimental details. More detailed technical and practical information on these instruments and other apparatus can be found in Hering (1989) and Baron and Willeke (1993). As already mentioned, lidar readings of the integrated aerosol backscatter have been measured in the Alpine region. Such systems fire light pulses from lasers, at a particular wavelength, into the atmosphere and then monitor the reflected signals. Complex algorithms allow the vertical aerosol structure, typically from near the surface to 30 km with a resolution of 600 m, to be

**Table 7.5** Details of typical aerosol instrumentation for the measurement of GAW-recommended parameters and previously used for the observation of aerosols in the Alpine region

| Instrument | Measured parameter | Typical background values (and resolution) |
|---|---|---|
| Lidar system | Column integrated aerosol backscatter (sr$^{-1}$) 532 nm (Nd:YAG laser) 694 nm (ruby laser) | $10^{-3}$–$10^{-4}$ sr$^{-1}$ ($\pm 10\%$) |
| Integrating nephelometer (TSI 3563) | $\sigma_{SP}$, $\sigma_{BSP}$ at 450, 550, 700 nm | $10^{-7}$–$10^{-5}$ at 550 nm ($> 1.0 \times 10^{-7}$ m$^{-1}$ for 30 s average) |
| Integrating plate method | $\sigma_{AP}$ | 2–16 $\times$ 10$^{-6}$ m$^{-1}$ ($\sim 1 \times 10^{-9}$ m$^{-1}$) |
| Aethalometer (Magee AE-10) | Black carbon mass concentration and $\sigma_{AP}$ by inference | 50–200 ng m$^{-3}$ ($\sim 5$ ng m$^{-3}$ 1 hour avg.) |
| Optical particle counter (PMS Las-X) | Optical equivalent diameter for $d_{OPT} = 0.1$–7.0 μm | $> 200$ cm$^{-3}$ for $d_{OPT} \sim 0.1$–0.2 ($< 1$ cm$^{-3}$) |
| Condensation nucleus counter (TSI 3025) | Number concentration for $d \sim 0.005$–1 μm | 500–1500 cm$^{-3}$ ($< 0.01$ cm$^{-3}$) |
| Berner impactor | Mass concentration for sizes $d = 0$–0.06, 0.06–0.125, 0.125, 0.25–0.5, 0.5–1, 1–2, 2–4, 4–8, 8–16, $> 16$ μm | Total mass concentration $\sim 5$–10 μg m$^{-3}$ ($\sim 2$ μg) |
| Epiphaniometer | Surface area concentration for $d \sim 0.01$–7 μm | $\sim 2$–24 μm$^2$ cm$^{-3}$ |

inferred. The aerosol scattering coefficient may be measured using an integrating nephelometer, giving $\sigma_{SP}$ and $\sigma_{BSP}$ simultaneously (at 450, 550, 700 nm), while the integrating plate method gives the absorption coefficient. Although the aethalometer is calibrated to measure the black carbon (BC) concentration, the absorption coefficient may also be derived. The measurement of scattering and absorption coefficients thus gives the extinction coefficient via equation (7.8). Measurement of the solar irradiance using sun photometers also allows the measurement of aerosol extinction, although it must be remembered that these measurements are column-integrated properties and cannot therefore be easily related to single-point measurements, such as those of the integrating nephelometer or aethalometer. More common instruments in use include optical particle counters, CN counters and impactors, for the measurement of the volume size distribution, the number concentration and the size-resolved mass concentration, respectively. Impactor data are particularly useful, as the size-resolved aerosol chemical composition can be subsequently determined from the collected samples. The epiphaniometer is a recently developed instrument for the measurement of the aerosol surface area concentration ($S$) and has been deployed at various high-alpine sites since 1988 on a continuous basis (Baltensperger et al., 1991). The instrument principle relies on the attachment of artificial radioactive $^{211}$Pb atoms to the aerosol surface, whose concentration is then proportional to the subsequently measured count rate. For the relatively stable aerosol size distribution measured at remote alpine sites, a Jungfraujoch site-specific calibration of 1 count per second has been found equivalent to a surface area concentration $S \sim 2.6\,\mu m^2\,cm^{-3}$ also corresponding to $\sim 1\,\mu g\,m^{-3}$.

### 7.6.3    Characterisation of the background aerosol

*7.6.3.1    Diurnal variations.* One aspect of studies in mountainous regions has concerned the transport of polluted PBL air by advective and convective processes to these sites, the latter often occurring in the spring–summer months as a consequence of increased solar irradiation. This phenomenon has long been recognized and corrected for in aerosol measurements, for instance at Mauna Loa (Bodhaine, 1983). Similar diurnal cycles also occur at the Jungfraujoch site, although on a more irregular basis due to synoptic weather patterns and the complexity of the local meteorology over the alpine terrain. Figure 7.11 illustrates the diurnal variation of epiphaniometer data averaged over the four seasons, at the Jungfraujoch and at Colle Gnifetti (4450 m). The latter site is on the Swiss–Italian border and is representative of the FT over the southern slopes of the Alps. It can be seen that a distinct diurnal pattern with a peak in the late afternoon is developed during the summer months, which is absent during the winter months as the air masses at high-alpine sites are decoupled from the polluted air masses below. Analyses of the data sets from a number of instruments have indicated that the 03:00–09:00 period (all times local standard time, GMT + 1 hour) is considered to be FT-influenced and the 09:00–03:00 period is often PBL-influenced. While Figure 7.11 suggests an FT-influenced period 01:00–11:00, the period 03:00–09:00 may be confidently assumed to be representative. By consideration of the regional meteorology, convective transport processes occurred 29% of the time during the period 1989–1993. Such diurnal convective patterns are not regularly evident in the aerosol signal and hamper the determination of whether the station is influenced by FT or PBL air masses. Similar observations have also been made at other alpine sites in Switzerland (Baltensperger et al., 1991) and at the Sonnblick. The available evidence suggests that remote high-alpine peaks are not necessarily decoupled from the PBL all the time. It is interesting to note that this process might be an important mechanism for the transport of anthropogenic pollution from the PBL to the FT.

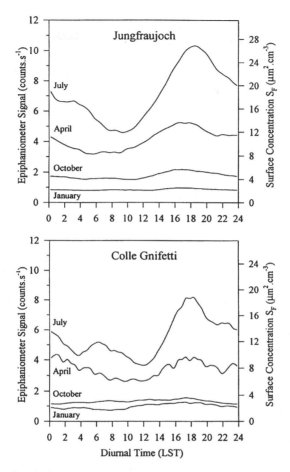

**Fig. 7.11** Diurnal variation of the epiphaniometer signal and surface area concentration $S$ for the four seasons at the Jungfraujoch (3450 m) and at Colle Gnifetti (4450 m, CH). Median values for the indicated months from 1988 to 1996, where the solid line indicates a running mean of 2.5 hours. Note that the surface area concentration axes have an altitude-dependent calibration. After Lugauer *et al.* (1997).

In contrast to the aerosol signal, the monthly averaged ozone concentration measured at a number of high-alpine sites does not show any distinct diurnal variation, which led to the conclusion that they were influenced by the FT most of the time. While these aspects remain unresolved, it has been suggested that photochemically induced gas-to-particle formation may be responsible for the observed diurnal variations. Evidence favouring transport-related processes comes from the monitoring of aerosol-borne radioactive radon and black carbon. Radon exhibits a low emission rate from snow-covered surfaces and may therefore be used as a tracer of polluted air masses, while black carbon aerosol is a good tracer for combustion processes, as mentioned previously. Both parameters exhibit similar diurnal variations, corroborating the fact that transport processes are the major reason for this diurnal signal. Hence, discrimination between FT and PBL conditions has to be performed carefully, based on a variety of parameters.

7.6.3.2 *Seasonal variations.* The annual record of aerosol parameters at the Jungfraujoch, principally due to sulphate aerosol, exhibits a seasonal cycle as a consequence of the solar irradiation cycle and the regional meteorology. Increased levels in the summer result from the complex interplay between a seasonal minimum in the $SO_2$ concentration from anthropogenic sources, a seasonal maximum in solar induced photochemistry, and a seasonal maximum in convective transport processes. Long-term records of sulphate at a Swiss midland site (Payerne, 490 m) indicate no significant seasonal cycle, which suggests that convection is principally responsible for elevated concentrations. Monthly median values for FT conditions at the Jungfraujoch exhibit the following summer maxima and winter minima, respectively (Nyeki *et al.*, 1997a): $\sigma_{SP}$ (550 nm) $\sim 16.1$ and $0.43 \times 10^{-6} \, m^{-1}$; $\sigma_{BSP}$ (550 nm) $\sim 210$ and $0.09 \times 10^{-6} \, m^{-1}$; $\sigma_{AP}$ (550 nm) $\sim 10.4$ and $0.76 \times 10^{-7} \, m^{-1}$, CN concentration $\sim 670$ and $280 \, cm^{-3}$; and $S \sim 24.1$ and $1.7 \, \mu m^2 \, cm^{-3}$. The seasonal variation is around an order of magnitude in each case, apart from the CN concentration which varies by a factor of 3. It may be concluded that the ratio of ultrafine ($d < 0.1 \, \mu m$) to total particle number concentration exhibits a seasonal variation. It is interesting to note that the seasonal variation is even more enhanced if all diurnal periods are considered. In comparison, NOAA baseline stations exhibit variations in $\sigma_{SP}$ from below $10^{-7}$ to above $10^{-5} \, m^{-1}$ and in CN concentration from 30 to $300 \, cm^{-3}$, according to season and location (NOAA, 1994). The limited data set for the Jungfraujoch site exhibits comparable background values to NOAA baseline stations and may therefore be designated under a clean continental category as in Table 7.4.

Long-term epiphaniometer data for the same site and for the FT-influenced diurnal period 03:00–09:00 illustrate an annual variation in aerosol surface area concentration (Figure 7.12). Analysis of the data from 1988 to 1996 indicates nearly a factor 10 difference between summer and winter monthly means. Even though such a data set is too short to estimate confidently any trend in surface area concentration, a trend of less than $0.01\% \, yr^{-1}$ compares with $-2$ to $+6.5\% \, yr^{-1}$ for $\sigma_{SP}$ and CN concentrations at NOAA baseline stations.

The analysis of air mass backtrajectories can aid in establishing the origins of aerosol background concentrations. Their implementation is, however, somewhat complex and time-consuming at present. Evidence of

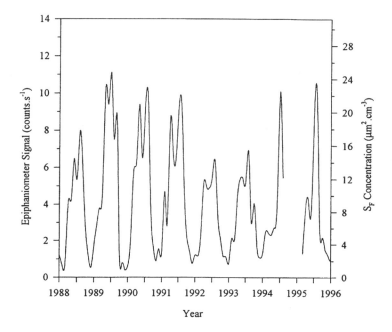

**Fig. 7.12** Annual variation of the epiphaniometer signal and surface area concentration $S$ at the Jungfraujoch (3454 m, CH) from monthly arithmetic mean data, for 1988–1996. Only data for the diurnal period 03:00–09:00 LST are included, representing free troposphere influenced air. After Lugauer *et al.* (1997).

long-range aerosol transport, such as Saharan dust episodes, can be ascertained from certain aspects of the measured aerosol data set. Figure 7.13 illustrates two size distributions obtained with an optical particle counter for FT- and PBL-influenced conditions at the Jungfraujoch. The volume size distributions are very similar in form in the size range 0.1–2 µm, but the real difference lies in the larger concentration of PBL aerosols in the accumulation mode. The trans-alpine transport of aerosols has also been observed in epiphaniometer measurements at the Jungfraujoch, Sonnblick and the Glacier de la Girose in France (3360 m). Similar episodic aerosol events were noted at all stations. Saharan dust episodes occur several times a year at such sites and are characterized by an increase in the background mass concentration from ∼ 5 to over 100 µg m$^{-3}$ (Schwikowski *et al.*, 1995).

### 7.6.4   *Recent trends in the background aerosol*

Seasonally varying convective and advective transport processes have been studied in terms of aerosol chemical composition. Numerous campaigns

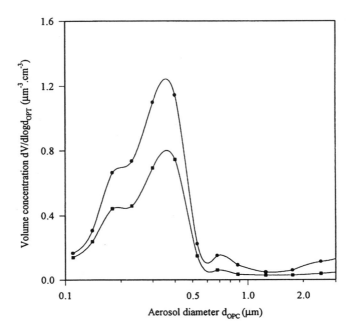

**Fig. 7.13** Average aerosol volume size distributions for free tropospheric (—■—; 03:00–09:00 LST) and PBL-influenced conditions (—●—; 09:00–03:00) at the Jungfraujoch (3454 m) during August 1995. After Nyeki *et al.* (1997b).

have investigated the composition of airborne species and their subsequent accumulation in snow and glaciers, to aid in the determination of past aerosol and gaseous concentrations in ice core samples. The first extended chemical composition measurements to be carried out in the Alps, near the Zugspitze, observed a seasonal variation in all measured soluble and insoluble species (Reiter *et al.*, 1985). The most abundant species in order of mass concentration were, sulphate, ammonium, nitrate and $SiO_2$. Emission sources of these species have already been discussed, where high concentrations of $SiO_2$ are evidence for the long-range transport of Saharan mineral dust. Figure 7.14 illustrates the sulphate aerosol data sets since 1972–1973 for a rural station on the Swiss plateau (Payerne), a pre-alpine station (Mt Wank, 1780 m) in Germany and a high-alpine station (Jungfraujoch). The seasonal variation in sulphate concentration, with a maximum in summer and minimum in winter, may be explained to a large extent by the seasonal variation of vertical transport processes. In order to assess the trends, 2-year running means are indicated as solid lines in Figure 7.14. Despite uncertainty in quality assurance procedures pre-1982 for the Payerne and the Jungfraujoch data, an overall downward trend in sulphate concentration is illustrated. The trend at the latter two sites,

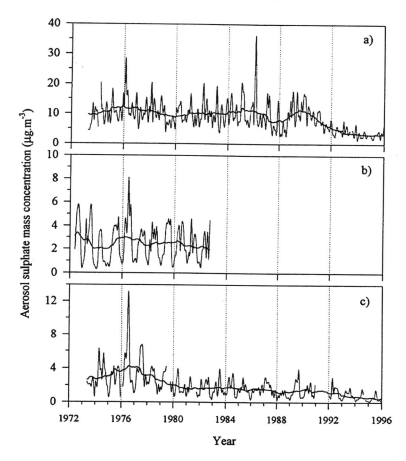

**Fig. 7.14** Annual variation in aerosol sulphate composition at: (a) a rural site on the Swiss plateau (Payerne, 490 m); (b) at a pre-alpine site (Mt Wank, 1780 m, Germany); and (c) at a high-alpine site (Jungfraujoch, 3454 m, CH). Solid lines represent 2-year running means. Note that the data include both free tropospheric and PBL-influenced conditions. Data from Payerne and the Jungfraujoch from Gehrig (1996).

$-2.8\%$ yr$^{-1}$ and $-5.7\%$ yr$^{-1}$ respectively, indicates a discernible decrease at the Jungfraujoch since the mid-1970s and the early 1990s at Payerne. In comparison, the $0.25\%$ yr$^{-1}$ trend at Mt Wank is not as evident, and may be attributed in part to the limited data set. However, it must be noted that by the limited time resolution of these samples, FT- and PBL-influenced conditions cannot be differentiated and hence the trend of the background FT aerosol is not obvious. All available evidence points to a reduction of anthropogenic sulphate emissions in PBL-influenced air masses, due mostly to abatement strategies for $SO_2$ release.

## 7.6.5   Glaciers as historical archives of past pollution

Ice core records from polar (including Greenland) or alpine (temperate latitudes) glaciers are commonly used in climatology to study environmental changes on a hemispheric or even global scale (Delmas, 1992). Many long- and short-term variations have been found preserved in ice cores such as climatic variations (ice ages and intermediate warm periods), single-event inputs to the atmosphere (e.g. volcanic eruptions), fallout from nuclear weapons tests, and also the generally observed increase over the last century of gaseous components like $CO_2$, $N_2O$, and $CH_4$ and aerosol components like $NO_3^-$ and $SO_4^{2-}$. While glaciers in very remote environments, such as Antarctica, have not exhibited an appreciable increase in nss sulphate and nitrate from increased anthropogenic emissions, the Greenland ice sheet and especially mid-latitude regions such as the Alps, are related more directly to local or regional environments (Oeschger and Langway, 1989). Atmospheric species accumulating in glaciers generally reflect four source categories: (i) long-range transport in the FT; (ii) the PBL via convective transport; (iii) local emissions close to glaciers; and (iv) emissions to the upper atmosphere, such as nuclear explosions, volcanic eruptions and desert dust injection. The mechanism in (ii) has already been mentioned in section 7.6.3.1 and results in summer glacier accumulation that is of regional influence. Winter glacier accumulation is, however, more likely to represent continental background values.

Measurements of the concentration of nss sulphate in snow from Greenland indicate an increase from about 20 to over $100\,ng\,g^{-1}$ during the last century. In comparison, the Alps, which are surrounded by densely populated regions, indicate sulphate increases from about 80 to $450\,ng\,g^{-1}$ (Wagenbach, 1991; Döscher et al., 1995) over the same time period. Such measurements are conducted on so-called cold glaciers generally above $4000\,m$ in the Alps, where the snow precipitation record is not disturbed by melt-water during the warmer seasons. However, the declining trend in aerosol sulphate, since the 1970 period, is also reflected in these studies and contrasts an ongoing increase in nitrate and ammonium species. The latter are responsible for the increasing acidification and eutrophication of natural ecosystems and therefore pose new environmental challenges.

To a first approximation, concentrations of glacier constituents reflect their atmospheric concentration. However, the deposition of atmospheric constituents to glaciers is dependent on, among other parameters, wet and dry deposition, the local topography, macro- and micrometeorology and the mechanisms of glacier formation. Hence, the reconstruction of past atmospheric concentration levels of aerosols and gases is no trivial problem. The complex topography, as well as the high snow accumulation rates, of alpine regions limits studies to essentially the last several thousand years, whereas the ice sheets of Greenland and Antarctica are suited to palaeoclimatic studies over hundreds of millennia. For these various reasons, polar glaciers are of hemispheric importance and alpine glaciers of regional importance.

In conclusion, the trend observed in Alpine glaciers is a close reflection of increased air pollution due to anthropogenic activity, where the variability in the ice core record can be mostly accounted for by the seasonal variation of aerosol concentration.

## Acknowledgements

This work was supported by the NERC. The support of the staff and facilities at the Jungfraujoch research station and the PSI are gratefully acknowledged. Thanks also go to M. Lugauer and M. Schwikowski for help in the preparation of several graphs.

## References

Albrecht B.A. (1989) Aerosols, cloud microphysics and fractional cloudiness. *Science*, **245**, 1227–1230.

Andreae M.O. (1991) Biomass burning, in *Global Biomass Burning* (ed. J.S. Levine). MIT Press, Cambridge, MA.

Andreae M.O. (1995) Climatic effects of changing atmospheric aerosol levels, in *World Survey of Climatology, Vol. XVI Future Climate of the World* (ed. A. Henderson-Sellers). Elsevier, Amsterdam.

Baltensperger U., Gäggeler H.W., Jost D.T., Emmenegger M. and Nägeli W. (1991) Continuous background aerosol monitoring with the epiphaniometer. *Atmos. Environ. A*, **25**, 629–634.

Baltensperger U., Gäggeler H.W., Jost D.T., Lugauer M., Schwikowski M. and Seibert P. (1997) Aerosol climatology at the high-alpine site Jungfraujoch, Switzerland. *J. Geophys. Res.*, **D102**, 19 707–19 715.

Baron P.A. and Willeke K. (eds) (1993) *Aerosol Measurement*. Van Nostrand Reinhold, New York.

Barry R.G. (1992) *Mountain Weather and Climate*. Routledge, London.

Berresheim H., Wine P.H. and Davis D.D. (1995) Sulfur in the atmospheric, in *Composition, Chemistry and Climate of the Atmosphere* (ed. H.B. Singh). Van Nostrand Reinhold, New York, pp. 251–307.

Blumthaler M. and Ambach W. (1994) Changes in solar radiation fluxes after the Pinatubo eruption. *Tellus B*, **46**, 76–78.

Bodhaine B. (1983) Aerosol measurements at four background sites. *J. Geophys. Res.*, **88**, 10 753–10 768.

Braga Marcazzan M.G., Bonelli P., Della Bella E., Fumagalli A., Ricci R. and Pellegrini U. (1993) Study of regional and long-range transport in an Alpine station by PIXE analysis of aerosol particles. *Nucl. Instr. Meth. Phys. Res.*, **75**, 312–316.

Cess R.D., Zhang M.H., Minnis P., Corsetti L., Dutton E.G., Forgan B.W., Garber D.P., Gates W.L., Hack J.J., Harrison E.F., Jing X., Kiehl J.T., Long C.N., Morcrette J.J., Potter G.L., Ramanathan V., Subasilar B., Whitlock C.H., Young D.F. and Zhou Y. (1995) Absorption of solar-radiation by clouds – observations versus models. *Science*, **267**, 496–499.

Chamberlain A.C. (1991) *Radioactive Aerosols.* Cambridge University Press, Cambridge.

Charlson R.J. and Heintzenberg J. (eds) (1995) *Aerosol Forcing of Climate.* Wiley, Chichester.

Charlson R.J., Lovelock J.E., Andreae M.O. and Warren S.G. (1987) Oceanic phytoplankton, atmospheric sulphur, cloud albedo and climate. *Nature,* **326,** 655–661.

Charlson R.J., Schwartz S.E., Hales J.M., Cess R.D., Coakley J.A. Jr, Hansen J.E. and Hofmann D.J. (1992) Climate forcing by anthropogenic aerosols. *Science,* **255,** 423–430.

Chazette P., David C., Lefrère J., Godin S., Pelon J. and Mégie G. (1995) Comparative lidar study of the optical, geometrical, and dynamical properties of stratospheric post-volcanic aerosols, following the eruptions of El Chichón and Mount Pinatubo. *J. Geophys. Res.,* **100,** 23 195–23 207.

Clarke A.D. (1989) Aerosol light absorption by soot in remote environments. *Aerosol Sci. Technol.,* **10,** 161–171.

Clarke A.D. and Charlson R.J. (1985) Radiative properties of the background aerosol: Absorption component of extinction. *Science,* **229,** 263–265.

Coakley J.A., Bernstein R.L and Durkee P.A. (1987) Effect of ship-stack effluents on cloud reflectivity. *Science,* **237,** 1020–1022.

d'Almeida G.A. (1989) Desert aerosol: Characteristics and effects on climate, in *Paleoclimatology and Paleometeorology: Modern and Past Patterns of Global Atmospheric Transport* (eds M. Leinen and M. Sarntheim), Kluwer Academic, Dordrecht, Netherlands.

d'Almeida G.A., Koepke P. and Shettle E.P. (eds) (1991) *Atmospheric Aerosols: Global Climatology and Radiative Characteristics.* Deepak, Hampton, VA.

Delmas R.J. (1992) Environmental information from ice cores. *Rev. Geophys.,* **30,** 1–21.

Dentener F.J. and Crutzen P.J. (1993) Reaction of $N_2O_5$ on tropospheric aerosols: impact on the global distributions of $NO_x$, $O_3$, and OH. *J. Geophys. Res.,* **98,** 7149–7163.

Döscher A., Gäggeler H.W., Schotterer U. and Schwikowski M. (1995) A 130 years deposition record of sulfate, nitrate, and chloride from a high-alpine glacier. *Water, Air and Soil Poll.,* **85,** 603–609,

Duce R.A. (1995) Sources, distributions and fluxes of mineral aerosols and their relationship to climate, in *Aerosol Forcing of Climate* (eds R.J. Charlson and J. Heintzenberg). Wiley, Chichester.

Finlayson-Pitts B. and Pitts J. (1986) *Atmospheric Chemistry: Fundamentals and Experimental Techniques.* Wiley Interscience, New York.

Gehrig R. (1996) Swiss EMEP data, partly published in Annual EMEP Reports, Chemical Coord. Centre, NILU, Kjeller, Norway.

Gillani N.V., Schwartz S.E., Leaitch W.R., Strapp J.W. and Isaac G.A. (1995) Field observations in continental stratiform clouds: Partitioning of cloud particles between droplets and unactivated interstitial aerosols. *J. Geophys. Res.,* **100,** 18 687–18 706.

Grasserbauer M., Paleczek S., Rendl J., Kasper A. and Puxbaum H. (1994) Inorganic constituents in aerosols, cloud water and precipitation collected at the high alpine measurement station Sonnblick: Sampling, analysis and exemplary results. *Fres. J. Anal. Chem.,* **350,** 431–439.

Hansen A.D.A., Bodhaine B.A., Dutton E.G. and Schnell R.C. (1988) Aerosol black carbon measurements at the South Pole: Initial results, 1986–7. *Geophys. Res. Lett.,* **15,** 1193–1196.

Hansen J.E., Lacis A., Ruedy R. and Sato M. (1992) Potential climate impact of Mount Pinatubo eruption. *Geophys. Res. Lett.,* **19,** 215–218.

Harshvardhan (1993) Aerosol–climate interactions, in *Aerosol–Cloud–Climate Interactions* (ed. P.V. Hobbs). Academic Press, San Diego, CA.

Heimann M., Monfray P. and Polian G. (1990) Modeling in the long-range transport of [222]Rn to subantarctic and antarctic areas. *Tellus B,* **42,** 83–99.

Heintzenberg J. (1982) Size-segregated measurements of particulate elemental carbon and aerosol light absorption at remote Arctic locations. *Atmos. Environ.,* **16,** 2461–2469.

Hering S.V. (1989) (ed.) *Air Sampling Instruments.* American Conference of Governmental Industrial Hygienists, Cincinnati, OH.

Hidy G.M. (1984) *Aerosols: An Industrial and Environmental Science.* Academic Press, Orlando, FL.

Hinds W.C. (1982) *Aerosol Technology – Properties, Behavior, and Measurements of Airborne Particles.* Wiley Interscience, New York.

Hobbs P.V. (1993) Aerosol–cloud interactions, in *Aerosol–Cloud–Climate Interactions* (ed. P.V. Hobbs). Academic Press, San Diego, CA.

Hobbs P.V. and McCormick M.P. (eds) (1988) *Aerosols and Climate*. Deepak, Hampton, VA.

IPCC (1992) *Climate Change 1992: The Supplementary Report to the IPCC Scientific Assessment* (eds J.T. Houghton, B.A. Callander and S.K. Varney). Cambridge University Press, Cambridge.

IPCC (1995) *Climate Change 1994* (eds J.T. Houghton, L.G. Meira Filho, J. Bruce, Hoesung Lee, B.A. Callander, E. Haites, N. Harris and K. Maskell). Cambridge University Press, Cambridge.

Jaenicke R. (1988) Atmospheric physics and chemistry, in *Meteorology: Physical and Chemical Properties of Air* (ed. G. Fischer), Landolt-Boernstein Series, Group V, Vol. 4b. Springer-Verlag, Berlin, pp. 391–457.

Jaenicke R. (1993) Tropospheric aerosols, in *Aerosol–Cloud–Climate Interactions* (ed. P.V. Hobbs), Academic Press, San Diego, CA.

Jäger H. (1992) The Pinatubo eruption cloud observed by Lidar at Garmisch-Partenkirchen. *Geophys. Res. Lett.*, **19**, 191–194.

Jennings S.G. (ed.) (1993) *Aerosol Effects on Climate*, Arizona University Press, Tucson.

Junge C.E. (1963) *Air Chemistry and Radioactivity*. Academic Press, New York.

Kasper A. and Puxbaum H. (1994) Determination of $SO_2$, $HNO_3$, $NH_3$ and aerosol components at a high alpine background site with a filter pack method. *Anal. Chim. Acta*, **291**, 297–304.

Kouimtzis T. and Samara T. (eds) (1995) *Airborne Particulate Matter*. Springer-Verlag, Berlin.

Langner J. and Rodhe H. (1991) A global three-dimensional model of the tropospheric sulfur cycle. *J. Atmos. Chem.*, **13**, 255–263.

Leinen M. and Sarnthein M. (eds) (1989) *Paleoclimatology and Paleometeorology: Modern and Past Patterns of Global Atmospheric Transport*. Kluwer Academic, Dordrecht, Netherlands.

Levine J.S. (ed.) (1991) *Global Biomass Burning*. MIT Press, Cambridge, MA.

Lodge J.P. Jr. (ed.) (1989) *Methods of Air Sampling and Analysis*. Lewis, Chelsea, Michigan.

Lugauer M., Baltensperger U., Furger M., Gäggeler H.W., Just D.T., Schwikowski M. and Wanner H. (1997) Aerosol transport to the high Alpine sites Jungfraujoch (3454 m asl) and Colle Gniffeti (4452 m asl), *Tellus B*, in press.

McCormick M.P., Wang P.H. and Poole L.R. (1993) Stratospheric aerosois and clouds, in *Aerosol–Cloud–Climate Interactions* (ed. P.V. Hobbs). Academic Press, San Diego, CA.

Möller D. (1995) Sulfate aerosols and their atmospheric precursors, in *Aerosol Forcing of Climate* (eds R.J. Charlson and J. Heintzenberg). Wiley, Chichester.

Montzka S.A., Butler J.H., Myers R.C., Thompson T.M., Swanson T.H., Clarke A.D., Lock L.T. and Elkins J.W. (1996) Decline in the tropospheric abundance of halogen from halocarbons: implications for stratospheric ozone depletion. *Science*, **272**, 1318–1322.

Nicholson K.W. (1988) The dry deposition of small particles: A review of experimental measurements. *Atmos. Environ.*, **22**, 2653–2666.

NOAA (1994) *Climate Monitoring and Diagnostics Laboratory*, No. 22 (eds J.T. Peterson and R.M. Rosson), Summary Report 1993. NOAA Environmental Research Laboratories, Boulder, CO.

Nyeki S., Baltensperger U., Colbeck I., Jost D.T. and Weingartner E. (1997a) The Jungfraujoch high-alpine research station (3454 m) as a background clean continental site for the measurement of aerosol parameters. *J. Geophys. Res.*, in press.

Nyeki S., Li F., Rosser D., Colbeck I. and Baltensperger U. (1997b) The background aerosol size distribution at a high-alpine site: An analysis of the seasonal cycle. *J. Aerosol Sci.*, **28**, S211–S212.

Oeschger H. and Langway C.C. Jr. (eds) (1989) *The Environmental Record in Glaciers and Ice Sheets*. Wiley, Chichester.

Penner J.E. (1995) Carbonaceous aerosols influencing atmospheric radiation: Black and organic carbon, in *Aerosol Forcing of Climate* (eds R.J. Charlson and J. Heintzenberg). Wiley, Chichester.

Penner J.E., Dickinson R.E. and O'Neill C.A. (1992) Effects of aerosol from biomass burning on the global radiation budget. *Science*, 256, 1432–1434.

Penner J., Charlson R.J., Hales J.M., Laulainen N.S., Leifer R., Novakov T., Ogren J., Radke L.F., Schwartz S.E. and Travis L. (1993) Quantifying and minimizing uncertainty of climate forcing by anthropogenic aerosols. *Bull. Am. Met. Soc.*, 75, 375–400.

Porstendörfer J., Röbig G., and Ahmed A. (1979) Experimental determination of the attachment coefficients of atoms and ions on monodisperse aerosols. *J. Aerosol Sci.*, 10, 21–28.

Prospero J.M., Uematsu M. and Savoie D.L. (1989) Mineral aerosol transport to the Pacific Ocean, in *Chemical Oceanography* (eds J.P. Riley, R. Chester and R.A. Duce), Vol. 10. Academic Press, London, pp. 188–218.

Pruppacher H.R. and Jaenicke R. (1995) The processing of water vapor and aerosols by atmospheric clouds, a global estimate. *Atmos. Res.*, 38, 283–295.

Pruppacher H.R., and Klett J.D. (1978) *Microphysics of Clouds and Precipitation*. Reidel, Dordrecht, Netherlands.

Pueschel R.F. (1995) Atmospheric aerosols, in *Composition, Chemistry and Climate of the Atmosphere* (ed. H.B. Singh). Van Nostrand Reinhold, New York, pp. 120–175.

Pye, K. (1987) *Aeolian Dust and Dust Deposits*. Academic Press, London.

QUARG (1993) *Urban Air Quality in the United Kingdom*, Quality of Urban Air Review Group. Department of Environment, London.

Radojevic M. and Harrison R.M. (1992) *Atmospheric Acidity*. Elsevier, London.

Ramanathan V., Subasilar B., Zhang G.J., Conant W., Cess R.D., Kiehl J.T., Grassl H. and Shi L. (1995) Warm pool heat-budget and shortwave cloud forcing – a missing physics. *Science*, 267, 499–503.

Reiter R., Sladkovic R. and Pötzl, K. (1985) Determination of the concentration of chemical main and trace elements (chemical matrix) in the aerosol from 1972 to 1982 at a north-alpine pure air mountain station at 1780 m a.s.l. Part IV: Analysis of the routes of transport of air masses with a high content of predominantly mineral as well as anthropogenic material. *Arch. Met. Geophys. Bioclim. B*, 36, 43–66.

Rowland F.S. and Isaksen I.S.A. (eds) (1988) *The Changing Atmosphere*. Wiley Interscience, Chichester.

Ruijgrok W., Davidson C.I., Nicholson K.W. (1995) Dry deposition of particles. *Tellus B*, 47, 587–601.

Schwikowski M., Seibert P., Baltensperger U. and Gäggeler H.W. (1995) A study of an outstanding Saharan dust event at the high-alpine site Jungfraujoch, Switzerland. *Atmos. Environ.*, 29, 1829–1842.

Seinfeld J.H. (1986) *Atmospheric Chemistry and Physics of Air Pollution*. Wiley, New York.

Singh H.B. (ed.) (1995) *Composition, Chemistry and Climate of the Atmosphere*. Van Nostrand Reinhold, New York.

Sturges W.T. (ed.) (1991) *Pollution of the Arctic Atmosphere*. Elsevier, London.

Tegen I., Lacis A.A. and Fung I. (1996) The influence on climate forcing of mineral aerosols from disturbed soils. *Nature*, 380, 419–422.

Tolbert M.A. (1996) Polar clouds and sulfate aerosols. *Science*, 272, 1597.

Tomasi C. and Vitale, V. (1984) Vertical variations of aerosol particle extinction in an alpine valley. *J. Aerosol Sci.*, 15, 413–416.

Twomey S.A. (1977) The influence of pollution on the short-wave albedo of clouds. *J. Atmos. Sci.*, 34, 1149–1152.

Van Roozendael M., de Maziere M. and Simon P.C. (1994) Ground-based visible measurements at the Jungfraujoch station since 1990. *J. Quant. Spectrosc. Radiat. Transfer*, 52, 231–240.

Wagenbach D. (1991) Environmental records in Alpine glaciers, in *The Environmental Record in Glaciers and Ice Sheets* (eds H. Oeschger and C.C. Langway Jr.). Wiley Interscience, Chichester, pp. 69–83.

Wallén C.C. (ed.) (1977) *Climates of Central and Southern Europe*, Vol. 6, World Survey of Climatology. Elsevier, Amsterdam.

WMO (1991) *Report on the WMO Meeting of Experts to Consider the Aerosol Component of GAW*, No. 79. World Meteorological Organization, Geneva.

## Nomenclature

| | |
|---|---|
| $A_0$ | cloud albedo |
| $\mathring{a}$ | Ångström exponent |
| $B(d)$ | the attachment coefficient |
| $\bar{c}$ | the mean thermal velocity of a diffusing molecule |
| $C_c$ | dimensionless Cunningham slip correction factor |
| $C_{SCAV}$ | mass scavenging coefficient of aerosols by cloud droplets |
| $d$ | aerosol geometric diameter |
| $d_{CRIT}$ | aerosol critical diameter at which activation occurs |
| $d_{OPT}$ | aerosol optical equivalent diameter |
| $D$ | diffusion coefficient of a diffusing molecule |
| $f(RH)$ | aerosol growth factor as a function of relative humidity |
| $g$ | acceleration due to gravity |
| $g_{ASY}$ | asymmetry factor of a scattering aerosol |
| $J$ | flux of gaseous molecules sticking to aerosol particles |
| $L$ | cloud liquid-water content |
| $M$ | fine mode aerosol mass concentration |
| $N$ | aerosol number concentration |
| $N_{CLOUD}$ | cloud droplet concentration |
| $N_i(d)$ | the particle number concentration with particle diameter $d$ |
| $N_{mol}$ | the molecular number concentration |
| $r$ | aerosol geometric diameter |
| $r_e$ | effective cloud droplet radius |
| $S$ | aerosol surface area concentration |
| $S_C$ | critical supersaturation at which aerosol activation into a cloud droplet occurs |
| $v_D$ | aerosol deposition velocity due only to gravitational sedimentation |
| $v_{DEP}$ | aerosol dry deposition velocity |
| $V$ | aerosol volume concentration |
| $z_0$ | roughness length of surface for aerosol dry deposition |
| | |
| $\alpha_{AP}$ | aerosol specific absorption efficiency |
| $\alpha_{SP}$ | aerosol specific scattering efficiency |
| $\beta(\phi)$ | angular scattering phase function |
| $\gamma$ | dimensionless sticking coefficient |
| $\delta$ | cloud optical thickness |
| $\eta$ | viscosity of air |
| $\lambda$ | wavelength of light |
| $\lambda_{MFP}$ | mean free path of a diffusing molecule |
| $v$ | Junge parameter |
| $\sigma_{AP}$ | aerosol absorption coefficient |
| $\sigma_{BSP}$ | aerosol backwards hemispheric scattering coefficient |

$\sigma_{EXT}$      aerosol extinction coefficient; the sum of $\sigma_{SP}$ and $\sigma_{AP}$

$\sigma_{SP}$      aerosol total scattering coefficient

$\tau$      atmospheric residence time

$\bar{\omega}$      aerosol single scattering albedo; the ratio of $\sigma_{SP}$ and $\sigma_{EXT}$

$\Omega$      aerosol optical depth; the vertical integral of $\sigma_{EXT}$ over the light beam path length $dl$

# 8  Synthesis and processing of nanoparticle materials

S. JAIN, T.T. KODAS and D. MAJUMDAR

## 8.1  Introduction

Nanoparticles have attracted considerable interest in the recent past due to their unique properties which differ from those of the bulk material (Gleiter, 1989). The properties of nanoparticles are unique because they have insufficient atoms to exhibit bulk properties. This chapter focuses on the synthesis of nanoparticles by aerosol routes and their processing.

Aerosol routes have been used to synthesize both nanoparticles and nanophase materials. The term 'nanoparticle' refers to single particles where the size of each particle is of the order of tens of nanometres or less (Figure 8.1). In some cases, agglomerates of these small particles are formed. If the size of an agglomerate is of the order of tens of nanometres or less the agglomerate can be considered a nanoparticle as well (Figure 8.1). Some aerosol routes result in particles each consisting of many small crystals. If the size of these crystallites is of the order of nanometres (though the particle size maybe of the order of a few micrometres or larger) then these particles can be termed as nanophase powder/material (Figure 8.1).

Spherical unagglomerated nanoparticles are the most desirable (compared to agglomerated nanoparticles) for processing to make bulk materials of different shapes and sizes because they sinter easily and exhibit other

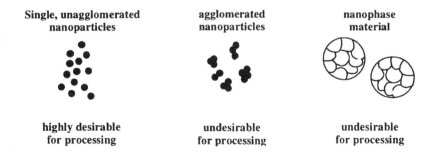

| Single, unagglomerated nanoparticles | agglomerated nanoparticles | nanophase material |
|---|---|---|
| highly desirable for processing | undesirable for processing | undesirable for processing |

Fig. 8.1 Nanoparticles and nanophase materials/powders.

unique properties. Agglomerated nanoparticles and large particles containing nanocrystallites are undesirable because it is difficult to eliminate the porosity between the particles and agglomerates to achieve the desired properties (Suraynarayan *et al.*, 1996b). Nanophase materials in the form of micrometre-size particles are used for processing when it is not possible to synthesize unagglomerated nanoparticles in sufficient amounts due to problems associated with the method of synthesis. Several examples exist: superconducting oxides such as $YBa_2Cu_3O_{7-x}$ (Setaka *et al.*, 1988; Tohge *et al.*, 1988; Kodas *et al.*, 1988), BiSrCaCuO (Pebler and Charles, 1988; Kodas *et al.*, 1989).

Aerosol routes for particle synthesis can be broadly classified into two categories: gas-to-particle conversion; and intraparticle reaction which is also termed spray pyrolysis (Gurav *et al.*, 1993a; Messing *et al.*, 1993), aerosol decomposition (Kodas, 1989), spray roasting (Ruthner, 1983), evaporative decomposition (Gardner *et al.*, 1984), spray calcination (Vollath, 1990), liquid aerosol thermolysis (Jayanthi *et al.*, 1993) and other names. Particles of the order of tens of nanometres or less can be synthesized by the gas-to-particle conversion route, but particles generated by spray pyrolysis are generally of the order of hundreds of nanometres or larger (submicrometre; Messing *et al.*, 1993) and therefore are not nanoparticles. Nevertheless, nanophase particles can be made by spray pyrolysis in the sense that the particles consist of smaller crystals (grains) which are of the order of few nanometres in size (Gurav *et al.*, 1993a; 1993b). Therefore. this chapter discusses gas-to-particle conversion as a route to nanoparticles and spray pyrolysis as a route to nanophase materials.

Powder processing is used for particles made by the gas-to-particle conversion route and by spray pyrolysis, but the conditions used for the powders made by different routes vary. This is because the particles made by these two routes have different particle size ranges and morphologies (hollow, solid or agglomerated particles). The two aerosol routes are described briefly at this point in order to explain the origin of the differences in the properties of the particles made by the two routes.

*Gas-to-particle* conversion involves the introduction of reactants into a reactor either in the vapour form or as liquid droplets or solid particles which are vaporized inside the reactor before chemical reaction takes place (Figure 8.2a). The gaseous precursors react in the gas-phase to form small product particles (hence the term gas-to-particle conversion). In the physical analogue of this process, the product vapour is formed by vaporization of a bulk material. The particles grow when they collide with other particles (collision) and fuse (coalescence) to form bigger particles (Friedlander, 1977). Particles can also grow by condensation, the consumption of vapour by the particles (Friedlander, 1977) when the particle number concentration is low enough so no particle–particle collisions

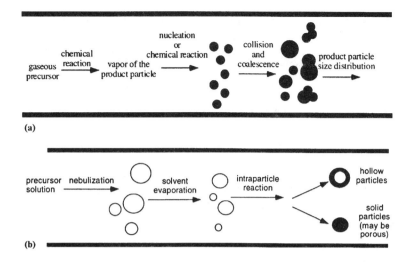

**Fig. 8.2** Schematic diagram of: (a) the gas-to-particle conversion route; (b) the liquid-to-particle route (spray pyrolysis).

occur in the reactor. During synthesis of nanoparticles it is usually growth by collision and coalescence that must be controlled to limit the particle size. This occurs because the particle concentration is usually high enough for appreciable particle–particle collisions to occur. Particle number concentrations of $10^{15} cm^{-3}$ are common.

Because of the random nature of the collision process the particles formed are not monodisperse and control of stoichiometry for multi-component systems can be difficult (Gurav *et al.*, 1993a). Further, the particles are frequently aspherical and agglomerated because of in-complete coalescence. But the particles formed are usually solid (rather than porous or hollow as sometimes occurs in spray pyrolysis). Conditions such as temperature, residence time and reactant concentration can be controlled so that it is possible to synthesize particles of only a few nanometres in size by this route (Kirkbir and Komiyama, 1987; Morooka *et al.*, 1991).

In *spray pyrolysis* the reactants are dissolved or suspended in a liquid which is atomized into droplets using an aerosol generator. A gas then carries these droplets into a reactor where the solvent evaporates and the precursor undergoes intraparticle chemical reaction in the liquid/solid state (rather than gas phase) to form the product particles (Figure 8.2b). Because of the low droplet number concentration ($10^6$–$10^7 cm^{-3}$; Messing *et al.*, 1993) produced by most atomizers, collisions between the droplets/particles do not take place to any significant extent, thereby allowing the formation of unagglomerated powders (in contrast with the particles formed by gas-to-particle conversion). Because the reactants are mixed in a common liquid this method offers greater versatility and control over stoichiometry for multi-component systems. However, hollow or porous particles can be formed by this route (Gurav *et al.*, 1993a; Messing *et al.*, 1993).

The final particle size for spray pyrolysis can be controlled by varying the initial reactant concentration. For example, an initial droplet size (containing the precursor and the solvent, which is assumed to be water) of 5 μm (typical value) and a concentration of 4 mol $AgNO_3$ results in a final Ag particle size of $\sim 1$ μm using an ultrasonic generator (Pluym *et al.*, 1993). In order to synthesise particles a few nanometres in size the reactant concentration has to be extremely low. The low reactant concentration and the low droplet number concentration together result in very low production rates for nanoparticles. For example, the production rate for 5 nm ($d_p$) particles of unit density ($\rho$) with $10^7 \, cm^{-3}$ ($N$) droplet concentration and gas flow rate of $5.0 \, l \, min^{-1}$ ($Q$) is $\sim 0.21 \, mg \, h^{-1}$ ($\frac{\pi}{6} d_p^3 \rho N Q$). This is the major reason why particles made by this route are generally in the submicrometre range or larger.

The above-mentioned aerosol routes form the subject matter of this chapter. Section 8.2 describes the gas-to-particle conversion route for the synthesis of nanoparticles, with emphasis on the fundamental processes involved and the technology currently used to synthesize different materials. Section 8.3 focuses on nanophase material synthesis by spray pyrolysis, with emphasis on the fundamentals and the technology. These two sections concentrate on the method of *synthesis* of nanoparticles and nanophase particles. The particles prepared by these routes must in many cases be subjected to various processes to form coatings or bulk materials and impart the final properties depending on their applications. This is broadly termed as *powder processing* and is discussed in detail in section 8.4. Thus, this chapter provides an overview of the complete process for making the final product beginning from the starting materials.

## 8.2  Gas-to-particle conversion

This section discusses the synthesis of nanoparticles by the gas-to-particle conversion route. The major chemical and physical steps in this process are: vaporization of the reactant (when the reactant is not a gas), gas-phase reaction, nucleation, collision and sintering. In physical processes, reactant vaporization and reaction are replaced by vaporization, ablation or sputtering of the material of which particles are desired. Condensation and surface reaction (Friedlander, 1977) are two chemical processes which can lead to an increase in particle size when particles are synthesized by the gas-to-particle conversion route. However, these steps are usually not important during synthesis of nanoparticles. The chemical reaction version of this process is used for industrial production of titania, fumed silica and carbon blacks (Ulrich, 1984). Nanoparticles of elements, alloys, compounds and mixtures of particles have been synthesized at laboratory scales by both the chemical and physical variations of this route (Gleiter, 1981; Kagawa et al., 1983; Rohlfing et al., 1984; Liu et al., 1986; Okuyama et al., 1986; Siegel and Eastman, 1989; Hahn and Averback, 1990; Hung and Katz, 1992).

### 8.2.1 Basic physicochemical processes

This section briefly discusses the fundamental processes such as vapour generation, chemical reaction, nucleation and aerosol/particle dynamics (collision and coalescence) separately to facilitate understanding of the basic concepts. These individual concepts are then integrated in order to describe qualitatively the overall behaviour of an aerosol reactor. The concept of characteristic times is also introduced to illustrate how each of the basic processes affects the particle size distribution and morphology. Various types of reactors are then discussed in terms of the fundamental processes.

8.2.1.1 *Generation of the precursor vapour.* A prerequisite for particle synthesis by gas-to-particle conversion is the generation of the precursor vapour. An important distinction is necessary at this point with respect to the nature of the precursor. In most cases, the precursor is a material different from the product material and in such cases a chemical reaction is necessary to convert the precursor to the final product. In other cases, the precursor is the same material as the final product and in such cases the process does not involve any chemical reaction but involves the processes of evaporation, ablation or sputtering of the starting material (which may be in the powder or bulk form) to generate the vapour.

The case where chemical reaction in the gas phase is necessary to convert the precursor to the final product is now discussed. For example, volatile precursors such as $SiCl_4$ and $TiCl_4$ (for $SiO_2$ and $TiO_2$ particle synthesis; see Powers, 1978; Pratsinis et al., 1990) are heated outside the reactor to generate sufficient vapour which is then transported to the reactor using a carrier gas (which may be nitrogen, oxygen or some other coreactant). In cases where the reactant has low volatility it can be dissolved in a solvent and the solution can be atomized to form droplets. The droplets are then carried into the reactor using a carrier gas where the solvent and the precursor evaporate (usually in less than 1% of the total residence time). This approach has been used in flow reactors (Kagawa et al., 1983; Powell et al., 1992; Weimer et al., 1991). Dry precursor powder delivery can also be used.

The case where chemical reactions are not necessary to convert the precursor into the final product usually involves evaporation by simple resistive heating (Gleiter, 1989; Siegel and Eastman, 1989). In some cases high-energy lasers such as excimer or $CO_2$ lasers (Gonsalves et al., 1991; Rohlfing et al., 1984: Haggerty, 1984; Liu et al., 1986) are used to provide the energy required to vaporize a solid material and the vapour is then used as the precursor for the synthesis of the desired material. In other cases, a plasma is generated which provides the energy to vaporize the precursor (Vissikov et al., 1988). Laser and plasma approaches can also be used for precursors where chemical reaction is needed to convert the precursor into the final product (Vissikov et al., 1981; Nakahigashi et al., 1992).

High-energy ions have also been used to sputter precursors where the ions dislodge atoms from the material (called the target) to form atoms of the precursor in the gas phase (Gleiter, 1989; Hahn and Averback, 1990; Chow et al., 1990; Ying, 1993). These modes of precursor generation are useful for non-volatile precursors.

### 8.2.1.2 Chemical reaction.

This step is applicable only to systems where the precursor undergoes a gas-phase chemical reaction to form the final product. The gaseous precursor is introduced into the reactor, where the high temperature of the reactor results in the homogenous reaction of the gaseous precursor. Chemical reaction results in the generation of molecules of the product material. The rate of chemical reaction ($r_A$, $mol\,cm^{-3}\,s^{-1}$) usually increases exponentially with temperature ($T$, in kelvin), depends on the reactant partial pressure ($C_A$, $mol\,cm^{-3}$) and is described by the following equation (Levenspiel, 1983) for a first-order reaction:

$$r_A = k_0 \exp\left(-\frac{E}{RT}\right) C_A, \qquad (8.1)$$

where $k_0$ is a constant ($s^{-1}$), $E$ is activation energy ($J\,mol^{-1}$) and $R$ is the molar gas constant ($8.3145\,J\,K^{-1}\,mol^{-1}$). In order to obtain a high reaction rate (and for other reasons discussed later) high temperatures are often necessary and therefore gas-to-particle conversion reactors frequently operate in the temperature range of 1000–2500 K. For example, the reaction rate for $TiCl_4$ is $\sim 6.7 \times 10^{-5}\,mol\,cm^{-3}\,s^{-1}$ at a temperature of 1500 K and $TiCl_4$ concentration of $1 \times 10^{-6}\,mol\,cm^{-3}$ (Pratsinis et al., 1990). This rate drops to $\sim 1 \times 10^{-5}\,mol\,cm^{-3}\,s^{-1}$ at a temperature of 1200 K. Frequently, the reaction goes to completion early in the reactor (10 ms) as in the oxidation of $TiCl_4$ and $SiCl_4$ to make $TiO_2$ and $SiO_2$, respectively. As is discussed later, control of reaction rate is essential for control of particle size distribution.

8.2.1.3  *Nucleation.*    This step is applicable to all systems irrespective of the nature of the precursor (i.e. whether it undergoes a chemical reaction or not). The chemical reaction of the precursor or vaporization/sputtering results in the formation of gas atoms or molecules of the product material. These atoms/molecules as a group lower their energy by undergoing the phase transformation to the liquid or the solid state, resulting in the formation of particles (McDonald, 1962). The driving force for and also the rate of this transformation depend on the ratio of the partial pressure of the product molecules to the saturated vapour pressure of the product material at that temperature; this quantity is termed the *saturation ratio.* The high partial pressure relative to the vapour pressure is generated by chemical reaction in cases where the precursor undergoes chemical reaction or by simple vaporization or sputtering followed by cooling in other cases. This is shown schematically in Figure 8.3.

Nucleation takes place when clusters of the gaseous molecules of the product material attain a certain critical size (McDonald, 1962; Friedlander, 1977) which depends on the properties of the material such as surface tension and density. This size may or may not be greater than the size of the individual molecule of the product material. If this critical size

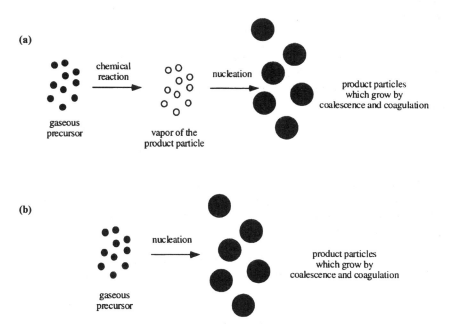

**Fig. 8.3** Schematic of the process when: (a) the precursor undergoes chemical reaction; (b) the precursor does not undergo chemical reaction.

is less than the size of a single molecule of the product material it implies that there is no barrier to nucleation and solid particles are formed directly from chemical reaction or by cooling of the vapour (Xiong and Pratsinis, 1991). In such cases chemical reaction and nucleation cannot be distinguished from each other. This is the case for $TiO_2$ formation (Xiong and Pratsinis, 1991). It should be noted that nucleation always follows chemical reaction for all systems. But in some cases (such as $TiO_2$) nucleation may not be the rate-limiting step for the generation of solid particles. In such systems formation of solid particles is limited by the rate of chemical reaction.

8.2.1.4 *Aerosol dynamics.* The generation of new particles (either by nucleation or by chemical reaction) rapidly increases the particle number concentration in the system. Typically, simulation studies show that the particle concentrations exceed $10^{12}\,cm^{-3}$ (Okuyama *et al.*, 1986; Xiong and Pratsinis, 1991; Xiong *et al.*, 1992). For example, the production of $1\,g\,s^{-1}$ of 10 nm particles of unity density requires particle concentration of $\sim 4 \times 10^{17}\,cm^{-3}$ at a gas flow rate of $5.01\,min^{-1}$ (based on a simple mass balance). The high particle number concentration leads to collisions between the particles. (*Collision* refers to the colliding of two particles. *Coagulation*, on the other hand, refers to collision of two particles along with complete coalescence of these particles to form a single bigger spherical particle.) The rate of collision can be quantified by the characteristic time for collision which is defined as the time required to halve the particle number concentration under a given set of conditions.

The characteristic time for collision $\tau_{coag}$ can be given by $1/KN$, where $K$ is the collision coefficient ($K = 4k_B TC_c/3\eta$), $N$ the particle number concentration, $k_B$ the Boltzmann constant, $T$ temperature, $\eta$ gas viscosity and $C_c$ the Cunningham correction factor (Hinds, 1982). The collision coefficient is independent of the particle size ($C_c = 1.0$) when the particles are much greater than the mean free path of the gas molecules. When the particle size is on the order of or smaller than the mean free path of the gas molecules, $K$ increases as particle size decreases (Friedlander, 1977; Hinds, 1982).

At standard conditions and a particle number concentration of $10^{12}\,cm^{-3}$ (a typical number for nanoparticle production) the characteristic time for collision is 0.15 ms (i.e. it takes 0.15 ms to halve the concentration of monodisperse particles from an initial value of $10^{12}\,cm^{-3}$) for particle size of 1 nm (Hinds, 1982). A typical residence time in an aerosol reactor is $\sim 1\,s$, and therefore collision can never be ignored (in fact, it is the *dominant growth mechanism*) in reactors used for nanoparticle production.

The particles that collide stick together and the random nature of the particle collision process leads to a distribution of particle sizes. These distributions are often approximately lognormal (Hinds, 1982) or self-

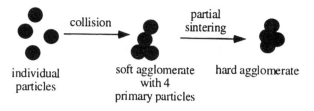

**Fig. 8.4** Formation of hard and soft agglomerates.

preserving size distributions (Friedlander, 1977) and therefore the particles made by this route are never monodisperse. The individual particles are called **primary particles** and the structures formed by collisions of these particles are called **agglomerates**. The agglomerates formed by particles simply sticking to each other through van der Waals force can be easily broken apart into the primary particles and are called **soft agglomerates**. When some of the particles fuse (coalescence) partially, **hard agglomerates** are formed. Incomplete sintering (fusing) of these particles results in the aspherical nature of these agglomerates. Figure 8.4 shows the two different types of agglomerate formed by this process.

8.2.1.5 *Sintering.* Sintering (or coalescence) is one of the most important processes taking place in aerosol reactors. Sintering is the fusion of two or more solid particles into a bigger single spherical particle. Sintering is a process by which the surface energy of the particles is minimized (Kingery *et al.*, 1976) and it can result in the formation of the geometrical shape with the lowest surface energy. This shape often has the lowest surface area for a given volume (i.e. a sphere). It is important to note that collisions between the particles alone are not sufficient to increase their size; it is essential that the particles fuse together in order to form a bigger particle. Thus, sintering is as important as collision for growth of nanoparticles. Incomplete sintering leads to the formation of aspherical hard agglomerates in the gas-to-particle conversion route.

The rate of sintering can be quantified by the characteristic time for sintering $\tau_{sint}$ and is generally of the form $cd^n \exp(-E/RT)$ where $c$ is a constant, $d$ the primary particle size, $n$ an integer (from 1 to 4), $E$ the activation energy, $R$ the molar gas constant and $T$ temperature (Kingery *et al.*, 1976). When $n$ is 4 the dependence on the primary particle size is strong; for example, 10 nm titania particles sinter in 1 ms at 1500 K but it takes 10 s for 0.1 μm particles to sinter at the same temperature (Kobata *et al.*, 1991). This means that if the residence time in an aerosol reactor is ~1 s then 10 nm particles will sinter completely while 0.1 μm particles will fuse partially.

Temperature has the most dominant effect on the sintering rate due to the exponential dependence. For example, it takes 10 nm titania particles 22 s to sinter at 1000 K, whereas it takes only 1 ms for these particles to sinter at 1500 K. For this reason (apart from the effect on reaction rate) the temperatures in these reactors must be kept as high as possible for as long as possible when micrometre-sized particles are desired and temperatures must be as low as possible for as short a time as possible when nanometre-sized particles are desired.

It is important to keep in mind that the sintering rate always decreases along the length of the reactor due to the increase in the particle size even if the temperature is constant (Xiong *et al.*, 1992). In practice, however, the temperature usually drops along the length of the reactor and the sintering rate decreases to a greater extent. This can be used to advantage in nanoparticle production where the temperature can be dropped rapidly to prevent further particle growth by collision and sintering and therefore limit the particle size to the nanometre scale.

8.2.1.6  *Particle deposition.*  The particles formed in the reactor spend a certain amount of time in the reactor during which they may deposit on the reactor walls by impaction, diffusion, electrophoresis, thermophoresis and gravitational sedimentation (Hinds, 1982) and may then be re-entrained by the fluid/particle flow. This process is complicated and few of the modelling studies of aerosol reactor behaviour so far have accounted for these phenomena (i.e. deposition and re-entrainment) (Gelbard and Seinfeld, 1979; Okuyama *et al.*, 1986: Xiong and Pratsinis, 1991; Xiong *et al.*, 1992). Gravitational sedimentation is almost never an important deposition mechanism for nanoparticles due to their very low settling velocities (Hinds, 1982). Thermophoresis is the motion of particles from a high-temperature region to a low-temperature region and is exploited for collecting product particles on cooler surfaces in some of the processes discussed below. As an example, the thermophoretic velocity for a 10 nm particle is $2.8 \, \mu m \, s^{-1}$ (unit temperature gradient, $1 °C \, cm^{-1}$) compared to the sedimentation velocity of $0.07 \, \mu m \, s^{-1}$ (Hinds, 1982). At higher temperature gradients thermophoretic deposition becomes more important and in fact is used to collect the particles made in certain types of reactor (Pfund, 1930; Burger and Cittert, 1930; Siegel *et al.*, 1988; Mayo *et al.*, 1990). The role of impaction and electrophoresis in aerosol reactors has not been analysed in the literature.

8.2.1.7  *Overall reactor behaviour.*  Thus far, the fundamental processes have been discussed separately. In reality, these steps take place simultaneously and the overall process is more complicated. However, considerable simplifications can be achieved by the calculation of the characteristic times discussed above: residence time, time for chemical reaction, time for collision and time for sintering. These simplifications allow an easier understanding of the interplay of the various processes and in many cases obviate detailed modelling computations.

The case of $TiO_2$ formation from $TiCl_4$ is discussed specifically. This is because it is one of the best-studied systems and therefore data on the reaction and sintering rates are available. The process is discussed in an integrated manner so that the interplay of the different processes such as chemical reaction, collision and sintering can be understood. The process is also discussed in terms of characteristic times to provide a qualitative as well as semi-quantitative analysis of the process.

The synthesis of $TiO_2$ from $TiCl_4$ in a tubular reactor is described in the schematic diagram of Figure 8.2a. The precursor, $TiCl_4$, is carried into the reactor using oxygen as the carrier gas which also is a coreactant in this process. One mole of $TiCl_4$ results in 1 mole of $TiO_2$. Therefore, a simple mass balance reveals that 0.0125 moles of $TiCl_4$ must be processed per hour to obtain a production rate of $1\,g\,h^{-1}$ of $TiO_2$. Therefore, $TiCl_4$ must be heated to a sufficiently high temperature to deliver this amount of precursor into the reactor if the goal is a production rate of $1\,g\,h^{-1}$.

The $TiCl_4$ that enters the tubular reactor (assumed to be at 1500 K) undergoes homogeneous gas-phase reaction to form $TiO_2$ molecules. The characteristic time for first-order chemical reaction ($TiCl_4$) is $\tau_{rxn} = 1/k_0 \exp(-E/RT)$. This means that at 1500 K the characteristic time for reaction is $\sim 15\,ms$ (Pratsinis *et al.*, 1990). If the residence time is assumed to be $\sim 1\,s$ (typical value) then it implies that the chemical reaction is complete very early in the reactor. However, if the temperature is maintained at 1000 K the characteristic time for chemical reaction is $\sim 0.5\,s$. Therefore, in the latter case the characteristic time for chemical reaction is comparable to the residence time (1 s) and therefore chemical reaction takes place to a significant extent all along the reactor.

The ratio of the characteristic time for chemical reaction and the residence time has a significant influence on the particle size distribution of the product particles. In the case where the characteristic time for chemical reaction is much shorter than the residence time, the particle size distribution can attain a unimodal self-preserving form (Landgrebe and Pratsinis, 1989). A self-preserving distribution means that if the particle size distribution function is normalized by the total particle number concentration and plotted as a function of the particle volume (which is normalized by the total volume) then the shape of the distribution remains constant with time (Friedlander, 1977); these distributions are roughly lognormal in shape.

On the other hand, if the characteristic time for chemical reaction is comparable to the residence time then the particle size distribution can be bimodal (Landgrebe and Pratsinis, 1989). Thus, a simple analysis of the characteristic times can reveal information about the *nature* of the particle size distribution.

When the reaction is complete very early in the reactor a simple mass balance can be used to obtain an idea of the particle size. For example, assume that $0.0125\,mol\,h^{-1}$ of $TiCl_4$ reacts completely very early in the reactor generating $0.0125\,mol\,h^{-1}$ of $TiO_2$. This corresponds to $7.5 \times 10^{16}$ $TiO_2$ molecules per cubic centimetre for a flow rate of $100\,l\,h^{-1}$. A single $TiO_2$ molecule has a size of $\sim 4\,Å$. If collision is allowed to proceed to the point where the particle number drops to about $5 \times 10^{12}\,cm^{-3}$ then the average size increases to $1.0\,nm$ based on a simple mass balance. Similarly, the particle number concentration for an average size of $100\,nm$ is $\sim 5 \times 10^9\,cm^{-3}$, for a flow rate of $100\,l\,h^{-1}$. The above calculation shows that the particle size can be controlled by controlling the initial reactant concentration (limiting the total number of $TiO_2$ molecules that are generated).

The above analysis demonstrates that when the characteristic time for chemical reaction is much shorter than the residence time, collision and sintering are the only significant processes taking place along most of the reactor length. In fact, these are the processes that lead to particle growth and must be controlled in order to achieve control over the particle size and morphology.

The particle size and morphology are strongly influenced by the process of collision and sintering. The processes of collision and coalescence (sintering) commence when the particles are being formed by chemical reaction and continue even when the chemical reaction is complete. The particle concentration increases rapidly due to chemical reaction (or nucleation for other systems). For example, at $1500\,K$ the reaction rate is $\sim 8.4 \times 10^{-6}\,mol\,cm^{-3}\,s^{-1}$ from equation (8.1). In the case of $TiO_2$, where the rate-limiting step is the chemical reaction, this rate can be multiplied by Avogadro's number to give the rate of generation of $TiO_2$ particles (Xiong and Pratsinis, 1991). This converts to $5 \times 10^{18}$ molecules per cubic centimetre per second, where each particle is a titania molecule. It is this high rate of generation of $TiO_2$ particles that leads to a rapid increase of the particle number concentration which results in a high collision rate.

Collision and coalescence lead to an increase in the particle size and their relative rates influence the morphology of the particles. When the particles are generated by chemical reaction their size is small and sintering takes place at a high rate. For example, the characteristic time for sintering of a $1\,nm$ $TiO_2$ particle is $\sim 0.1\,ms$ (Kobata et al., 1991). This is much shorter than the characteristic time for collision of $\sim 0.15\,ms$, assuming a particle number concentration of $10^{12}\,cm^{-3}$ and particle size of $1\,nm$. This means that the particles that collide have time to fuse completely before the next collision and therefore the particles remain spherical (Figure 8.5). This approach of estimating relative rates of collision and coalescence based on characteristic times was first proposed by Koch and Friedlander (1990).

| $\tau_{coal}/\tau_{coll} < 1$ | collision limited | spheres |  |
|---|---|---|---|
| $\tau_{coal}/\tau_{coll} \sim 1$ | intermediate | hard agglomerates |  |
| $\tau_{coal}/\tau_{coll} > 1$ | coalescence limited | aggregates "soft" | 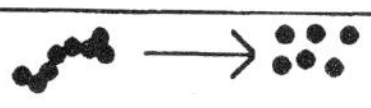 |

**Fig. 8.5** Effect of the collision and coalescence rate on the morphology of the particles.

As a particle travels downstream, the particle size increases and therefore the characteristic time for sintering increases (rate of sintering slows down). The particle number concentration also drops and increases the characteristic time for collision (rate of collision decreases). However, due to the strong dependence of the sintering characteristic time on the particle size the net effect is that the rate of sintering becomes lower than the collision rate. This means that the particles that collide do not have time to fuse completely before the next collision. This results in the formation of agglomerates. This is shown in Figure 8.5, where the ratio of the characteristic times for coalescence and sintering is used to describe how formation of agglomerates and aspherical particles takes place in these reactors.

The formation of spherical particles is favoured by high sintering rates (higher temperature and smaller particle sizes) and by low collision rate (lower reactant or particle concentration) as shown in Figure 8.5. Further, sintering as well as collision rate need to be controlled in order to limit particle size to the nanometre range. Although the above example focused on the $TiO_2$ system in a tubular reactor, the analysis presented above applies to all material systems.

### 8.2.2  Gas-to-particle conversion technology

This section discusses the established and emerging aerosol routes for synthesis of nanoparticles. Many of the technologies for nanoparticle synthesis developed from the understanding of conventional aerosol routes for synthesis of micrometre-sized particles. This is the reason why most of the existing and emerging technologies are variations of the conventional aerosol flow reactor shown in Figure 8.2a. Only the manner in which the precursor is vaporized or the way the heat for chemical reaction and sintering is introduced into the system varies based on materials, production rate and particle size distribution considerations. Thus numerous types of reactor exist in which particles are produced by gas-to-particle conversion.

Some of the types of gas-to-particle conversion are: flame reactors (Matsoukas and Friedlander, 1991; Zachariah *et al.*, 1989; Hung and Katz, 1992; Hung *et al.*, 1992), gas condensation (Granqvist and Buhrman, 1976; Gleiter, 1989), plasma reactors (Vissikov *et al.*, 1981; 1988; Ohno and Uda, 1984; Nakahigashi *et al.*, 1992), tube furnaces (Okuyama *et al.*, 1986), lasers (Haggerty and Flint, 1990) and sputtering (Ichinose *et al.*, 1992). These are discussed below.

8.2.2.1 *Flame reactors.* Precursors are introduced into a flame generated in most cases by the oxidation of hydrocarbons or hydrogen in the reactors. The exothermicity of the oxidation (of the hydrocarbons) reactions results in temperatures as high as 2000–3000 K. The high temperature drives the chemical reactions which form product particles (different from the oxidation reaction of the hydrocarbons). The precursors may be introduced into the system as a vapour or as solution droplets (for low-volatility precursors) (Ulrich, 1971; 1984; Shon *et al.*, 1979; Zachariah *et al.*, 1989). Therefore particles can be synthesized from even non-volatile (at room temperature) precursors which may be cheap compared to the more volatile precursors. The scale-up of this process for production of carbon black, silica and titania has been demonstrated (Gurav *et al.*, 1993a).

Due to the higher temperatures (compared to other routes), chemical reaction usually goes to completion very early in the reactor because of rapid reactant consumption. Particle formation may take place by either nucleation or direct chemical reaction (Ulrich, 1984). The usual steps of collision and coalescence lead to particle growth. The residence time in these reactors is short (between 1 ms and 1 s) and so, despite the high temperatures (which favour high growth rates of the particles), nanometre-sized particles can be synthesized.

In these systems there exist steep temperature gradients and it is difficult accurately to determine the temperature profile. This complicates control over the particle size and morphology. The product particles are usually hard agglomerates with a broad size distribution. Additives provide limited control over particle morphology and distribution by modifying the rates of sintering and collision (Mezey, 1966; Haynes *et al.*, 1979). This process has been used primarily for synthesis of ceramic powders which require high temperatures for the formation of desired phases.

8.2.2.2 *Furnace reactors.* These are tubes which are heated externally and therefore provide excellent control of the temperature profile. The precursors are introduced into the reactor as vapours or as droplets

**Fig. 8.6** Titania particles made in a furnace reactor (courtesy of Powell, 1995).

(containing low-volatility precursors) along with a carrier gas in a controlled manner where they undergo gas-phase reaction (and nucleation), collision and coalescence to form the product particles (Kagawa *et al.*, 1983; Powell *et al.*, 1992). The mass and heat transfer processes can be accurately determined in such systems and good control over particle morphology and size distribution is possible. The residence time usually varies in the 1–10 s range at the laboratory scale and is controlled usually by varying the flow rate of the carrier gas. Figure 8.6 is a micrograph of titania particles synthesised in a furnace reactor and shows non-spherical particles with a size distribution that is not mono-disperse (Powell, 1995).

The temperatures in such reactors are usually limited to below 1800 K by practical constraints for heating elements in furnaces. Ultrafine

(17–80 nm) titanium dioxide (amorphous and crystalline) particles have been synthesized in such reactors by various researchers (Kirkbir and Komiyama, 1987; Morooka *et al.*, 1991). Other compounds that have been synthesized in nanoparticle form by this route are GaAs and ZnS (Sercel *et al.*, 1992; Powell *et al.*, 1992). Reviews by Kodas (1989) and Gurav *et al.* (1993a) discuss the synthesis of other materials in such reactors.

8.2.2.3 *Gas-condensation method.* This method is the most extensively studied route for nanoparticle production. Initially this method was limited to the synthesis of nanoparticles of volatile elements (Pfund, 1930; Burger and Cittert, 1930) but was later extended to the synthesis of alloys, compounds and a mixture of particles (Yukawa *et al.*, 1972; Kashu *et al.*, 1984; Hirayama, 1987; Siegel *et al.*, 1988; Mayo *et al.*, 1990). This method, followed by *in-situ* compaction has been used to produce bulk nanophase materials (Gleiter, 1981; Siegel and Eastman, 1989; Averback *et al.*, 1990). There are excellent reviews in the literature on this method (Gleiter, 1989; Granqvist and Buhrman, 1976; Uyeda, 1991).

Gas condensation is versatile and can be used to produce high-purity particles. The process may be operated in continuous mode (Koizumi *et al.*, 1989) or batch mode (Siegel and Eastman, 1989). The greatest disadvantage of the batch version of this process is the low production rates of particles. Also, for the batch as well as the continuous version it is difficult to synthesize multi-component particles with the appropriate stoichiometries (Gurav *et al.*, 1993a).

A schematic of a typical batch reactor is shown in Figure 8.7. The main assembly consists of an ultrahigh vacuum chamber, evaporation sources, a liquid-nitrogen-cooled finger and scraper and a compaction unit to consolidate the powders (Siegel and Eastman. 1989). The whole process can be classified into three parts (Edelstein, 1993): generation of high partial pressure to achieve supersaturation; gas-phase nucleation, growth and agglomeration; and collection of particles.

The operation begins with the evacuation of the vacuum chamber to pressures less than $10^{-5}$ Pa followed by the introduction of an inert gas (He or Ar) with a pressure of few hundred pascal. The material is evaporated by resistive heating to generate sufficient vapour pressure. The free convective plume above the source cools as it rises, leading to supersaturation of the vapour and nucleation to form particles. These particles are carried by the gas current to the nitrogen-cooled finger where the particles deposit by thermophoresis. The particle size distribution in such systems is approximately lognormal (Granqvist and Buhrman, 1976). In many cases, the collected particles are scraped from the cold finger and directed into a piston and anvil device where they are compacted under high pressures (1–5 GPa) to form a bulk nanophase material.

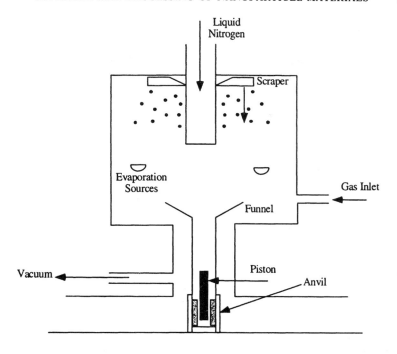

**Fig. 8.7** Schematic diagram of the gas-condensation route.

Generation of high partial pressures of materials can be achieved by thermal evaporation, sputtering or laser ablation (Mandich *et al.*, 1987; Hahn and Averback, 1990). Thermal evaporation suffers from the temperature limitations imposed by the crucible material and reaction of the material with the crucible. This is the reason why thermal evaporation has been largely limited to synthesis of metals. Sputtering has the advantage that it can be used for a wide variety of metals and oxides. It is important to inhibit the gas-phase transport of the vaporized atoms or extensive film formation occurs on the walls.

The inert gas pressure, the evaporation rate and gas composition can be varied to control the particle characteristics in batch systems. The particle size decreases with decrease in the evaporation rate and decrease in the gas pressure or by using a lighter gas such as He instead of Xe (Granqvist and Buhrman, 1976). Production of small clusters while maintaining a reasonable production rate is difficult in the batch approach and is a drawback of this method.

8.2.2.4   *Plasma reactors.*   This method can be considered as a variation of the continuous gas-condensation method. In these reactors a plasma is used to vaporize the precursor to generate a high partial pressure of the precursor. Precursors in powder form can be carried into the plasma zone where they are completely vaporized. Therefore, cheap and non-volatile precursors can be used in this case. After the plasma zone the vapour cools rapidly and chemical reactions may occur in the gas phase (if the precursor material is different from the product). This results in supersaturation of the vapour, leading to subsequent nucleation and particle formation (see reviews by Kodas, 1989; Kodas and Sood, 1990).

This process has been used to synthesize nanoparticles of different elements and compounds. For example, metal particles of Mn, Fe, Co and Zn in the range of 5–50 nm have been made from their respective oxides (Vissikov et al., 1988). Nitrides of Al, Si and Mg have been synthesized in an $Ar/N_2$ atmosphere (Vissikov et al., 1981; Vissikov, 1992) and oxides of Si, Al, Fe and Co have been made in an oxidizing atmosphere using $SiCl_4$, Al, $FeCl_2$ and CoS as precursors (Vissikov et al., 1988). Ultrafine metal, alloy and ceramic–metal composite particles have also been made by this route (Ohno and Uda, 1984; Nakahigashi et al., 1992). A major drawback of this process is the formation of hard agglomerates at high production rates (Gurav et al., 1993a).

### 8.2.2.5 *Sputtering.*

This method is similar to the gas-condensation route except that ions are used to sputter the starting material from the surface. In this process ions, often of Ar or Kr, are accelerated to high energies and then directed towards the surface of the starting material. The high energy of the ions causes atoms and clusters to be ejected from the surface into the gas phase (Gleiter, 1989). Thus this method is especially useful for synthesis of particles where the starting substance is non-volatile. Further, the atoms and clusters in the gas phase have the same composition as that of the starting material. In some cases, the initial starting material is an element which is then reacted with oxygen to produce the oxide (Ying, 1993; Chow et al., 1990). The oxidation may be carried out after the particles of the elements have formed or in the gas phase itself. The major limitations of this process are poor lack of control over the size distribution and low production rates.

Some of the materials synthesized by this route consist of particles of Al, Mo, Cu and W in the size range of 7–50 nm (Hahn and Averback, 1990), oxides of Zr (Hahn and Averback, 1990), and composites of $Pt–TiO_{2-x}$, Al/Mo, $Cu_{91}Mn_9$ and $Al_{52}Ti_{48}$ (Chow et al., 1990; Hahn and Averback, 1990; Ying, 1993).

### 8.2.2.6 *Laser ablation.*

In this process the energy to vaporize the reactant is supplied by high-power pulsed lasers. These lasers, when directed to the target substrates, vaporize the target material (which may be the same material as that of the product particles). This creates a high vapour pressure of the material, leading to high supersaturations. To obtain pure metal particles this process is carried out under an inert atmosphere. The laser wavelength is an important consideration and is dictated by the choice of the target material (Gleiter, 1989). For example, ultraviolet lasers are used for metal targets.

In some cases where the target material is not the same as that of the product particles, an oxidizing atmosphere (for oxide particle synthesis) is used (Johnston *et al.*, 1992). In this case the vaporized material reacts in the gas phase with the oxygen to form the metal oxides which form small particles by homogeneous nucleation or direct chemical reaction. The laser energy is sometimes used to facilitate gas-phase reaction of a gaseous precursor to form metal clusters of 3–10 nm (Jasinski and LeGoues, 1991).

This process can be used to generate nanoparticles of metals (Dietz *et al.*, 1981; LaiHing *et al.*, 1987), alloys (Rohlfing *et al.*, 1984), semiconductors (Liu *et al.*, 1986) and ceramics (Gupta *et al.*, 1987; Borsella *et al.*, 1992) by using appropriate target substrates and is therefore a versatile route for nanoparticle synthesis. However, the production rates are low, the production costs are high and scale-up is a problem. This is a relatively new method for nanoparticle synthesis and there are problems and challenges of control of particle size distribution and scale-up that need to be resolved.

## 8.3 Liquid-to-particle conversion

Numerous liquid-phase (solution) techniques are available for the preparation of metal and ceramic powders. Solution-based methods can be broadly classified into two groups: precipitation (direct strike precipitation, precipitation from homogeneous solution, sol-gel (Pugh and Bergstrom, 1994) and solvent-evaporation techniques. Liquid-to-particle conversion aerosol routes are a subset of solvent-evaporation techniques. This section will compare and contrast this route with conventional liquid-phase routes and discuss the scientific and technological aspects of powder generation by liquid-phase aerosol routes. We begin with a brief overview of conventional liquid-phase routes.

In **precipitation** methods, the particles of the product material are precipitated by the addition of a precipitating agent to a solution of a soluble salt to form a sparingly soluble salt (direct strike precipitation) or by supersaturating a solution to cause nucleation and consequent particle formation (precipitation from homogeneous solution, PFHS). In the direct strike method, the addition of the precipitating agent leads to copious precipitation at a rate faster than the time-scale required for mixing of the reactants, which is in contrast to the homogeneity of the reaction mixture and controlled release of the precipitating species during PFHS. The precipitated solid is then filtered from the reaction mixture. Precipitation routes are widely employed for industrial-scale production of powders. For example, submicrometre metal powders (Ag, Ag/Pd) are made by precipitation routes for thick-film microelectronic applications (Wang *et al.*, 1994). Lead zirconate titanate (PZT) powders are used in piezoelectric actuators. Lanthanum-modified PZT powders have been prepared by inducing precipitation from an aqueous solution containing zirconium oxychloride, titanium tetrachloride, hydrogen peroxide and the nitrates of lead and lanthanum by the addition of ammonium hydroxide and maintaining a pH of ~9 (Murata *et al.*, 1976). For the production of submicrometre or nanosized particles, solution concentration, pH, heating rate, temperature, choice of precipitation reaction and other conditions must be carefully controlled (Pugh and Bergstrom, 1994). These routes usually require heating of the powder to improve crystallinity. This heating can result in agglomeration, which is reduced by subsequent milling.

**Sol-gel** processing involves particle growth from a liquid solution or colloidal suspension of silica, metallic acids or chlorides and metal alkoxides ($M(OR)_n$), where R is an alkyl and M is a metal with valency $n$ (Millberg, 1987). For metal alkoxides, the reaction involves several steps. The alkoxides mix because the OR groups provide solubility (though alcohol may be necessary). Mixing is followed by hydrolysis where the OR groups become OH groups, resulting in chain-like or three-dimensional polymers. Hydrolysis rates are dependent on temperature, pH and the concentrations of the starting materials. Sol-gel routes are used mainly for glassy materials; formation of highly crystalline material usually requires subsequent heating which can result in agglomeration.

In **solvent-evaporation** techniques for powder generation, the desired product is formed by the decomposition of precursor that precipitates from a solution due to vaporization of the solvent. For non-aerosol solvent-evaporation techniques, this is often carried out in large trays resulting in some degree of agglomeration of the product powder, thereby necessitating milling to reduce agglomeration.

*Spray* and *freeze drying* and *spray pyrolysis* are solvent-evaporation aerosol techniques for powder generation. The common step in all of the processes is the generation of droplets from a liquid (Johnson, 1981; Masters, 1985; Kodas, 1989; Messing *et al.*, 1993). In the case of spray pyrolysis, droplets are dried and reacted at elevated temperatures in the gas phase. Because of its inherent advantages over conventional liquid-phase routes, *spray pyrolysis* has emerged as a promising technique for powder generation (Kodas, 1989; Messing *et al.*, 1993). Spray pyrolysis involves the atomization of a precursor solution to form droplets which

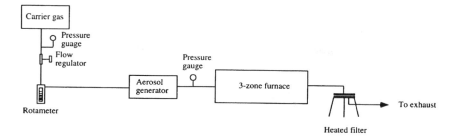

**Fig. 8.8** Experimental set-up for spray pyrolysis.

are then directed through a heated furnace (Figure 8.8). During passage through the furnace, the droplets are converted to solid product particles by the processes of solvent evaporation and solute precipitation and decomposition. The particles formed undergo thermally driven processes like sintering and may densify within the furnace, after which they are collected on a filter. A large number of materials have been synthesized at the laboratory scale by this route (Kodas, 1989; Messing *et al.*, 1993).

The attractiveness of spray pyrolysis lies in its overall simplicity. Droplets of the precursor solution are generated using a wide variety of atomizers such as pressure atomizer, rotating disc, electrostatic and ultrasonic (Messing *et al.*, 1993). These generators yield product particle size distributions with Sauter mean diameter (SMD) values of 0.1–10 µm, roughly lognormal distributions and geometric standard deviations greater than 1.4. The reactor tube is often a ceramic cylinder (mullite, alumina) with flanges at its open ends. It is heated by a furnace with multiple heated zones which are maintained at a desired temperature. Oxygen or air-fuel flames have also been used instead of a reactor to synthesize metal oxides, mixed metal oxides and ferrite powders from metal chloride, sulphate, acetate, and nitrate solutions (Neilson *et al.*, 1963). The flow rate of the carrier gas can be adjusted. The residence time of the aerosol stream in the furnace/flame depends on the carrier gas flow rate and furnace/flame temperature. The powder is collected in a filter. Membranes filters, made of a wide variety of synthetic material such as PTFE and polythene, are commonly used for laboratory scale operations. Bag filters are also used.

This process was first commercialized by Ruthner to produce metal oxide powders from metal chloride solutions (Ruthner, 1983). The precursor solutions were atomized into the top of a vertical furnace held at 700–1000°C. The generated particles were separated from the carrier gas stream at the reactor exit by a cyclone. However, the powders consisted of 25–400 µm particles which required milling when used for fine-grain ceramic production. Of late, SSC, Inc. (now Praxair Specialty Ceramics) in Seattle, WA, USA, Scimarec, Inc., in Tokyo, Japan and E. Merck, Inc., in Darmstadt, Germany have started marketing ceramic powders produced by processes similar to spray pyrolysis (Messing *et al.*, 1993).

In the case of spray drying the droplets are exposed to lower temperatures and dried, while in freeze drying the droplets are frozen by contact with a cold liquid and subsequently dried. Spray drying and freeze drying differ from spray pyrolysis by the absence of the occurrence of any chemical reactions. The temperatures used are also much lower (below 500°C for spray drying). Spray drying is often used to granulate very fine powder which cannot be handled in its original state. For the synthesis of some powdered materials, the dried precursors are collected and subsequently reacted and milled to form the final powder.

The general advantages of aerosol routes for liquid-to-particle conversion over other powder generation methods are:

- simplicity and cheapness;
- no waste streams;
- for spray pyrolysis, the ability to generate pure, micrometre-scale, unagglomerated, spherical powders with molecular-level compositional homogeneity for many multi-component powders;
- wide range of available precursors;
- scale-up demonstrated.

However, there are a few potential disadvantages:

- formation of hollow or porous particles which are not desired for some powder applications;
- wide particle size distribution due to limitations of aerosol generator;
- improper stoichiometry due to loss of volatile intermediates.

There follows a description of the different stages of the morphological evolution of particles and the basic physicochemical phenomena encountered during spray pyrolysis. Spray drying (Masters, 1985) and freeze drying are discussed elsewhere (Johnson, 1981; 1987).

### 8.3.1 *Morphological evolution during spray pyrolysis*

The term 'morphology' is used here to refer to particle shape and internal structure (solid or hollow particle). Figure 8.9 shows a possible sequence of steps for spray pyrolysis. A liquid precursor or solution is atomized into small droplets (usually of the order of 1 μm) and carried through a heated furnace by a carrier gas. The atomized liquid may also be a colloidal suspension of precursor particles. It is also possible to have a colloidal precursor suspended in a solution of another precursor(s). A droplet loses solvent to form a dry particle of precursor which then decomposes in the solid state to form the product composition in cases where the precursor does not melt before complete reaction. In some cases, the dry precursor particle melts before

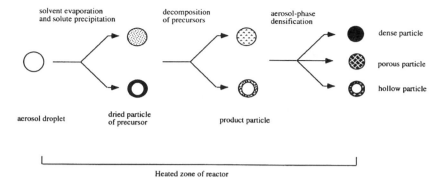

**Fig. 8.9** Morphological evolution during spray pyrolysis.

decomposing and reaction occurs partially in a melt. The product particle subsequently undergoes aerosol-phase densification to form the final particle morphology. The final morphology obtained (hollow, porous or solid) is dependent on factors such as reactor temperature, residence time of the particle in the reactor, the melting point (more specifically, sintering temperature) of the product composition and precursor characteristics (discussed in section 8.3.1.2). The following paragraphs describe the morphological evolution of the particle during the process.

8.3.1.1 *Aerosol droplet.* Droplet generation is the first stage in the process of particle generation by spray pyrolysis and takes place in all systems irrespective of the precursor nature (colloid or solution, organic or inorganic). A pure liquid, solution or colloidal suspension is atomized to form aerosol droplets which are then carried into a furnace by a flowing gas stream. Each droplet loses solvent and eventually yields a product particle. Due to loss of solvent the final particle size is always smaller than the droplet size and is a function of the precursor solution concentration, the size of the initial droplet, the density of the product particle and the molar weight of the product particle composition. Pluym *et al.* (1993) compared measured particle sizes with estimated values for silver powders. For the generation of silver from silver nitrate, the relation between the droplet and final particle size ($d_{\mathrm{drop}}$ and $d_{\mathrm{Ag}}$, respectively) reduced to (assuming that the silver particle had reached its theoretical density),

$$\frac{d_{\mathrm{Ag}}}{d_{\mathrm{drop}}} = \left\{ \frac{M_{\mathrm{Ag}}[\mathrm{AgNO_3}]}{1000\rho} \right\}^{1/3} \qquad (8.2)$$

where $[\mathrm{AgNO_3}]$ is the molar concentration of silver nitrate solution ($\mathrm{mol\,l^{-1}}$), $\rho$ the density of silver ($\mathrm{g\,cm^{-3}}$) and $M_{\mathrm{Ag}}$ the molar weight of silver (g).

This is a general relationship and is valid for particle formation from any soluble precursor. As an example, a 1 μm droplet of 1 mol silver nitrate would produce a 0.22 μm particle of silver. Because aerosol generators usually produce a distribution of droplet sizes, the final particle size distribution is directly affected by the size distribution of the aerosol droplets entering the furnace. However, because this relationship assumes the formation of one particle from every droplet, it is not valid when particle size and its distribution are affected by such phenomena as bursting of droplets within the reactor (Messing *et al.*, 1993).

8.3.1.2 *Particles containing precipitated solute.* This stage is reached due to partial evaporation of the solvent in a droplet which results in increasing solute concentration and solute precipitation. Depending on solution concentration and ambient conditions, the precipitation can occur throughout the particle volume or at the droplet periphery (Jayanthi *et al.*, 1993). The manner in which precipitation occurs can decide the morphology of the particle formed due to solvent evaporation; this intermediate morphology can in turn affect the morphology of the final product particle. For example, if precipitation occurs at the droplet surface, it is possible that the final particle morphology will be hollow. Similarly, uniform precipitation throughout the droplet volume can yield a solid or porous particle. However, subsequent phenomena (e.g. precursor melting, intraparticle sintering and densification) can completely mask any effects of these intermediate morphologies (Figure 8.10).

Gases evolved within a particle with a surface of precipitated solids (due to solvent evaporation and/or reaction of the precursor) can cause an increase of pressure which can exert an explosive force on the particle. The permeability of the precipitated solids decides the ease with which these gases can escape, which in turn decides the magnitude of the outward

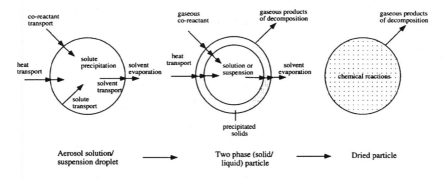

**Fig. 8.10** Basic physicochemical phenomena during spray pyrolysis.

force exerted on the solid skeleton of the particle. If the force is large enough, the particle may rupture and form a hollow shell and a family of smaller droplets (Gardner and Messing, 1984).

The effect of solvent evaporation from solution droplets (~100–1000 μm) on the morphology of dried particles has been investigated in great detail in a classic paper by Charlesworth and Marshall (1959). They identified six different routes of morphological evolution depending on *precursor characteristics* and *ambient conditions*. With the surrounding air temperature below the boiling point of the solution, some solutes formed a porous and rigid crust which did not rupture during the drying process. Other solutes formed pliable and impervious skins which caused the dried particle to dimple or wrinkle due to a decrease in volume caused by solvent removal.

Different results were observed when the surrounding air temperature was above the boiling point of the solution. If the crust was very porous, the pressure exerted by solvent evaporation within the particle forced solution out through the pores to the surface where the solvent quickly evaporated. The pressure of evaporated solvent caused impermeable and rigid crusts to rupture (e.g. sodium chloride) while plastic ones inflated (e.g. ammonium sulphate). Such morphologies can be obtained after the completion of solvent removal during spray pyrolysis (droplet drying) and can affect the morphology of the final product particle.

8.3.1.3 *Dried particles.* The two-phase (solid and liquid) particle becomes a dry particle of solute after losing all the solvent, after which intraparticle chemical reactions form the final product composition. The final morphology of a particle can be influenced by the intraparticle reactions. However, if the precursor melts before the start of the intraparticle reactions, the dry particle of precursor formed after complete solvent removal will convert to a sphere of molten liquid from which the product particle evolves. In such a case, the final morphology is not expected to be affected by the intermediate morphology obtained before melting of the precursor. In cases where the reactants do not melt, solid-state reaction of a hollow particle can yield a hollow particle of the product.

8.3.1.4 *Product particles.* Chemical reaction in the dry particles results in the final product composition. After the product composition has been attained, intraparticle sintering and densification processes can influence particle morphology. Hollow and dense are the two main types of product particle morphology that are broadly recognized (Messing *et al.*, 1993). Particles with porosity distributed uniformly throughout their volume can also be obtained (porous particle in Figure 8.9). Figure 8.11 shows hollow and solid copper(I) oxide particles (Majumdar *et al.*, 1996a).

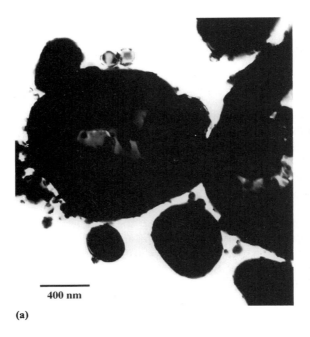

400 nm

(a)

400 nm

(b)

**Fig. 8.11** (a) Hollow copper(I) oxide particles; (b) dense copper(I) oxide particles.

Having looked at the sequential stages of the evolution of particle morphology in the reactor, the following subsection will discuss the basic physicochemical phenomena that characterize them.

### 8.3.2   Basic physicochemical phenomena associated with spray pyrolysis

The different physicochemical processes (Figure 8.10) that occur during spray pyrolysis can be categorized as: droplet generation; heat and mass transfer in carrier gas, droplets and particles; chemical reactions and phase transformations; and intraparticle sintering and densification. The following paragraphs will describe these phenomena with reference to when they occur during spray pyrolysis as well as their effect, if any, on particle characteristics (morphology, phase composition).

#### 8.3.2.1   Droplet generation.

Droplet generation is the first step in spray pyrolysis. Various types of atomizer are used for droplet generation (Pluym et al., 1993). The size of the droplets generated is affected by the type of generator and by liquid properties such as surface tension and density. The TSI-3076, Collison CN-25, an ultrasonic home humidifier and other atomizer have been used for spray pyrolysis at a laboratory scale (Pluym et al., 1993). For example, the droplet size generated by an ultrasonic atomizer is given by (Ogihara et al., 1991)

$$d_p = 0.34 \left( \frac{8\pi\gamma}{\rho f^2} \right)^{1/3} \tag{8.3}$$

where $d_p$ is droplet size ($\mu$m), $\rho$ solution density (g cm$^{-3}$), $\gamma$ surface tension (dyn cm$^{-1}$) and $f$ the frequency of the piezoelectric transducer of the atomizer (MHz). The surface tension and density of the solution are strongly influenced by the type of solvent and the solute concentration. For dilute solutions, the surface tension and density of the solution can be assumed to be that of the solvent. The surface tension of organic-based systems (less than 40 dyn cm$^{-1}$) is generally less than those of water-based ones ($\sim$70 dyn cm$^{-1}$). For an operating frequency of 1.6 MHz, the droplet size generated is 3 $\mu$m for an aqueous solution ($\rho = 1, \gamma = 70$) and 2.7 $\mu$m when ethanol ($\rho = 0.8, \gamma = 40$) is used as the solvent. As discussed in an earlier section, the final particle size can be determined by the droplet size in some cases.

#### 8.3.2.2   Heat transfer.

The temperature of the droplet/particle is an important parameter because rates of solvent evaporation, chemical reactions, intraparticle densification and phase transformations are strongly affected by it. The heat needed to raise the gas and particle temperatures is provided by the hot reactor walls. Heat transfer from the walls to the droplet/particle occurs through the carrier gas by convection and conduction. During the initial stages of spray pyrolysis, heat transfer from the gas to the droplet supplies the energy for solvent evaporation which is usually completed in a very short time ($\sim 1$ ms for $\sim 1$ μm droplet) compared to typical residence times for spray pyrolysis ($\sim 1$–10 s).

Because the gas provides the latent heat for solvent evaporation, it may be expected to cool. A calculation based on an energy balance for a spray pyrolysis system with $10^7$ droplets per cubic centimetre of water and droplet size of 1 μm (typical for many atomizers) shows that the gas cools by less than 10°C due to evaporation of the droplets. Therefore, the gas temperature is the same as the reactor wall temperature during all the stages of spray pyrolysis. This is only true beyond the length of the reactor required by the gas stream in order fully to develop its temperature profile and reach steady-state conditions, known as the **thermal entrance length** (Bird *et al.*, 1960).

The time required to heat a droplet/particle by conduction from the surrounding gas to the gas temperature is roughly $d_p^2/\alpha_g$, where $d_p$ is droplet/particle diameter (cm) and $\alpha_g$ the thermal diffusivity of the surrounding gas (cm$^2$ s$^{-1}$). For solid particles, their high thermal diffusivities ($\sim 1$–10 cm$^2$ s$^{-1}$) compared to those for gases ($\sim 0.1$ cm$^2$ s$^{-1}$) implies that the only resistance to the heating of a particle is the resistance to conductive heat transfer from the gas to the particle. Considering the thermal diffusivity of air to be roughly 0.1 cm$^2$ s$^{-1}$ (Bird *et al.*, 1960), the time for a particle with a 1 μm diameter to reach thermal equilibrium with its surroundings is about $10^{-7}$ s. This suggests that the particle can heat at extremely high rates. For the particle to be unable to reach the temperature of the surrounding gas by conduction, the gas itself must be heated at rates greater than millions of kelvin per second. Therefore, for most practical situations, the particle temperature is the same as that of the reactor walls. Again, this is true only beyond the thermal entrance length.

Two situations are known when droplet/particle temperature may deviate from the gas temperature: during solvent evaporation and in the case of an exothermic reaction of the precursor. Solvent evaporation can result in a depression of the droplet temperature, whereas an exothermic reaction can cause a temperature elevation of a particle relative to the gas temperature. The lowering of droplet temperature during evaporation can be quantified by an energy balance and is given by (Hinds, 1982)

$$T_\infty - T_d = \frac{D_v H M}{R k_v} \left( \frac{p_d}{T_d} - \frac{p_\infty}{T_\infty} \right), \tag{8.4}$$

where $T_\infty$ is the gas temperature, $T_d$ the particle temperature, $p_d$ the partial pressure of the solvent vapour at the particle surface, $p_\infty$ the partial pressure of the solvent vapour far away from the particle, $D_v$ the diffusivity of the solvent vapour in the surrounding gas, $H$ the latent heat of evaporation of the solvent, $M$ the molar weight of the solvent, $R$ the molar gas constant and $k_v$ the thermal conductivity of the surrounding gas.

As an example, for a temperature of 40°C and relative humidity of 50% for the gas, the droplet temperature is depressed by less than 10°C. This is a realistic temperature for solvent evaporation during spray pyrolysis since evaporation begins and may be complete below the boiling point of the solvent (usually water) in the precursor solution. Because a particle usually heats up to a much higher temperature during subsequent stages of spray pyrolysis, for practical systems droplet temperature depression does not play an important role in deciding the characteristics of the final product particle.

The elevation of particle temperature due to an exothermic chemical reaction (e.g. the decomposition of magnesium acetate to magnesium oxide in an oxidizing atmosphere; Gardner and Messing, 1984) is handled by balancing the rate of heat release due to the reaction with the rate of heat transfer in the gas phase away from the particle. If the characteristic time for the exothermic chemical reaction is much smaller than that for heat transfer away from the particle, the temperature of the particle rises compared to the gas temperature. The elevation of the particle temperature with respect to the gas phase is given by

$$T_d - T_\infty = \frac{d_p H_r r_s}{6 h_g},$$  (8.5)

where $T_d$ is the particle temperature, $T_\infty$ the gas temperature, $d_p$ the particle diameter, $H_r$ the enthalpy of the reaction, $r_s$ the intraparticle reaction rate and $h_g$ the heat transfer coefficient in the surrounding gas. This equation assumes a constant reaction rate, $r_s$.

This shows that the elevation of the particle temperature with respect to the gas phase is favoured by larger particle sizes, high exothermicity, higher rate of chemical reaction and slower rates of heat transfer in the gas phase. For example, consider an exothermic chemical reaction with an enthalpy of 30 kcal mol$^{-1}$. The heat transfer coefficient in air is about $10^{-5}$ cal K$^{-1}$ cm$^{-2}$ s$^{-1}$ (Perry et al., 1984). Assuming $r_s = 10^{-3}$ mol cm$^{-3}$ s$^{-1}$, a 1 μm particle can experience a temperature elevation of 50°C.

### 8.3.2.3 *Mass transfer.*

Like heat transfer, mass transfer occurs between the gas and a droplet/particle and also within a droplet/particle. Intraparticle mass transfer includes diffusion of solvent in a droplet, diffusion of solute and gaseous reactants and products of decomposition within a droplet/particle, and solid-state diffusion during densification of a product particle. Mass transfer between a droplet/particle and the surrounding gas includes diffusion of gases (solvent vapour and decomposition products) into the surrounding gas away from the droplet/particle surface (Figure 8.10). It may also include diffusion of a gaseous coreactant from the gas to a droplet/particle surface.

Mass transport processes can influence particle morphology through their influence on the solute precipitation process. Diffusion inside droplets/particles and in the gas phase occur in series with each other. Both are driven by concentration gradients. During spray pyrolysis, as a droplet travels along the reactor, the solvent starts evaporating. This tends to reduce the concentration of the solvent and increase the solute concentration at the droplet surface. This sets up movement of solvent and solute within the droplet, with solvent molecules diffusing towards the surface from the interior and solute molecules diffusing in the opposite direction. However, the rates are not rapid enough to keep up with the loss of solvent from the surface of the droplet into the gas phase (because the transport rate is proportional to the diffusion coefficient which is much higher in gases $(0.15 \, cm^2 \, s^{-1})$ than the effective diffusion coefficients in liquids $(10^{-5} \, cm^2 \, s^{-1})$, leading to solute and solvent concentration gradients within the droplet. A simplified analysis with typical numbers shows that the solute concentration at the droplet surface can be 25% greater than at the centre of the droplet. Such gradients can ultimately lead to solute precipitation at the droplet periphery (Gardner, 1965).

The time for droplet drying depends on the rate of solvent evaporation which in turn depends on the temperature and relative humidity of the surrounding gas. A 1 μm droplet of pure water takes about 2 ms to evaporate at 20°C and 50% relative humidity (Hinds, 1982). This time increases at lower temperatures and higher values of relative humidity. It also increases with the concentration of the solute in the droplet. Charlesworth and Marshall (1959) showed that evaporation rates from water droplets and dilute solution droplets were nearly identical until the completion of crust formation for droplet diameters of several hundred micrometres. The evaporation rate after complete crust formation depended on the nature of the solute and the temperature and humidity of the surrounding air. In general, it is only for large droplets of the order of 100 μm, for very high values of relative humidity (near 100%), and when solvent evaporation is significantly slowed down by impermeable crusts on the particles that the droplet evaporation time must be taken into account when determining the reactor residence time (typically ∼1–10 s for spray pyrolysis) required for generation of product particles by spray pyrolysis.

In the case where a liquid precursor reacts with (a) gaseous coreactant(s), the gas has to diffuse into liquid-phase droplets (Visca and Matijevic, 1979). The characteristic time for diffusion is given by $d_p^2/D$, where $D$ is the diffusivity of the gaseous species in the droplet $(cm^2 \, s^{-1})$ and $d_p$ is the droplet diameter (cm). A typical value for diffusion coefficients of gases in liquids ($10^{-5}\,\text{cm}^2\,\text{s}^{-1}$) yields a characteristic time for diffusion of 1 ms for a 1 μm droplet. Thus, the time required for a gaseous species to diffuse into a 1 μm droplet is of the order of a millisecond, which implies that gaseous reactants such as hydrogen can diffuse into droplets very rapidly relative to typical reactor residence times for spray pyrolysis.

For the case of a solid with (a) gaseous coreactant(s), the coreactant(s) must diffuse into solid particles. For example, during the generation of Pd particles at temperatures above 900°C in an aerosol reactor, PdO can form by oxidation of Pd as the particles cool while exiting the reactor. However, due to the low diffusion coefficient of oxygen in Pd metal, $4.43 \times 10^{-10}\,\text{cm}^2\,\text{s}^{-1}$ at 900°C (Park and Altstetter, 1985), the amount of oxidation is low enough (characteristic time for diffusion much larger than residence time) that nearly pure Pd particles emerge from the reactor.

8.3.2.4 *Intraparticle reaction and phase transformation.*     Intraparticle chemical reactions can influence particle morphology. Gardner and Messing (1984) showed that MgO powder characteristics were determined by the decomposition path of the precursor salt. The product particles were micrometre-sized and consisted of smaller submicrometre particles when chloride and nitrate precursors were used, while much smaller, submicrometre particles were formed from magnesium acetate in an oxidizing atmosphere. This difference in morphology was due to the exothermic oxidation of the precursor during decomposition which caused fragmentation of the large particles formed from individual droplets into finer particles.

The final powder composition is determined by the kinetics or thermodynamics of the intraparticle reactions. As an example, during the generation of copper(I) oxide, increasing the residence time from 3 s to 5 s with $N_2$ as the carrier gas changed the composition of the product powder from CuO to $Cu_2O$, the thermodynamic product (Majumdar *et al.*, 1996a). This is an example of product composition being limited by reaction kinetics.

Phase transformations can also play a role in deciding final product composition. The phase transformations of titania from anatase to rutile is a well-known example for aerosol processes. Such transformations occur primarily by the processes of nucleation and growth or spinodal decomposition and are discussed elsewhere (Kingery *et al.*, 1976). As in the case of chemical reactions, sufficiently high temperatures are needed to initiate and complete phase transformations within the residence times available during spray pyrolysis.

8.3.2.5 *Intraparticle sintering and aerosol-phase densification.*     Intraparticle mass transfer processes also influence phenomena such as sintering

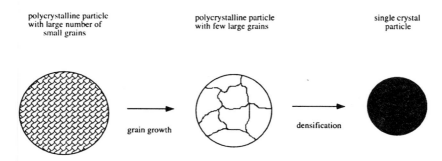

polycrystalline particle with large number of small grains

polycrystalline particle with few large grains

single crystal particle

grain growth

densification

**Fig. 8.12** Schematic of the progressive reduction in the crystallite boundary area which provides the driving force for intraparticle densification during spray pyrolysis.

of crystallites within a particle. After the intraparticle reactions are complete and the product phase has been formed, a particle can be composed of a large number of small grains or crystallites (in the case of a crystalline product). These crystallites eliminate surface area and porosity by the processes that occur during sintering of powders. Some of these processes result in densification of particles which can cause morphological changes such as increased sphericity. Densification can also be accompanied by grain growth through which larger crystallites grow at the expense of smaller ones, which disappear. If grain growth is fast enough, single-crystal particles may be formed (Figure 8.12). As an example, single-crystal particles of copper(I) oxide were formed at 1200°C by spray pyrolysis of copper nitrate with residence times of approximately 3 s (Majumdar *et al.*, 1996a).

Densification is favoured by high temperatures and long residence times. For densification by solid-state sintering (discussed later) of crystallites within a particle, the closer the temperature is to the melting point the greater is the chance of obtaining complete densification of the particle. If reactor temperatures are high enough, particles can melt to form perfectly spherical particles. Figure 8.13 shows gold particles made at 1200°C (Majumdar *et al.*, 1996b). The particles were spherical because gold melts at 1064°C. However, melting cannot always be predicted as in multicomponent systems. In such cases, a phase diagram can provide guidelines for the temperature required for densification. Incongruent melting of one or more phases can cause liquid-phase sintering in multi-component particles (Hingorani *et al.*, 1995).

The densification of particles can also be affected by the chemical composition of the gaseous environment in the reactor. Gas trapped in closed pores in a particle will limit pore shrinkage unless the gas is soluble in the boundaries between the crystallites and can diffuse from the pores (Kingery *et al.*, 1976). Alumina doped with MgO can be sintered to

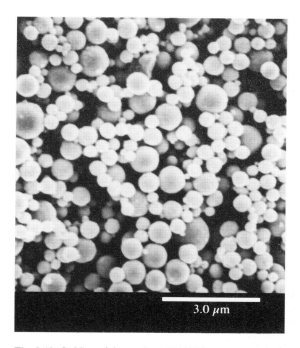

**Fig. 8.13** Gold particles made at 1200°C by spray pyrolysis.

essentially zero porosity in an atmosphere of $H_2$ or $O_2$, which are soluble, but not in air, which contains insoluble nitrogen (Kingery *et al.*, 1976). The infiltration of particles by a reactive gas may produce reaction bonding, resulting in significant densification (Reed, 1988).

## 8.4  Powder processing

A large number of methods (aerosol and non-aerosol) are currently available to prepare powders consisting of dense, spherical particles of desired size and purity (Johnson, 1987). While many of these processes yield micrometre-sized particles, there has been increasing interest in nanosized powders that have the potential to yield sintered products with improved properties at lower sintering temperatures.

Several advantages arise from the extremely high atomic diffusivities and driving forces available for sintering of nanoparticles. Theoretical predictions indicate that the densification rate of 14 nm titania particles is eight orders of magnitude greater than that for samples with 1.3 μm particles (Hahn *et al.*, 1990), which allows for quicker processing at a given temperature compared to micrometre-sized particles. Problems such as decomposition of constituent phases that are frequently encountered in the sintering of metal nitrides due to extremely high sintering temperatures (above 2000°C) can be avoided by resorting to nanoparticles because they can be processed at much lower temperatures (Skandan *et al.*, 1994). The formation of micrometre-scale cracks in zirconia ceramics due to the dilatational strain (volume expansion) associated with the monoclinic to tetragonal phase transformation has been prevented by using nano-powders which sinter below the phase transformation temperature of 1170°C (Skandan *et al.*, 1994).

Perhaps the biggest potential impact of nanoparticles is the possibility of processing brittle (fracturing with little deformation) crystalline materials using techniques developed for materials like metals that deform more easily (Anon. 1995). Synthesis of ceramics with nanosized grains has resulted in materials that can deform at relatively low temperatures by mechanisms (e.g. diffusional creep) which usually become prominent only at high temperatures in materials with larger grains. These ceramics may deform plastically (permanently) under an applied stress due to the ease with which nanosized grains slide past each other.

Another way to view this behaviour of nanocrystalline materials is through an analogy to the Hall–Petch (Petch, 1953) relationship for metals, which states that the yield stress for brittle failure is related to the grain size by

$$\sigma = \sigma_0 + fd^{-1/2} \tag{8.6}$$

where $\sigma$ is the yield stress, $\sigma_0$ the lattice friction stress required to move individual dislocations, $f$ a constant and $d$ the spatial grain size. According to this relationship, the smaller the grain size the larger the yield stress, which in turn implies a material with greater resistance to brittle failure. This higher resistance to brittle failure, combined with the ease of plastic deformation by diffusional creep, provides the opportunity for shaping ceramic objects without catastrophic damage.

Nanograined materials can also exhibit improved final properties compared to their micrograined counterparts. For example, Table 8.1 lists

**Table 8.1** Selected properties of WC composites with different grain sizes (adapted from Raghunathan *et al.*, 1996)

| Property | 0.2 μm diameter | 0.8 μm diameter |
|---|---|---|
| Fatigue strength (MPa) | 2200 | 500 |
| Hardness (HV 0.5 g) | 2200 | 1800 |
| Wear loss (g) | 0.3 | 1.4 |
| Crack resistance (kg mm$^{-1}$) | 60 | 30 |

the effect of grain size on selected mechanical properties of tungsten carbide (WC) composites used to make cutting tools (Raghunathan *et al.*, 1996). Some other advantages of nanoparticles are as follows:

1. Their small particle size during synthesis can permit normal restrictions to phase equilibria to be overcome by the combination of high driving forces and short diffusion distances. This allows the formation of new metastable structures, as has been demonstrated for bulk metallic glasses (Schwarz and Johnson, 1983).
2. The large fraction of atoms residing in the interfaces between grains (almost one-half in the case of a 5 nm size grain) provides the opportunity for new atomic arrangements with novel materials properties. Remarkable linear and nonlinear optical phenomena have been predicted, and to some extent observed, for composite materials made from semiconductor crystallites with diameters of 5–15 nm dispersed in dielectric media. High third-order nonlinear optical coefficients and picosecond response times have been reported (Brus, 1986). These nonlinear optical properties may ultimately find use in optical logic elements (Andres *et al.*, 1989).
3. The possibilities for reacting, coating and mixing *in situ* various types of particle create the potential for synthesising new multi-component composites with nanometre-sized microstructures and engineered properties (Andres *et al.*, 1989). Composite materials with ultrafine grain size and highly controlled compositions may be synthesized by physical (e.g. mechanical alloying-forging) rather than chemical processing, thereby providing greater flexibility in designing advanced materials (Andres *et al.*, 1989). It should be possible to produce composite materials consisting of ultrafine grains with two elements that are normally immiscible. With 30% of the atoms lying in the grain boundaries, a nanophase, binary composite material has almost as many alloy neighbour atoms as in a completely mixed binary system (Andres *et al.*, 1989).

Nanoparticles are rarely used in particulate form; in order to be useful, they need to be processed further. In many cases, the nanopowder must be transformed into a coherent, consolidated body by processes which are collectively grouped under **forming** and **sintering** (Figure 8.14). In other cases, nanoparticles may be incorporated in polymeric or dielectric films to impart unique mechanical, electrical or optical properties (Brus, 1986).

Forming refers to the consolidation of powder into a desired shape which is usually close to that of the finished product. The forming methods which presently exist for powders were developed by industry for powders with particle sizes greater than roughly 0.1 μm. The interest in and the ability to generate nanosized powders being relatively new, forming techniques exclusively designed for nanopowders are still in the nascent

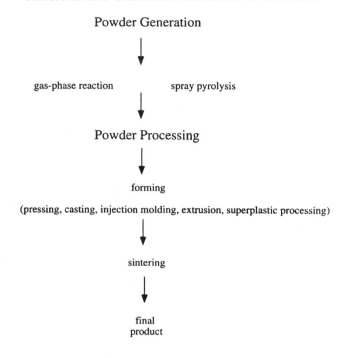

Powder Generation

gas-phase reaction          spray pyrolysis

Powder Processing

forming

(pressing, casting, injection molding, extrusion, superplastic processing)

sintering

final
product

**Fig. 8.14** Flow diagram of product fabrication from powder.

stage and consequently available methods developed for micrometre-sized powders are generally used. These forming methods, used for ceramic and metal powders, are **pressing**, **casting**, **injection moulding** and **extrusion**. While these methods are potentially applicable for nanometre- to micrometre-sized powders, a very different method, called **superplastic processing**, has emerged as a novel technique for the net shape forming of nanocrystalline materials. This technique exploits the unique deformation behaviour exhibited by nanocrystalline materials. After the powder has been consolidated into the desired shape, the formed object is usually subjected to carefully designed thermal treatment which is referred to as **firing** (Kingery *et al.*, 1976; Reed, 1988). In many modern powder processing sequences, forming and sintering are carried out simultaneously (section 8.4.1.4). When powders of metals, ceramics and glasses are formed into shapes, the resultant green (unsintered) bodies are porous due to voids between individual particles in the compact. Upon firing, sintering occurs during which the individual particles fuse together and eliminate the porosity, which results in an increase in the density and consequent decrease in volume of the compact. The thermodynamic driving force for sintering is the reduction in total energy by the reduction of surface area or replacement of higher-energy surfaces (interfaces) with lower-energy ones. Sintering can be broadly categorized into *solid-state* and *liquid-phase* sintering depending on whether or not liquid phase(s) assist the microstructural changes occurring during the firing process.

In this section we will introduce the different forming methods and sintering mechanisms.

### 8.4.1  *Forming*

8.4.1.1  *Pressing.*  **Die** and **isostatic pressing** are the two principal methods of consolidating powders by the application of an external stress. **Uniaxial die pressing** has been the predominant forming method used to fabricate nanostructured materials in the laboratory for scientific studies because of its simplicity and inexpensiveness (Suryanarayanan *et al.*, 1996a). In this process, powder is placed in a mould in the shape of the final object and pressed with a metal die (which fits the mould) to form a green body. In industry, it is used to fabricate ceramic and metal parts from powder feeds which are in the form of granules ($\sim$100 µm spherical agglomerates of particles usually formed by spray drying) containing processing additives. Pressures in the range of 3000–30 000 psi are commonly used (Figures 8.15 and 8.16). This method can only be used to fabricate simple shapes (e.g. cylinders). Another limitation is that for a shape with a high length-to-diameter ratio, the frictional forces of the powder, particularly against the die wall, lead to pressure gradients and

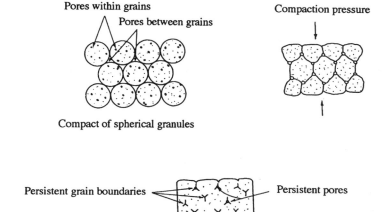

**Fig. 8.15** Transformation of pore size distribution and granule shape during pressing. Adapted from Reed (1988).

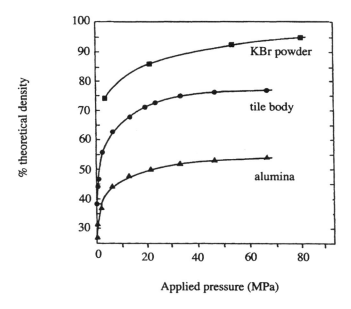

**Fig. 8.16** Change in density of powder compacts of a few selected materials in response to applied pressure. Adapted from Reed (1988).

the resulting variation of density within a piece causes shrinkage variations and loss of tolerances during firing. However, because of the high productivity and the ability to produce parts (to close tolerances) ranging widely in size and shape with essentially no drying shrinkage, pressing is the most widely used industrial forming method.

Pressed products with one elongated dimension, a complex shape, or a large volume are not easily die-pressed and are formed by isostatic pressing or isopressing (Figure 8.17). In this method the pressure is applied from all directions (as opposed to uniaxially during die pressing) on the powder mass by a liquid through a flexible mould. The main advantage of this forming method is the uniform application of pressure throughout the sample volume which minimizes non-uniform packing densities due to pressure gradients. Machining is often used to perfect the final shape. Isostatic pressing, together with machining, is well suited to prototype production and can be automated for large-scale production as for insulators for spark plugs (Pugh and Bergstrom, 1994). Combination and roll pressing are other variations of powder forming by pressing (Reed, 1988).

One of the obstacles in fabricating products by pressing nanoparticles is their tendency to agglomerate. Agglomerates in a mass of nanopowder are resistant to densification during sintering and can therefore be sites for

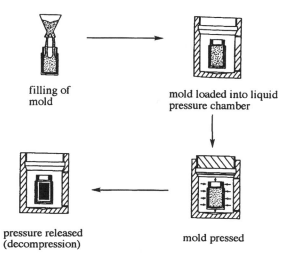

filling of
mold

mold loaded into liquid
pressure chamber

pressure released
(decompression)

mold pressed

**Fig. 8.17** Sequential steps in wet bag isopressing. Adapted from Reed (1988).

mechanical defects in parts formed by pressing. A theoretical study of nanoparticle systems based on an established model for hot isostatic pressing (see section 8.4.2.3 for hot isostatic pressing) has attempted to account for the effects of agglomeration, bulk and surface impurities and other factors on densification (Suryanarayanan *et al.*, 1996b). The investigation revealed that agglomeration and impurities can be an impediment to densification despite the expectations of high densification rates due to high atomic diffusivities. *This is one of the main motivations for development of aerosol processes that can generate unagglomerated nanoparticles.*

8.4.1.2 *Casting.*   Complex shapes cannot be formed by pressing (Pugh and Bergstrom, 1994). Casting processes can then be used to produce a self-supporting shape called a **cast** from a specially formulated slurry. An aqueous slurry containing fine clay has been traditionally called a **slip** and conventional casting is therefore called **slip casting**, the most common forming method under this category (Figure 8.18). A low-viscosity, high-solids-content slurry is poured into a mould (usually made of plaster). The solids form a body against the mould wall as the liquid in the slip drains out. The liquid removal may be hastened by applying a pressure to the slip (**pressure casting**) or pulling a vacuum on the mould (**vacuum casting**). The solid body is then dried in the mould to improve its green strength, after which it is removed and the residual moisture evaporated by slow drying. The advantage of this forming method lies in the ease and inexpensiveness

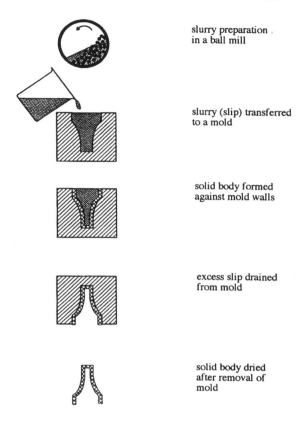

slurry preparation
in a ball mill

slurry (slip) transferred
to a mold

solid body formed
against mold walls

excess slip drained
from mold

solid body dried
after removal of
mold

**Fig. 8.18** Flow diagram for slip casting. Adapted from Pugh and Bergstrom (1994).

with which prototypes can be made. The disadvantages are imprecision in the formed shape and the long time required for casting.

The difficulties of finding proper dispersing agents can be another disadvantage in slip casting of nanoparticles because of their inherent tendency to agglomerate. Agglomeration may lead to non-uniform densities in the body of an object made by slip casting, thereby causing problems during its subsequent processing by firing. Proper choice of the dispersion medium and deflocculating agents, therefore, is critical to the success of ceramic fabrication by slip casting of nanoparticles. Some progress has been made in this direction. Bossel *et al.* (1995) investigated the deagglomeration of nanosized silicon powders in different dispersing media. The extent of deagglomeration of the particles in different solvents depended on the surface properties of the powders and on the interparticle interactions. Ethanol was a suitable dispersing medium for silicon with which the average aggregate size of the powders could be reduced to 80 nm.

Nanoparticle slurries used for slip casting can exhibit other unique characteristics. Shan and Zhang (1996) processed nanosized tetragonal zirconia powders by slip casting. The zirconia powders were suspended in water using appropriate dispersants. Compared with micrometre-sized powders, the slurries had lower solids content for a given viscosity. They also exhibited thixotropic behaviour (important for the smooth flow of the slurries) because of their decreasing apparent viscosity with time under constant shear stress (pressure casting).

Although the small particle size of nanopowders is one of the biggest advantages relative to micrometre-sized powders, the large surface area associated with the small particles can be a detriment during powder processing by slip casting. The slip cast specimens in the previous example had a low relative green density (36.4%) and underwent a large linear shrinkage (34%) on sintering at 1530–1600°C for 4 hours. Large shrinkages are undesirable because of the difficulty in achieving stipulated tolerances in the final product. The observed characteristics were attributed to the large specific surface area of the powders leading to a high absorbed water content between the nanoparticles and the consequent lowering of solids content and green density.

A special form of casting used by the hybrid microelectronics industry is **tape casting** (Reed, 1988). A film of controlled thickness is formed by pouring a slip onto a ceramic tape and laying it out in a thin layer using a blade (also called 'doctor blading'). Organics in the slip help to optimize its rheological (flow) properties and also bind the solid particles together after removal of the liquid phase of the slip by drying. Further firing results in a dense, sintered film. The process is generally referred to as **continuous tape casting** when the blade is stationary and the substrate moves and **doctor blade casting** when the blade moves across a stationary substrate (Figure 8.19). Ceramic sheets with thicknesses of 0.025–1.0 mm

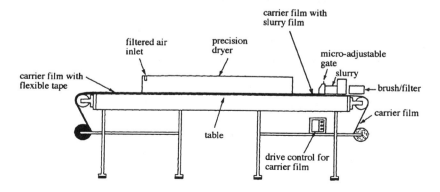

**Fig. 8.19** Set-up for continuous tape casting. Adapted from Reed (1988).

can be produced with tolerances of ±10% or better (Hyatt, 1986). In precision tape casting, sheets with thicknesses of 0.025–0.25 mm and with tolerances of ±0.008 mm can be produced (Williams, 1986).

Agglomeration of nanoparticles during tape casting can pose problems as in the case of pressing and slip casting. Agglomerated powders are difficult to disperse in slurry form, a requirement for tape casting. This leads to non-uniform slip consistencies which can degrade the quality of the film (non-uniform film thickness) formed during tape casting. This is a probable reason for the paucity of experimental work on tape casting of nanoparticle slurries. Also, the possible advantage of being able to fabricate very thin films by using small particles would be mitigated by the mechanical fragility of such films.

8.4.1.3 *Injection moulding.*    Forming methods used with polymers are also applicable to powders. In injection moulding, a heated mixture of a powder and 35–50 vol% of a polymeric binder is injected into a cooled metal die through a screw feeder. The major component of the binder is a thermoplastic polymer or wax which is removed by controlled firing of the moulded body in air or an inert atmosphere. Processing additives are also added to decrease the viscosity of the moulding batch and therefore allow moulding with higher solids content. The advantages of injection moulding are the ability to form complicated shapes and the ease of automation of large-scale production. The high cost and long processing times are the primary disadvantages. Injection moulding is used for fabricating complex shapes such as multi-vaned turbine rotors and small electronic components. Very few examples of the application of this technique to process nanoparticles are currently available. However, the use of injection moulding in the production of nanostructured magnetic components is increasingly improving properties of the end product (Anon., 1996).

8.4.1.4 *Extrusion.*    Extrusion is the shaping of the cross-section of materials plasticized with an organic binder by forcing it through a rigid die orifice, a method commonly used for objects which have an axis normal to a fixed cross-section (Kingery *et al.*, 1976; Reed, 1988). Extrusion can also be done in water systems. In these cases, a water-soluble polymeric binder is added to increase plasticity during extrusion and to improve the green strength of the extruded material. The most widely practised method is to use a vacuum auger to eliminate air bubbles, thoroughly mix the body with 12–20 vol% water and force it through a hardened steel or carbide die. Hydraulic piston extruders are also widely used. Extrusion is used to fabricate a wide variety of parts of complex shape which include hollow furnace tubes, honeycomb catalyst supports, transparent alumina tubes for lamps, very small electronic and magnetic ceramics and graphite electrodes which can exceed 1 ton in size.

Extrusion is particularly useful for attaining improved structural properties in a final product compared to uniaxial die pressing which can cause embrittlement due to delamination microcracks produced during sintering (due to the establishment of non-uniform density gradients during pressing). Sometimes, during extrusion, objects are heated to cause densification during the forming process. Extrusion in the presence of a hot gas (high-temperature gas extrusion, HTGE) produces homogeneous stress gradients resulting in more uniform density distributions than can be achieved by uniaxial pressing. HTGE of iron and nickel nanopowders (30 nm particles) at 1120 K using a 0.5 GPa pressure produced almost fully dense specimens in a matter of seconds with mean grain sizes of 100 nm (Alymov and Leontieva, 1995). The smaller grain size compared to hot isostatic pressing (1000 nm) is advantageous from the standpoint of mechanical strength properties.

8.4.1.5 *Superplastic forming.* Superplasticity is phenomenologically defined as the ability of a material to exhibit exceptionally large tensile elongation (without failure) during stretching (Chen and Xue, 1990). This property is commonly found in many metals and alloys when the grains are nanosized and the deformation due to applied pressure occurs at temperatures above two-thirds of the melting point. The small grains can flow, much like sand particles in a water-saturated slip, by way of atomic diffusion along grain boundaries. For refractory materials like SiC which sinter at high temperatures, the maximum temperature for superplastic deformation is in practice restricted to below the conventional sintering temperature to be able to utilize superplasticity as a practical shaping method for complex components (Mitomo *et al.*, 1996).

Superplastic forming is an attractive method for the processing of nano-particles, especially for material systems that undergo large shrinkages (in excess of 15%) during sintering, because of its ability for near-net shaping. Plastic (permanent) deformation in crystalline materials generally occurs by the movement of one-dimensional crystal defects (known as dislocations) or diffusion of atoms under stress. Most ceramics and intermetallics cannot be deformed plastically at ambient temperatures and consequently undergo brittle failure. The discovery that superplastic behaviour under tensile loading is possible in nanograined yttria-containing tetragonal zirconia polycrystals (Y-TZP) (Wakai *et al.*, 1986) and a covalent crystal composite, $Si_3N_4/SiC$, at elevated temperatures (Wakai *et al.*, 1990) generated great interest in developing this capability for other ceramic materials. Several commercial processes in metal industries have taken advantage of these high ductilities to form intricate, large-scale components directly into their net shape (Chen and Xue, 1990).

Such deformation processes have also been demonstrated for ceramics. Nanocrystalline titania powders were superplastically formed, in one step, into a final ceramic product at temperatures below 900°C (much lower than ~1200°C, the sintering temperature of micrometre-sized titania). The mechanical strength (fracture toughness) of these nanocrystalline ceramics was twice that of conventional polycrystalline ceramics with micrometre-sized grains (Karch and Birringer, 1990). Compared with the conventional forming processes, which are conducted at low temperatures before firing, superplastic forming has the advantages of greater shape flexibility and better dimensional accuracy.

Nanoparticles are not always required for the formation of materials with nanosized grains. Mechanical attrition has been developed as a versatile alternative to other processing routes for preparing nanocrystalline materials. In this process, bulk powder is mixed together in desired proportions and subjected to high-energy milling. During this step, the grain sizes are reduced by $10^4$ by rearrangement of the internal microstructure of the powder particles (creation and self-organization of dislocations to form high-energy grain boundaries). Tatsuhiko et al. (1995) investigated a repeated mechanical alloying-forging method to fabricate non-equilibrium and nanostructured solid materials and near-net-shaped nanograined materials for binary systems. Si–Al and Bi–Al were chosen as the experimental systems because Si and Bi are difficult to disperse in aluminium by conventional powder metallurgical techniques due to their low solubility in aluminium which causes them to separate from the aluminium matrix. Cyclic applications of high-pressure pressing and severe shear deformation produced Si–Al and Bi–Al mixtures with fine Si and Bi particles dispersed in an Al matrix. The most attractive aspect of this work was the production of near-net-shaped bodies of Si–Al and Bi–Al in the same mechanical alloying-forging machine by only changing die-sets, thereby making the process amenable for adoption by industry.

### 8.4.2 Sintering

Objects made by consolidating powders (called 'green products') are heated in a kiln or furnace to develop the desired microstructure necessary for them to exhibit desired properties. This process, called **firing**, causes further consolidation by the elimination of porosity or voidage through sintering. In structural materials, the densification of fired bodies increases mechanical strength (Kingery et al., 1976; Reed, 1988). In conductive microelectronic thick films, solid–solid contact, continuity and homogeneity of the metal phase are the critical microstructural characteristics for better conductivity (Chung and Kim, 1988). In such films, densification (sintering) during firing lowers the resistance to the flow of electricity by increasing the contact area of the metal particles and thereby decreasing the length of the current path.

Most sintering operations are carried out without subjecting the object to any externally applied pressure. This is called **pressureless sintering**. The major distinction among pressureless sintering techniques is between *solid-state* and *liquid-phase* processes. The following sections will discuss and differentiate these processes and also introduce **pressure sintering** where the driving force for sintering is enhanced by the application of external pressure.

8.4.2.1  *Solid-state sintering.*  As the name implies, solid-state sintering refers to sintering processes occurring in compacts of solid particles without the presence of any liquid phases. The microstructure of such a compact is characterized by grains and pores and the microstructural changes associated with solid-state sintering are characterized by changes in the size and shape of these grains and pores. These transformations can cause densification of the powder body which is necessary for maximizing properties such as strength, translucency, thermal conductivity and electrical conductivity. In solid-state sintering, the free-energy change that gives rise to densification is due to the decrease in surface area by the elimination of solid–vapour interfaces with the coincidental formation of new but lower-energy solid–solid surfaces (Kingery *et al.*, 1976). The net decrease in free energy occurring on sintering a 1 μm particle size material corresponds to an energy decrease of about $1 \, cal \, g^{-1}$. This decrease is of the order of $10 \, cal \, g^{-1}$ or more for a 1 nm grain-size aggregate, which upon sintering would undergo much larger changes in surface area per unit mass (Andres *et al.*, 1989). This implies a much higher driving force for sintering in nanopowders.

Temperature and particle size are the most important parameters in solid-state sintering. On a microscopic scale, material transfer is affected by the pressure difference across the surfaces of the particles (Kingery *et al.*, 1976). For particles of small sizes, this effect becomes prominent due to their large curvatures resulting in larger driving forces for phenomena such as densification and grain growth. Nanosized copper (Sakka, 1989) and gold (Sakka *et al.*, 1989) particles exhibited very low densification and grain growth temperatures. In the case of ultrafine gold powder, these phenomena occurred at 110°C, whereas copper powders exhibited grain growth at 200°C when heated in an $H_2$ atmosphere. The explicit dependence of sinterability of SiC (produced by a gas-phase aerosol route) on particle size is evident from Figure 8.20 (Kato, 1987). The sinterability decreases appreciably with increasing particle size due to the decreasing driving force for sintering. Porat *et al.* (1996) found that while micrometre-sized powder (1.8 μm) of WC–Co sintered to full-density mainly by liquid-phase sintering at 1360°C, nanosized particles (30 nm) of

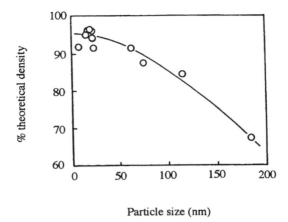

Particle size (nm)

**Fig. 8.20** Effect of particle size on sinterability of SiC powder made by a gas-phase aerosol route, sintered at 2050°C for 30 min under 300 mmHg pressure in an argon atmosphere. Adapted from Kato (1987).

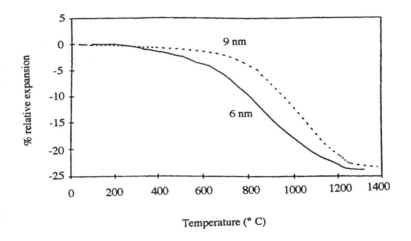

Temperature (° C)

**Fig. 8.21** Shrinkage data for compacts of zirconia nanoparticles with different average particle sizes. Adapted from Skandan (1995).

the same material densified in the solid state at 1250°C. Skandan *et al.* (1994) showed that even at the nanometre scale, particle size can affect densification, as is evident from the comparison of shrinkage data of 6 and 9 nm $n$-ZrO$_2$ powders (Figure 8.21).

The lowering of sintering temperature is due to the higher diffusion coefficients found in nanocrystalline powders (Andres *et al.*, 1989). In a compact of nanocrystalline particles, a high percentage of atoms are in

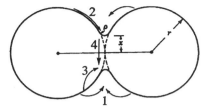

1. evaporation-condensation
2. surface diffusion
3. volume diffusion
4. grain-boundary diffusion

**Fig. 8.22** Paths for matter transport during solid-state sintering. Adapted from Kingery *et al.* (1976).

grain boundaries which are in an unrelaxed state. The diffusion coefficients of these atomic species are much higher than normal grain-boundary diffusion, presumably resulting from the unrelaxed nature of the grain boundaries (Andres *et al.*, 1989). This rapid atomic diffusion makes possible the lower processing temperatures.

Though the driving force for sintering is the reduction of surface area in all systems, considerable differences in behaviour occur among the different systems. The understanding of sintering phenomena obtained with the help of simple models has allowed these differences to be traced to different mechanisms of mass transfer (Kuczynski, 1949). The different mechanisms of solid-state sintering are evaporation and condensation, surface diffusion, grain-boundary diffusion, volume diffusion and viscous flow (Figure 8.22). In the following we discuss these mechanisms with an emphasis on how the small particle size of nanoparticles influences these mechanisms.

8.4.2.1.1 *Evaporation–condensation.* Consider the interaction of the two particles in Figure 8.22 in the initial stages of sintering. At the particle surface there is a positive radius of curvature so that the vapour pressure is somewhat higher than would be observed for a flat surface. However, just at the junction between the particles there is a neck with a small radius of curvature corresponding to a vapour pressure an order of magnitude lower than above the particle itself. Due to this pressure difference, material can vaporize from the particle surface and condense in the neck region between the two particles. Quantitative expressions relating the diameter of the contact area between the particles and the variables influencing its growth rate are available (Kingery *et al.*, 1976). The distance between the particle centres is not affected by such mass transfer. Therefore, densification of a powder mass cannot occur by vapour-phase material transfer. Materials must be heated to a temperature sufficiently high for the vapour pressure to be appreciable. For micrometre-scale particles vapour pressures of the order of $1$–$10\,Pa\,(10^{-2}$–$10^{-1}\,torr)$ are required. In nanoparticle systems, similar vapour pressures can be obtained at lower temperatures or higher vapour pressures can be achieved at the same temperature.

The equilibrium vapour pressure over curved surfaces is given by the Kelvin equation (Reed, 1988):

$$\ln\frac{P}{P_0} = \frac{2\gamma V_{mol}}{rRT},\qquad(8.7)$$

where $P$ is the vapour pressure over the curved surface, $P_0$ the vapour pressure over a flat surface, $\gamma$ the surface energy, $V_{mol}$ the molar volume, $r$ the particle radius (m), $T$ the temperature (K) and $R$ the molar gas constant. For example, at 1850°C a 10 nm alumina particle has a relative vapour pressure $(P/P_0)$ of 1.2, while for a 1 μm particle the relative vapour pressure is only 1.002 ($\gamma = 905\,erg\,cm^{-2}$) (Kingery et al., 1976). This implies that initial contact formation and necking between individual particles in a compact due to evaporation–condensation can occur much more readily in nanoparticles.

8.4.2.1.2 *Surface diffusion.* This phenomenon is characterized by the migration of atoms along the particle surface to the neck between the particles. The rounding-off of the internal and external surfaces and pores in powder compacts is also produced by surface diffusion, which has the lowest activation energy of all types of diffusion and thus is favoured at low temperatures (Thummler and Thomma, 1967). The transfer of material by this method, like vapour transport, does not lead to any decrease in the distance between particle centres and therefore cannot cause densification. It can, however, cause coarsening or grain growth.

The interactions of two idealized nickel nanoparticles were simulated at the atomic scale using molecular dynamics (Heinisch, 1996). The spherical particles with face-centred cubic crystal structure had 767 atoms each. During the interaction time of 500 ps at 945 K in the simulation, necking occurred by surface diffusion of atoms in the surface layers. The short time needed for necking indicated the ease with which surface diffusion processes can be activated.

Bonevich and Marks (1992) found evidence of surface diffusion during their investigation of the sintering behaviour of ultrafine (20–50 nm) alumina particles at 1000°C for a dwell time of 100 ms. Some particles had adhered to each other and sintered with no apparent reorientation to form low-energy interfaces, which indicated that surface diffusion was sufficiently fast to 'lock in' the initial contact orientation before significant neck growth occurred. During the low-temperature (below 500°C) sintering of nanocrystalline titania (50 nm), the necessary mass transport required for elimination of small pores (2–20 nm) presumably occurred by surface diffusion from the surface of the nanoparticles to the neck regions at the particle contacts (Hahn *et al.*, 1990).

These examples suggest that surface diffusion is common during the early stages of solid-state sintering of nanoparticles.

8.4.2.1.3    *Grain-boundary diffusion.*    The transport of matter along or across the region between two grains is called grain-boundary diffusion. Since this kind of transport can remove material from the particle volume or from the grain boundaries and thereby reduce the distance between particle centres, it can cause shrinkage and pore elimination, which result in densification. This mode of diffusion is especially important for nanoparticles because basic deformation characteristics during superplastic forming are dominated by atomic diffusion along grain boundaries.

Several examples of grain-boundary diffusion in nanoparticle systems exist. The onset of significant densification during the sintering of nanosized titania at around 600°C coincided with grain-boundary diffusion becoming sufficiently large to influence sintering (Hahn *et al.*, 1990). Venkatachari *et al.* (1995) determined that densification of nanocrystalline yttria-stabilized zirconia (average crystallite size 8–12 nm) could be enhanced by suppressing coarsening or grain growth and enhancing grain-boundary diffusion. This was done by heating the powder compacts rapidly through the temperature regime in which coarsening mechanisms (surface diffusion) predominate to the temperature regime in which densification mechanisms are active (grain-boundary diffusion). Thus they were able to obtain sintered bodies with densities greater than 99% of theoretical and final average grain sizes less than 200 nm. The successful exploitation of grain-boundary diffusion phenomena in nanoparticle systems can therefore be useful in making better products.

Solid-state sintering by volume diffusion and viscous flow are also known. However, studies of these phenomena in nanoparticle systems are almost non-existent. For a general discussion of these phenomena the interested reader is referred to other sources (Kuczynski, 1949; Kingery *et al.*, 1976).

8.4.2.2    *Liquid-phase sintering.*    Liquid-phase sintering involves coexisting liquid and solid phases during some part of the firing step (German, 1985). Depending on whether the liquid phase is constantly present during sintering or not, the process may be broadly subdivided into **persistent** and **transient** liquid-phase sintering. When the solid phase dissolves in the liquid the process is called **sintering with a reactive liquid,** while in the absence of any dissolution it is known as **vitrification**. The other major factors involved in liquid-phase sintering pertain to the interfacial energies between the liquid and solid phase (wetting versus non-wetting liquids) and the extent to which the liquid-phase penetrates along solid–solid grain boundaries (German, 1985).

The interaction of these and other factors such as particle size, sintering temperature, time, atmosphere and green density provides a variety which makes for a highly flexible and useful manufacturing technique with widespread application to both metals and ceramics. For example, in the fabrication of nanoceramics of refractory materials such as SiC, liquid-phase sintering and nanosized particles (90 nm) can be used to retard grain growth (grain growth is undesirable from the standpoint of development of mechanical properties) that is typical of bodies made by liquid-phase sintering of conventional SiC powders (0.28 μm) (Mitomo *et al.*, 1996). This is possible because of the greater densification caused by the presence of a liquid phase at a lower processing temperature for nanosized powders. The lowering of the processing temperature reduces the driving force for grain growth, which helps retain a nanostructured microstructure. This example also underlines the important role of particle size and temperature during sintering in general and liquid-phase sintering in particular.

Liquid-phase sintering can be conceptually divided into three stages (Figure 8.23). As the object to be sintered is heated, a liquid phase appears. This causes rapid densification by particle rearrangement brought about by capillary forces exerted by the liquid on the solid particles due to spreading (the extent of spreading is decided by the wettability of the solid phases by the liquid). The densification rate slows down continuously as the surface energy of the object is reduced due to decreasing porosity. This

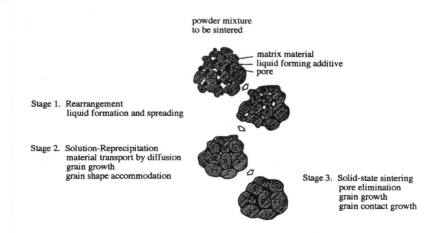

**Fig. 8.23** Stages of liquid-phase sintering. Adapted from German (1985).

stage typically describes the *vitrification* process. The amount of densification obtained by rearrangement is decided by the size and shape of the solid particles, liquid viscosity and surface tension and amount of liquid phase. The rate of densification during this stage is inversely proportional to the particle size (Kingery and Narasimhan, 1959). In changing from a 1 μm to a 10 nm particle, this rate is increased by a factor of 100. Rearrangement processes can be inhibited by irregular particle shape.

Full density (zero porosity) is possible by rearrangement if enough liquid is formed. It is estimated that 35 vol% liquid is needed for complete densification by particle rearrangement (Kingery, 1959). However, this is not usually achieved in practical systems.

Various other events characteristic of *reactive liquid-phase sintering* may occur simultaneously with rearrangement. However, these become dominant only when the rearrangement process slows down. The effect of the solubility of the solid particles in the liquid phase and the diffusivity of the dissolved atoms in the liquid characterize the second stage of classic liquid-phase sintering. The solubility of a grain (or particle) in a surrounding liquid is inversely proportional to its size (German, 1985). This results in the faster dissolution of the smaller grains. The difference in solubilities sets up a concentration gradient in the liquid which results in material transport from the smaller to the larger grains by diffusion through the liquid. This process is termed **coarsening** or **Ostwald ripening** and results in the progressive growth of larger grains. This stage of solution–reprecipitation also causes densification (only in conjunction with the rearrangement process).

Because the solubility of the solid particles is inversely proportional to their size, nanoparticle compacts can be expected to show enhancements in grain growth compared to those of micrometre-sized particles under identical liquid-phase sintering conditions. For example, Hingorani *et al.* (1995) observed significant enhancement in grain growth of ZnO nanoparticles (40 nm) when doped with 0.5–4 wt% $Bi_2O_3$. They argued that the dopant concentrations used were above the solidus boundary in the phase diagram of the $ZnO$–$Bi_2O_3$ system which led to the formation of a $Bi_2O_3$-rich liquid phase at the grain boundaries. The kinetics of grain growth was influenced by liquid-phase sintering which increased the grain-boundary mobility and the rate of grain growth. However, the activation energy for grain growth ($148 \pm 5 \, kJ \, mol^{-1}$) was unaffected by the amount of the dopant. Thus, by properly manipulating the conditions under which liquid-phase sintering occurs, contradictory phenomena of grain growth enhancement and retardation can be induced in nanoparticle compacts.

If a high liquid content is present, complete densification can be reached by the rearrangement process and no need arises for the solution–precipitation phenomena which would also be expected to change grain shapes. Consequently, the greatest uniformity in grain shape is expected for systems at high liquid contents. Grains in iron-copper pellets with different amounts of liquid phase and sintered for 4 hours were characterized by a 'sphericity index' (Kingery and Narasimhan, 1959). The maximum departure from spherical shape occurred at small liquid contents. This suggested that at low liquid contents, maximum solution–reprecipitation was required to achieve densification by alteration of spherical grain shapes to allow tighter packing, thereby resulting in irregularly shaped grains. This phenomenon is referred to as **grain shape accommodation**.

Grain shape accommodation was also observed during the liquid-phase sintering of thick-film resistor compositions of lead ruthenate (pyrochlore structure, 34 nm particle size) and lead aluminosilicate glass (Chiang *et al.*, 1994). Though the actual reason behind this observation could not be determined due to lack of sequential observations of microstructure with time, solution–precipitation in response to van der Waals interparticle forces was determined as the cause behind flattening of the pyrochlore phase particles at their points of contact. These observations reinforce the presence and role of these phenomena (grain shape accommodation and solution–precipitation) during liquid-phase sintering.

The final stage of classic liquid-phase sintering is referred to as **solid-state controlled sintering**. The existence of a solid skeleton slows down the rate of densification in this stage. The rigidity of the solid skeleton inhibits further rearrangement. The residual pores enlarge if they contain trapped gases leading to expansion of the compact. Contacts between solid grains are formed by solution–reprecipitation, coalescence of grains and solid-state diffusion. The grain contacts allow densification by solid-state sintering. However, microstructural coarsening continues to occur by diffusion. In general, properties of most liquid-phase sintered materials are degraded by prolonged final-stage sintering.

The different phenomena that characterize liquid-phase sintering of conventional powders are also manifested during the sintering of nano-particle systems. The known advantages of liquid-phase sintering can therefore be exploited during the processing of nanosized powders.

### 8.4.2.3 *Pressure sintering.*

Pressureless sintering depends on the capillary pressures resulting from surface energy to provide the driving force for densification. In pressure sintering an external pressure is applied at an elevated temperature which provides an additional driving force to enhance the rate and extent of densification. Techniques such as *hot pressing*, *hot isostatic pressing*, *sinter forging* and *hot extrusion* use a combination of temperature, stress and strain rate to densify powder compacts (Reed, 1988). These methods are fairly old and were developed in order fully to densify objects made from coarse (of the order of a few micrometres) powders available in the days before sophisticated routes to

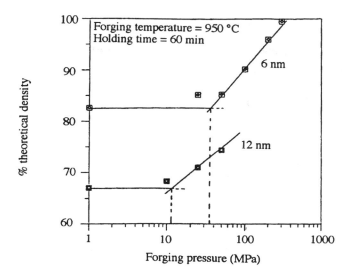

**Fig. 8.24** Change in relative density of nanoparticle compacts as a function of forging pressure. Adapted from Skandan (1995).

making submicrometre and nanosized powders were developed. Covalent materials such as boron carbide, silicon carbide and silicon nitride can be hot-pressed to nearly complete density. Another advantage is that in some cases densification can be obtained at a lower temperature than pressure-less sintering, so that excessive grain growth does not occur. Since the mechanical properties of many ceramic systems are maximized with high density and small grain size, optimum properties can be obtained by hot-pressing techniques. The main disadvantages of hot pressing for oxides are the unavailability of inexpensive and long-life dies for high temperatures and the difficulty in automating the process for high productivity.

The advantage of using nanopowders is that they can be hot-pressed to much higher densities than micrometre-sized powders at the same temperature and pressure. Figure 8.24 shows the effect of applied pressure on the relative density of two powder compacts made from nanopowders with different particle sizes (Skandan, 1995). Powder A has an average particle size of 6 nm and specific surface area of $136 \, m^2 \, g^{-1}$, powder B is coarser, with an average particle size of 12 nm and a specific surface area of $65 \, m^2 \, g^{-1}$. However, the relative density versus applied pressure plots for these powders indicate the presence of a *threshold stress* (beyond which rapid densification occurs, Figure 8.24) which increases with decreasing particle size.

This unique behaviour of nanoparticle ceramics under an applied stress is due to the fact that the *driving force* for densification is composed of an *intrinsic* force due to particle curvature and an *extrinsic* force due to applied stress (Figure 8.25). This can be mathematically represented as (Skandan, 1995)

$$DF = \gamma \Omega K + g P_a \tag{8.8}$$

where $DF$ is the driving force for densification, $\gamma$ is the surface energy of the material, $\Omega$ is the molecular volume, $K$ is proportional to the curvature of the shrinking pore which in turn is approximately inversely proportional to the grain size, $g$ is a geometrical constant and $P_a$ is the applied pressure.

Figure 8.25 is a representation of this equation. At small particle sizes, the contribution of the applied stress to the driving force is far less significant than that due to the curvature. At large particle sizes, the contribution due to the applied stress becomes predominant due to the smaller extrinsic driving force resulting from lower particle curvatures. Therefore, due to the large intrinsic driving force for sintering in nanoparticle ceramics, very high pressures are required to cause additional densification during hot pressing and sinter-forging which explains the high threshold stresses. On the other hand, in micrograined materials, the threshold stress is so small that a small amount of applied pressure causes noticeable change in densification.

Densification during pressure sintering can occur by all the mechanisms which have been discussed for solid-state sintering. It is often advan-

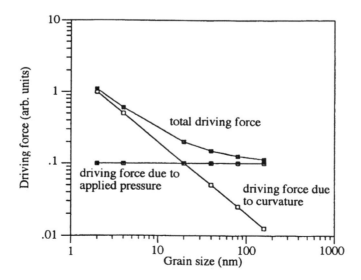

**Fig. 8.25** Functional dependence of driving force for densification on grain size. Adapted from Skandan (1995).

tageous to add a small fraction of a liquid-phase-forming material (e.g. LiF to MgO) to allow pressure-induced liquid-phase sintering to occur (Kingery *et al.*, 1976).

Pressure sintering (useful for the successful processing of micrometre-sized powders) of nanoparticles is advantageous only when sintering highly refractory materials. This is because the high intrinsic driving forces in nanoparticle systems necessitate very high pressures to cause additional densification. Nevertheless, it is useful for sintering materials which require very high temperatures for densification because the added driving force for sintering due to the applied pressure reduces the temperature required to achieve the required amount of densification. This in turn helps to restrict the grain size in the final product.

## References

Alymov M.I. and Leontieva O.N. (1995) Synthesis of nanoscale Ni and Fe powders and properties of their compacts. *Nanostructured Materials*, **6**, 393–395.

Andres R.P., Averback R.S., Brown W.L. *et al.* (1989) Research opportunities on clusters and cluster-assembled materials – A Department of Energy, Council on Materials Science Panel Report. *J. Mater. Res.*, **4**, 704–736.

Anon. (1995) New developments in powder metallurgy. *Adv. Mater. Process.*, **148**, 28–32.

Anon. (1996) Trends in the processing of soft magnetic materials. *Met. Powder Rep.*, **51**, 18–19.

Averback R.S., Hahn H., Hofler H.J. and Logas J.C. (1990) Processing and properties of nanophase metallic alloys: Ni-Ti. *Appl. Phys. Lett.*, **57**, 1745–1747.

Bird R.B., Stewart W.E. and Lightfoot E.N. (1960) *Transport Phenomena*. Wiley, New York.

Bonevich, J.E. and Marks, L.D. (1992) The sintering behaviour of ultrafine alumina particles. *J. Mater. Res.*, **7**, 1489–1500.

Borsella E., Botti S., Fontoni R., Alexandrescu R., Morjan I., Popescu C., Dikonimos-Markis T., Giorgi R. and Enzo S. (1992) Composite silicon/carbon/nitrogen powder production by laser induced gas-phase reaction. *J. Mater. Res.*, **7**, 2257–2268.

Bossel C., Dutta J., Houriet R., Hilborn J. and Hofmann H. (1995) Processing of nano-scaled silicon powders to prepare slip cast structural ceramics. Paper presented to *1994 Symposium on Engineering of Nanostructured Materials*, Boston, MA.

Brus L.E. (1986) Electronic wave functions in semiconductor clusters: experiment and theory. *J. Phys. Chem.*, **90**, 2555–2560.

Burger H.C. and Cittert P.H.V. (1930) Die Herstellung von Wismut-Antimon-Vakuum-thermoelementen durch Verdampfung. *Z. Phys.*, **66**, 210–217.

Charlesworth D.H. and Marshall W.R. Jr. (1960) Evaporation from drops containing dissolved solids. *AIChE Journal*, **6**, 9–23.

Chen I.-W. and Xue L.A. (1990) Development of superplastic structural ceramics. *J. Am. Ceram. Soc.*, **73**, 2585–2609.

Chiang Y.-M., Silverman L.A., French R.H. and Cannon R.M. (1994) Thin glass film between ultrafine conductor particles in thick-film resistors. *J. Am. Ceram. Soc.*, **77**, 1143–1152.

Chow G.M., Holtz R.L., Pattnaik A. and Edelstein A.S. (1990) Alternative approach to nanocomposite synthesis by sputtering. *Appl. Phys. Lett.*, **56**, 1853–1855.

Chung Y.S. and Kim H. (1988) Effect of oxide glass on the sintering behavior and electrical properties in Ag thick films. *IEEE Trans. Comp., Hybrids, Manuf. Technol.*, **11**, 195-199.

Dietz T.G., Duncan M.A., Powders D.E. and Smally R.E. (1981) Laser production of supersonic metal cluster beams. *J. Chem. Phys.*, **74**, 6511–6512.

Edelstein A.S. (1993) in G.C. Hadjipanayis and R.W. Siegel (eds), *Nanophase Materials*. Kluwer Academic Publishers, London.

Friedlander S.K. (1977) *Smoke, Dust and Haze*. Wiley Interscience, New York.

Gardner G. (1965) Asymptotic concentration distribution of an involatile solute in an evaporating drop. *Int. J. Heat Mass Transfer*, **8**, 667–668.

Gardner T.J. and Messing G.L. (1984) Magnesium salt decomposition and morphological development during evaporative decomposition of solutions. *Thermochim. Acta.*, **78**, 17–27.

Gardner T.J., Sproson D.W. and Messing G.L. (1984) Precursor chemistry effects on development of particulate morphology during evaporative decomposition of solutions. *Mater. Res. Soc. Symp. Proc.*, **32**, 2227–2232.

Gelbard F. and Seinfeld J.H. (1979) The GDE for aerosols – theory and application to aerosol formation and growth. *J. Colloid Interface Sci.*, **68**, 363–382.

German R.M. (1985) *Liquid Phase Sintering*. Plenum Press, New York.

Gleiter H. (1981) in N. Hansen, A. Horsewell, T. Leffers and H. Lilhot (eds), *Deformation of Polycrystals: Mechanisms and Microstructures*. Ris National Lab., Rackilde, Denmark, pp. 15–19.

Gleiter H. (1989) Nanocrystalline materials. *Prog. Mater. Sci.*, **33**, 223–228.

Gonsalves, K.E., Strutt P.R. and Xio T.D. (1991) Synthesis of ceramic nanoparticles by the ultrasonic injection of an organosilazane precursor. *Adv. Mater.*, **3**, 202–204.

Granqvist C.G. and Buhrman R.A. (1976) Ultrafine metal particles. *J. Appl. Phys.*, **47**, 2200–2219.

Gupta A., Beeson K.W., Donlan J.P. and West G.A. (1987) Titanium silicide ultrafine powder: $CO_2$ laser generation and thin film applications. *J. Appl. Phys.*, **61**, 1162–1167.

Gurav A., Kodas T., Pluym T. and Xiong Y. (1993a) Aerosol processing of materials. *Aerosol Sci. Technol.*, **19**, 411–452.

Gurav A.S., Duan Z., Wang L., Hampden-Smith M.J. and Kodas T.T. (1993b) Synthesis of fullerene–rhodium nanocomposites via aerosol decomposition. *Chem. Mater.*, **5**, 214–216.

Haggerty J.S. (1984) Synthesis of powders and thin films by laser induced gas-phase reactions. *Mater. Sci. Res.*, **17**, 137–154.

Haggerty J.S. and Flint J.H. (1990) Synthesis of Si, SiC and $Si_3N_4$ powders under high density conditions, in Z.A. Munir and J.B. Holt (eds), *Combustion and Plasma Synthesis of High Temperature Materials*. VCH, New York, pp. 399–405.

Hahn H. and Averback R.S. (1990) The production of nanocrystalline powders by magnetron sputtering. *Appl. Phys. Lett.*, **67**, 1113–1115.

Hahn H., Logas J. and Averback R.S. (1990) Sintering characteristics of nanocrystalline $TiO_2$. *J. Mater. Res.*, **5**, 609–614.

Haynes B.S., Jander H. and Wagner H.G. (1979) The effect of metal additives on the formation of soot in pre-mixed flames. in *Seventeenth Symposium (International) on Combustion*, Pittsburg, PA, The Combustion Institute, 1365–1367.

Heinisch H.L. (1996) Computer simulations of nanoparticle interactions, *1996 125th TMS Annual Meeting*, in D.L. Burrell (ed.) *Synthesis and Processing of Nanocrystalline Powder*, Anaheim, CA, Minerals, Metals & Materials Soc (TMS).

Hinds W.C. (1982) *Aerosol Technology*. Wiley, New York.

Hingorani S., Shah D.O. and Multani M.S. (1995) Effect of process variables on the grain growth and microstructure of $ZnO–Bi_2O_3$ varistors and their nanosize ZnO precursors. *J. Mater. Res.*, **10**, 461–467.

Hirayama T. (1987) High temperature characteristics of transition $Al_2O_3$ powder with ultrafine particles. *J. Am. Ceram. Soc.*, **70**, C122–C124.

Hung C.H. and Katz J.L. (1992) Formation of mixed-oxide powders in flames: 1. $TiO_2–SiO_2$. *J. Mater. Res.*, **7**, 1861–1869.

Hung C.H., Katz J.L. and Miguel P.F. (1992) Formation of mixed-oxide powders in flames: $2SiO_2–GeO_2$ and $Al_2O_3–TiO_2$. *J. Mater. Res.*, **7**, 1870–1875.

Hyatt E.P. (1986) Making thin flat ceramics – A review. *Am. Ceram. Soc. Bull.*, **65**, 637–638.

Ichinose N., Ozaki Y. and Kashu S. (1992) *Superfine Particle Technology*. Springer-Verlag, London.

Jasinski J.M. and LeGoues F.K. (1991) Photochemical preparation of crystalline silicon nanoclusters. *Chem. Mater.*, **3**, 989–992.

Jayanthi G., Messing G.L. and Zhang S.-C. (1993) Modeling of solid particle formation during spray pyrolysis. *J. Aerosol Sci. Tech.*, **19**, 478–490.

Johnson D.W. Jr. (1981) Nonconventional powder preparation techniques. *Am. Ceram. Soc. Bull.*, **66**, 221–224.

Johnson D.W. Jr. (1987) Innovations in ceramic powder preparation, in K.S.M.G.L. Messing, J.W. McCauley and R.A. Haber (eds) *Advances in Ceramics*. The American Ceramic Society, Westerville, OH, pp. 3–21.

Johnston G.P., Muenchausen R., Smith D.M., Farenholtz W. and Foltyn S. (1992) Reactive laser ablation synthesis of nanosized aluminum nitride. *J. Am. Ceram. Soc.*, **75**, 3465–3468.

Kagawa M., Honda F., Onodera H. and Nagae T. (1983) The formation of ultrafine $Al_2O_3$, $ZrO_2$ and $Fe_2O_3$ by the spray-ICP technique. *Mater. Res. Bull.*, **18**, 1081–1087.

Karch J. and Birringer R. (1990) Nanocrystalline ceramics. Possible candidates for net-shape forming. *Ceram. Int.*, **16**, 291–294.

Kashu S., Fuchita E. and Manabe T. (1984) Deposition of ultrafine particles using a gas jet. *J. Appl. Phys. II*, **23**, L910–L912.

Kato, A. (1987) Recent production methods of ultrafine powders. *Ind. Ceram.*, **7**, 105–108.

Kingery W.D. (1959) Densification during sintering in the presence of a liquid phase. I. Theory. *J. Appl. Phys.*, **30**, 301–306.

Kingery W.D. and Narasimhan M.D. (1959) Densification during sintering in the presence of a liquid phase. II. Experimental. *J. Appl. Phys.*, **30**, 307–310.

Kingery W.D., Bowen H.K. and Uhlmann D.R. (1976) *Introduction to Ceramics*. Wiley, New York.

Kirkbir F. and Komiyama H. (1987) Formation and growth mechanisms of porous, amorphous and fine particles by chemical vapour deposition. Titania from titanium isopropoxide. *Can. J. Chem. Eng.*, **65**, 759–766.

Kobata A., Kusakabe K. and Morooka S. (1991) Growth and transformation of $TiO_2$ crystallites in aerosol reactors. *AIChE J.*, **37**, 347–359.

Koch W. and Friedlander S.K. (1990) Particle growth by coalescence and agglomeration. *J. Aerosol Sci.*, **21**, S73–S76.

Kodas T.T. (1989) Generation of complex metal oxides by aerosol processes: Super-conducting ceramic particles and films. *Adv. Mater.*, **28**, 794–806.

Kodas T.T. and Sood A. (1990) Alumina powder production by aerosol process, in L. Hart (ed.) *Alumina Chemicals: Science and Technology Handbook*. American Ceramic Society, Westerville, OH, pp. 375–402.

Kodas T.T., Engler E.M., Lee V., Jacowitz R., Baum T.H., Roche K., Parkin S.S.P., Young W.S., Hughes S., Kleder J. and Auser W. (1988) Aerosol flow reactor production of fine $YBa_2Cu_3O_{7-x}$ powder: Fabrication of superconducting ceramics. *Appl. Phys. Lett.*, **52**, 1622–1624.

Kodas T.T., Engler E. and Lee V. (1989) Generation of thick $Ba_2YCu_3O_7$ films by aerosol deposition. *Appl. Phys. Lett.*, **54**, 1923–1925.

Koizumi T., Yokota S., Matsumura S. and Inoue Y. (1989) Apparatus for producing superfine particles. US Patent 336 560.

Kuczynski G.C. (1949) Self-diffusion in sintering of metallic particles. *Metals Trans.*, Feb., 169–178.

LaiHing K., Wheeler R.G., Wilson W.L. and Duncan M.A. (1987) Photoionization dynamics and abundance patterns in laser vaporized tin and lead clusters. *J. Chem. Phys.*, **87**, 3401–3409.

Landgrebe J.D. and Pratsinis S.E. (1989) Gas-phase manufacture of particulates: Interplay of chemical reaction and aerosol coagulation in the free-molecular regime. *Ind. Eng. Chem. Res.*, **28**, 1474–1481.

Levenspiel, O. (1983) *Chemical Reaction Engineering*. Wiley, New Delhi.

Liu Y., Zhang Q.L., Tittel F.K., Curl R.F. and Smally R.E. (1986) Photodetachment and photo-fragmentation studies of semiconductor cluster ions., *J. Chem. Phys.*, **85**, 7434–7441.

Majumdar D., Shefelbine T.A., Kodas T.T. and Glicksman H.D. (1996a) Copper (I) oxide powder generation by spray pyrolysis. *J. Mater. Res.*, **11**, 2861–2868.

Majumdar D., Kodas T.T. and Glicksman H.D. (1996b) Gold particle generation by spray pyrolysis. *Adv. Mater.*, **8**, 1020–1022.

Mandich M.L., Bondybey V.E. and Reents W.D. (1987) Reactive etching of positive and negative silicon cluster ions by nitrogen dioxide. *J. Chem. Phys.*, **86**, 4245–4257.

Masters K. (1985) *Spray Drying Handbook*. Wiley, New York.

Matsoukas T. and Friedlander S.K. (1991) Dynamics of aerosol agglomerates. *J. Colloid Interface Sci.*, **146**, 495–506.

Mayo M.J., Siegel R.W., Narayanasamy A. and Nix W.D. (1990) Mechanical properties of nanophase TiO$_2$ as determined by nanoindentation. *J. Mater. Res.*, **5**, 1073–1082.

McDonald J.E. (1962) Homogeneous nucleation of vapor condensation I. Thermodynamic aspects. *Am. J. Phys.*, **30**, 870–877.

Messing G.L., Zhang S.-C. and Jayanthi G. (1993) Ceramic powder synthesis by spray-pyrolysis. *J. Am. Ceram. Soc.*, **76**, 2707–2726.

Mezey E.J. (1966) Pigments and reinforcing agents, in C.F. Powell, J.H. Oxley and J.M. Blocher, Jr. (eds), *Vapor Deposition*. Wiley, New York, pp. 423–438.

Millberg L.S. (1987) The synthesis of ceramic powders. *JOM*, Aug., 9–13.

Mitomo M., Kim Y.-W. and Hirotsuru H. (1996) Fabrication of silicon carbide nano-ceramics. *J. Mater. Res.*, **7**, 1601–1604.

Morooka S., Kobata A. and Kusakabe K. (1991) Growth and crystal transformation of TiO$_2$ particles in a plug flow CVD reactor, in S.I. Hirano, G.L. Messing and H. Hausner (eds), *Ceramic Powder Science IV*. American Ceramic Society, Westerville, OH, pp. 59–63.

Murata M., Wakino K., Tanaka K. and Hamakawa Y. (1976) Chemical preparation of PLZT powder from aqueous solution. *Mat. Res. Bull.*, **2**, 323–328.

Nakahigashi K, Ishibashi H., Minamigawa S. and Kogachi M. (1992) X-ray diffraction study on ultrafine particles of Ag–Cu alloys prepared by hydrogen plasma–metal reactor method. *Jap. J. Appl. Phys.*, **31**, 2293–2298.

Neilson M.L., Hamilton P.M. and Walsh R.J. (1963) Ultrafine metal oxides by decomposition of salts in a flame, in C.S.W.E. Kuhn (ed.), *Ultrafine Particles*. Wiley, New York, pp. 181–195.

Ogihara T., Ookura T., Yanagawa T., Ogata N. and Yoshida K. (1991) Preparation of submicrometre spherical oxide powders and fibres by thermal spray decomposition using an ultrasonic mist atomiser. *J. Mater. Chem.*, **1**, 789–794.

Ohno S. and Uda M. (1984) Room temperature oxidation of ultrafine iron particles at low oxygen partial pressure. *Nippon Kagaku Kaishi.*, **6**, 924–929.

Okuyama K., Kousake Y., Toghe N., Yamamoto S., Wu J.J., Flagan R.C. and Seinfeld J.H. (1986) Production of ultrafine metaloxide particles by thermal decomposition of metal alkoxide vapors. *AIChE J.*, **32**, 2010–2019.

Park J.-W. and Altstetter C.J. (1985) The diffusivity and solubility of oxygen in solid palladium. *Scr. Metall.*, **19**, 1481–1485.

Pebler A. and Charles R.G. (1988) Synthesis of small particle size YBa$_2$Cu$_3$O$_{7-x}$ by a vapor phase process. *Mater. Res. Bull.*, **23**, 1337–1344.

Perry R.H., Green D.W. and Maloney J.O. (eds) (1984) *Perry's Chemical Engineer's Handbook*. Wiley, New York.

Petch N.J. (1953) Cleavage strength of polycrystals. *J. Iron Steel Inst.*, **174**, 25–28.

Pfund A.H. (1930) Bismuth-black and its applications. *Phys. Rev.*, **35**, 1434.

Pluym T.C., Ward T.L., Powell Q.H., Gurav A.S., Kodas T.T., Wang L.M. and Glicksman H.D. (1993) Solid silver particle production by spray pyrolysis. *J. Aerosol Sci.*, **24**, 383–392.

Porat R., Berger S. and Rosen A. (1996) Dilatometric study of the sintering mechanism of nanocrystalline cemented carbides. *Nanostructured Mat.*, **7**, 429–436.

Powell Q. (1995) Synthesis of titania particles in tubular flow reactors. MS thesis, University of New Mexico, Albuquerque, NM.

Powell Q., Garvey J., Gurav A., Hampden-Smith M. and Kodas T.T. (1992) Multi-component particle formation via gas-phase reactions of single-component precursors. *J. Aerosol Sci.*, **23**, S249–S252.

Powers D.R. (1978) Kinetics of SiCl$_4$ oxidation. *J. Am. Ceram. Soc.*, **61**, 295–297.

Pratsinis S.E., Bai H., Biswas P., Frenklach M. and Masterangelo S.V.R. (1990) Kinetics of titanium(IV) chloride oxidation. *J. Am. Ceram. Soc.*, **73**, 2158–2162.

Pugh R.J. and Bergstrom L. (eds) (1994) *Surface and Colloid Chemistry in Advanced Ceramics Processing*, Surfactant Science Series. Marcel Dekker, New York.

Raghunathan S., Caron R., Friederichs J. and Sandell P. (1996) Tungsten carbide technologies. *Adv Mater. Process.*, **4**, 21–23.

Reed J.S. (1988) *Introduction to the Principles of Ceramic Processing.* Wiley, New York.

Rohlfing E.A., Cox D.M., Petrovic-Luton R. and Kaldor A. (1984) Alloy cluster beams: Nickel/chromium and nickel/alumina. *J. Phys. Chem.*, **88**, 6227–6231.

Ruthner M. (1983) Industrial production of multicomponent ceramic powders by means of the spray roasting technique, in P. Vincenzin (ed.), *Ceramic Powders.* Elsevier, Amsterdam, pp. 515–518.

Sakka Y. (1989) Reduction and sintering of ultrafine copper particles. *J. Mater. Sci. Lett.*, **8**, 273–276.

Sakka Y., Uchikoshi T. and Ozawa E. (1989) Low-temperature sintering and gas desorption of gold ultrafine powders. *J. Less-Common Met.*, **147**, 89–96.

Schwarz R. and Johnson W.L. (1983) Formation of an amorphous alloy by solid-state reaction of the pure polycrystalline metals. *Phys. Rev. Lett.*, **51**, 415–418.

Sercel P.C., Saunders W.A., Atwater H.A. and Flagan R.C. (1992) Nanometer-scale GaAs clusters from organometallic precursors. *Appl. Phys. Lett.*, **61**, 696–698.

Setaka R., Komatsu W., Shibata T. and Nakajima M. (1988) Preparation of single-crystalline powder of superconducting $YBa_2Cu_3O_{7-x}$ by gas-phase solidification method. *Jap. J. Appl. Phys.*, **61**, L2100–L2102.

Shan H. and Zhang Z. (1996) Slip casting of nanometer sized tetragonal zirconia powder. *Brit. Ceram. Trans.*, **95**, 35–38.

Shon S.N., Kasper G. and Shaw D.T. (1979) *Aerosole Naturwissenschaften, Med. Tech.* (eds W. Stoeber and R. Janicke). Gesellshaft für Aerosolforschung, Mainz, Germany, pp. 34–39.

Siegel R.W. and Eastman J.A. (1989) Synthesis, characterization and properties of nanophase ceramics. *Mater. Res. Soc. Symp.*, **132**, 3–14.

Siegel R.W., Ramaswamy S., Hahn H., Zongquan L. and Ting L. (1988) Synthesis, characterization and properties of nanophase $TiO_2$. *J. Mater. Res.*, **3**, 1367–1372.

Skandan G. (1995) Processing of nanostructured zirconia ceramics. *Nanostructured Mat.*, **5**, 111- 126.

Skandan G., Hahn H., Roddy M. and Cannon W.R. (1994) Ultrafine-grained dense monoclinic and tetragonal zirconia. *J. Amer. Ceram. Soc.*, **77**, 706–710.

Suryanarayanan R., Frey C.A., Sastry S.M.L., Waller B.E. and Buhro W.E. (1996a) Deformation, recovery and recrystallization behavior of nanocrystalline copper produced from solution-phase synthesized nanoparticles. *J. Mater. Res.*, **11**, 449–457.

Suryanarayanan R., Sastry S.M.L., Jerina K.L. Trentler T.J., Waller B.E. and Buhro W.E. (1996b) Densification of nanocrystalline powder produced by solution phase synthesis: theoretical modeling and comparison with experiments. *1996 125th TMS Annual Meeting*, Anaheim, CA, Minerals, Metals & Materials Soc (TMS).

Tatsuhiko A., Osamu K., Kiyohiko T. and Junji K. (1995) Net-shape forming of particulate materials by repeated MA-forming. *1995 International Conference & Erhibition on Powder Metallurgy and Particulate Materials*, Seattle, WA, ASM International.

Thummler F. and Thomma W. (1967) The sintering process. *Metallurgical Reviews*, 69–108.

Tohge N., Tatsumigaso M., Minami T., Okuyama K., Adachi M. and Kousaka Y. (1988) Direct preparation of uniformly distributed $YBa_2Cu_3O_{7-x}$ powders by spray pyrolysis. *Jap. J. Appl. Phys.*, **27**, L1068–L1088.

Ulrich G.D. (1971) Theory of particle formation and growth in oxide synthesis flames. *Combust. Sci. Technol.*, **4**, 47–57.

Ulrich G.D. (1984) Flame synthesis of fine particles. *Chem. Eng. News*, **62**, 22–26.

Uyeda R. (1991) Studies of ultrafine particles in Japan: Crystallography. Methods of preparation and technological applications. *Prog. Mat. Sci.*, **35**, 1–96.

Venkatachari K.R., Huang D., Ostrander S.P., Schulze W.A. and Stangle G.C. (1995) Preparation of nanocrystalline yttria-stabilized zirconia. *J. Mater. Res.*, **10**, 756–761.

Visca M. and Matijevic E. (1979) Preparation of uniform colloidal dispersions by chemical reaction in aerosols. I. Spherical particles of titanium dioxide. *J. Colloid Interface Sci.*, **68**, 308–319.

Vissikov G.P. (1992) Structural, phase and morphological features of plasmachemically synthesized ultradispersed particles. *J. Mater. Sci.*, **27**, 5561–5568.

Vissikov G.P., Manolova K.D. and Brakalov L.B. (1981) Chemical preparation of ultrafine aluminium oxide by electric arc plasma. *J. Mater. Sci.*, **16**, 1716–1719.

Vissikov G.P., Stefanov B.I., Gerasimov N.T., Oliver D.H., Enikov R.Z., Vrantchev A.I., Balabanova E.G. and Pirgov P.S. (1988) Photochemical technology for high-dispersion products. *J. Mater. Sci.*, **23**, 2415–2418.

Vollath D. (1990) Pyrolytic preparation of ceramic powders by spray calcination technique. *J. Mater. Sci.*, **25**, 2227–2232.

Wakai F., Sakaguchi S. and Matsuno Y. (1986) Superplasticity of yttria stabilized tetragonal zirconia polycrystals. *Adv. Ceram. Mater.*, **1**, 259–263.

Wakai F., Kodama Y., Sakaguchi S., Murayama N., Izaki K. and Nihara K. (1990) A superplastic covalent crystal composite. *Nature*, **344**, 421–423.

Wang S.F., Dougherty J.P., Huebner W. and Pepin J.G. (1994) Silver-palladium thick-film conductors. *J. Am. Ceram. Soc.*, **77**, 3051–3072.

Weimer A.W., Roach R.P. and Hanle C.N. (1991) Rapid carbothermal reduction of boron oxide in graphite transport reactor. *AIChE J.*, **37**, 759–768.

Williams J.C. (1986) Doctor-blade processes, in F.F.Y. Wang (ed.), *Treatise on Materials Science and Technology, Vol. 9: Ceramic Fabrication Processes*. Academic Press, New York, pp. 173–198.

Xiong Y. and Pratsinis S.E. (1991) Gas-phase production of particles in reactive turbulent flows. *J. Aerosol Sci.*, **22**, 637–655.

Xiong Y., Akhtar K. and Pratsinis S.E. (1992) Modeling the formation of boron carbide particles in an aerosol flow reactor. *AIChE J.*, **38**, 1685–1671.

Ying J.Y. (1993) Structure and morphology of nanostructured oxides synthesized by thermal vaporization/magnetron sputtering and gas condensation. *J. Aerosol Sci.*, **24**, 315–318.

Yukawa N., Hida M., Imura T., Kawamura M. and Mizuno Y. (1972) Structure of chromium rich Cr–Ni, Cr–Fe, Cr–Cu and Cr–Ni–Fe alloy particles made by evaporation in argon. *Met. Trans.*, **3**, 887–895.

Zachariah M.R., Chin D., Semerjian H.G. and Katz J.L. (1989) Silica particle synthesis in a counter diffusion flame. *Combust. Flame*, **78**, 287–298.

# 9 Pharmaceutical aerosols for delivery of drugs to the lungs

C.B. LALOR and A.J. HICKEY

## 9.1 Introduction

The accessibility of the lungs, as a target organ, predispose locally acting pharmacological agents to superior performance in comparison with other conventional routes. The respiratory tract has been recognized as a site for deposition and action of bronchodilator, anti-inflammatory and antibiotic agents and some products of biotechnology (Hubbard *et al.*, 1989). In addition, the exceedingly large surface area (143 m$^2$; Hollinger, 1985) of the lungs and slow mucociliary clearance from the lung periphery, compared to the upper airways, renders this a viable route of adminis-tration (Lansley, 1993; Byron, 1986a; Yeates *et al.*, 1981 ) for systemically acting agents. The potential of the lungs as a route for systemic administration will depend upon the deposition and clearance of aerosol particles or droplets. Initial deposition depends upon the physicochemical properties of therapeutic aerosol particles or droplets. Clearance mech-anisms include the mucociliary transport, cell-mediated transport, absorption and metabolism in the lungs. It is essential that inhalation products generate an optimum therapeutic dose in a narrow particle size range (less than 5 μm; Hickey, 1992a). The physicochemical properties of the candidate compound will significantly influence the choice of in-halation device based upon these performance features.

### 9.1.1 *Historical perspective*

Aerosols have been used for the treatment of lung diseases for thousands of years. Several thousand years ago the inhabitants of the Indus Valley and China used smokes from roots and leaves of plants in the genera *Datura* and *Ephedra* that relieved the symptoms of respiratory distress and asthma. It is now known that these smokes contained stramonium alkaloids, specifically atropine, and ephedrine which have been used in modern medicine in their pure form (Goodman and Gilman, 1985). Forerunners of the modern nebulizers and dry powder inhalers were in use in the late nineteenth and early twentieth centuries (Clark, 1993). However, the invention of the metering valve for the delivery of respirable aerosol particles from a propellant solution or suspension represents the most significant advance in pharmaceutical aerosol development in recent history. Propellant-driven metered dose inhalers are the standard for unit-dose delivery of potent therapeutic agents to the lungs against which all other systems are measured. This packaging and delivery system was first employed for the delivery of epinephrine (adrenaline) in the 1950s and has since been expanded to more specific beta-adrenergic agonists, such as albuterol (Sciarra and Cutie, 1990).

## 9.1.2   Use of aerosols in the treatment of disease

### 9.1.2.1   Asthma.   Asthma affects as many as 10 million people in the United States and is a major cause of emergency hospitalizations (Frew and Holgate, 1993). Manifestations of this disease are regionally displayed in the lungs and commonly affect the small lower tracheobronchial airways (Smith and Bernstein, 1996). The most widespread symptom of asthma is the bronchospasm which can be treated with topically applied bronchodilators: beta-adrenergic agonists (albuterol, metaproterenol and terbutaline), corticosteroids (beclomethasone, budesonide, triamcinolone) and anticholinergic drugs (ipratropium), all of which are available in aerosol form (Hickey, 1992a). Improved therapeutic approaches utilizing these compounds have been reported in the literature. Recently, a number of new drugs have been discovered and developed to treat asthma. Long-acting potent $beta_2$-adrenergic agonists are currently being marketed (salmeterol, formoterol) (Wallin et al., 1993; Kemp et al., 1993; D'Angio, 1993). These have the advantage over earlier marketed $beta_2$-adrenergic agonists because they can be administered in very small doses and less frequently, improving patient compliance. They are not likely to replace the first-generation beta agonists as first-line agents because of their slower onset of action. The role they may play in asthma therapy will be for treating those patients who experience nocturnal and exercise-induced asthma (Wallin et al., 1993; Kemp et al., 1993).

The most effective therapeutic approach for treatment of acute asthma depends largely upon the optimal administration of pharmacological agents. The treatment of therapeutically controlled chronic asthma depends to a greater extent than acute cases on the arsenal of agents available and combinations. Continuous aerosol administration and frequent intermittent administration are two approaches to treat acute onset asthma attacks. Each method has been studied extensively (Idris et al., 1993; Lin et al., 1993).The utility of $beta_2$-adrenergic agonist aerosol agents and their place in treating acute episodes of asthma will for now remain a first-line defence. The cost-effectiveness of multiple-drug therapies for the treatment of stable asthma has been considered. The addition of a steroid to bronchodilator therapy has been evaluated. It was concluded that the cost of treatment and overall health care were dramatically reduced (van Molken et al., 1993).

The future developments for asthma therapy may include leucotriene receptor antagonists. The proposed target pathway for leucotriene receptor antagonists action is arachidonic acid metabolism. The 5-lipooxygenase metabolites (leucotrienes), are potent constrictors of airway smooth muscle. Inhibition of this pathway by candidate compounds is being explored (Snyder and Fleisch, 1989; Gaddy et al., 1992; Gupta et al., 1995; Altiere and Thompson, 1996). Heparin, a compound usually useful as an anticoagulant, has shown protective action in exercise-induced asthma in high inhalation doses (Lane and Adams, 1993; Ahmed et al., 1993).

9.1.2.2 *Infectious diseases.* There is an urgent need to prevent and treat opportunistic infections in acquired immune deficiency syndrome patients, the leading micro-organism being *Pneumocystis carinii* (Montgomery, 1992). Oral agents are recommended as first-line defence and for prophylaxis, but aerosolized pentamidine remains a proven though less effective alternative prophylactic regimen for those who cannot tolerate or are allergic to the oral trimethoprim/sulphamethoxazole.

Prevention and treatment of invasive pulmonary aspergillosis in neutropenic patients has been attempted using an aerosolized amphotericin B. This was shown to be very well tolerated because a minimal amount was systemically absorbed (Beyer et al., 1993).

9.1.2.3 *Cystic fibrosis.* Cystic fibrosis attacks the lung airways compartmentally but it is displayed throughout the whole tracheobronchial tree (Smith and Bernstein, 1996). The promotion of mucus secretion poses a barrier to the delivery of compounds to the lungs. Excessive mucus in the airways will affect the deposition pattern of aerosols and act as a barrier to drug diffusion (Martonen et al., 1995; Lethem, 1993; Edwards, 1978). After depositing on the epithelial surface of the upper respiratory tract, particles and debris are swept away by ciliated cells of the mucous blanket and transported to the throat to be swallowed or expectorated (Smaldone et al., 1979; Yeates, 1991). Mucociliary clearance reduces the amount of time a formulation is in contact with an absorbing membrane, which may affect the maximum bioavailability of the delivered drug.

Enzymes which are present in the diseased lung of cystic fibrosis patients are likely to be potential problems for peptide-based drugs which may be delivered by aerosols. A promising compound in the arsenal for treatment of stable cystic fibrosis is aerosolized deoxyribonuclease (Ranasinha et al., 1993). It has the ability to improve lung function and has been shown to be efficacious and safe (Smith and Bernstein, 1996). Cystic fibrosis patients experience a variety of pulmonary infections and targeted aerosolization of antimicrobial agents may prove to be beneficial in treating these infections. Establishing the efficacy and safety of these agents will be a key factor for their eventual pharmaceutical production and marketing (Ramsey *et al.*, 1993).

### 9.1.3 *Opportunities of development of this dosage form*

The pulmonary epithelium provides an excellent absorptive surface for aerosolized compounds and may be more suitable for selected compounds than the gastrointestinal tract. Novel compounds most widely regarded as potential agents for delivery via the pulmonary route are the peptide and protein formulations (Adjei and Garren, 1990; Patton and Platz, 1992; Byron and Patton, 1994). Epidemiologically significant disease states may one day be treated using a therapeutic approach which combines the products of biotechnology and aerosol technology. The lung may even be used as a route of administration to enhance the action of traditional drugs. For example, the delivery of morphine to the lung may result in advances in pain control for severely debilitated patients (Tooms, 1993). Advances in pharmaceutical aerosol science in the context of drug disposition in/from the pulmonary region of the lungs will play a major role in development of future drug delivery systems. The physicochemical properties of the aerosols, such as particle size and distribution, density, morphology, hygroscopicity and, in the case of liquids, viscosity, surface tension and evaporation–condensation play a significant role in effective drug delivery and efficacy. Pharmaceutical aerosols can be distinguished from most occupational, environmental and industrial aerosols in that they are generally not stable or in equilibrium with the surrounding atmosphere. The unique properties of these aerosols require that close attention is given to the interpretation of any data collected characterizing their physical properties.

## 9.2  Particulate product manufacture

Particle production occurs by a number of constructive and destructive methods (Jimbo, 1990; Hickey, 1991). The methods most frequently employed by pharmaceutical scientists and engineers are destructive methods. Milling shatters solid particles to reduce their size (Van Cleef, 1991). Solution droplets are generated from the fluid motion in high-velocity airstreams. These droplets may then be dried to form solid particles (Masters, 1991). Recently, an alternative constructive method of particle production has been evaluated for drug particle formulation

(Sacchetti and Van Oort, 1996). Supercritical fluid manufacture, while varying in the precise nature of the process, consists of precipitating particles from solution in liquefied gases.

### 9.2.1 *Milling*

Milling is generally carried out by one of a number of procedures depending upon the required product (Ansel *et al.*, 1995). Hammer milling and ball milling reduce particles to $\sim$50 μm. Air jet or attrition milling reduces particles to $\sim$1–10 μm.

#### 9.2.1.1 *Hammer and ball milling.*   The act of milling particulate products requires the use of tumbling bodies which are usually held in large numbers within a closed container and are moved as the container rotates, shakes, or stirs. Material is introduced between the tumbling bodies and direct pressure is applied by impaction or shear to pulverize particulates using their relative motion.

Figure 9.1a illustrates an impact grinding mill known as a hammer mill. The impact stress on the solid surface is achieved using implements which swing from a rotor. The stress is determined by kinetic energy of motion (velocity 20–200 m s$^{-1}$; Rumpf, 1990). Hammer mills are capable of continuous milling, limited only by the feed and milling rates.

Ball mills, as shown schematically in Figure 9.1b, are used for fine comminution of hard and moderately hard materials and produce high outputs of up to $200 \times 10^3$ kg h$^{-1}$. Ball mills vary in size, ranging from laboratory size to industrial sizes. The grinding chamber consists of a cylindrical vessel fitted with grinding bars on the inside. It is filled 25–45% with balls of the same or various sizes. The extent of filling is determined by the ratio of the bulk volume of the balls to the volume of the grinding

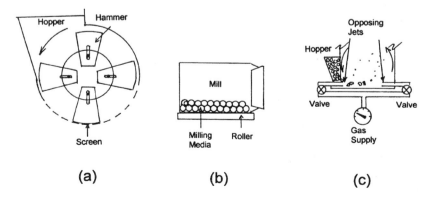

(a)    (b)    (c)

**Fig. 9.1** Schematic diagrams of: (a) hammer; (b) ball; (c) attrition mills.

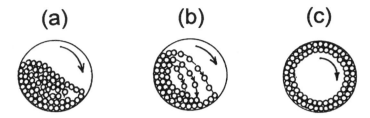

**Fig. 9.2** Schematic diagrams of states of motion in a ball mill: (a) cascading; (b) cataracting; (c) centrifuging.

chamber (Rumpf, 1990). Three types of motion can be displayed by the balls inside a chamber based on the rotational speed and the extent of the filling. These three states of motion are cascading, cataracting and centrifuging, and are depicted schematically in Figure 9.2. If the extent of filling is too small, the balls swing back and forth in a compact mass and no practical comminution will occur. Transitions between the three states of motion occur, usually with all types of motion occurring in the same chamber. Centrifugal motion can be achieved by altering the speed of the rotation to 75% of its critical value. The critical speed ($V$) is defined by the inner diameter ($D$) of the mill and can be determined as follows (Rumpf, 1990; Carstensen, 1993):

$$V = [2g/D]^{0.5}, \tag{9.1}$$

where $g$ is the acceleration due to gravity. The usual rotation speeds of ball mills are 50–80% of the critical speed, $V$. The optimum diameter, $d_b$, of a ball in a mill is

$$d_b = K[D]^{0.5}, \tag{9.2}$$

where $K$ is a constant related to equipment parameters. A higher degree of comminution, a finer powder, is achieved by the use of long milling times, rapid rotation speeds (within the critical limit) and large ball–powder charge ratio. Optimization of the energy required to achieve comminution requires adapting the mill to the material to be ground. Impact energy from the balls must be sufficient to crush the coarse material, but the resultant particles should not be subject to compression or further impaction resulting in fines or aggregates outside the desired particle size range. Therefore, the energy can be adjusted according to the size and number of balls, rotation speed, the charge of material being milled and the milling time. The energy expenditure ($E$) in milling has been defined by Kick's law,

$$E = K \ln[d'/d''], \tag{9.3}$$

where $d'$ and $d''$ are the diameters of the milled powder and the starting bulk powder respectively, and $K$ is a constant. In a system consisting of two fractions $W_a$ [of diameter $d'$] and $W_b$ [of diameter $d''$] where $W$ is the total charge (weight) in the mill, the following relationship is true:

$$\ln(W/W_a) = -(q/C)t \tag{9.4}$$

where $q = E/t$ is constant and $C$ is a constant related to equipment parameters; thus the fraction unmilled follows an exponential decay in time.

9.2.1.2 *Attrition jet milling.* Attrition milling involves particle–particle impaction that results in reduction in particle size, as shown in Figure 9.1c (Hickey, 1991). Opposing jet milling brings about particle impaction by introducing particles on a high-velocity air jet in a direction opposed by an equally large opposing air jet. The resulting turbulence and reduction in particle momentum brings particles together at high velocity with forces capable of producing very small particle sizes. Most commercially available systems employ a cyclone separator that redirects large particles back to the impinging jets and allows small particles to pass into a collection vessel. Jet energy milling is designed to grind crystalline or friable materials to an average particle size of less than 5 µm without contamination or harm to the product. The quantities that can be milled are limited only by the collection vessel and the feed hopper dimension. The rate of milling depends upon the physicochemical properties of the material and the capacity for which the mill was designed. Commercially available attrition mills range in capacity from bench size, which may reasonably be employed to mill gram quantities, up to manufacturing scale, which may produce kilogram quantities. These systems are suitable for most materials including some heat-sensitive substances. Generally, the product obtained requires no further sieving or classifying.

### 9.2.2 *Sieving/cyclone separation*

Separation of dry powders is performed by sifting using a single sieve or a nest of sieves whose apertures decrease from top to bottom (Allen, 1990). Sieves for fine-particle analysis are pans with bottoms of electroformed grids. The cut size is determined essentially by the width of the sieve apertures, as shown in Table 9.1. The size distribution of particles in bulk form can be obtained down to about 25 µm by sieving. The distribution so obtained will be that of the ultimate particles (assuming complete deagglomeration) and not necessarily that of the primary particles. The particle mass retained on each sieve is usually determined by weight. The fraction of particles by weight on each sieve is presented in tabular or graphic form.

**Table 9.1** Sieve standards, indicating mesh size (micrometres) and US and British standards. Values in parentheses are closest approximation to US standard

| Micrometre | US Stand. ASTME 11–61 | Brit. Stand. BS 410: 1969 | Micrometre | US Stand. ASTME 11–61 | Brit. Stand. BS 410: 1969 |
|---|---|---|---|---|---|
| 3360 | 6 | (5) | 297 | 50 | (52) |
| 2830 | 7 | (6) | 250 | 60 | 60 |
| 2380 | 8 | (7) | 210 | 70 | 72 |
| 2000 | 10 | 8 | 177 | 80 | (85) |
| 1680 | 12 | 10 | 149 | 100 | (100) |
| 1410 | 14 | (12) | 125 | 120 | 120 |
| 1190 | 16 | (14) | 105 | 140 | 150 |
| 1000 | 18 | 16 | 88 | 170 | (170) |
| 840 | 20 | (18) | 74 | 200 | (200) |
| 707 | 25 | (22) | 63 | 230 | 240 |
| 595 | 30 | (25) | 53 | 270 | 300 |
| 500 | 35 | 30 | 44 | 325 | (350) |
| 420 | 40 | 36 | 37 | 400 | 400 |
| 354 | 45 | (44) | – | – | – |

The probability of an undersized particle passing a sieve is not unity (Rumpf, 1990). Particle sizes which are nearly equivalent to the mesh width have a greater obstacle to overcome passing through the mesh and most likely will stick in the mesh. If the ratio of the particle size to mesh size lies between 0.8 and 1.0, the aperture of the mesh has a greater chance of being obstructed. Below 0.8 and greater than 1.5, the possibility of blockage is reduced. Large-particle sifting is faster since to some extent these particles can aid the release of trapped granules in the mesh. Feeding too much material onto the sieve causes migration of the fines toward the mesh within the heap of material moving around and can increase the time to achieve sharp cuts. Theoretically, the longer sifting times do improve the sharpness of cut, and raise the cut point. The use of vibration or tapping to aid in passage of particles through a sieve carries a risk of changing the particle size distribution. Particles may be broken by the energy input or aggregate under the influence of vibration (Pietsch, 1991).

Cyclones are devices that accomplish a solid gas separation by imparting a spinning motion to a two-phase system, thereby removing particles and droplets from gas or liquid streams. The particle passes through the cyclone, and from the rotating motion a centrifugal force is created; then the particles travel downward and enter a discharge hopper, as illustrated schematically in Figure 9.3. The movement of gas to entrain particulate materials is required in the spray-drying operation to achieve gas–solid product separation. The gas leaves the cyclone through a top central outlet. For small-particle collection, the diameter of the cyclones is an inch or less but the

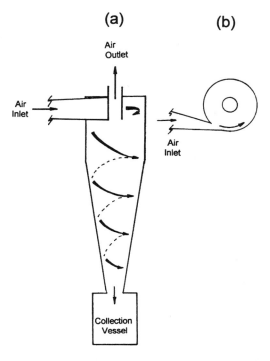

**Fig. 9.3** Schematic diagram of cyclone separator: (a) transverse; (b) horizontal sections.

efficiency of the cyclone will fall off rapidly as the particles become less than 5 μm in diameter (Brookman *et al.*, 1963; Richard, 1976). Figure 9.3 shows a typical cyclone where the solid–gas matter enters the cyclone body tangentially and moves downward in a vortex motion. Cyclones generate much lower radial acceleration than centrifuges; therefore, the larger the cyclone, the less efficient it is. The pressure drop is equal for cyclones if their geometric ratios are equal. Therefore, there is some incentive to arrange several cyclones in parallel operation rather than one. Inlet velocities are normally around 10–25 m s$^{-1}$, and efficiency increases when inlet velocity increases. Entrance velocities in excess of 30 m s$^{-1}$ cause turbulence, leading to bypassing and re-entrainment of separated particles and decrease in efficiency. Cyclones are capable of size-segregating particles in smaller size ranges than sieves down to micrometre-sized material.

### 9.2.3  *Spray drying*

The spray dryer operates on the principle of nozzle spraying in parallel flow (spray and drying air flow in the same direction) or countercurrent flow (spray and drying air flow in opposite directions) (Sacchetti and Van

Oort, 1996). This process is a convection-type drying process (Cook and DuMont, 1988). It is designed to dry the product down, micronize and produce structural changes (crystalline to amorphous) in the powder. Today the most frequently used method for drying aqueous extracts or solutions is freeze drying (lyophilization; Pikal, 1990) which, while effective and frequently appropriate, is rather time-consuming. The result is a soft cake which is usually ground to give a dry but very hygroscopic powder (Schreier *et al.*, 1994). When spray drying is used, the dried powder is obtained within a few minutes in the form of a uniform powder. It is much less hygroscopic than a lyophilized product. The powder obtained should be free flowing to be used directly in dosing, packing or transporting. The moisture content may be required to be in a specified range for quality reasons (shelf life, dispersability and flow).

Both time and temperature are important in thermal degradation of the final product. Ingredients which are sensitive to heat, such as enzymes and antibiotics, retain their activity when spray dried. In convection drying the product adopts the inlet temperature only for a relatively short time (seconds or less). On the other hand, conduction drying allows the product to achieve the hot metal temperature and it may remain at this temperature for several hours. In spray drying, the countercurrent flow pattern can expose the dried material to high temperatures but it can provide the lowest final moisture content for the material. Countercurrent flow is able to overheat materials. Cocurrent or parallel flow exposes the feed to high temperatures and establishes high initial drying rates, but if evaporation occurs too rapidly the product can be damaged.

Particles comprising sprays and dried products are never monodisperse. An atomizer cannot form totally homogeneous sprays. Spray droplets are subjected to different shape distortions depending upon their drying characteristics and travel within the dryer. A number of factors contribute to the final morphology and physicochemical properties of the product. The air inlet temperatures may vary in a wide range, 100–800°C, but the outlet temperatures usually vary in the range 50–150°C (Rumpf, 1990). In addition to temperature the method of atomization may influence the initial droplet size (Lalor *et al.*, 1996). Pneumatic high shear may be contrasted with ultrasonic droplet production and certain benefits may accrue depending upon the method selected. Not only will these methods give rise to different droplet sizes but also the mechanism of production of the droplets may result in effects on drug stability. Drugs that are susceptible to degradation by mechanical shear may be affected by the pneumatic method, while those that are heat liable may be affected by the ultrasonic method. The drying process is the last stage in the production of solid particles and the final size, shape, density, crystallinity and solvent content of the particles are all affected by the conditions in the drying chamber.

Finally, the powder may be collected by a number of methods including cyclone separation, bag filtration and electrostatic precipitation.

9.2.3.1 *Aqueous.* The liquid feed to a spray drier can contain drug in solution or suspension including emulsions and pastes. The resulting dried product will depend upon the physical and chemical properties of the liquid feed and, most importantly, the dryer design and its operation conditions. The technique of spray drying materials has many advantages for material preparation in the pharmaceutical industry. Spray drying has been used to prepare spherical particles loaded with fine drug crystals such that no further processing (crushing or sieving) is necessary. The fine crystals show no crystal growth because of the rapid solvent evaporation in the spray drying process. Spray drying can be widely used in preference to other drying methods for heat-sensitive materials. The foremost advantage of this technique is that both drying and individual particle formation or microencapsulation of the drugs can be accomplished simultaneously.

Atomized liquid will break up into a large number of individual droplets and form a spray (Lefebvre, 1989; Niven, 1996). The energy necessary for this is supplied by Venturi pressure or ultrasonic effects directly supplied from the selected nozzle. During spray–air contact, droplets meet hot air and a rapid moisture evaporation takes place. The instrument must satisfactorily dry the product and remove the particles efficiently. The evaporation of spray droplets is rapid and by the time removal of the product from the chamber occurs, the drying air has cooled sufficiently. The drying chamber and air disperser must create a flow pattern which prevents deposition of partially dried product on the walls and the atomizer. Wall deposition occurs when rapid movement of droplets occurs with insufficient drying times and may also be affected by the spray or plume angle of the nozzle. The particle droplet size, form and density will be the ultimate determinants of the fall characteristics through the air chamber. Small droplets will move with the air current in the dryer and larger particles will be independent of the flow characteristics.

The plume angle from the nozzles employed has significance for efficient drying and collection of the sprayed droplets. The selection of dimensions for a drying chamber for each nozzle are dictated largely by the plume dimensions of the emitted spray. In broad terms, a narrow spray angle will allow a narrow long tube to achieve drying. A broad spray angle will require a tube with large cross-sectional diameter to reduce impaction losses, but this may be short because much of the energy of the spray will be dissipated near the nozzle.

Greater liquid rotation creates a wider spray angle. Increase in spray angle signifies decrease in nozzle capacit. Increase in energy for atomization will create smaller droplets sizes at constant feed rates. Increase in feed viscosity by suspension methods will produce coarser sprays on atomization at fixed atomizer operating conditions. This will also affect evaporation rates by increasing the particle and bulk density. The simplest atomizer is a jet issuing from a round orifice into still air (Lefebvre, 1989). In a swirl nozzle the liquid is given a rotation prior to its emergence from an orifice which aids in droplet dispersion.

9.2.3.2 *Non-aqueous.* In the case of non-aqueous spray drying, which may be required for compounds of low aqueous solubility, the organic vapour must be removed from the solid. Using highly volatile organic liquids facilitates drying unless a solvate is formed or trapping of the solvent in the matrix occurs. Depending upon the nature of the solvent, operating conditions may be restricted by the risk of flammability or explosion, and solvent recovery equipment may be required.

### 9.2.4 *Supercritical fluid manufacture*

Supercritical fluids may be employed to perform a controlled recrystallization of a substance (Phillips and Stella, 1993; Sacchetti and Van Oort, 1996). Supercritical fluids occur in high-pressure and high-temperature regions of the phase diagrams for gases such as carbon dioxide and alkanes. These fluids are characterized by their similarity to liquids under these conditions. The most popular medium for pharmaceutical purposes is carbon dioxide, which has a critical temperature of 31°C. Supercritical fluids may be employed as solvents (rapid expansion supercritical fluid solutions, RESS) or antisolvents (gas antisolvent, GAS) for the substance being recrystallized (Bruno and Ely, 1991; Tom and Debenedetti, 1991). The process of RESS involves preparing a solution in supercritical fluid and rapidly evaporating by pressure reduction, resulting in reduced solubility and precipitation of the drug or excipient. GAS involves adding the supercritical antisolvent to a solution or compound, or vice versa, to induce precipitation. More recently, an interesting approach has been suggested of polymerization in supercritical fluids (Chaffer and DeSimone, 1995). This method appears to offer excellent control of particle size and distribution, although it has yet to be employed to produce large numbers of pharmaceutical products (Schaffer *et al.*, 1996; Canelas *et al.*, 1996).

## 9.3   Pressure packaged metered dose inhalers

### 9.3.1   *Inhalers*

The most readily available commercial aerosol generation device is the metered dose inhaler (MDI). The MDI is made up of five main components, schematically illustrated in Figure 9.4a: a container, a

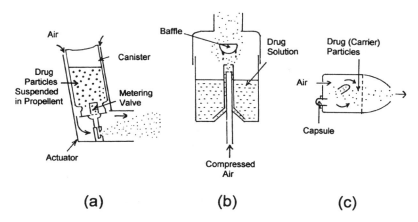

**Fig. 9.4** Schematic diagrams of: (a) pressurized metered dose inhaler; (b) air jet nebulizer; (c) powder inhaler.

metering valve, the liquefied propellants with excipients, the active compound in a suspension or solution form and the actuator which provides the mouthpiece for oral inhalation (Atkins *et al.*, 1992; Hallworth, 1987). MDIs are designed to be very compact and repeatedly to deliver accurate doses ranging from 50 µg to 5 mg (Hallworth, 1987). The volume delivered upon actuation is 25–100 µl in a finely atomized spray over approximately 100–200 ms (Hallworth, 1987).

9.3.1.1 *Non-aqueous suspension/solutions.* The formulation of drug, propellant and excipients in an MDI can exist in two forms, suspension or solution. The suspension is preferred because few compounds are fully soluble in propellants and suspensions exhibit long-term chemical stability. In addition, since cosolvents are not employed the full propellant force is retained for dispersing the aerosol droplets. Suspensions are heterogeneous dispersed systems that may exhibit physical instability (Carstensen, 1993; Patel *et al.*, 1986). Physical stability is very important to the performance of suspension-type MDIs. In dilute suspensions, separation of the phases will occur proportionally to the square of the particle size. Therefore, coarse suspensions are more unstable than dilute suspensions. Increasing the concentration may improve the quality of the suspension, but it will reduce the fine particle fraction. Efficient size reduction and classification of the dry powder component are required for suspension formulations to achieve microfine particles that will deliver the highest dose in a respirable size range.

Suspension-based MDIs must be vigorously shaken before actuation to provide homogeneous, uniform medication delivery. The drug particles are suspended in the propellant by hydrophobic surfactants (oleic acid, sorbitan trioleate, lecithin) and the amount of surfactant is kept extremely low (0.1–2.0% w/w) to avoid particle growth (Hickey et al., 1988). The surfactant stabilizes the suspension by preventing caking, slowing the rate of sedimentation and maximizing sedimentation volume. The extent of surfactant adsorption onto various compounds has been shown to be inversely proportional to the equilibrium moisture content and hydrophobicity (Clarke et al., 1993). This is the main reason why moisture is considered a source of instability of a suspension-based MDI. Excessive amounts of surfactant can adversely affect the particle size of the aerosol cloud by preventing the evaporation of the propellant and may cause undue side effects in the patient (Hickey et al., 1988; Johnson, 1996). Surfactant addition decreases the vapour pressure of the aerosol and lowers the energy required for drying the drug particle (droplet volatilization rate). A coarse aerosol will result and this then may reduce the fine particle fraction (Hallworth, 1987). Surfactant molecules have been implicated in crystallization phenomena (Canselier, 1993). Evidence of crystal formation in a suspension-type MDI will alter the bio-availability of the formulation. Crystal growth can be detected early in formulation development. To maintain a viable product, detection of crystal growth should be performed relatively early in the development stages of the formulation (Phillips et al., 1993). Physical stability can also be impacted by the viscosity of the formulation. This parameter can be a useful indicator of flocculation in aerosol suspensions (Sidhu et al., 1993). Early detection of physical instability can improve the bio-availability, delivered dose and particle size distribution of a suspension-based MDI.

In a solution formulation, the compound must be freely soluble in either volatile propellant or a combination of propellants and acceptable cosolvents (i.e. ethanol). Chlorofluorocarbon (CFC) propellants are considered to be poor solvents for both non-polar and polar compounds (Sanders, 1974). Therefore, it is necessary to add cosolvents to increase their solvent properties and use different blends of propellants to achieve the desired vapour pressure (Byron et al., 1994). Ethanol is used as a cosolvent in solution formulations because it has a high affinity for both non-polar and polar compounds alike and it is miscible with propellants (Herzka, 1966). High concentrations are often necessary when large therapeutic doses are required. This can adversely affect the particle size of the aerosol since a concommitant reduction in vapour pressure occurs. In addition, ethanol is known to act as an irritant to the airways. Ethanol can also cause corrosion of aluminium containers and coated canisters may be required (Sanders, 1974). By varying the amounts of propellant and cosolvents in a formulation, the aerosol quality may change. Cosolvents can alter the vapour pressure and stability of the formulation (Gorman and Carroll, 1993). A higher respirable fraction is obtained from solutions because the aerodynamic size of suspension aerosols cannot be smaller than the initial geometric size of the primary particles (Dalby and Byron, 1988).

Complete delivery in a respirable size range has never been achieved with the current MDIs in solution or in suspension form. Development of a dilute solution formulation in a highly volatile propellant blend has been shown to result in a 40% respirable fraction as compared to only approximately 10% respirable output from a suspension formulation (Dalby and Byron, 1988; Newman *et al.*, 1981b). Use of hydrofluoro-alkane propellants (HFAs) is being considered instead. The two leading candidates from this class of compounds are HFA-134a (1,1,1,2-tetrafluoroethane) and HFA-227 (1,1,1,2,3,3,3-heptafluoropropane) (Dalby *et al.*, 1990; Whitman and Eagle, 1994). These two compounds have similar physicochemical properties to CFC-12. Additionally, they have the most acceptable toxicological and environmental characteristics (Daly, 1992). However, even with these new propellants, a host of new challenges to the formulation of MDIs will ensue because of the significant solubility issues compared to present-day propellants and the incompatibility with surfactants (Dalby *et al.*, 1990). HFA-134a has a similar vapour pressure to CFC-12 which results in high velocity upon expulsion, but its action must be optimized during formulation to minimize upper airway and maximize peripheral lung deposition. It is imperative for future development to test the specific interactions of these propellants with components of the product.

9.3.1.2 *Packaging components.* The drug in the propellant–excipient blend is expelled from the pressurized canister by the metering valve through the actuator. A decline in the effective delivered dose can be related to the valve function. The metering valve is the most critical part in the design of an MDI. Many valve designs have been proposed to achieve efficient actuation (Sanders, 1979; Schultz *et al.*, 1994). The valve must meter doses repeatedly and accurately, providing a good vapour seal throughout the shelf life of the product. It should allow complete opening and closing to the reservoir and atmosphere, alternately as required, and have the capacity to provide an accurate dose upon priming (Fiese *et al.*, 1988; Cyr *et al.*, 1991). The materials that make up the valve and container must be physically and chemically compatible with the active compound, propellants and any necessary excipients.

9.3.1.3 *Droplet formation.* Only a small fraction of the metered dose of aerosol will deposit in the lungs, as the particles are not all in an appropriate or uniform size range and have high inertia. Most of the output at the orifice of the actuator consists of droplets which are relatively large (25 μm) and have high velocities (30–70 m s$^{-1}$) resulting in oropharyngeal deposition (Hickey and Evans, 1996; Byron, 1986b; 1987; Newman, 1983). Generation occurs explosively (at over $30\,\mathrm{m\,s^{-1}}$), and the propellant containing the therapeutic substance disintegrates while passing out of the metering device at high velocity and moves the evaporating particle through the air away from the valve. When the propellant evaporates, the particles undergo very rapid decrease in diameter (Clark, 1993). New MDIs have been introduced which deliver the aerosol at much lower velocity (Gentlehaler, Schering Corporation; see Newman and Clarke, 1993). The difference is found in the actuator, which includes a vortex chamber immediately upstream of the actuator nozzle together with a narrow air inlet in the rear of the device and a horn-shaped internal surface in the mouthpiece. The device is $7\,\mathrm{cm}$ long. These features provide for a spray that emerges at a velocity of less than $2\,\mathrm{m\,s^{-1}}$. The advantage of this device, compared to the use of auxiliary spacers, is its manageable size and retention drug delivery characteristics (it decreases oropharyngeal deposition, and maintains total lung deposition and regional lung deposition). Various approaches have been taken to predicting the performance of aerosol systems each of which is based on a knowledge of the vapour pressure of the propellant formulation, the metered volume and the actuator orifice dimensions (Clark, 1991; Dunbar, 1996; Hickey and Evans, 1996).

9.3.1.4 *Characteristic dose and droplet size output.* Placement of the MDI upon actuation is a determining factor in the amount of drug delivered to the patient. Many studies have been devoted to optimizing the delivery of aerosols from MDI by considering the placement and the proper usage of these devices (Dolovich *et al.*, 1989). It is recognized that as little as 10–20% of the dose from an MDI actually enters the lungs (Dolovich, 1993) even when the particle size emitted is only 2–3 μm (Fults *et al.*, 1991).

9.3.2 *Spacers*

Spacer devices and tube extensions have been developed to allow more time for evaporation of the liquid droplet and for reduction of the velocity of the particles (Newman *et al.*, 1984; 1986a; Dolovich, 1993; Hickey, 1992a). This can increase the airway penetration because the droplet size may fall within a respirable range. Additionally, a spacer device, in combination with the MDI, aids children and the elderly because less hand–breath coordination is necessary. It is important to consider the characteristic aerosol output delivered from an MDI in combination with a spacer device if this is recommended clinically. The spacer will retain a portion of the aerosol and modify the particle size distribution and the dose emitted from the MDI (Newman *et al.*, 1981a). Azmacort® (Rhone-

Poulenc Rorer, Collegeville, PA) and Pulmicort Spacer® (Nebulator®, Astra, Lund, Sweden) are the only devices currently marketed with spacer devices permanently attached to the actuator. Modification of the spacer device may be used in patients with arthritis or an artificial airway (Meeker and Stelmach, 1992; Dolovich et al., 1983; Newman et al., 1989). Physicians have long been aware that metered aerosols are often incorrectly used because of lack of understanding, poor instruction, or difficulty in coordinating respiratory manoeuvres with delivery from these devices. Administration of inhalation drugs for the treatment of asthma and other chronic pulmonary diseases requires the most efficient deposition of the drug in the lungs. Since even the best inhalation devices can deliver only 10–15% (Jenkins et al., 1987; Summer et al., 1989; Turner et al., 1988; Morley et al., 1988) of the aerosol to the lungs, it is clinically important that the patient is instructed and can use the MDI properly. It is not uncommon for the position of the MDI with respect to the mouth to have the greatest influence on the amount of aerosol delivered to the lungs. The trained patient will benefit from proper use of the spacer and the MDI. The combined use of a spacer and an MDI can be demonstrated to be useful in treating acute asthma in emergency settings. Comparison of spacer device administration and nebulizer delivery of albuterol has shown that the use of the spacer devices is as effective in delivery of beta agonists to children with acute asthma as the nebulizer was (Kerem et al., 1993).

9.3.2.1 *Droplet velocity.* The droplet velocity emitted from a spacer approximates the entrained velocity of the patient inspiratory flow. An inspiratory flow rate of less than $1 \, \mathrm{l \, s^{-1}}$ is easily taught and manageable by most patients. The reduced velocity in comparison with the droplet velocity delivered directly from the MDI results in reduced oropharyngeal deposition.

9.3.2.2 *Droplet evaporation.* Evaporation takes place as propellant flash evaporates. This is a complex process that cannot be modelled adequately. The multi-component nature of the heat and mass transfer phenomenon may be ultimately too complex to be fully understood. For example, the motion of the droplets with respect to the ambient air must be known to account for convective heat and mass transport. It is reasonable to suggest that the droplets exhibit little motion relative to their dispersing vapour at the orifice but as the flow is disrupted by interaction with the surrounding air the boundary vapour layer around the droplets must become smaller. This simple view of the plume development is complicated by diffusive mass transfer which depends upon the vapour gradient and thickness of the boundary layer, both of which will change as a function of time and distance from the actuator orifice.

9.3.2.3 *Droplet sedimentation.* Large particles or droplets sediment through the plume and are deposited in the spacer device (Hickey and Evans, 1996). This sedimentation reduces oropharyngeal deposition. This is particularly important for drugs such as corticosteroids which may give rise to local immunosuppression with consequent risk of infection with *Candida albicans*, giving rise to oral candidiasis.

9.3.2.4 *Characteristic dose and droplet size output.* The characteristic dose delivered to the lungs from a spacer accessory to an MDI is similar to that from an MDI alone. The total emitted dose (leaving the spacer) is smaller than that leaving an MDI, since large particles otherwise deposited in the oropharynx are retained in the spacer.

## 9.4  Nebulizers

Nebulizers (air jet or ultrasonic) are used to produce aerosols of pure liquid or solid particles (Niven, 1996). They are most frequently used for acute care of non-ambulatory patients and in infants and children. Solutions of drug candidates in nebulized aerosol droplet form are frequently employed in early formulation development and may ultimately result in marketed products, especially if solid particles are shown to be unstable upon storage. Air jet nebulizers employ the Bernoulli effect to draw solution from a reservoir through a capillary to a region of low pressure induced by a high airflow. Figure 9.4b illustrates a common arrangement of concentric liquid and air ducts. Ultrasonic nebulizers generate aerosols using high-frequency ultrasonic waves (100 kHz and higher) focused in the liquid chamber by a ceramic piezoelectric crystal that mechanically vibrates upon stimulation (Dennis and Hendrick, 1992; O'Doherty *et al.*, 1992). In some instances, an impellor blows the particles out of the nebulizers or the aerosol is inhaled directly by the patient. The ultrasonic nebulizer is capable of greater output than the air jet nebulizer and for this reason is used frequently in aerosol drug therapy. Unfortunately, these devices are expensive and may pose problems when nebulizing certain suspensions and protein formulations (Sato *et al.*, 1992).

### 9.4.1  *Solution/suspension delivery*

The choice between solution and suspension formulations in nebulizers is similar to that in MDIs. Nebulization of suspensions occurs when a salt solution or particles in suspension are used in the generator reservoir. The droplets produced evaporate and the salt residue or the particles suspended in the drops remain to form the aerosol. However, solution formulations are the most common and the only approved products for use in nebulizers. Nebulizer formulations contain water with cosolvents, such as ethanol, glycerin and propylene glycol (Davis, 1978) and surfactants added to improve solubility and stability. An osmotic agent is also added into nebulizers to prevent bronchoconstriction from hypo-osmotic or hyperosmotic solutions (Witeck and Schacter, 1984; Desager *et al.*, 1990). The formulation chosen will affect total mass output and particle size (Masinde and Hickey, 1993; Hickey *et al.*, 1994b). The addition of any agent will have a dramatic effect upon droplet formation, and therefore judicious selection of additives is required.

The initial solution temperature and environmental temperature can have a dramatic effect upon the efficient delivery of compound from a jet nebulizer. There are certain solution compounds (pentamidine) which can precipitate during the atomization process. This is directly caused by the temperature decrease during atomization. If the initial solution temperature is adjusted above room temperature, the problem is not likely to occur (Stelliou *et al.*, 1993).

### 9.4.2  *Spray formation*

Liquid aerosols are typically droplets of spherical shape. The air jet nebulizer operates to produce a thin film of liquid that breaks up into droplets (Mercer *et al.*, 1968) The air enters the nebulizer at a flow rate which depends on the diameter of the orifice and the pressure drop across it. The air expands and passes the mouth of the liquid inlet tube at high speed. A drop in pressure brings about the flow of liquid into the airstream. Acceleration of the droplets causes large ones to impinge on the walls of the nebulizer and return to the bulk solution with a decrease in the temperature of the solution. Water evaporation causes a continuous increase in the concentration of the solute in the liquid remaining in the nebulizer. The small droplets evaporate to form a small particle of solute which is entrained in the airstream. It is this final particle of solute which has a high probability of lung deposition in inhalation therapy. These particles can be sized and an estimate of the initial droplet size ($d_D$) can be obtained from the relation between the initial droplet size and the final particle size ($d_p$) if the density of the droplet ($\rho_D$) and particle ($\rho_p$) and solution concentration ($C$, mass fraction) are known:

$$\tfrac{1}{6}\pi d_p^3 \rho_p = C(\tfrac{1}{6}\pi d_D^3 \rho_D) \tag{9.5}$$

$$d_p = d_D(C\rho_D/\rho_p)^{1/3}. \tag{9.6}$$

The dispersing force in the ultrasonic nebulizer is mechanical energy produced by a piezoelectric crystal vibrating in an electric field produced by an electronic high-frequency oscillator. The energy imparted to the solution of an ultrasonic nebulizer is used to overcome surface tension, disperse the solution droplets and heat the solution. A molecule experiences no net force while in the interior of the bulk phase, but when it is brought to the interface the forces become unbalanced. The forces keeping the molecule in the bulk are known as van der Waals forces.

### 9.4.3   *Characteristic dose and droplet size output*

Doses administered by nebulization are much larger than those from MDIs. The metered doses from a pressure packaged inhaler are limited in size, resulting in short, single-duration therapy (Atkins *et al.*, 1992; Clay *et al.*, 1983a; 1983b). Nebulizers provide very small droplets and high mass output. Baffles within nebulizers remove larger droplets. The droplet size in the airstream is influenced by the compressed air pressure (Newman *et al.*, 1987b). Mass median diameters are normally 2–5 μm with air pressures of 20–30 pounds per square inch gauge. Commercially available air jet nebulizers do not perform equally. This will affect the clinical efficacy of nebulized aerosols which depends primarily on the amount of active substance depositing at various sites in the respiratory tract. Deposition depends on the droplet size (Newman *et al.*, 1986b; Stahlhofen *et al.*, 1983) and total output from the nebulizer (Newman *et al.*, 1983; Faurisson *et al.*, 1995; Clay *et al.*, 1983a; 1983b; Sterk *et al.*, 1984; Phipps and Gonda, 1990; Smye *et al.*, 1992; Langford and Allen, 1993; Matthys and Kohler, 1985) as well as patient parameters such as inspiratory flow rate, respiratory tract morphology and disease state of the lungs. The wide variations in performance from marketed nebulizers are the primary reason why patients experience poor response to the therapy. The performance of nebulizers has been reviewed thoroughly (Dahlback *et al.*, 1986; Phipps and Gonda, 1994; Nerbrink *et al.*, 1994). It has been concluded that nebulizers give rise to various aerosol size distributions, solution temperatures (ultrasonic), solution concentrations and rates of output. Improved construction and closer attention to engineering may result in better output and particle size characteristics (Nerbrink *et al.*, 1994).

The observed evaporation of liquid droplet particles before the aerosol leaves the device is caused by a significant drop in gas temperature (Porstendörfer *et al.*, 1977). For jet nebulizers, the fall in temperature is approximately 5–6 degrees below the ambient temperature at low flow rates and approximately 11–15 degrees at higher flow rates. This decrease in temperature occurs rapidly and is complete in about 4 minutes (Dennis *et al.*, 1991; Ferron and Soderholm, 1990). Ultrasonic nebulizers show an increase in temperature by approximately 18 degrees and this occurs over a much longer period of time – approximately 20 minutes (Phipps and Gonda 1990). The aerosol entering the respiratory tract is saturated with water vapour at the temperature in the nebulizer and must eventually come to equilibrium within the airways (Phipps *et al.*, 1987). Final size will depend on the solute concentration of the droplet solution and the vapour pressure effect associated with small particle curvature (Kelvin effect; Mercer *et al.*, 1968).

The change in concentration of solution in the nebulized aerosol droplets is significantly increased with jet nebulizers because of the fall in temperature and the reduction in dilution air humidity. The ultrasonic nebulizers have little effect on droplet solution concentration. For jet nebulizers the total output of these systems falls during nebulization, this being dependent on the flow rate. Output from ultrasonic nebulizers is much greater than from jet nebulizers. In this case the output can remain relatively constant even though the temperature increases. The droplets formed using ultrasonic nebulizers, which depend upon the frequency, are coarser (i.e. have higher mass median diameter) than those delivered by air jet nebulizers (Newman *et al.*, 1987a; Smaldone *et al.*, 1988).

## 9.5   Dry powder inhalers

The typical dry powder inhaler device consists of two elements: the inhalation appliance to disperse unit doses of the powder formulation into the inspired airstream; and a reservoir of the powder formulation to dispense these doses. An example of dry powder device is shown schematically in Figure 9.4c. The dosing unit can be of three different types. A bulk reservoir allows a precise quantity of powder to be dispensed upon individual dose delivery up to approximately 200 doses, and the unit dose and multiple unit dose provide individual doses (in blister packaging or in gelatine capsule form) for inhalation as required. Dry powder inhalers are known to perform well in comparison with MDIs (Biddiscombe *et al.*, 1993; Borgstrom and Newman, 1993). Nevertheless, the dry powder inhaler has yet to meet its full potential and there is optimism that it may eventually replace MDIs for some indications in response to the international control of CFCs in these latter products. The device does have shortcomings because it can only deliver a fraction of its load in the respirable size range.

### 9.5.1   *Interparticulate forces*

The most important mechanisms of adhesion between particles or between particles and a wall arise through liquid bridges, van der Waals and electrostatic interactions (Rietema, 1991). Liquid that is present between surfaces of two solid bodies develops boundary forces as a result of the surface tension of the liquid (Kiesvaara and Yliruusi, 1993; Kiesvaara *et al.*, 1993). Moreover, a difference in pressure, called the capillary pressure, is established. Adhesive forces always consist of boundary forces and capillary forces. Van der Waals interactive energy is of the order of 0.1 eV and decreases with the sixth power of the distance between the molecules (Fuchs, 1964). This range of interaction is large compared with that of the chemical bond. When two solid surfaces come into contact electrostatic forces of attraction arise as a result of the contact potential generated. Surplus charges can cause attractive or repulsive forces. Additionally, a force that is associated with adhesive forces between particles is the solid bridge. The solid bridge is formed between particles by melting or by a crystallization process.

The surface roughness is influenced by the van der Waals, liquid bridges and electrostatic forces. Electrostatic forces remain with the particles at farther separations than the other two forces do. For ideally smooth surfaces the adhesive forces due to liquid bridges and van der Waals and electrostatic interactions increase linearly with the radius of the sphere for particle–particle interactions. The effect of roughness on adhesive forces is most pronounced in van der Waals force because of their unique short range. This phenomenon is often used in practice when fine particles are blended with larger particles to reduce adhesive forces and reduce agglomeration.

There are other important forces that can act on particles in flow. Field forces are important and include gravitational, electrical and magnetic forces. Aerodynamic and hydrodynamic forces are based on the motion of other particles. These are not constant values and will depend upon the flow being at steady state, the Reynolds number, the degree of turbulence in the fluid, its compressibility, mean free path, proximity of boundary walls and other particle properties (i.e. surface roughness and shape of the particle). Pressure forces – inertial forces, diffusion forces – arise through the direct molecular bombardment of particles (Brownian movement, radiation pressure).

### 9.5.2  *Powder flow properties*

Cohesion is a measure of the stickiness of the solids or a measure of the ratio of adhesive forces between the particles to their weight. The angle of friction can be demonstrated using a container filled to the rim with cohesionless material. Tipped over gradually, the granules begin to flow over the slope as soon as a certain angle of inclination is reached. If the tilt of the container is increased, the angle of inclination of the granular material (i.e. the angle of repose) is maintained. The angle of repose is equal to the angle of friction for cohesionless solids. The gravitational component of force exerted by the uppermost layer of particles on the sloping surface is small. The tangential shear stress is correspondingly small. It is a matter of experience that for cohesive solids the angle of repose is less reproducible and is not a measure of the angle of friction. This is shown when a cohesive material is packed into a container which is then tilted 90° without any of the material falling out. This angle is not possible otherwise in a shear test the shear stress would have to become infinitely large. Flow properties (angle of repose and flow rate) and packaging characteristics were influenced by particle micromorphology, as demonstrated by normalized amplitudes of higher harmonics and fractal dimension (Cartilier and Tawashi, 1993). Micromorphology is as important as shape descriptors in assessment of close packing arrangements.

### 9.5.3  Metering of powders

Dry powder generators are subject to variability because of the physical and chemical properties of the powder. These inhalers are designed to meter doses ranging from 200 µg to 20 mg (Bell and Treneman, 1994). The preparation of drug powder in these devices is very important. The powder in these inhalers requires the same efficient size reduction that has been discussed for suspensions in MDIs. Micronized particles flow and are dispersed more unevenly than coarse particles. Therefore, the micronized drug powder may be mixed with an inert carrier (Byron et al., 1990; Concessio et al., 1996). This carrier is usually α-lactose monohydrate because this product is available in a variety of well-characterized particle size ranges (Byron et al., 1990). The carrier particles have a larger particle size than the therapeutic agent to prevent the excipient from entering the airways. Segregation of the two particles will occur when turbulent airflow is created upon patient inhalation through the mouthpiece. This turbulence of inspiration will provide a certain amount of energy to overcome the interparticulate cohesive and particle surface adhesive forces for the micronized particles to become airborne (Hickey et al., 1994a). High concentrations of drug particles in air are easily attained using dry powder generation, but stability of the output and the presence of agglomerated and charged particles are common problems. With very small particles, dispersion is difficult because of electrostatic, van der Waals, capillary and mechanical forces that increase their energy of association. Powder inhalers will only usually disperse about 10–20% of the contained drug into respirable particles (Bell and Treneman, 1994).

### 9.5.4  Powder dispersion

Particles for investigation can be individual particles or agglomerates. In order fully to disperse a powder as an aerosol it is necessary to supply sufficient energy to a small volume of the bulk powder to separate the particles by overcoming the attractive forces between them. These forces, which depend on the surface properties of the material, are proportional to the first power of the particle diameter but because the separating forces are proportional to the cube of the diameter for vibration and the diameter squared for drag, dispersibility increases rapidly with particle size. There is a size below which particles cannot be satisfactorily dispersed with a given generator (Fuchs, 1964).

Attraction between similar solid particles and electrostatic forces which can act over large distance are very common problems. Dispersion of solids in gases can be facilitated by compensating the charges with ionized gases. Alternatively, the addition of dispersants that carry charge can also disperse the solids. The addition of dispersants can also reduce van der Waals forces.

The hand-held dry powder device is designed to be manipulated to break open the capsule/blister package or to load bulk powder followed by dispersion from the patient's inspiration. Airflow will deaggregate and aerosolize the powder. Many of the devices are designed to pass over baffles, or through tortuous channels that disturb flow properties and further disperse the powder. In most cases, the patient's inspiratory airflow provides the energy to disperse and deagglomerate the dry powder and also determines the amount of medicament that will reach the lungs. There is little coordination involved in using the powder inhalers because the breath of the patient activates the device.

### 9.5.5   *Characteristic dose and particle size output*

The characteristic dose from commercially available dry powder inhalers is dependent upon the patient's inspiratory flow rate (Clark and Egan, 1994). Although the stated dose may range from 200 µg to 5 mg, it is clear that the proportion of this mass reaching the lungs is dependent upon the force of dispersion which is derived in all currently available dry powder inhalers from the breath of the inspiratory flow of the patient. There are other inhalers currently under development that supply the force necessary to generate the powder independently of the patient (suppliers include Dura, San Diego, CA, and Inhale, Palo Alto, CA) by a variety of mechanical methods.

### 9.6   Summary of methods of characterization

### 9.6.1   *Total emitted dose*

The total emitted dose may be obtained by a number of methods which predominantly involve filtration, elutriation and impinger techniques (Hickey, 1992b). Traditionally aerosols have been sampled by absolute,

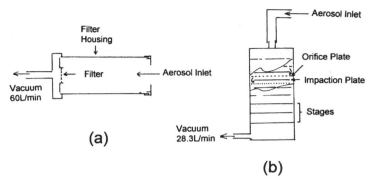

**Fig. 9.5** Schematic diagrams of: (a) a unit dose sampler; (b) a multi-stage inertial sampler.

fibre-glass filters, under airflow conditions that either dictate the sampled volume or are of relevance to the performance of the device. A filter may be mounted in a housing constructed of stainless steel or PTFE through which air carrying the aerosol can be drawn at a fixed flow rate. This type of system is suitable for sampling any of the aerosols described above. One version of this sampler is shown in Figure 9.5a (USP Advisory Panel on Aerosols, 1994). Nebulized droplets present a polar medium to any surface encountered, dry powder aerosols present a range of charge and dielectric composition and MDIs present a predominantly non-polar medium. Since each of the aerosols being characterized is physicochemically unique it may be necessary to consider the composition of the sampling tube and the filter to allow extraction of the drug following administration.

### 9.6.2  Dynamic particle size analysis

Extraction of a representative sample of an aerosol is subject to the influence of inertial, diffusional, gravitational, thermal and electrical forces as well as the heterogeneity of dispersion. It is easier to correct for sampling error in a moving flow than in a stagnant air mass. Terminal settling velocities, at 20°C and atmospheric pressure, of respirable spherical particles in air (10, 1, 0.1 µm) range over four orders of magnitude (18.11, 0.208 and 0.00514 cm min$^{-1}$, respectively). Obtaining representative size analysis for suspended particles is difficult. It will depend upon the material to be analysed and on the purpose of the measurement. The size range, physical and chemical composition need to be known.

9.6.2.1  *Inertial impaction.*  Impaction is ordinarily defined as the deposition of particles from a gas stream on obstacles or surfaces in the path of the stream (Hinds, 1982). Collision arises when the inertia of the particle carries it across the lines of gas flow so that it collides with the obstruction. Collision is not always followed by adhesion or sticking. Impaction occurs primarily at high velocities, but then the chances are much greater that the deposit will be swept away. Particle size, density, gas velocity and obstacle size are the principal factors influencing the efficiency of collection. The impaction device is the primary device to sample and size analyse particles (Andersen, 1958; Hickey, 1992b; Cohen, 1986; Mitchell et al., 1988). Impingers are impaction devices in which collection occurs in a liquid rather than on a solid surface. Particle-laden air flows through a series of orifices, or jets, arranged in order of decreasing size and directed against impaction plates which results in the deposition of particles in overlapping size fractions, as illustrated in Figure 9.5b. Fractions collected by multi-stage impactor can be analysed to determine the size distribution since particles are collected in approximate proportion to size at each stage. This quantity of particles at each stage is usually defined in terms of mass. Calibration is performed with known size and density particles. The size distribution curve on a mass basis is obtained by plotting the sum of one-half the mass collected on a particular stage plus the mass collected on all succeeding stages as a percentage of the total mass collected against the mass mean diameter for that stage. The accuracy of this procedure depends on the magnitude of the overlap of particles sized on one stage and that passed on to succeeding stages. In most cases it is assumed that the data describing the particle size and distribution can be approximated by a lognormal distribution (Raabe, 1971). However, it has been noted that not all particle size distributions can be accurately approximated using this approach (Graham et al., 1994; Thiel, 1994). If a lognormal distribution is assumed the mass median aerodynamic diameter and geometric standard deviation can be derived for the distribution from a graphical fit to the data (Hickey, 1990b; Hickey, 1992b). The advantage of a lognormal distribution is that conversion to other distributions (defined by number or surface) is possible by applying the Hatch–Choate equations (Hinds, 1982).

Pharmaceutical aerosols are particularly sensitive to sampling conditions (Dalby and Tiano, 1993; Hickey, 1995). The geometry of the sampling inlet, sometimes referred to as the induction port (Fults et al., 1991), impaction plate collection surfaces (Hickey, 1990a; 1990b) and airflow sampling conditions (Van Oort, 1996) all affect the calibration of the device and estimates of the particle size.

### 9.6.2.2 Laser diffraction. 

Laser diffraction monitors the characteristic diffraction pattern of particles of specific sizes. According to the principle of Fraunhofer diffraction, particles diffract light to a greater or lesser extent based on their size. By introducing particles into a viewing volume and placing concentric detectors in front of the illuminating laser,

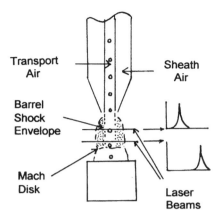

**Fig. 9.6** Schematic diagram of the principles of laser diffraction.

diffracted light can be collected and segregated into regions attributable to various particle sizes. These regions can be defined by calibration with particle standards or using reticles. Figure 9.6 illustrates the general principle of the technique.

Laser light diffraction (LLD) methods are conventionally employed to estimate the size of particles of spherical morphology. The use of this method to estimate the size of irregularly shaped particles is not generally accepted although it has been attempted (Kanerva *et al.*, 1993). In this case LLD was compared with image analysis for the determination of particle size distributions of pharmaceutical powders. In all determinations the LLD method broadened size distributions compared to image analysis. Furthermore, the LLD method overestimated the particle size of spherical particles by 40%. The error increased clearly with increasing particles size. Differences between the methods increased with increasing differences in particle dimensions. It could be concluded that because of significant errors, LLD should be restricted to preliminary studies of particle size distributions and cannot be substituted for other methods.

9.6.2.3 *Laser velocimetry.* Particles are sized by laser velocimetry on the basis of their inertia by accelerating them through a well-defined nozzle and timing their flight between two portions of a split laser beam (Hindle and Byron, 1995). The instrument detects the light scattered by the particles that cross the laser beams, and the signal starts and stops a timer. The time for a particle to travel a known distance at a fixed flow rate is measured and relates to the aerodynamic size. Figure 9.7 illustrates the arrangement of the apparatus for velocimetry measurements. This method is sensitive to particle concentration and samples should be diluted to avoid coincidence errors.

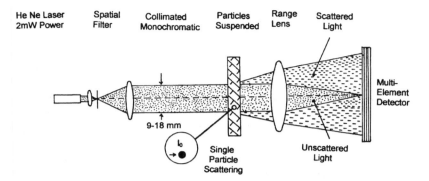

**Fig. 9.7** Schematic diagram of the sensing zone of a laser velocimeter.

#### 9.6.2.4 *Phase Doppler techniques.*

The SPART (single particle aerodynamic relaxation time) analyser measures the aerodynamic diameter of single particles while they are still in the aerosol phase, in real time (Hiller *et al.*, 1980; McCusker *et al.*, 1982). Acoustic Doppler effects are employed to estimate the relaxation time of the particles. The relaxation time is directly related to the terminal settling velocity and in turn the particle diameter.

Phase Doppler Anemometry (PDA) allows simultaneous measurement of particle size and velocity (Ranucci and Chen, 1993; Dunbar, 1996). The laser light is split using a Bragg cell and the two beams are impinged on a viewing volume where they illuminate particles. The presence of the moving particle gives rise to a frequency shift in the intensity pattern of scattered light which is proportional to the particle velocity. In addition, the phase difference in the interference pattern of the light scattered by particles and detected by two photodetectors is proportional to the particle size.

A variety of other laser sizing techniques can be used to assess the *in vitro* performance of aerosol devices. These aerosol analysers operate on the basis of light's interaction with particles making them have the disadvantage of not requiring chemical analysis. The potential for droplets or particles that do not contain drug to be included in estimates of size using these techniques makes it unlikely that they will be adopted for regulatory purposes at this time. A twin-beam phase Doppler analyser has been used to size the cloud delivered from a number of commercially available devices (Clifford *et al.*, 1990). This technique was developed from laser Doppler anemometry, and information about individual particle velocity and size can be obtained simultaneously. There are a number of instruments that use light scattering techniques, and they can characterize particles in the 0.1–100 µm size range. Laser diffraction has been used to evaluate the aerosol clouds emitted from medical nebulizers (Clark, 1995; Ho *et al.*, 1986; Kaye, 1979). The size distribution of a population of particles is determined by weighted volumes rather than counting individual particles entering a fixed measured volume (Kaye *et al.*, 1979). This method offers some advantages over the impaction approach. Sizing can be done directly at the exit of the actuator mouthpiece before appreciable evaporation occurs and in a non-intrusive manner (e.g. Malvern® 2600, Malvern Instruments, MA). This provides a more accurate initial droplet size, facilitating evaluation of the performance differences of devices. The dosimetry of inhaled particles can be determined and the amount of aerosol in each fraction of inhaled and expired air can be measured (Gebhart, 1989). Laser diffraction requires less time because it is fully automated and chemical analysis is not performed. This technique can be used in combination with gamma scintigraphy and in some circumstances compares well with impaction methods. Using a non-volatile aerosol, it was demonstrated that the diffraction technique measures a size parameter that correlates with *in vivo* deposition patterns (Clark, 1995). Many of the single-particle analysers are subject to error as a function of particle concentration. The necessity to dilute aerosols to obtain accurate particle or droplet size estimates brings into question the relevance of these techniques for characterizing the concentrated inhalation aerosols, at least from a regulatory perspective.

### 9.6.3  *Fine particle fraction of the dose*

The fine particle fraction has been used to establish the quality of the aerosol with regard to the proportion of the particles that may be therapeutic. Since this proportion of particles below a certain size can be stated quantitatively, this has historically been treated as good indicator of product uniformity in a potentially therapeutic range. In recent years the implication that this number reflects lung deposition and therapeutic effect has come under attack, particularly from those practising in other health-related fields for which there are current definitions of 'respirable fractions'. This criticism and the increasing concern regarding the particle size distribution within the fine particle fraction may ultimately lead to greater use of data from multi-stage impingers or impactors.

### 9.6.4  *Pharmacopoeial methods*

Current pharmacopoeial methods include a filter system used for unit-dose analysis two-stage impinger and multi-stage impactor methods (USP Advisory Panel on Aerosols, 1994). None of the optical sizing methods is considered suitable for regulatory purposes at this time.

## 9.7 Conclusions

This chapter is intended to give an overview of the rationale for the use of inhalation aerosols in therapy and the key methods of administration and characterization, with some description of the underlying physico-chemical considerations of importance. Some important topics, such as hygroscopic growth (Hickey *et al.*, 1990; Hickey and Martonen, 1993) and particle shape (Hickey, 1990c; Hickey *et al.*, 1992), have not been discussed in this chapter; refer to Chapter 1 for shape and Chapter 5 for hygroscopic growth.

Prediction of the behaviour of an aerosol system usually requires the simultaneous consideration of interactions of the particles with the gas phase, interactions of particles, kinetics of particle behaviour in a specified flow and the motion of the particles with respect to constraining flow boundaries or obstacles in the flowstream.

The criteria for selection of a particular method of delivery of drugs to the lungs depends mainly on physicochemical considerations. In recent times sociopolitical issues have also become a consideration. The current concern over the potential for CFC propellants to deplete atmospheric ozone and their involvement in global warming subjects them to close scrutiny when regulatory approval is sought for products in which they are major components. Although an MDI product may be the first approach to delivering a drug since it is capable of delivering bolus doses (at least 100 µg, 2–3 µm in median diameter) it is unlikely to receive regulatory approval unless a good case for the inclusion of CFC propellants can be made. The recent approval of the alternative propellant HFA-134a in products in Europe and the US (3M Pharmaceuticals, St Paul, MN) makes it likely that the future of MDIs depends on the success of alternative propellant formulations.

Nebulizers produce aqueous droplets 1–2 µm in diameter which are passively inhaled by patients. These aerosols are particularly suitable for treating compromised patients as the droplets can enter the peripheral regions of the lungs readily. It must be noted that these aerosols are usually relatively dilute solutions and since the droplet size is small it requires a considerable period of time to deliver a therapeutic dose (15–30 min). Currently nebulizers are not conveniently used outside the home or clinic. However, potential hand-held systems are currently under development that would expand the use of aqueous systems (Farr *et al.*, 1996; Zierenberg, 1996).

The dry powder inhaler is the logical alternative to the MDI as it delivers a bolus of drug in approximately the same particle size range, 1–5 µm. The production of a stable pounder formulation of peptides or proteins will no doubt promote the development of this method of delivery. The limitations of this approach are the poor dispersability of some powders and their susceptibility to moisture ingress and physico-chemical instability. However, these are areas in which significant progress is being made as the search for alternatives to propellant-driven MDIs continues. The development of aerosol systems with desired physico-chemical properties to achieve optimal therapeutic effect remains a high priority in the fields of pharmaceutics and pharmaceutical technology. Undoubtedly, important fundamental observations regarding the physico-chemical properties of these extremely heterogeneous systems will continue to occur and improve the prospects for future drug delivery to the lungs.

# References

Adjei A. and Garren J. (1990) Pulmonary delivery of peptide drugs: Effect of particle size on bioavailability of leuprolide acetate in healthy male volunteers. *Pharm. Res.*, 7, 565–569.

Ahmed T., Garrigo J. and Danta I. (1993) Preventing bronchoconstriction in exercise induced asthma with inhaled heparin. *N. Engl. J. Med.*, 329, 90–95.

Allen T. (1990) *Particle Size Measurement*, 4th edn. Chapman & Hall, New York

Altiere R.J. and Thompson D.C. (1996) Physiology and pharmacology of the airways, in A.J. Hickey (ed.), *Inhalation Aerosols: Physical and Biological Basis for Therapy*. Marcel Dekker, New York, pp. 85–137.

Andersen A.A. (1958) A new sampler for the collection, sizing and enumeration of viable airborne particles. *J. Bacteriol.*, 76, 471–484.

Ansel H.C., Popovich N.G. and Allen L.V. (1995) *Pharmaceutical Dosage Form and Drug Delivery Systems*, 6th edn. Lea and Febiger, Philadelphia.

Atkins P.J., Barker N.P. and Mathisen D. (1992) The design and development of inhalation drug delivery systems, in A.J. Hickey (ed.), *Pharmaceutical Inhalation Aerosol Technology*. Marcel Dekker, New York, pp. 155–185.

Bell J. and Treneman B. (1994) Design and engineering of a DPI, in P.R. Byron, R.N. Dalby and S.J. Farr (eds), *Proceedings of Respiratory Drug Delivery IV*. Interpharm Press, Buffalo Grove, IL, pp. 93–98.

Beyer J., Barzen G., Risse G., Weyer C., Siegert W. *et al.* (1993) Aerosol amphotericin B for prevention of invasive pulmonary aspergillosis. *Antimicrob. Agents Chemother.*, 37, 1367–1369.

Biddiscombe M.F., Melchor R., Mak V.H., Marriott R.J. and Spiro S.G. (1993) Lung deposition of salbutamol directly labeled with technetium-99m, delivered by pressurized metered dose and dry powder inhalers. *Int. J. Pharm.*, 91, 111–121.

Borgstrom L. and Newman S. (1993) Total and regional lung deposition of terbutaline sulphate inhaled via a pressurized MDI or via Turbuhaler. *Int. J. Pharm.*, 97, 47–53.

Brookman R.S., Phillippi J.F. and Maisch C.L. (1963) Small diameter cyclones, *Chem. Eng. Prog.*, 59(11), 66–69.

Bruno T.J. and Ely J.F. (1991) *Supercritical Fluid Technology: Reviews in Modern Theory and Applications*. CRC Press, Boca Raton, FL.

Byron P.R. (1986a) Prediction of drug residence times in regions of the human respiratory tract following aerosol inhalation. *J. Pharm. Sci.*, 77, 693–695.

Byron P.R. (1986b) Some future perspectives for unit dose inhalation aerosols. *Drug Dev. Ind. Pharm.*, 12, 993–1015.

Byron P.R. (1987) Pulmonary targeting with aerosols. *Pharm. Tech.*, 11 (May), 42–56.

Byron P.R. and Patton J.S. (1994) Drug delivery to the respiratory tract. *J. Aerosol Med.*, 7, 49–75.

Byron P.R., Jashnani R. and Germain S. (1990) Efficiency of aerosolization from dry powder blends of terbutaline sulphate and lactose NF with different particle size distribution. *Pharm. Res.*, 7, S81.

Byron P.R., Miller N.C., Blondino F.E., Visich J.E. and Ward G.H. (1994) Some aspects of alternative propellant solvency, in P.R. Byron, R.N. Dalby and S.J. Farr (eds), *Proceedings of Respiratory Drug Delivery IV*. Interpharm Press, Buffalo Grove, IL, pp. 231–242.

Canelas D.A., Betts D.E. and DeSimone J.M. (1996) Dispersion polymerization of styrene in supercritical carbon dioxide: Importance of effective surfactants. *Macromolecules*, **29**, 2818–2821.

Canselier J.P. (1993) Effects of surfactants on crystallization phenomena. *J. Disper. Sci. Tech.*, **14**, 625–644.

Carstensen J.T. (1993) *Pharmaceutical Principles of Solid Dosage Forms*. Technomic Publishing Co., Lancaster, PA.

Cartilier L.H. and Tawashi R. (1993) Effect of particle morphology on the flow and packing properties of lactose. *STP Pharma. Sci.*, **3**, 213–220.

Clark A.R. (1991) Metered atomization for respiratory drug delivery. PhD thesis, Loughborough University of Technology, UK.

Clark A.R. (1993) Medical aerosol inhalers: past, present and future. *J. Aerosol Med.*, **6**, 224.

Clark A.R. (1995) The use of laser diffraction for the evaluation of the aerosol clouds generated by medical nebulizers. *Int. J. Pharm.*, **115**, 69–78.

Clark A.R. and Egan M. (1994) Modeling the deposition of inhaled powdered drug aerosols. *J. Aerosol Sci.*, **25**, 175–186.

Clarke J.G., Wicks S.R. and Farr S.J. (1993) Surfactant mediated effects in pressurized metered dose inhalers formulated as suspensions. Part I. Drug surfactant interactions in a model propellant system. *Int. J. Pharm.*, **93**, 221–231.

Clay M.M., Pavia D., Newman S.P., Lennard-Jones T. and Clarke S.W. (1983a) Assessment of jet nebulizers for lung aerosol therapy. *Lancet*, **2**, 592–594.

Clay M.M., Pavia D., Newman S.P., Lennard-Jones T. and Clarke S.W. (1983b) Factors influencing the size distribution of aerosols from jet nebulizers. *Thorax*, **38**, 755–759.

Clifford R.H., Ishii I. and Montaser A. (1990) Dual beam light-scattering interferometry for simultaneous measurements of droplet-size and velocity distributions of aerosols from commonly used nebulizers. *Chemistry*, **62**, 309–394.

Cohen B.S. (1986) Introduction: The first 40 years, in J.P. Lodge and T.L. Chan (eds) *Cascade Impactor, Sampling and Data Analysis*. American Industrial Hygiene Association, Akron, Ohio.

Concessio N.M., Van Oort M.M. and Hickey A.J. (1996) Comparison of inertial impaction methods to determine the efficiency of delivery of powder blends from the Rotahaler®. *Pharm. Res.*, **13**, S174.

Cook E. and DuMont H. (1988) Start up of drying systems, *Chem. Proc.*, **11**, 54–59.

Cyr R.D., Graham S.J., Li K.Y.R. and Lovering E.G. (1991) Low first spray drug content in albuterol metered dose inhalers. *Pharm. Res.*, **8**, 658–660.

D'Angio R. (1993) Salmeterol: Long acting inhaled beta adrenergic receptor agonist for the treatment of asthma. *Hosp. Formul.*, **28**, 741–744.

Dahlback M., Nerbrink O., Arborelius M. and Hansson H.C. (1986) Output characteristics from three medical nebulizers. *J. Aerosol Sci.*, **17**, 563–574.

Dalby R.N. and Byron P.R. (1988) Comparison of output particle size distributions from pressurized aerosols formulated as solutions or suspensions. *Pharm. Res.*, **5**, 36–39.

Dalby R.N. and Tiano S.L. (1993) Pitfalls and opportunities in the inertial sizing and output testing of nebulizers. *Pharm. Tech.*, **17**(9), 144–156.

Dalby R.N., Byron P.R., Shepard H.R. and Papadopoulos E. (1990) CFC propellant substitution: P-134a as a potential replacement for P-12 in MDIs. *Pharm. Tech.*, **14**(3), 26.

Daly J.J. (1992) Replacement for CFC propellants: A technical and environmental overview. *J. Biopharm. Sci.*, **3**, 265.

Davis S.S. (1978) Physico-chemical studies on aerosol solutions for drug delivery: I. Water-propylene glycol systems. *Int. J. Pharm.*, **1**, 71–83.

Dennis J.H. and Hendrick D.J. (1992) Design characteristics for drug nebulizers. *J. Med. Eng. Tech.*, **16**, 63–68.

Dennis J.H., Stenton S.C., Beach J.R., Avery A.J., Walters E.H. and Hendrick D.J. (1991) Jet and ultrasonic nebulizer output: Use of a new method for direct measurement of aerosol output. *Thorax*, **46**, 151–152.

Desager K.N., Van Bever H.P. and Stevens W.J. (1990) Osmolality and pH of antiasthmatic drug solutions. *Agents and Actions*, 31, 225–228.

Dolovich M. (1993) Lung dose, distribution, and clinical response to therapeutic aerosols. *Aerosol Sci. Tech.*, 18, 230–240.

Dolovich M., Ruffin R., Corr D. and Newhouse M.T. (1983) Clinical evaluation of a simple demand inhalation MDI aerosol delivery device. *Chest*, 84, 36–41.

Dolovich M., Ruffin R.E., Robers R. and Newhouse M.T. (1989) Optimal delivery of aerosols from MDI. *Chest*, 80, 911–915.

Dunbar C.A. (1996) An experimental and theoretical investigation of the spray issued from a pMDI. PhD thesis, UMIST, Manchester, UK.

Edwards P.A.W. (1978) Is mucus a selective barrier to macromolecules? *Br. Med. Bull.*, 34, 55–56.

Farr S.J., Schuster J.A., Lloyd P., Lloyd L.J., Okikawa J.K. and Rubsamen R.M. (1996) Aer$_x$-Development of a novel liquid aerosol delivery system: Concept to clinic, in R.N. Dalby, P.R. Byron and S.J. Farr (eds), *Respiratory Drug Delivery V*. Interpharm Press, Inc., Buffalo Grove, IL, pp. 175–185.

Faurisson F., Dessanges J.F., Grimfeld A., Beaulieu R., Kitzis M.D., Peytavin G., Lefebvre J.P., Farinotti R., Sautegeau A. (1995) Nebulizer performance: AFLM study. *Respiration*, 62, Suppl., 13–16.

Ferron G.A. and Soderholm S.C. (1990) Estimation of the times for evaporation of pure water droplets and for stabilization of salt solution particles. *J. Aerosol Sci.*, 21, 415–429.

Fiese E.F., Gorman W.G., Dolinsky D., Harwood R.J., Hunke W.A., Miller N.C., Minzer H. and Harper N.J. (1988) Test method for evaluation of loss of prime in metered dose aerosols. *J. Pharm. Sci.*, 77, 90–93.

Frew A.J. and Holgate S.T. (1993) Clinical pharmacology of asthma: Implications for treatment. *Drugs*, 46, 847–862.

Fuchs, N.A. (1984) *The Mechanics of Aerosols*. Pergamon Press, New York.

Fults K., Cyr T. and Hickey A.J. (1991) The influence of sampling chamber dimensions on aerosol particle size measurement by cascade impactor and twin impinger. *J. Pharm. Pharmacol.*, 43, 726–728.

Gaddy J.N., Margolskee D.J., Bush R.K., Williams V.C. and Busse W.W. (1992) Bronchodilation with a potent and selective leukotriene D4 (LTD4) receptor antagonist in patients with asthma. *Am. Rev. Respir. Dis.*, 146, 358–363.

Gebhart J. (1989) Dosimetry of inhaled particles by means of light scattering, in J.D. Crapo *et al.* (eds) *Extrapolation of Dosimetric Relationships for Inhaled Particles and Gases*. Academic Press, New York, pp. 235–245.

Goodman L.S. and Gilman A.G. (1985) *The Pharmacological Basis of Therapeutics*, 7th edn. (eds. A.G. Gilman, L.S. Goodman, T.W. Rall and F. Murad). MacMillan Publishing Company, New York, pp. 169–170 and 131–138.

Gorman W.G. and Carroll F.A. (1993) The internal pressure of small volume metered dose aerosols. *Pharm. Tech.*, 17(8), 24–58.

Graham S.J., Lawrence R.C. and Hickey A.J. (1994) Aerosol particle size statistics. *Pharm. Res.*, 11, S39.

Gupta P., Cheskin H. and Adjei A. (1995) Pulmonary bioavailability and absorption characteristics of the 5-lipoxygenase inhibitor, Abbott-79175, in beagle dogs. *Int. J. Pharm.*, 115, 95–102.

Hallworth G.W. (1987) The formulation and evaluation of pressurized metered-dose inhalers, in D. Ganderton and T. Jones (eds), *Drug Delivery to the Respiratory Tract*. Ellis Horwood Inc., VCH Publishers, New York, pp. 87–118.

Herzka A. (1986) *International Encyclopedia of Pressurized Packaging (Aerosols)*. Pergamon Press, New York.

Hickey A.J. (1990a) An investigation of size deposition upon individual stages of cascade impactor. *Drug Dev. Ind. Pharm.*, 16, 1911–1929.

Hickey A.J. (1990b) Factors influencing aerosol deposition in inertial impactors and their effect on particle size characterization. *Pharm. Tech.*, 14(9), 118–130.

Hickey A.J. (1990c) The effect of hydrophobic coatings upon the behavior of pharmaceutical aerosol powders, in S. Masuda and K. Takahashi (eds), *Aerosols: Science, Industry, Health and Environment*. Pergamon Press, Oxford, pp. 1315–1318.

Hickey A.J. (1991) Lung deposition and clearance of pharmaceutical aerosols: What can be learned from inhalation toxicology and industrial hygiene? *Aerosol Sci. Tech.*, **18**, 290–304.

Hickey A.J. (1992a) Summary of common approaches to pharmaceutical aerosol administration, in A.J. Hickey (ed.), *Pharmaceutical Inhalation Aerosol Technology.* Marcel Dekker, New York, pp. 155–185.

Hickey A.J. (1992b) Methods of aerosol particle size characterization, in A.J. Hickey (ed.), *Pharmaceutical Inhalation Aerosol Technology.* Marcel Dekker, New York, pp. 219–254.

Hickey A.J. (1995) Sampling inlet considerations in inertial particle-size analysis of inhalation aerosols. *Pharm. Tech.*, **19**(3), 58–68.

Hickey A.J. and Evans R. (1996) Aerosol generation from propellant driven metered dose inhalers, in A.J. Hickey (ed.), *Inhalation Aerosols; Physical and Biological Basis for Therapy.* Marcel Dekker, New York, pp. 417–439.

Hickey A.J. and Martonen T.B. (1993) Behavior of hygroscopic pharmaceutical aerosols and the influence of hydrophobic additives. *Pharm. Res.*, **10**, 1–7.

Hickey A.J., Dalby R.N. and Byron P.R. (1988) Effects of surfactants on aerosol powders in suspension. Implications for airborne particle size. *Int. J. Pharm.*, **42**, 267–270.

Hickey A.J., Gonda I., Irwin W.J. and Fildes F.J.T. (1990) Effect of hydrophobic coating on the behavior of a hygroscopic aerosol powder in an environment of controlled temperature and relative humidity. *J. Pharm. Sci.*, **79**, 1009–1014.

Hickey A.J., Fults K.A. and Pillai R.S. (1992) Use of particle morphology to influence the delivery of drugs from dry powder aerosols. *J. Biopharm. Sci.*, **3**, 107–113.

Hickey A.J., Concessio N.M., Van Oort M.M. and Platz R.M. (1994a) Factors influencing the dispersion of dry powders as aerosols. *Pharm. Tech.*, 18(Aug.), 58–64.

Hickey A.J., Kuchel K. and Masinde L.E. (1994b) Comparative efficiency of solution and suspension output from jet nebulizers, in P.R. Byron, R.N. Dalby and S.J. Farr (eds), *Proceedings of Respiratory Drug Delivery IV.* Interpharm Press, Buffalo Grove, IL, pp. 259–263.

Hiller F.C., Mazumder M.K., Wilson J.D. and Bone R.C. (1980) Effect of low and high relative humidity on metered dose bronchodilator solution and powder aerosols. *J. Pharm. Sci.*, **69**, 334–337.

Hindle M. and Byron P.R. (1995) Size distribution control of raw materials for dry-powder inhalers using the Aerosizer with the Aero-disperser. *Pharm. Tech.*, **19**(6), 64–78.

Hinds W.C. (1982) *Aerosol Technology,* Wiley, New York.

Ho K.K.L., Kellaway I.W. and Tredree R.L. (1986) Particle size analysis of nebulized aerosols using Fraunhofer laser diffraction and inertial impaction methods. *J. Pharm. Pharmacol.*, **38**, Suppl., p. 26.

Hollinger M.A. (1985) Basic lung structure and function, in *Respiratory Pharmacology and Toxicology.* W.B. Saunders, Philadelphia, pp. 1–20.

Hubbard R.C., Casolaro M.A., Mitchell M., Sellers S.E., Arabia R., Matthay M.A. and Crystal R.G. (1989) Fate of aerosolized recombinant DNA-produced α1-antitrypsin: Use of the epithelial surface of the lower respiratory tract to administer proteins of therapeutic importance. *Pro. Nat. Acad. Sci. USA*, **86**, 680–684.

Idris, A.H., McDermott M.F., Raucci J.C., Morrabel A., McGorray S. and Hendeles L. (1993) Emergency Department treatment of severe asthma. Metered dose inhaler plus holding chamber is equivalent in effectiveness to nebulizer. *Chest*, **103**, 665–672.

Jenkins S.C., Healtin R.W., Fulton T.J., Maxham J. (1987) Comparison of domiciliary use of a nebulizer and MDI to deliver salbutamol in stable chronic airflow limitation. *Chest*, **91**, 804–807.

Jimbo G. (1990) Funtai, powder, particle and beyond, in *Second Congress on Particle Technology,* Society of Powder Technology of Japan, pp. 1–23.

Johnson K.A. (1996) Interfacial phenomena and phase behavior in metered dose inhaler formulations, in A.J. Hickey (ed.), *Inhalation Aerosols: Physical and Biological Basis for Therapy.* Marcel Dekker, New York, pp. 385–415.

Kanerva H., Kiesvaara J., Muttonen E. and Yliruusi J. (1993) Use of laser light diffraction in determining the size distributions of different shaped particles. *Pharm. Ind.*, **55**, 849–853.

Kaye B.H. (1979) Characterization of fine particle systems by utilizing diffraction pattern analysis, in *Proceedings of the 2nd European Symposium Particle Characterization*, Nuremberg, Germany, pp. 626–644.

Kemp J.P., Bierman C.W., Cocchetto D.M. (1993) Dose response study of inhaled salmeterol in asthmatic patients with 24 hour spirometry and holter monitoring. *Ann. Allergy*, **70**, 316.

Kerem K., Levison H., Schuh S., O'Brodovich H., Canny G. *et al.* (1993) Efficacy of albuterol administered by nebulizer versus spacer device in children with acute asthma. *J. Pediatr.*, **123**, 313–317.

Kiesvaara J. and Yliruusi J. (1993) Use of the Washburn method in determining the contact angles of lactose powders. *Int. J. Pharm.*, **92**, 81–88.

Kiesvaara J., Yliruusi J. and Ahomaki E. (1993) Contact angles and surface free energies of theophylline and salicylic acid powders determined by the Washburn method. *Int. J. Pharm.*, **97**, 101–109.

Lalor C.B., Concessio N.M. and Hickey A.J. (1996) Spray drying performance characteristics of airblast and ultrasonic nozzles. *Pharm. Res.*, **13**, S179.

Lane D.A. and Adams L. (1993) Non-anticoagulant uses of heparin. *N. Engl. J. Med.*, **329**, 129–130.

Langford S.A. and Allen M.B. (1993) Salbutamol output from two jet nebulizers. *Resp. Med.*, **87**, 99–103.

Lansley A.B. (1993) Mucociliary clearance and drug delivery via the respiratory tract. *Adv. Drug Delivery Rev.*, **11**, 299–327.

Lefebvre A. (1989) *Atomization and Sprays*. Hemisphere, New York.

Lethem M.I. (1993) The role of tracheobronchial mucus in drug administration to the airways. *Adv. Drug Del. Rev.*, **11**, 271–298.

Lin R.Y., Smith A.J. and Hergenroeder P. (1993) High serum albuterol levels and tachycardia in adult asthmatics treated with high dose continuously aerosolized albuterol. *Chest*, **103**, 221–225.

Martonen T.B., Katz, I. and Cress W. (1995) Aerosol deposition as a function of airway disease: Cystic fibrosis. *Pharm. Res.*, **12**, 96–102.

Masinde L.E. and Hickey A.J. (1993) Aerosolized aqueous suspensions of poly(L-lactic acid) microspheres. *Int. J. Pharm.*, **100**, 123–131.

Masters K. (1991) *Spray Drying Handbook*, 5th edn. Wiley, New York.

Matthys H. and Kohler D. (1985) Pulmonary deposition of aerosols by different mechanical devices. *Respiration*, **48**, 269–276.

McCusker K., Hiller F.C., Wilson J.D., McLeod P., Sims R. and Bone R.C. (1982) Dilution of cigarette smoke for real time aerodynamic sizing with a SPART analyser. *J. Aerosol Sci.*, **13**, 103–110.

Meeker D.P. and Stelmach K. (1992) Modification of the spacer device: use in the patient with arthritis or an artificial airway. *Chest*, **102**, 1243–1244.

Mercer T.T., Tillery M.I., Chow H.Y. (1968) Operating characteristics of some compressed air nebulizers. *Am. Ind. Hyg. Assoc. J.*, **29**, 66–78.

Mitchell J.P., Costa P.A. and Waters S. (1988) An assessment of an Andersen Mark-II Cascade Impactor. *J. Aerosol Sci.*, **19**, 213–221.

Morley T.F., Marozsan E., Zappasodi S., Gordon R., Griesback R. and Giudice J. (1988) Comparison of beta adrenergic agents delivered by nebulizer vs metered dose inhaler with InspirEase in hospitalized asthmatic patients. *Chest*, **94**, 1205–1210.

Montgomery A.R. (1992) Aerosolized pentamidine for the treatment and prophylaxis of *Pneumocystis carinii* pneumonia in patients with acquired immunodeficiency syndrome, in A.J. Hickey (ed.), *Pharmaceutical Inhalation Aerosol Technology*. Marcel Dekker, New York, pp. 307–320.

Nerbrink O., Dahlback M. and Hansson H.C. (1994) Why do medical nebulizers differ in their output and particle size characteristics? *J. Aerosol Med.*, **7**, 259–276.

Newman S.P. (1983) *Deposition and Effects of Inhalation Aerosols*. Astra Pharmaceuticals, Lund, Sweden, pp. 63–72.

Newman S.P. and Clarke S.W. (1993) Bronchodilator delivery from Gentlehaler, a new low velocity pressurized aerosol inhaler. *Chest*, **103**, 1442–1446.

Newman S.P., Moren F., Pavia D., Little F. and Clarke S.W. (1981a) Deposition of pressurized suspension aerosols inhaled through extension devices. *Am. Rev. Res. Dis.*, **124**, 317–320.

Newman S.P., Pavia D., Moren F., Sheahan N.F. and Clarke S.W. (1981b) Deposition of pressurized aerosols in the human respiratory tract. *Thorax*, **36**, 52–55.

Newman S.P., Pellow P.G., Clay M.M. and Clarke S.W. (1983) Therapeutic aerosols: Physical and practical considerations. *Thorax*, **38**, 755–759.

Newman S.P., Millar A.B., Lennard-Jones T.R., Moren F. and Clarke S.W. (1984) Improvement of pressurized aerosol deposition with Nebuhaler™ spacer device. *Thorax*, **39**, 935–941.

Newman S.P., Woodman G., Clarke S.W. and Sackner M.A. (1986a) Effect of InspirEase on the deposition of metered dose aerosols in the human respiratory tract. *Chest*, **89**, 551–556.

Newman S.P., Pellow P.G.D. and Clarke S.W. (1986b) Droplet size distributions of nebulized aerosols for inhalation therapy. *Clin. Phys. Physiol. Meas.*, **7**, 139–146.

Newman S.P., Pellow P.G.D. and Clarke S.W. (1987a) In vitro comparison of Devilbiss jet and ultrasonic nebulizers. *Chest*, **92**, 991–994.

Newman S.P., Pellow P.G.D. and Clarke S.W. (1987b) Flow pressure characteristics of compressors used for inhalation therapy. *Eur. J. Respir. Dis.*, **71**, 122–126.

Newman S.P., Clark A.R., Talalee N. and Clarke S.W. (1989) Pressurized aerosol deposition in the human lung with and without an 'open' spacer device. *Thorax*, **44**, 706–710.

Niven R.W. (1996) Atomization and nebulizers, in A.J. Hickey (ed.), *Inhalation Aerosols; Physical and Biological Basis for Therapy*. Marcel Dekker, New York, pp. 273–312.

O'Doherty M.J., Thomas S.H., Page C.J., Treacher D.F. and Nunan T.O. (1992) Delivery of a nebulized aerosol to a lung model during mechanical ventilation: Effect of ventilator settings and nebulizer type, position, and volume fill. *Am. Rev. Resp. Dis.*, **146**, 383–387.

Patel N.K., Kennon L. and Levinson R.S. (1986) Pharmaceutical suspensions, in L. Lachman, H.A. Lieberman and J.L. Kanig (eds), *The Theory and Practice of Industrial Pharmacy*, 3rd edn. Lea and Febiger, Philadelphia, pp. 479–502.

Patton J.S. and Platz R.M. (1992) Pulmonary delivery of peptides and proteins for systemic action. *Adv. Drug Del. Rev.*, **8**, 179–196.

Phillips E.M. and Stella V.J. (1993) Rapid expansion from supercritical solutions: Application to pharmaceutical processes. *Int. J. Pharm.*, **94**, 1–10.

Phillips E.M., Byron P.R. and Dalby R.N. (1993) Axial ratio measurements for early detection of crystal growth in suspension type metered dose inhalers. *Pharm. Res.*, **10**, 454–456.

Phipps P.R. and Gonda I. (1990) Droplets produced by medical nebulizers: Some factors affecting their size and solute concentration. *Chest*, **97**, 1327–1332.

Phipps P.R. and Gonda, I. (1994) Evaporation of aqueous aerosols produced by jet nebulizers: Effects on particle size and concentration of solution in the droplets. *J. Aerosol Med.*, **7**, 239–258.

Phipps P., Borham P., Gonda, I., Bailey D., Bautovich G. and Anderson S. (1987) A rapid method for the evaluation of radioaerosol delivery systems. *Eur. J. Nuc. Med.*, **13**, 183–186.

Pietsch W. (1991) *Size Enlargement by Agglomeration*. Wiley, New York.

Pikal M. (1990) Freeze-drying of proteins, Part 1: Process design. *BioPharm.*, **3**(8), 18–27.

Porstendörfer J., Gebhart J. and Röbig G. (1977) Effect of evaporation on the size distribution of nebulized aerosols. *J. Aerosol Sci.*, **8**, 371–380.

Raabe O.G. (1971) Particle size analysis using grouped data and the lognormal size distribution. *J. Aerosol Sci.*, **2**, 289–303.

Ramsey B.W., Dorkin H.L., Eisenberg J.D., Gibson R.L. and Smith A.L. (1993) Efficacy of aerosolized tobramycin in patients with cystic fibrosis. *N. Engl. J. Med.*, **328**, 1740–1746.

Ranasinha C., Assoufi B., Shak S., Christiansen D. and Hodson M. (1993) Efficacy and safety of short term administration of aerosolized recombinant human DNAse I in adults with stable stage cystic fibrosis. *Lancet*, **342**, 199–202.

Ranucci J.A. and Chen F.C. (1993) Phase doppler anemometry: A technique for determining aerosol plume-particle size and velocity. *Pharm. Tech.*, **17**(6), 62–74.

Richard D. (ed.) (1976) *Handbook on Aerosols*. Technical Information Center, USERDA, Oak Ridge, TN.

Rietema, K. (1991) *The Dynamics of Fine Powders*, Elsevier Science, New York.

Rumpf H. (1990) *Powder Technology* (transl. F.A. Bull). Chapman & Hall, New York.

Sacchetti M. and Van Oort M.M. (1996) Spray drying and supercritical fluid particle generation techniques, in A.J. Hickey (ed.), *Inhalation Aerosols; Physical and Biological Basis for Therapy*. Marcel Dekker, New York, pp. 337–384.

Sanders P.A. (1974) Propellants and solvents, in J.J. Sciarra and L. Stoller (eds), *The Science and Technology of Aerosol Packaging*. Wiley, New York, pp. 97–150.

Sanders P.A. (1979) *Handbook of Aerosol Technology*. Van Nostrand Reinhold Co., New York, pp. 85–115.

Sato Y., Sato M., Kita M. and Kishida T. (1992) A comparison of 28 kHz and 160 kHz ultrasonic aerosolization of interferon-alpha. *J. Aerosol Med.*, **5**, 59.

Schaffer K.A. and DeSimone J.M. (1995) Chain polymerizations in inert near- and supercritical fluids. *TRIP*, **3**, 146–153.

Schaffer K.A., Jones T.A., Canelas D.A. and DeSimone J.M. (1996) Dispersion polymerizations in carbon dioxide using siloxane-based stabilizers. *Macromolecules*, **29**, 2704–2706.

Sciarra J.J. and Cutie A.J. (1990) Pharmaceutical aerosols, in G.S. Banker and C.T. Rhodes (eds), *Modern Pharmaceutics*. Marcel Dekker, New York, pp. 605–634.

Schreier H., Mobley W.C., Concessio N.M., Hickey A.J. and Niven R.W. (1994) Formulation and in-vitro performance of liposome powder aerosols. *s.t.p. pharma sciences*, **4**, 38–44.

Schultz R.K., Dupont R.L. and Ledoux K.A. (1994) Issues surrounding metered dose valve technology: past, present and future perspectives, in P.R. Byron, R.N. Dalby and S.J. Farr (eds), *Proceedings of Respiratory Drug Delivery IV*. Interpharm Press, Buffalo Grove, IL, pp. 211–219.

Sidhu B.K., Washington C., Davis S.S. and Purewal T.S. (1993) Rheology of model aerosol suspensions. *J. Pharm. Pharmacol.*, **45**, 597–600.

Smaldone G.C., Itoh H., Swift D.L. and Wagner H.N. (1979) Effect of flow-limiting segments and cough on particle deposition and mucociliary clearance in the lung. *Am. Rev. Resp. Dis.*, **120**, 747–758.

Smaldone G.C., Perry F.J. and Deutch D.G. (1988) Characteristics of nebulizers used in the treatment of AIDS-related *Pneumocystis carinii* pneumonia. *J. Aerosol Med.*, **1**, 113–126.

Smith S.J. and Bernstein J.A. (1996) Therapeutic uses of lung aerosols, in A.J. Hickey (ed.), *Inhalation Aerosols: Physical and Biological Basis for Therapy*. Marcel Dekker, New York, pp. 233–269.

Smye S.W., Jollie M.I., Cunliffe H. and Littlewood J.M. (1992) Measurement and prediction of drug solvent losses by evaporation from jet nebulizers. *Clin. Phys. Physiol.*, **13**, 129–134.

Snyder D.W. and Fleisch J.H. (1989) Leukotriene receptor antagonists as potential therapeutic agents. *Annual Rev. Pharmacol. Toxicol.*, **29**, 123–143.

Stahlhofen W., Gebhart J., Heyder J. and Scheuch G. (1983) Deposition pattern of droplets from medical nebulizers in the human respiratory tract. *Bull. Eur. Physiopath. Resp.*, **19**, 459–463.

Stelliou I., Venthoye G. and Taylor K.M. (1993) Manipulating the temperature of pentamidine isethionate solutions in jet nebulizers. *Int. J. Pharm.*, **99**, R1–R3.

Sterk P.J., Plomp A., Van de Vate J.F. and Quanjer P.H. (1984) Physical properties of aerosols produced by several jet and ultrasonic nebulizers. *Clin. Respir. Physiol.*, **20**, 65–72.

Summer W., Elston R., Tharpe L., Nelson S. and Haponik E. (1989) Aerosol bronchodilator delivery methods. *Arch. Intern. Med.*, **149**, 618–623.

Thiel C.G. (1994) A pitfall in the use of the log-probability curve for particle size calculations and a solution to the problem. *Pharm. Res.*, **11**, S39.

Tom J.W. and Debenedetti P.G. (1991) Particle formation with supercritical fluids – a review. *J. Aerosol Sci.*, **22**, 555–584.

Tooms A., McKenzie A. and Grey H. (1993) Nebulized morphine. *Lancet*, **342**, 1123–1124.

Turner J., Corkery K., Eckman D., Gelv A., Lipavasky A. and Sheppard D. (1988) Equivalence of continuous flow nebulizer and MDI with reservoir bag for treatment of acute airflow obstruction. *Chest*, **93**, 476–481.

USP Advisory Panel on Aerosols (1994) Recommendations on the USP general chapters on aerosols (601) and uniformity of dosage units (905). *Pharm. Forum*, **20**, 7477–7504.

Van Oort M.M., Downy B. and Roberts W. (1996) Verification of operating the Andersen cascade impactor at different flow rates. *Pharm. Forum*, **22**, 2211-2215.

Van Cleef J. (1991) Powder technology. *Am. Scientist*, **79**, 304–315.

Van Molken M.P., van Doorslaer E.K., Jansen M.C., van Essenzandvliet E.E. and Rutten F.F. (1993) Cost effectiveness of inhaled corticosteroid plus bronchodilator therapy versus bronchodilator monotherapy in children with asthma. *PharmacoEcon*, **4**, 257–270.

Wallin A., Sandstrom T., Rosenhall L. and Melander B. (1993) Time course and duration of bronchodilation with formoterol in patients with stable asthma. *Thorax*, **48**, 611–614.

Whitman M.E. and Eagle A.M. (1994) Alternative propellants: Proprietary rights, toxicological issues and projected licensing problems, in P.R. Byron, R.N. Dalby and S.J. Farr (eds), *Proceedings of Respiratory Drug Delivery IV*. Interpharm Press, Buffalo Grove, IL, pp. 203–209.

Witeck T.J. and Schacter E.N. (1984) Detection of sulfur dioxide in bronchodilator aerosols. *Chest*, **86**, 592–594.

Yeates D.B. (1991) Mucus rheology, in R.G. Crystal and J.B. West (eds), *The Lung: Scientific Foundations. Volume 1*. Raven Press, New York, pp. 197–203.

Yeates D.B., Gerrity T.R. and Garrard C.S. (1981) Particle deposition and clearance in the bronchial tree. *Ann. Biomed. Eng.*, **9**, 577–592.

Zierenberg B., Eicher J., Dunne S. and Freund B. (1996) Boehringer Ingelheim nebulizer Bineb®: a new approach to inhalation therapy, in P.R. Byron and S.J. Farr (eds), *Respiratory Drug Delivery V*. Interpharm Press, Inc., Buffalo Grove, IL, pp. 1887–1893.

# 10  Computational fluid dynamics in particle sampling

D.B. INGHAM

## 10.1  Introduction

Particle samplers are widely used in workplaces in order to determine the concentration of airborne particles in the atmosphere. They generally operate by drawing air, with the aid of a pump, through one or more orifices in the sampler body, and housed within the sampler is a filter through which the air is subsequently drawn. The airborne particles are collected on the filter and their concentration is determined. Various samplers have been designed for this purpose, including static samplers which are located in fixed positions in a working environment and determine the dust concentration averaged over a prescribed period of time at that one point, and 'personal' samplers which are mounted on a working person near the breathing zone. A typical static sampler is the ORB sampler, which was designed by Ogden and Birkett (1978) to have the same entry efficiency, for particles with aerodynamic diameter up to at least 25 μm, as a human head equally exposed to all wind directions for wind speeds between 0 and 2.75 m s$^{-1}$. There are numerous other static samplers, and for a fuller discussion of these see, for example, Vincent (1989). Typical personal samplers are the single 4 mm hole sampler, the seven-hole sampler and the 25 mm open face holder sampler; see Vincent (1989). Although the computational techniques discussed in this chapter apply equally to both static and personal samplers, the examples used are for the personal sampler.

It is important that the particle sample collected on the filter within the sampler is representative of the particle concentration upstream of it; such a sample is known as a true sample. If changes do occur they should be known or predicted as accurately as possible.

The physical presence of the sampler and the action of the sampling will cause a disturbance to the ambient atmosphere, and particles possessing inertia may not follow the distorted fluid flow. Hence the sample collected on the filter may, and probably will, not be a true sample. In the case of an infinitely thin-walled sampler which faces the direction of the oncoming flow, and with the fluid being withdrawn through the sampler at the same velocity as the freestream in the absence of the sampler, no disturbance to the flow will occur. Hence, the particle sample collected will be a true sample. However, even for thin-walled samplers, if the velocity of

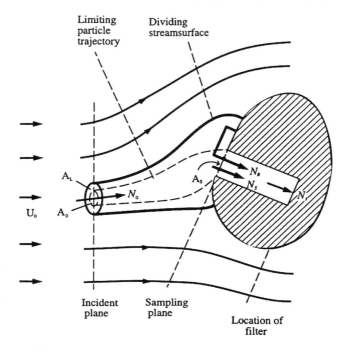

**Fig. 10.1** A schematic diagram of the flow past a sampler of arbitrary shape.

sampling and the freestream velocity are not equal the flow will be distorted and the sample collected will differ from the true concentration. In addition, it is impossible for a truly infinitely thin-walled sampler to be manufactured and *all* samplers are in fact 'blunt body' samplers.

The collection efficiency of a sampler can be defined in many different ways, and in order to illustrate some of these conventions we consider the uniform flow, with undisturbed speed $U_0$, past a sampler of arbitrary shape with a uniform suction speed $U_s$. Figure 10.1, which is a schematic diagram of the flow past an arbitrary shaped sampler, shows

- the dividing steam surface, which separates the sampled air from that which is not sampled;
- the limiting particle surface, which separates the sampled particles which pass into the orifice of the sampler from those which do not directly enter the sampler;
- the location of the filter in the sampler;
- the area of the plane of the sampling orifice ($A_s$), of the plane perpendicular to the oncoming flow contained by the limiting stream surface ($A_L$) and of the plane perpendicular to the oncoming flow contained by the limiting particle surface ($A_0$).

For a given particle size and airflow conditions, we assume that $N_0$, $N_S$, $N_B$ and $N_F$ are respectively the number of particles passing through the area $A_0$; passing through the orifice directly; passing through the orifice having first come into contact with the sampler; and sampled on the filter. We therefore define the aspiration efficiency,

$$A = \frac{N_S}{N_0};$$  (10.1)

the entry efficiency,

$$A_E = \frac{N_S + N_B}{N_0};$$  (10.2)

and the overall sampling efficiency,

$$A_0 = \frac{N_F}{N_0}.$$  (10.3)

In general, these quantities will be different depending on the nature of the particles and the surface of the sampler, and this will be described in more detail in section 10.4. In addition to these parameters, the sampling efficiency depends on the geometry of the sampler, the ratio of the sampling speed, $U_s$, to the oncoming unperturbed speed, $U_0$, and the Stokes number, $St$, which is a measure of how the particle will move relative to the fluid, and is defined to be

$$St = (d^2 \rho_p U_0)/(18\eta D_c)$$  (10.4)

where $d$ is the particle diameter, $\rho_p$ the particle density, $\eta$ the viscosity of the fluid and $D_c$ a characteristic dimension of the sampler. It will also depend on other parameters, which we will identify later.

The majority of research work performed on thin-walled samplers has investigated the errors involved in sampling particles suspended in a moving fluid, although some work has been performed on the sampling of particles in calm or near-calm conditions. Most work, however, has concentrated on the idealized conditions of sampling isokinetically, i.e. the mean velocity of sampling is equal to the freestream velocity, and the nozzle of the sampler is aligned parallel to the oncoming flow.

There have been numerous experimental investigations into thin-walled sampling, both isokinetic and not, and with the sampler both aligned and not aligned with the oncoming flow. One of the earliest investigators of thin-walled samplers was May and Druett (1953), but there have also been substantial contributions by Watson (1954), Badzioch (1959), Vitols (1966), Sehmel (1967), Davies (1968), Belyeav and Levin (1972; 1974), Jayasekera and Davies (1980), Durham and Lundren (1980), Davies and Subari (1982), Vincent et al. (1985) and Vincent (1989). Most of these investigators have then correlated their results and obtained empirical expressions for one of the definitions of the efficiency of the collection as given in equations (10.1), (10.2) and (10.3), although it is not always evident which. Usually these formulae depend on $U_0/U_s$, $St$ and $\beta$ (the angle of orientation of the oncoming flow to the axis of the sampler). Vincent (1987) has developed a semi-theoretical and empirical approach, and his results are in a reasonable agreement with most of the available experimental data. However, because of the large range of possible parameters involved in this problem it is impossible to obtain an empirical expression for the collection coefficients as defined in expressions (10.1), (10.2) and (10.3) which covers all aspects of thin-walled sampling. Thus it is necessary to develop a more thorough and accurate mathematical investigation using computational fluid dynamics (CFD).

Thin-walled samplers cannot exist in reality, and therefore, for the purpose of this chapter, we will concentrate on an example of blunt body sampling. Some of the early work on such samplers has been carried out by Davies and Peetz (1954). Davies (1967a; 1967b), Ogden and Wood (1975), Ogden and Birkett (1978), Vincent and Amburster (1981), Mark *et al.* (1985) and Mark and Vincent (1986). Most of this research again involved much experimental work and then fitting empirical formulae to their data. It was not until the pioneering research of Vincent *et al.* (1982) that this work was put on a more scientific basis with the introduction of his semi-theoretical and empirical approach. He found that, in addition to the parameters mentioned earlier, the collection efficiency of a sampler also depends on numerous other parameters. However, in order for his theory to agree adequately with the experimental data he had to determine some empirical constants that he had introduced into his theory. Clearly these constants are different for different sampler geometries. He found that his results were in good agreement with most of the available experimental data. Based on the approach of Vincent and his co-workers, Dunnett and Ingham (1988a) developed a more mathematically rigorous formula for the collection efficiency of a blunt body sampler.

Prior to the mid-1980s virtually all the research work on aerosol samplers had been done experimentally and empirical formulae were being proposed which fitted the particular experimental configuration under investigation. If the geometry was changed then new experimental investigations had to be performed and the resulting new empirical formulae developed. It was not until the exciting new developments introduced by Vincent and his co-workers that a more robust technique was developed for determining the collection efficiency of aerosol samplers. In the late 1980s, Dunnett and Ingham (1986; 1987; 1988a; 1988b; 1988c) addressed the problem of aerosol sampling using CFD techniques. However, their work was only valid for situations where the orifice of the sampler was either directly facing, or at a relatively small angle to, the oncoming airflow. Certainly the method was not able accurately to predict the situation when the orifice was rear-facing the oncoming flow. In order to overcome this problem Ingham, Wen and their co-workers, in a series of papers (1993, 1994, 1995, 1996) have addressed it using sophisticated mathematical and numerical techniques.

## 10.2   A simple mathematical model

As shown in Figure 10.1, if the orifice of the sampler is facing either directly upstream, or at a relatively small angle to the oncoming flow, and the sampling rate is not too large, then the fluid within the limiting stream surface behaves as an ideal fluid. This means that the airflow may be assumed to be inviscid since the flow comes from a region of low viscosity and does not pass regions where there are solid boundaries which develop boundary layers or turbulent wakes. Further, the limiting particle trajectory surface also lies within the limiting streamsurface and hence the particles collected come from a region where the flow may be considered to be inviscid. Hence it is only necessary accurately to predict the airflow within the limiting streamsurface and the nature of the flow outside this surface is relatively unimportant. In such circumstances, within the limiting streamsurface the airflow is essentially laminar and inviscid and therefore in this section we concentrate on the potential flow past a sampler with suction.

Because of the complex geometries and suction orifices of aerosol samplers it is not in general possible to obtain analytical solutions of the governing flow equations, even though this equation is relatively simple. Thus numerical schemes have to be employed, and the most popular techniques for solving such problems are the finite-difference and the finite-element methods. Although these two methods work extremely well and efficiently for the determination of the flow, the disadvantages of using such methods are that once the velocity potential has been obtained the air velocity has still to be obtained and also the treatment of irregular shaped boundaries is not easy. The need to evaluate the air velocity many times is a severe limitation of using the finite-difference and finite-element methods, but it is essential that this is done in order to track particle trajectories. To overcome the above two problems, the boundary-element method (BEM) is found to be an ideal technique for solving for the fluid velocity at numerous given arbitrary points in space.

Although the BEM works equally well in two and three dimensions, for ease in presenting the technique only a two-dimensional situation in which the plane of the sampling orifice is perpendicular to the oncoming airflow will be examined. Thus we consider the airflow past a two-dimensional horizontal cylinder of radius $a$ and centre $O$, in a uniform stream with an unperturbed constant air velocity, $U_0$, in the positive $x$

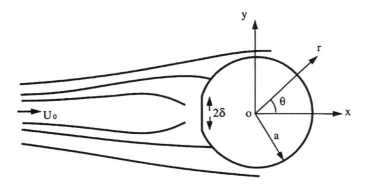

**Fig. 10.2** Coordinate system and notation for flow past a two-dimensional circular cylinder.

direction; see Figure 10.2. The cylinder has a finite inlet, of width $2\delta$ on the upstream side of the cylinder, through which an amount $Q$ of air per unit time per unit length is withdrawn. In normal operating conditions, the air may be considered to be incompressible so that the fluid velocity $\mathbf{q}_f$ satisfies the continuity equation

$$\nabla \cdot \mathbf{q}_f = 0. \tag{10.5}$$

Also, we are assuming that the fluid is inviscid and irrotational, so we have

$$\nabla \times \mathbf{q}_f = 0. \tag{10.6}$$

It is therefore possible to introduce the velocity potential $\Phi$ such that

$$\mathbf{q}_f = \nabla\Phi, \tag{10.7}$$

which on substitution into equation (10.4) gives the Laplace equation

$$\nabla^2\Phi = 0. \tag{10.8}$$

From equation (10.7), the $x$ and $y$ components of the air velocity, $U$ and $V$, can be found to be given by

$$U = \frac{\partial\Phi}{\partial x}, \qquad V = \frac{\partial\Phi}{\partial y}. \tag{10.9}$$

At large distances from the sampler the velocity potential due to the uniform flow is $U_0 r \cos\theta$ and that due to the withdrawal of fluid is $-(Q/2\pi)\ln r$, since at a large distance from the sampler the airflow will behave as if there is a point sink at the origin. Therefore we define $\Phi_\epsilon(r, \theta)$ such that

$$\Psi(r, \theta) = U_0 r \cos\theta - \frac{\theta}{2\pi}\ln r + \Phi_\epsilon(r, \theta) \tag{10.10}$$

where $\Psi_\epsilon(r, \theta) \to 0$ as $r \to \infty$.

The 'perturbation' potential $\Phi_\epsilon$ also satisfies the Laplace equation and decays sufficiently rapidly in the domain outside the sampler to apply the exterior form of Green's integral formula (see Dunnett and Ingham, 1988c),

$$\eta(\mathbf{p})\Phi_\epsilon(\mathbf{p}) = \int_{\partial\Omega} [\Phi_\epsilon(\mathbf{q})\ln' |\mathbf{p} - \mathbf{q}| - \Phi'_\epsilon(\mathbf{q})\ln |\mathbf{p} - \mathbf{q}|]\,dq, \qquad (10.11)$$

where $\partial\Omega$ is the boundary of the solution domain $\Omega$ (in this case, $\Omega$ is the area outside the sampler and $\partial\Omega$ is the boundary of the sampler); $\mathbf{p} \in \Omega \cup \partial\Omega, \mathbf{q} \in \partial\Omega$; $'$ denotes differentiation with respect to the outward normal $\mathbf{n}$ to $\partial\Omega$ at $\mathbf{q}$; $\eta(\mathbf{p})$ is defined by

$$\eta(\mathbf{p}) = \begin{cases} 0 & \text{if} \quad \mathbf{p} \in \Omega \cup \partial\Omega \\ \alpha_t & \text{if} \quad \mathbf{p} \in \partial\Omega \\ 2\pi & \text{if} \quad \mathbf{p} \in \Omega, \end{cases} \qquad (10.12)$$

in which $\alpha_t$ is the angle included between the tangents to $\partial\Omega$ on either side of $\mathbf{p}$; and $d\mathbf{p}$ is the differential increment of $\partial\Omega$ at $\mathbf{q}$. To solve for $\Phi_\epsilon(\mathbf{p})$ requires knowledge of $\Phi_\epsilon(\mathbf{q})$ and $\Phi'_\epsilon(\mathbf{q})$ everywhere on $\partial\Omega$.

There is no flow across the boundary apart from at the inlet. So the boundary condition on the circular cylinder is that there is no normal velocity, i.e.

$$\frac{\partial\Phi}{\partial r} = 0 \qquad \text{on } r = a, \qquad (10.13)$$

which gives

$$\frac{\partial\Phi_\epsilon}{\partial n} = U_0 \cos\theta - \frac{Q}{2\pi a} \qquad \text{on } r = a \qquad (10.14)$$

where $\mathbf{n}$ is the outward normal.

On the sampling head the mean velocity of sampling, $v_m = Q/2\delta$, is directed in the positive direction, so that the boundary condition on the inlet is given by

$$\frac{\partial\Phi}{\partial x} = v_m, \qquad (10.15)$$

i.e.

$$\frac{\partial\Phi_\epsilon}{\partial n} = -U_0 + \frac{Qx}{2\pi(x^2 + y^2)} + \frac{Q}{2\delta}. \qquad (10.16)$$

Hence from the boundary conditions we have a value for $\Phi_\epsilon(\mathbf{q})$ at each $\mathbf{q} \in \partial\Omega$. The first step is to determine the missing boundary information, e.g. find $\Phi_\epsilon(\mathbf{q})$ at each $\mathbf{q} \in \partial\Omega$. To do this let $\mathbf{p} \to \mathbf{p}_1 \in \partial\Omega$. This gives Green's integral formula as

$$\int_{\partial\Omega} \Phi_\epsilon(\mathbf{q}) \ln' |\mathbf{q} - \mathbf{p}_1| \, dq - \int_{\partial\Omega} \Phi_\epsilon'(\mathbf{q}) \ln |\mathbf{q} - \mathbf{p}_1| \, dq - \eta(\mathbf{p}_1)\Phi_\epsilon(p_1) = 0, \quad (10.17)$$

where $\mathbf{q}, \mathbf{p} \in \partial\Omega$. The boundary is now discretized so that

$$\partial\Omega = \bigcup_1^N \partial\Omega_j, \quad (10.18)$$

where each $\partial\Omega_j$ is a straight line segment. A 'node' $\mathbf{q}$ is located at the end of each segment. Using the linear boundary integral equation method (LBIEM) on each interval $\partial\Omega_j$, $j = 1, \ldots, N$, $\Phi_\epsilon$ and $\Phi_\epsilon'$ are approximated by piecewise linear functions

$$\begin{aligned}
\Phi_\epsilon &= (1 - \zeta_f)\Phi_\epsilon(\mathbf{q}_j) + \zeta_f\Phi_\epsilon(\mathbf{q}_{j+1}) \\
\Phi_\epsilon' &= (1 - \zeta_f)\Phi_\epsilon'(\mathbf{q}_j) + \zeta_f\Phi_\epsilon'(\mathbf{q}_{j+1}),
\end{aligned} \quad (10.19)$$

where $\mathbf{q}_j$ and $\mathbf{q}_{j+1}$ are the endpoints of $\partial\Omega_j$ and $\zeta_f$ is a linear function which increases from zero at $\mathbf{q}_j$ to unity at $\mathbf{q}_{j+1}$. It was found that it was essential to use the linear form of the BEM method in order to accurately position the inlet. Correspondingly, Green's integral formula becomes

$$\sum_1^N \left\{ \Phi_{\epsilon j} \int_{\partial\Omega_j} (1 - \zeta_f) \ln' |\mathbf{q} - \mathbf{p}_1| \, dq + \Phi_{\epsilon j+1} \int_{\partial\Omega_j} \zeta_f \ln' |\mathbf{q} - \mathbf{p}_1| \, dq \right\}$$

$$- \sum_1^N \left\{ \Phi_{\epsilon j}' \int_{\partial\Omega_j} (1 - \zeta_f) \ln |\mathbf{q} - \mathbf{p}_1| \, dq + \Phi_{\epsilon j+1}' \int_{\partial\Omega_j} \zeta_f \ln |\mathbf{q} - \mathbf{p}_1| \, dq \right\}$$

$$- \eta(\mathbf{p}_1)\Phi_\epsilon(\mathbf{p}_1) = 0, \quad (10.20)$$

where $\Phi_{\epsilon j}$ and $\Phi_{\epsilon j}'$ denote $\Phi_\epsilon(\mathbf{q}_j)$ and $\Phi_\epsilon'(\mathbf{q}_j)$, respectively. As $\Phi_{\epsilon j}'$ is known everywhere on the boundary, from the boundary conditions, a system of $N$ equations in the $N$ unknowns $\Phi_{\epsilon j}$ can be generated by taking $\mathbf{p}_1$ at each of the points $q_i$ for $i = 1, \ldots, N$. Solving these equations gives $\Phi_{\epsilon j}$ and $\Phi_{\epsilon j}'$ everywhere on the boundary. Then

$$\Phi_\epsilon(\mathbf{p}) = \frac{1}{\eta(\mathbf{p})} \left\{ \sum_1^N \left\{ \Phi_{\epsilon j} \int_{\partial\Omega_j} (1 - \zeta_f) \ln' |\mathbf{p} - \mathbf{q}| \, dq \right. \right.$$

$$+ \Phi_{\epsilon j+1} \int_{\partial\Omega_j} \zeta_f \ln' |\mathbf{p} - \mathbf{q}| \, dq \right\} - \sum_1^N \left\{ \Phi_{\epsilon j}' \int_{\partial\Omega_j} (1 - \zeta_f) \ln |\mathbf{p} - \mathbf{q}| \, dq \right.$$

$$\left. \left. + \Phi_{\epsilon j+1}' \int_{\partial\Omega_j} \zeta_f \ln |\mathbf{p} - \mathbf{q}| \, dq \right\} \right\} \quad (10.21)$$

is the discretized form of Green's integral formula which is used to evaluate $\Phi_\epsilon(\mathbf{p})$ at the general field point $\mathbf{p} \in \Omega \cup \partial\Omega$. Before performing these calculations all quantities were non-dimensionalized. Then

$$\nabla^2 \Phi_\epsilon^* = 0 \qquad (10.22)$$

$$\frac{\Phi_\epsilon^*}{\partial r^*} = -\cos\theta + 4F \qquad \text{on } r^* = 1 \qquad (10.23)$$

$$\frac{\partial \Phi_\epsilon^*}{\partial x^*} = -1 + \frac{4Fx^*}{(x^{*2} + y^{*2})} + \frac{4\pi F}{\text{DELTA}} \qquad (10.24)$$

on the inlet $\Phi_\epsilon^* \to 0$ as $r^* \to \infty$. * refers to the non-dimensional quantities

$$\Phi_\epsilon^* = \frac{\Phi_\epsilon}{aU_0}, \qquad x^* = \frac{x}{a}, \qquad y^* = \frac{y}{a}, \qquad r^* = \frac{r}{a}, \qquad (10.25)$$

and

$$F = \frac{Q}{8\pi aU_0}, \qquad \text{DELTA} = \frac{\delta}{a}. \qquad (10.26)$$

Now the velocity potential can be found at any point outside the sampler. An important characteristic of the flow which needs to be calculated in order to determine the particle trajectories is the flow velocity. This is done using Richardson's extrapolation; see, for example, Smith (1978). Considering the component of velocity in the positive $x$ direction, $U$, at the general field point **p**, we obtain

$$U = \frac{U_2^2 - U_1 U_3}{2U_2 - U_3 - U_1}, \qquad (10.27)$$

where $U_1$, $U_2$ and $U_3$ are the first, second and third approximations to $U$, respectively, using mesh sizes $h_1, h_1/2, h_1/4$, where $h_1$ is of $O(10^{-3})$. A similar expression to (10.27) can be obtained for the component of velocity in the positive $y$ direction, $V$.

All the details of the flow, in particular the velocity components $U$ and $V$, are now known. This method can easily be extended to two-dimensional aerosol samplers of any other shape; all that is involved is a change in the conditions when setting up the boundaries of the sampler. The method described in this section can also be extended to general three-dimensional samplers; for a full description of this technique see, for example, Dunnett and Ingham (1988c).

Now that the air velocity can be easily determined it is only necessary to work out the sampling efficiency of the sampler by particle tracking techniques. A full description of this will be given in section 10.4, but before doing this it is essential to investigate more general sampling conditions. In particular, if the sampled air goes through a region of turbulence then the above theory breaks down and a fully turbulent flow model has to be developed.

## 10.3 A sophisticated mathematical model

Although theoretical studies based on analytical techniques and BEM have increased the understanding of the sampling characteristics of samplers, they are limited by the assumption of the inviscid flow model. Further, the air that carries the particulate may be laminar or turbulent, according to the operating conditions of the sampler and the oncoming flow, and although there have been some investigations on laminar flows there is very little published work on turbulent flows in sampling. Therefore a direct simulation of the flow of real fluids will provide a rigorous description of the airflow and the sampling mechanics and this will assist in the designing of new sampling systems without the need to perform too many extensive and expensive experiments.

In order to illustrate the numerical method we will consider the simplest form of blunt sampler, the disc-shaped sampler with a central orifice, extensively investigated by Chung and Ogden (1986) who obtained aspiration coefficients over a wide range of operating conditions. In particular, they investigated the case when the axis of the disc is aligned with the uniform flow at large distances from the sampler and it is on this situation that we will concentrate our attention.

The sampler under investigation is therefore that studied experimentally by Vincent *et al.* (1985) and Chung and Ogden (1986) and is shown schematically in Figure 10.3. It is assumed to take the form of an axisymmetric disc with diameter $D$ which faces the wind, and the air is aspirated through a circular central orifice with diameter $D_0$. Cylindrical coordinates are used in which $r$ is the coordinate in the radial direction

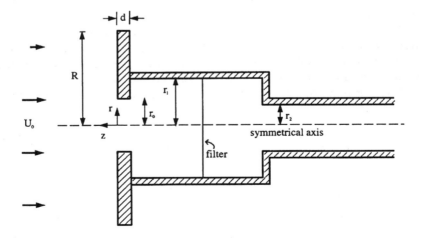

**Fig. 10.3** Schematic diagram of the sampler under investigation.

and $z$ is aligned with the axis of symmetry of the sampler and is measured positively in the opposite direction to the direction of the freestream. The magnitude of the freestream velocity is $U_0$ and the average sampling velocity is $U_s$. Thus the velocity ratio $R = U_0/U_s$ is a very important parameter which characterizes the nature of the flow. When dealing with sampling problems in turbulent flows, the steady, Reynolds averaged turbulent Navier–Stokes equations have to be solved and the key to the success of CFD lies with the accurate description of the turbulent behaviour of the flow. A number of turbulence models are available, ranging from the industry standard $k$–$\epsilon$ to the more complicated Reynolds stress models.

In most experimental situations, and in particular those investigated by Vincent *et al.* (1985) and Chung and Ogden (1986), the flows typically have Reynolds number of the order of 4000, and therefore the thickness of the turbulent boundary layer is of the same order as that of the diameter of the orifice of the sampler. In this situation the standard $k$–$\epsilon$ model cannot reveal the true nature of the fluid flow by using wall functions very close to the wall of the sampler. Therefore in this chapter we use a 'low Reynolds number turbulent $k$–$\epsilon$ model'; for further details see, for example, Launder and Sharma (1974).

For an incompressible fluid, the momentum and the continuity equations for turbulent fluid flow, in vector notation, are given by

$$\mathbf{V} \cdot \nabla \mathbf{V} = -\frac{1}{\rho}\nabla p + \nabla \cdot (v_\epsilon \nabla \mathbf{V}) \tag{10.28}$$

$$\nabla \cdot \mathbf{V} = 0, \tag{10.29}$$

where $\mathbf{V} = u\mathbf{e}_r + w\mathbf{e}_z$, $u$ and $w$ are the mean values of the turbulent components of the air velocity in the radial and axial directions respectively, and $\mathbf{e}_r$ and $\mathbf{e}_z$ are the unit vectors in the radial and axial directions respectively, $\rho$ is the density of the fluid, and $v_\epsilon$ is the effective kinematic viscosity of the fluid and consists of the sum of the laminar kinematic viscosity $v$ and the turbulent kinematic viscosity $v_t$, i.e. $v_\epsilon = v + v_t$.

A low Reynolds number turbulent $k$–$\epsilon$ model, developed by Launder and Sharma (1974), is given by

$$(\mathbf{V} \cdot \nabla)k = \nabla \cdot \left[\left(v + \frac{v_t}{\sigma_k}\right)\nabla k\right] + \phi - \epsilon + E_1, \tag{10.30}$$

$$(\mathbf{V} \cdot \nabla)\epsilon = \nabla \cdot \left[\left(v + \frac{v_t}{\sigma_\epsilon}\right)\nabla\epsilon\right] + C_1 f_1 \frac{\epsilon}{k}\phi - C_2 f_2 \frac{\epsilon^2}{k} + E, \tag{10.31}$$

where $k$ is the turbulent kinetic energy, $\epsilon$ is the turbulent energy dissipation, $\phi$ is the generation of the turbulent energy which is caused by turbulent stresses, and

$$f_1 = 1.0, \tag{10.32}$$

$$f_2 = 1 - 0.3\exp(-R_T^2), \tag{10.33}$$

$$R_T^2 = k^2/(v_\epsilon), \tag{10.34}$$

$$E = 2v\left(\frac{\partial k^{1/2}}{\partial x_j}\right)^2, \tag{10.35}$$

$$E_1 = -2.0vv_t\left(\frac{\partial^2 U_i}{\partial x_j \partial x_i}\right). \tag{10.36}$$

The turbulent viscosity $v_t$ is given by

$$v_t = C_\mu f_\mu \frac{k^2}{\epsilon}, \tag{10.37}$$

where

$$f_\mu = \exp\left(\frac{-3.4}{(1 + R_T/50)^2}\right). \tag{10.38}$$

Further, the coefficients which occur in equations (10.32)–(10.38) should be determined by performing an experimental investigation on the flow around the sampler. However, there is a severe lack of detailed measurements of the turbulent velocity distribution, and of other turbulence quantities, for flows around samplers. Therefore, the values of the unknown coefficients used in this paper are those suggested by Launder and Sharma (1974), namely,

$$C_\mu = 0.09, \quad \sigma_k = 1.0, \quad \sigma_\epsilon = 1.3, \quad C_1 = 1.44, \quad C_2 = 1.92. \tag{10.39}$$

These values are based on a very extensive examination of various fluid flow and form the best available data.

Equations (10.28)–(10.31) now have to be solved subject to the appropriate boundary conditions. Since the sampler is axisymmetric, we need only consider the solution in the semi-infinite domain $r > 0, -\infty < z < \infty$ (see Figure 10.4). However, in the numerical calculations we have to approximate the location of the boundary conditions, which are at $r \to \infty, -\infty < z < \infty$, to be at a finite radius, i.e. on $AB$, and at $z \to +\infty$ to be at finite distances, i.e. on $AA'$ and $BB'$. Equations (10.28)–(10.31) now have to be solved subject to the following boundary conditions.

On the upstream boundary $AA'$ the air velocity takes the constant value $U_0$ in the negative $z$ direction. For the freestream turbulence we use the experimental data obtained by Vincent et al. (1982) who used an adjustable system of square-mesh, biplanar-lattice grids. They give empirical expressions for the freestream turbulent intensity $I$ and the

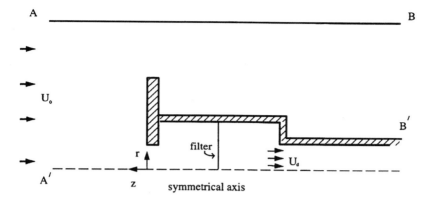

**Fig. 10.4** The computational domain for the sampler under investigation.

turbulent length scale $L$ as

$$I = 1.59\left(\frac{x}{b}\right)^{-0.7} \tag{10.40}$$

$$\frac{L}{b} = 0.02\left(\frac{x}{b}\right)^{1.03}, \tag{10.41}$$

where $b$ and $x$ are the width of the bars in the grid system and the distance downstream from the grid system, respectively.

*On the freestream boundary AB,*

$$\frac{\partial u}{\partial r} = \frac{\partial w}{\partial r} = \frac{\partial k}{\partial r} = \frac{\partial \epsilon}{\partial r} = 0. \tag{10.42}$$

*On the downstream boundary BB',*

$$\frac{\partial u}{\partial z} = \frac{\partial w}{\partial z} = \frac{\partial k}{\partial z} = \frac{\partial \epsilon}{\partial z} = 0. \tag{10.43}$$

*On the axis of symmetry,*

$$u = 0, \qquad \frac{\partial w}{\partial r} = \frac{\partial k}{\partial r} = \frac{\partial \epsilon}{\partial r} = 0. \tag{10.44}$$

*At the entrance of the exit pipe of the sampler* it is assumed that there is a uniform velocity $U_\infty$ which corresponds to a sampling flow rate $Q$ and $\partial k/\partial z = \partial \epsilon/\partial z = 0$.

Near the walls of the sampler the wall function method (see Wen and Ingham, 1993), is used to bridge the fully turbulent region and the flow in the vicinity of the wall. When the grid node $P$, which is nearest to the wall of the sampler, is located at a distance $y$ from the wall, the value of the quantity $y^+$ at the point $P$ is defined as

$$y^+ = \left(\frac{C_\mu^{1/4} k^{1/2} y}{v}\right)_{\mathrm{p}}.$$  (10.45)

For the momentum equations the wall shear stress $\tau_w$ is calculated using the linear or the logarithmic law of the wall, i.e.

$$\tau_w = \left(v\frac{U}{y}\right)_{\mathrm{p}} \qquad \text{for } y^+ \le 11$$  (10.46)

$$\tau_w = \left(\frac{\rho \kappa C_\mu^{1/4} k^{1/2} U}{\ln(Ey^+)}\right)_{\mathrm{p}} \qquad \text{for } y^+ \ge 11,$$  (10.47)

where the Karman constant $\kappa = 0.4$, for a smooth wall $E = 9.0$ and $U$ is the component of the fluid velocity parallel to the wall.

In the $k$-equation (10.30), the source term $S_k = \phi - \epsilon$ at the point $P$ nearest to the wall is modified by use of the wall function. The generation term $\phi$ is calculated by use of the wall shear stress expression (10.46) or (10.47), while the dissipation term $\epsilon$ is calculated from

$$\epsilon_{\mathrm{p}} = \left(\frac{C_\mu^{3/4} k^{3/2} y^+}{y}\right)_{\mathrm{p}} \qquad \text{for } y^+ \le 11$$  (10.48)

$$\epsilon_{\mathrm{p}} = \left(\frac{C_\mu^{3/4} k^{3/2} \ln(Ey^+)}{\kappa y}\right)_{\mathrm{p}} \qquad \text{for } y^+ \ge 11.$$  (10.49)

In the $\epsilon$-equation (10.31), the value of $\epsilon$ at the point $P$ is calculated from

$$\epsilon_{\mathrm{p}} = \left(\frac{C_\mu^{3/4} k^{3/2}}{\kappa y}\right)_{\mathrm{p}}$$  (10.50)

Pressure correction methods have been widely used to solve the Navier–Stokes equations for complex fluid flow problems ever since the pioneering work of Patankar and Spalding (1972) who developed the SIMPLE (Semi-Implicit Methods for Pressure Linked Equations) algorithm. Several authors have produced variants of this method in order to improve the rate of convergence of the algorithm, e.g. the SIMPLER and SIMPLEC methods. The mathematical development of SIMPLE-like algorithms concentrates on using the pressure–velocity relationship from the linearized momentum equations. In the SIMPLE and SIMPLEC algorithms the pressure correction equations are derived from the continuity equation for each control volume, and the solution is achieved by successively predicting and correcting the velocity components and pressure. In fact, they provide a pressure correction by use of the continuity equation, which is a Poisson-type pressure correction equation. The SIMPLER algorithm starts with an estimate for the velocity field rather than for the pressure field. This change is significant, since guessing an initial condition for the velocity rather than the pressure is generally much easier. However, the pressure correction equation of SIMPLE and SIMPLEC and the pressure equation of SIMPLER are derived only at a small control volume. In the SIMPLE and SIMPLEC algorithms the pressure correction at each grid point is constrained by the local mass conservation, which can be satisfied by an incorrect local velocity. For example, if all the initial velocity components take zero values, then the local mass conservation is satisfied everywhere, except at the control volumes on the boundary. Therefore, the global mass conservation is only propagated from the boundary to the whole computational domain by an iterative procedure. Because of the elliptic nature of the pressure correction equation, the larger the computational domain, or the finer the grid used, the slower the rate of convergence. Further, in complex fluid flows the pressure correction equation is not very sensitive to large pressure drops, which results in a slow rate of convergence.

In order to illustrate the methods used we only consider equations (10.28) and (10.29). We now consider an arbitrary volume of fluid contained in a volume $\Omega$ which has an outer surface $S$, and the unit outward normal to the surface is $n$. We employ the Gaussian theorem, i.e.

$$\int_s \int \mathbf{n} \cdot \mathbf{V} \, dS = 0 \tag{10.51}$$

$$\int_s \int (\mathbf{V} \cdot \mathbf{n}) \mathbf{V} \, dS = - \int_s \int \mathbf{n} p \, dS + \int_s \int \mathbf{n} \cdot (v \nabla \mathbf{V}) \, dS, \tag{10.52}$$

in order to transform the differential forms of equations (10.28) and (10.29) into their integral forms.

In order to obtain accurate results, more mesh points should be employed in, and in the vicinity of, the sampler, and hence non-uniform grids in both the $r$ and $z$ directions are used by means of the following coordinate transformation:

$$r = f(\eta), \tag{10.53}$$

$$z = g(\zeta), \tag{10.54}$$

where $\eta$ and $\zeta$ are new independent radial and axial variables and $f$ and $g$ are two functions which may be chosen in order to produce the required distribution of grids. In order to discretize the solution domain, a staggered grid which covers the physical domain is used. Finally, the finite-difference equation for the velocity component $u$ is obtained using equation (10.52) in the form

$$a_p u_p = a_E u_E + a_W u_W + a_N u_N + a_S u_S + (p_w - p_e)(h_2 h_3)_p \Delta\zeta \tag{10.55}$$

where

$$a_p = a_E + a_W + a_N + a_S + \left(\frac{vh_1 h_3}{h_2}\right)_p \Delta\eta\Delta\zeta \qquad (10.56)$$

$$a_E = \left(\frac{vh_2 h_3}{h_1}\right)\frac{\Delta\zeta}{\Delta\eta} + (h_2 h_3)_e \hat{u}_e \Delta\zeta \qquad (10.57)$$

$$a_W = \left(\frac{vh_2 h_3}{h_1}\right)\frac{\Delta\zeta}{\Delta\eta} + (h_2 h_3)_w \hat{u}_w \Delta\zeta \qquad (10.58)$$

$$a_N = \left(\frac{vh_1 h_2}{h_3}\right)\frac{\Delta\eta}{\Delta\zeta} - (h_1 h_2)_n \hat{w}_n \Delta\eta \qquad (10.59)$$

$$a_S = \left(\frac{vh_1 h_2}{h_3}\right)\frac{\Delta\eta}{\Delta\zeta} - (h_1 h_2)_s \hat{w}_s \Delta\eta, \qquad (10.60)$$

and $h_1 = f'(\eta)$, $h_2 = f(\eta)$, $h_3 = g'(\zeta)$, and $\Delta\eta$ and $\Delta\zeta$ are the mesh sizes in the $\eta$ and $\zeta$ directions, respectively. The quantities $\hat{u}_e$, $\hat{u}_w$, $\hat{w}_n$ and $\hat{w}_s$ are the velocities on the surfaces of the control volume and they should be replaced by schemes such as upwind, hybrid and so on. A similar expression exists for the $w$ component of velocity.

An under-relaxation parameter $E$ is introduced into the momentum equations, and in the radial direction this may be expressed in the form

$$a_p\left(1 + \frac{1}{E}\right)u_p = \sum a_{nb}u_{nb} + (p_w - p_e)(h_2 h_3)_p\Delta\zeta + \frac{a_p}{E}u_p^0 \qquad (10.61)$$

Let $p^*$ be the current pressure distribution and $u^*$ and $w^*$ be the velocities resulting from solving the radial and axial momentum equations with this pressure distribution. Therefore we have

$$a_e u_p^* = \sum a_{nb}^* + (p_w - p_e)(h_2 h_3)_p\Delta\zeta + \frac{a_p}{E}u_p^0 \qquad (10.62)$$

where $a_e = a_p(1 + 1/E)$.

If $p^*$ is the correct pressure distribution, then the velocities $u^*$ and $w^*$ will satisfy the continuity equation (10.51). However, in general this will not be the case. We therefore introduce corrections $u'$, $w'$ and $p'$ as follows:

$$u = u' + u^*, \qquad w = w' + w^*, \qquad p = p' + p^*. \qquad (10.63)$$

The relations between the velocity and pressure corrections can be seen by substituting equations (10.63) into equation (10.62), which gives

$$\left(a_e - \sum a_{nb}\right)u_p' = \sum a_{nb}(u_{nb}' - u_e) + (p_w' - p_e')(h_2 h_3)_p\Delta\zeta. \qquad (10.64)$$

In the SIMPLEC algorithm, the term $\sum a_{nb}(u_{nb}' - u_e)$ is omitted and thus equation (10.64) becomes

$$u_e' = d_e(p_p' - p_E'), \qquad (10.65)$$

where

$$d_e = \frac{(h_2 h_3)_e \Delta\zeta}{a_e - \sum a_{nb}}.$$

(10.66)

The pressure correction can then be obtained by substituting all the velocity components into the continuity equation for the control volume, and this yields

$$a_p p'_p = a_E p'_E + a_W p'_W + a_N p'_N + a_S p'_S + b,$$

(10.67)

where

$$b = (h_2 h_3)_e u_e^* \Delta\zeta - (h_2 h_3)_w u_w^* \Delta\zeta + (h_1 h_2)_n w_n^* \Delta\eta - (h_1 h_2)_s w_s^* \Delta\eta.$$

(10.68)

The SIMPLEC algorithm consists of the following steps:

(a)  guess the velocity and pressure fields $u^*$, $w^*$ and $p^*$;
(b)  solve the momentum equations to obtain the new values of $u^*$ and $w^*$;
(c)  solve for the pressure correction $p'$ and update the pressure using $p = p^* + p'$;
(d)  update the velocity components $u$ and $w$ by using the velocity correction equations;
(e)  repeat steps (b) to (d) until convergence has been reached.

The mass residual of every control volume is given by

$$R_{mass}^k = C_e - C_w + C_n - C_s,$$

(10.69)

where $C_e$, $C_w$, $C_n$ and $C_s$ represent the convection of mass through each face of the control volume which surrounds the point where the pressure is located. A measure of the convergence used in this paper is the sum of the mass residuals over all the control volumes, namely

$$R_{mass} = \sum_k |R_{mass}^k|.$$

(10.70)

When using the SIMPLEC algorithm as described, it was found that the numerical technique did not work well on this complex airflow problem. The rate of convergence of the iterative scheme was extremely slow no matter what value of the relaxation factor, $E$, or distribution of the grids was used. Even when a very good guess was prescribed for the initial pressure and velocity fields, the rate of convergence was still extremely slow and it would have taken many hours, or even days, of computing time on a large-scale computer in order to obtain a convergent solution. The difficulty is that it is not possible to obtain accurate velocity and pressure distributions in the vicinity of the orifice of the sampler or near the entrance to the exit pipe of the sampler. In fact the rate of convergence was much slower in the vicinity of the orifice than near the entrance to the exit pipe. This is because in the vicinity of the orifice the shape of the sampler changes very rapidly. The air which enters the sampler is accelerated rapidly in order to pass through the orifice, and this is accompanied by a sharp decrease in pressure. Then, immediately after the air has entered the sampler chamber, there is a sudden expansion which causes the air to decelerate rapidly. This contrasts with the air motion at the entrance of the exit pipe, where the air simply accelerates. The pressure correction equation is not sufficiently sensitive to detect this rapid variation in the geometry of the sampler, and the iterative procedure spends many iterations building up the steep pressure gradients and large pressure drops. A specialized correction technique has been introduced to overcome this problem; for more details, see Wen and Ingham (1993; 1996).

In conclusion, the inclusion of the average pressure correction in the SIMPLEC algorithm has successfully led to a substantial enhancement in the rate of convergence of the global mass conservation, and has also produced an acceleration in the rate of convergence in the local mass residual. Therefore, a substantial reduction in the cost of obtaining numerical solutions of the Navier–Stokes equations is possible.

## 10.4    Particle dynamics

Dispersed two-phase flow consists of a particle phase flowing in a continuous fluid; how to calculate the fluid flow in the continuous state, using CFD techniques, has been described in sections 10.2 and 10.3. Extension of single-phase CFD techniques to multi-phase flow calculations has many difficulties and, in general, there are two methodologies for extending single-phase techniques to two-phase and multi-phase flows. First there is the situation where the two or more phases are assumed to coexist in the flow domain, each occupying a fraction of the flow volume. The phases are assumed to have their own physical properties, flow velocities, etc., and the equations of conservation of mass, momentum, energy, etc. can be written separately for each phase. This approach of dealing with two- and multi-phase flows is important when one of the phases does not dominate the other phases. However, when considering the sampling of particles of low concentration from air, the flow field of the continuous phase may first be calculated for a single-phase flow. Then the transport of the particular phase is calculated by solving the equations

$$\frac{dU}{dt} = F, \qquad \frac{dX}{dt} = U, \qquad (10.71)$$

where $X$ is the position of the particle, $U$ the particle velocity and $F$ the force on the particle. Given the force $F$, the motion of the particle can be tracked given its initial position and velocity. Thus the difficulty of using this method is in the determination of the force **F**, and it will be this aspect that we shall consider further in this section.

A particle flowing in a fluid is subjected to several forces resulting from the motion of the fluid around it, such as the drag, the lift, the 'virtual mass' and the Basset force, in addition to its gravitational force. In most applications – in particular, to the sampling of aerosol particles – it is the drag force (which acts in the direction of the relative velocity between the fluid and the particle) which is the most important force acting on the particle. Further, in many situations the particles are sufficiently close to being spherical that they may be assumed to be spherical and studies of the drag on a sphere date back to the eighteenth century when Newton performed his famous experiments.

In general the drag force, $F_D$, on a body is related to a drag coefficient by the expression

$$C_D = F_D / \left( \frac{1}{2} \rho_c U_r^2 A \right), \tag{10.72}$$

where $A$ is the projected cross sectional area of the particle, $\rho_c$ is the density of the continuous phase and $U_r$ is the relative velocity of the sphere and the fluid. For a sphere of diameter $d_p$, we have

$$C_D = 8 F_D / (\pi d_p^2 \rho_c U_r^2), \tag{10.73}$$

and the drag coefficient will depend on the particle Reynolds number,

$$Re_p = (\rho_f d_p U_r) / \eta_f. \tag{10.74}$$

Stokes (1851) found that for $Re_p \ll 1$,

$$C_D = 24 / Re_p. \tag{10.75}$$

Since the work of Stokes numerous investigators have studied the drag coefficient on a sphere. They have found that expression (10.73) is valid for $Re_p < 0.1$, but for $0.1 < Re_p < 10$ the flow tends to oscillate. For $Re_p \geq 10$ the boundary layer separates, and a stationary ring vortex forms behind the sphere which grows in size and begins to oscillate when $Re_p \approx 150$. These oscillations became more violent until $Re_p \approx 500$. For $500 \leq Re_p < 2 \times 10^5$ the drag coefficient remains fairly constant, after which the boundary layer becomes turbulent. This results in a sudden drop in the drag coefficient but as the Reynolds number increases further the drag coefficient starts to increase. Numerous expressions which relate $C_D$ and $Re_p$ for a sphere have been postulated based on detailed experimental data; for an excellent description see, for example, Koncar-Djurdjevic *et al.* (1986). These results may be summarized as follows: Langmuir and Blodgett (1946) give the expression

$$C_D = \frac{24}{Re_p}(1 + 0.197 \, Re_p^{0.63} + 0.0026 \, Re_p^{1.38}), \qquad Re_p \leq 100; \qquad (10.76)$$

Schiller and Nauman (1933) have

$$C_D = \frac{24}{Re_p}(1 + 0.15 \, Re_p^{0.687}), \qquad Re_p \leq 700; \qquad (10.77)$$

and, according to Ihme et al. (1972),

$$C_D = \frac{24}{Re_p} + \frac{5.48}{Re_p^{0.573}}, \qquad Re_p \leq 30\,000. \qquad (10.78)$$

A particle may also experience a lift force which may result from the particle rotating or being in a shear flow. The force due to the rotation of a particle is usually referred to as the Magnus lift force (Magnus, 1853) and it has since been investigated by numerous authors; see, for example, Boothroyd (1971). When a particle is in a shear flow this causes the particle to rotate and a lift force is generated; see, for example, Saffman (1965) and Govan et al. (1989).

The force on a particle is also influenced by several other factors:

- *Non-sphericity.* It has been found that the drag coefficient of particles that are 'almost' spherical remains approximately the same as that for a sphere up to a particle Reynolds number of the order of 10. For more details see Becker (1959) and Govier and Aziz (1972).
- *Surface roughness.* When the boundary layer on the sphere is laminar, the effect of surface roughness on $C_D$ increases; see Torobin and Gauvin (1961).
- *Freestream turbulence.* Increasing the freestream turbulence decreases the Reynolds number at which the boundary layer becomes turbulent, see 'Surface roughness' above.
- *Virtual mass force.* Some additional energy has to be spent on accelerating particles through the fluid because they experience a resistance due to their acceleration relative to the fluid at infinity. A full explanation of virtual mass can be found in Cook and Harlow (1984).
- *The Basset force.* This takes into account the history of the motion of the particle and for more details see Basset (1961).

Fortunately, most aerosol particles of general interest are approximately spherical and have a small particle Reynolds number. Therefore only the drag force, given by expression (10.75), is important.

In sampling, particle–particle interactions should also be taken into account. However, for the purpose of this chapter we have assumed that the particle concentration is so small that these effects may be ignored. In contrast, particle–wall interactions cannot be ignored in sampling and the impact and rebounding processes involve the loss or gain in momentum by the particles. Depending on the nature of the particles and the surface, particles may:

- stick to the surface and not be re-entrained into the fluid flow, i.e. the coefficient of restitution of the particle is zero;
- bounce off the surface, i.e. the coefficient of restitution is greater than zero;
- initially stick to the surface but have the possibility of being re-entrained into the flow, i.e. we have blow-off or saltation (see Ingham and Yan, 1994);
- shatter into two or more particles on impact, with these smaller particles then behaving in any of the ways described above.

The above gives a brief review of several aspects of what, at a minimum, should be investigated before particle collection efficiencies can be calculated. We now investigate two methods of calculating particle collection efficiencies of the disc-shaped sampler described in section 10.3. In order to simplify the description only spherical particles with small particle Reynolds number are considered.

### 10.4.1  Stochastic model

When the particle density is much larger than that of air and the particle Reynolds number is much less than unity, the $i$ component of the Lagrangian particle equation of motion takes the form

$$\frac{dx_i}{dt} = u_{pi} \tag{10.79}$$

$$\frac{\rho_w d_a^2}{18\eta} \frac{du_{pi}}{dt} = \bar{u}_i + u_i' - u_{pi} \tag{10.80}$$

where $\rho_w$ is the density of water, $d_a$ is the particle aerodynamic diameter, $\eta$ is the viscosity of the fluid, $x_i$ normalized by $\delta/2$ is the position of the particle, $u_{pi}$ is the $i$ component of the velocity of the particle, $\bar{u}_i$ is the mean fluid velocity of the fluid and $u_i'$ is the fluctuating component of the fluid velocity; $u_{pi}$, $\bar{u}_i$ and $u_i'$ are normalized by $U_0$. The Stokes number is defined as

$$St = \rho_w d_a^2 U_0 / (18\eta\delta/2). \tag{10.81}$$

The motion of the particles is not equally affected by the different scales of turbulence, but rather is mainly governed by the interaction of the particle with a succession of large eddies, each of which is assumed to have constant flow properties. A method for tracking the particle motion was developed by Gosman and Ioannides (1981) assuming that the velocity fluctuations are isotropic and have a Gaussian distribution with a standard deviation given by

$$u_e = (2k/3)^{1/2}. \tag{10.82}$$

Thus the fluctuating velocity components are given by

$$u_i = \Psi_i u_e / U_0 \tag{10.83}$$

where $\Psi_i$ are normally distributed pseudo-random numbers and $u_i$ is used in the particle equation (10.80) to evaluate the instantaneous drag force on the particle. The length scale $L_e$ and the lifetime $\tau_e$ of the scale eddies (see, for example, Gosman and Ioannides, 1981; Shuen *et al.*, 1983) are also given by

$$L_e = C_\mu^{3/4} k^{1.5} / \epsilon, \qquad \tau_e = L_e / u_e, \tag{10.84}$$

respectively.

The transit time required for the particle to cross the eddy was determined by Shuen *et al.* (1983) for a particle in a uniform flow, i.e.

$$\tau_t = -\tau \ln\{1 - L_e / [\tau |(u_1^2 + u_2^2)^{1/2} - (u_{p1}^2 + u_{p2}^2)^{1/2}|]\}, \tag{10.85}$$

where $\tau$ is the relaxation time of the particle,

$$\tau = d_a^2 \rho_w / 18\eta. \tag{10.86}$$

Equations (10.79) and (10.80), which govern the motion of the particles, were integrated over the time $T$, which is the minimum of the eddy lifetime and the transit time, i.e. $T = \min\{\tau_e, \tau_f\}$.

In order to obtain the dispersion properties of the particles a number $M$ (a value of $M = 1000$ can be taken in order to obtain a stable statistic value of $P$) of particles of a given size were released at the same point at a large distance upstream of the sampler. Then the probability that the particle is sampled for this size of particles is defined as

$$P = \frac{N}{M}, \tag{10.87}$$

where $N$ is the number of sampled particles. Clearly, $P$ is a function of $Q_r$, where $Q_r$ is the flux of fluid across the area enclosed by the circle in the plane of constant $z$ on which the particles start. It is assumed that far upstream of the sampler the particle has the same velocity as that of the air and the concentration of particles of a given size is constant. Thus the aspiration efficiency of the sampler is given by the expression

$$A = \frac{\int_0^\infty P(Q_r) \, dQ_r}{Q} \tag{10.88}$$

where $Q$ is the sampled flux of air which enters the sampling probe.

### 10.4.2  *Mean motion model*

In general, the turbulent fluctuating air velocity is much smaller than the average air velocity, and, under the assumption of local isotopic turbulence, a particle has an equal chance of diffusing in the two directions which are normal to the mean particle motion. Therefore, when there exists a uniform concentration of particles at large distances from the sampler it is expected that the inertia of the particle will be more important than the particle diffusion in the determination of the aspiration efficiency Therefore, if we neglect the effect of particle diffusion on the aspiration efficiency of the sampler, i.e. we neglect the fluctuations in the air velocity, then equation (10.80) simplifies to

$$\frac{\rho_w d_a^2}{18\eta}\frac{du_{pi}}{dt} = \bar{u}_i - u_{pi}. \tag{10.89}$$

Equations (10.79) and (10.89) may now be solved and the different aspiration efficiencies, as described in equations (10.1)–(10.3), may be determined.

The stochastic model and the mean motion model have been used to evaluate the aspiration efficiency for numerous samplers. It has been found that the stochastic model for particle diffusion gives results for the aspiration efficiency of the sampler which are not substantially different from those predicted using the mean motion model. However, it is interesting to note that the stochastic model, which is very expensive in computing time, predicts an aspiration efficiency which is slightly greater than unity when the Stokes number is very small. In general, both the models numerically predict aspiration efficiencies which agree reasonably well with all the available experimental data.

## 10.5  Conclusions

It is useful to consider both the advantages and disadvantages of using CFD.

### 10.5.1  *Advantages of CFD*

- The most important advantage of numerical predictions is in its relatively low cost in comparison to experimental investigations. In most applications, including the sampling of aerosols, the cost of a computer run is substantially lower than performing the corresponding experimental investigation. Further, whereas the cost of labour and most materials is rising, computational costs are falling, and this trend is likely to continue for many more years.

- It is much faster to perform a computation than to set up and complete the corresponding experiment. Thus CFD can study numerous computational results, and suggest better designs and operating conditions, while the experimentalist is still performing one experiment.
- Computational fluid dynamists can simulate full-scale conditions without the need to do small-scale models, as is frequently necessary in order to perform the necessary experiments in a laboratory. For example, if personal sampling is to be performed under controlled conditions, this frequently requires the use of a wind tunnel. However, in order to simulate real practical situations either the wind tunnel has to be very large (and hence the cost of the experimentation is very large) or small-scale experiments have to be performed (and then there are problems associated with scale-up).
- When using CFD the solution is automatically calculated at all points in space and time and therefore the solution procedure produces all the relevant variables throughout the domain of interest. In contrast, the experimentalist is only able to make a relatively small number of measurements, often in a very restrictive number of locations.

### 10.5.2 *Disadvantages of CFD*

- The computed results are only as good as the mathematical model. Although our knowledge of turbulence has substantially improved over the last few years, it is still far from being well understood. Much more theoretical work on this aspect is required before the computational results can be used with confidence.
- Even if a very good mathematical model is available the results obtained are very much dependent on the numerical method. However, over recent years much progress has been made in this area.
- For problems which involve complex geometries, strong nonlinearities, or fluid property variations which are very sensitive to the variations of some of the other properties, such as temperature, a numerical solution may be hard to obtain and even if it can then it is computationally expensive. Even the relatively simple problem considered here, namely the problem of air being sucked through a small orifice, causes most commercial software packages difficulties. Hence the need for the specialized treatment of the problem rather than the use of the available commercial software packages.
- Even well-developed commercial software packages should only be used with great care. It is easy to produce 'numbers' rather than results from such packages.

### 10.5.3  *Recommendations*

Performing carefully controlled full-scale experiments of good quality under real-life operating conditions is the only way to ensure that the results on the problem of interest are accurate and reproducible. However, if improvements to the design of the equipment are to be made, or changes in the operating conditions are to be recommended, then this usually requires a very extensive range of parameters to be investigated. Thus accurate experimental solution of such problems is very expensive and time-consuming.

In aerosol sampling, often some very simple mathematical modelling, as performed for example by Dunnett and Ingham (1988c), will give some very valuable insight into the changes in design and operating conditions. Based on these predictions, further experimentation should be performed but on a much narrower range of parameters. It is at this stage that a CFD dynamical investigation should be performed and the mathematical model fully validated against the experimental data.

As indicated in section 10.3, commercial software packages are very useful for performing such calculations. However, the results obtained from these packages should only be used if validated by an experienced user. For too frequently numerical 'results' and 'graphical' data are produced which bear no resemblance to the solution of the problem being considered. Even experienced users, on occasions, can produce spurious results. Thus the results obtained by any numerical predictions should be very carefully checked before they are accepted.

### 10.5.4  *Outlook for CFD*

Rapid growth in CFD activities is now changing the role of experimental testing to one of verification, although there will be a continuing need for experimental research to improve the understanding of the physics of complex flows, such as turbulent flows. Lack of real-time and data visualization capabilities are noticeable deficiencies of most commercial CFD codes. CFD, by definition, generates millions of bytes of simulated data which, even with powerful interactive graphic workstations, would be very time-consuming to visualize. This is likely to become even more of a problem as three-dimensional time-dependent simulation is integrated into design systems. There are separate programs, such as AVS, Data Visualiser, IRIS Explorer and PV-Wave, which are designed specifically for the post-processing of complex data, but the associated disadvantages are the required extra training in the use of such additional codes and the expenditure in their procurement.

Mesh generation is the most tedious and time-consuming activity in the problem description and is by no means straightforward, especially in the case of three-dimensional modelling. Implementation of an adaptive grid strategy, i.e. dynamically adjusted grid points, will add an important dimension to the maturity of CFD. Three-dimensional modelling is an area that is not tested adequately and further advances in numerical techniques should broaden the capabilities of CFD codes, particularly in relation to problems where three-dimensional description is an essential prerequisite.

The computational time and memory requirements of current CFD codes are very high and for realistic problems this necessitates very expensive hardware. Better iterative solution algorithms and their implementation on parallel processing computers should produce greater computational efficiency and reduce the burden on the requirements. Another important issue which is vital to raising the confidence in the use of CFD is code validation or quality assurance. Improved code reliability and a reduction in the expertise required to exploit these codes will result if the codes are well validated.

# References

Badzioch S. (1959) Collection of gas-borne dust particles by means of an aspirated sampling nozzle. *Brit. J. Appl. Phys.*, **10**, 26–32.

Bassett A.B. (1961) *Hydrodynamics*. Dover, New York.

Becker H.A. (1959) Effects of shape and Reynolds number on drag in the motion of a freely oriented body in an infinite fluid. *Can. J. Chem. Eng.*, **37**, 85–91.

Belyaev S.P. and Levin L.M. (1972) Investigation of aerosol aspiration by photographing particle tracks under flash illumination. *J. Aerosol Sci.*, **3**, 127–140.

Belyaev S.P. and Levin L.M. (1974) Techniques for collection of representative aerosol samples. *J. Aerosol Sci.*, **5**, 325–338.

Boothroyd R.G. (1970) *Flowing Gas-Solids Suspension*. Chapman & Hall, London.

Chung K.Y.K. and Ogden T.L. (1986) Some entry efficiencies of disklike samplers facing the wind. *Aerosol Sci. Technol.*, **5**, 81–91.

Cook T.L. and Harlow F.H. (1984) Los Alamos National Laboratory, USA Report No. LA-10021-MS.

Davies C.N. (1967a) Movement of dust particles near a horizontal cylinder containing a sampling orifice. *Brit. J. Appl. Phys.*, **18**, 653–656.

Davies C.N. (1967b) The effect of a crosswind when sampling dust particles through an orifice in the base of a horizontal cylinder. *Brit. J. Appl. Phys.*, **18**, 1787–1792.

Davies C.N. (1968) The entry of aerosols into sampling tubes and heads. *J. Phys. D, Ser. 2*, 921–932.

Davies C.N. and Peetz V. (1954) Dust shadows below a cylinder containing a suction orifice and deposition of particles upon the cylinder. *Brit. J. Appl. Phys.*, S17–S20.

Davies C.N. and Subari M. (1982) Aspiration above wind velocity of aerosols with thin-walled nozzles facing and at right angles to the wind direction. *J. Aerosol Sci.*, **13**, 59–71.

Dunnett S.J. and Ingham D.B. (1986) A mathematical-theory to two-dimensional blunt body sampling. *J. Aerosol Sci.*, **17**, 839–853.

Dunnett S.J. and Ingham D.B. (1987) The effects of finite Reynolds number on the aspiration of particles into a bulky sampling head. *J. Aerosol Sci.*, **18**, 553–561.

Dunnett S.J. and Ingham D.B. (1988a) The human head as a blunt aerosol sampler. *J. Aerosol Sci.*, **19**, 365–380.

Dunnett S.J. and Ingham D.B. (1988b) An empirical model for the aspiration efficiencies of blunt aerosol samplers orientated at an angle to the oncoming flow. *Aerosol Sci. Technol.*, **8**, 245–264.

Dunnett S.J. and Ingham D.B. (1988c) *The Mathematics of Blunt Body Sampling.* Lecture Notes in Engineering. Springer-Verlag, London.

Durham M.D. and Lundgren D.A. (1980) Evaluation of aerosol aspiration efficiency as a function of Stokes number, velocity ratio and nozzle angle. *J. Aerosol Sci.*, **11**, 179–188.

Gosman A.D. and Ionnides E. (1981) *Aspects of Computer Simulation of Liquid Fueled Combustors*, AIAA paper no. 81–0323.

Govan A.H., Hewitt G.F. and Nyan C.F. (1989) Particle motion in a turbulent pipe flow. *Int. J. Multiphase Flow*, **15**, 471–481.

Govier G.W. and Aziz K. (1972) *The Flow of Complex Mixtures in Pipes.* Krieger Publishing Company, Florida.

Ihme F., Schmidt-Traub H. and Brauer H. (1972) Theoretical studies on mass transfer at and flow past spheres. *Chemie-Ing-Tech.*, **44**, 306–313.

Ingham D.B. and Wen X. (1993) Disklike body sampling in a turbulent wind. *J. Aerosol Sci.*, **24**, 629–642.

Ingham D.B. and Yan B. (1994) Re-entrainment of particles on the outer wall of a cylindrical blunt sampler. *J. Aerosol Sci.*, **25**, 327–340.

Ingham D.B., Wen X., Dombrowski N. and Foumeny E.A. (1995) Aspiration efficiency of a thin-walled shallow-tapered sampler rear-facing the wind. *J. Aerosol Sci.*, **26**, 933–944.

Jayasekera P.N. and Davies C.N. (1980) Aspiration below wind velocity of aerosols with sharp edged nozzles facing the wind. *J. Aerosol Sci.*, **11**, S35–547.

Koncar-Djurdjevic S., Zdanski F. and Dudukovic A. (1986) in N.P. Cheremissionoff (ed.), *Encyclopaedia of Fluid Mechanics*, Vol. 4. Gulf Publishing Corporation, USA.

Langmuir I. and Blodgett K.B. (1946) A mathematical investigation of water droplet trajectories. US Army Forces Technical Report No. 5418.

Launder B.E. and Spalding D.B. (1974) Numerical computation of turbulent flows. *Comp. Meth. Appl. Mech. Eng.*, **3**, 269–289.

Launder B.E. and Spalding D.B. (1972) *The Numerical Methods of Turbulence.* Academic Press, New York.

Launder B.E. and Sharma B.I. (1974) Application of the energy-dissipation model of turbulence to the calculation of flow near a spinning disc. *Lett. Heat and Mass Transfer*, **1**, 131–137.

Magnus G. (1853) Ueber die Abweichung der Geschosse, und ueber eine auffallende Erscheinung bei Rotirenden Körpern. *Ann. Phys. Chem.*, **88**, 1–29.

Mark D. and Vincent J.H. (1986) A new personal sampler for airborne total dust in workplaces. *Ann. Occup. Hyg.*, **30**, 89–102.

Mark D., Vincent J.H. and Witherspoon W.A. (1982) Particle blow-off – a source of error in blunt dust samplers. *Aerosol Sci. Technol.*, **17**, 463–469.

Mark D., Vincent J.H., Gibson H. and Lynch G. (1985) A new static sampler for airborne total dust in workplaces. *Am. Indust. Hygiene Assoc. J.*, **46**, 127–133.

May K. and Druett H. (1953) The pre-impinger: a selective aerosol sampler. *Brit. J. Ind. Med.*, **10**, 142–152.

Ogden T.L. and Birkett J.L. (1978) An inhalable dust sampler for measuring the hazard from total airborne particulate. *Ann. Occup. Hyg.*, **21**, 41–50.

Ogden T.L. and Wood J.D. (1975) Effects of wind on the dust and benzene-soluble matter captured by a small sampler. *Ann. Occup. Hyg.*, **17**, 187–196.

Patankar S.V. and Spalding D.B. (1972) A calculation procedure for heat, mass and momentum transfer in three dimensional parabolic flows. *Int. J. Heat Transfer*, **15**, 1787–1806.

Patankar S.V. (1980) *Numerical Heat Transfer and Fluid Flow.* Hemisphere, Washington.

Saffman P.G. (1965) The lift on a small sphere in slow shear flow. *J. Fluid Mech.*, **22**, 385–400.

Schiller L. and Naumann A. (1933) Mechanics of liquids and gases. Part I. *Die Physik*, **1**, 101–122.

Sehmel G.A. (1967) Errors in the subisokinetic sampling of an airstream. *Ann. Occup. Hyg.*, **10**, 73–82.

Shuen J.S., Chen C.D and Faeth G.M. (1983) Evaluation of a stochastic model of particle dispersion in a turbulent round jet. *AIChE J.*, **29**, 167–170.

Smith G.D. (1978) *Numerical Solution of Partial Differential Equations: Finite Difference Methods*, Oxford University Press, Oxford.

Stokes G.G. (1851) On the effect of the internal friction of fluids on the motion of a pendulum. *Trans. Cambridge Phil. Soc.*, **1**, 8–106.

Torobin L.B. and Gauvin W.H. (1961) Fundamental aspects of solids-gas flow. *Can. J. Chem. Eng.*, **38**, 142–153.

Vincent J.H. (1987) Recent advances in aspiration theory for thin-walled and blunt aerosol sampling probes. *J. Aerosol Sci.*, **18**, 487–498.

Vincent J.H. (1989) *Aerosol Sampling: Science and Practice*. Wiley, Chichester.

Vincent J.H. and Armbruster L. (1981) On the quantitative definition of the inhalability of airborne dust. *Ann. Occup. Hyg.*, **24**, 245–248.

Vincent J.H. and Mark D. (1982) Applications of blunt sampler theory to the definition and measurement of inhalable dust, in W.H. Walton (ed.), *Inhaled Particles*, Vol. V. Pergamon, Oxford, pp. 3–19.

Vincent J.H., Hutson D. and Mark D. (1982) The nature of air-flow near the inlets of blunt dust sampling probes. *Atmos. Environ.*, **16**, 1243–1249.

Vincent J.H., Emmett P.C. and Mark D. (1985) The effects of turbulence on the entry of airborne particles into a blunt dust sampler. *Aerosol Sci. Technol.*, **4**, 17–29.

Vitols V. (1966) Theoretical limits of error due to anisokinetic sampling of particle matter. *J. Air. Pollut. Control Assoc.*, **2**, 79–84.

Watson H.H. (1954) Errors due to anisokinetic sampling of aerosols. *Am. Ind. Hyg. Assoc. Quart.*, **15**, 21–25.

Wen X. and Ingham D.B. (1993) A new method for accelerating the rate of convergence of the SIMPLE-like algorithm. *Int. J. Numer. Methods Fluids*, **17**, 385–400.

Wen X. and Ingham D.B. (1995) Aspiration efficiency of a thin-walled cylindrical probe rear-facing the wind. *J. Aerosol Sci.*, **26**, 95–107.

Wen X. and Ingham D.B. (1996) A note on the application of the average correction technique. *Int. J. Numer. Methods Fluids*, **23**, 811–817.

## Nomenclature

| | |
|---|---|
| $A$ | aspiration coefficient |
| $A_E$ | entry efficiency |
| $A_s$ | area of sampling orifice |
| $A_0$ | overall sampling efficiency |
| $C_D$ | drag coefficient |
| $d$ | particle diameter |
| $d_a$ | aerodynamic diameter |
| $D$ | disc diameter |
| $D_c$ | characteristic dimension |
| $D_0$ | diameter of orifice in disc |
| $I$ | freestream turbulent intensity |
| $k$ | turbulent kinetic energy |
| $N$ | particle concentration |
| $P$ | probability |
| $Q$ | flow rate, sampled flux of air |
| $R$ | velocity ratio $(U_0/U_s)$ |
| $Re_p$ | particle Reynolds number |

| | |
|---|---|
| $St$ | Stokes number |
| $U$ | $x$ component velocity |
| $U_s$ | suction speed |
| $U_0$ | undisturbed speed |
| $v$ | laminar kinematic viscosity |
| $v_t$ | turbulent kinematic viscosity |
| $v_\epsilon$ | effective kinematic viscosity |
| $V$ | $y$ component velocity |
| $\beta$ | angle of orientation of sampler |
| $\epsilon$ | turbulent energy dissipation |
| $\eta$ | viscosity |
| $\kappa$ | Karman constant |
| $v_m$ | mean velocity of sampling |
| $\rho$ | density |
| $\rho_w$ | density of water |
| $\tau$ | relaxation time |
| $\Phi$ | velocity potential |
| $\Phi_\epsilon$ | perturbation potential |

# Index